GEOPHYSICAL FLUID DYNAMICS

Geophysical Fluid Dynamics
Understanding (Almost) Everything with Rotating Shallow Water Models

Vladimir Zeitlin

Laboratory of Dynamical Meteorology, Sorbonne University and École Normale Supérieure Paris, France

OXFORD

UNIVERSITY PRESS

OXFORD
UNIVERSITY PRESS

Great Clarendon Street, Oxford, OX2 6DP,
United Kingdom

Oxford University Press is a department of the University of Oxford.
It furthers the University's objective of excellence in research, scholarship,
and education by publishing worldwide. Oxford is a registered trade mark of
Oxford University Press in the UK and in certain other countries

First Edition published in 2018

Impression: 1

Published in the United States of America by Oxford University Press
198 Madison Avenue, New York, NY 10016, United States of America

British Library Cataloguing in Publication Data
Data available

Library of Congress Control Number: 2017959942

ISBN 978–0–19–880433–8

DOI 10.1093/oso/9780198804338.001

Printed and bound by
CPI Group (UK) Ltd, Croydon, CR0 4YY

To the memory of G. Zaslavsky

Preface

The book explains the key notions and fundamental processes of the dynamics of the fluid envelopes of the Earth (transposable to other planets) from the unifying viewpoint of rotating shallow-water model (RSW). The model, in its one- or two-layer versions, plays a distinguished role in geophysical fluid dynamics. It has been used now for about a century to aid conceptual understanding of various phenomena, for elaboration of approaches and methods to be used later in more complete models, for development and testing of numerical codes and schemes of data assimilations, and for many other purposes. In spite of its simplicity, the model grasps the essential features of the complete 'primitive equations' models of large- and medium-scale atmospheric and oceanic motions. Although RSW, most often, can not give a full quantitative explanation of observations, it provides qualitative (and, in many cases, semi-quantitative) understanding. It gives simple and clear representation of the principal dynamical processes and helps to develop physical intuition. In addition, it allows for efficient high-resolution numerical methods which achieve, with modest computational resources, resolutions and time spans hardly possible with the full primitive equations. The quasi-geostrophic reduction of the model is no less celebrated, having been used, for example, for the first successful numerical weather prediction.

After deriving one- and two-layer versions of the RSW model directly from the primitive equations and exploring its properties, we will explain and illustrate the fundamentals of geophysical fluid dynamics with its help, and treat traditional and recently arisen applications. We will be explaining both mathematics and physics underlying dynamical phenomena and the methods used to analyse them, with necessary demonstrations. However, most often we will remain on a heuristic level and will be, frequently in the first place, looking for qualitative insights. Hence, we will illustrate dynamical processes under consideration with abundant figures, and often sacrifice technical details or relegate them to exercises.

The book targets fluid dynamicists, physicists, and applied mathematicians interested in the dynamics of the climate system and the modelling of atmospheric and oceanic phenomena, mathematically minded meteorologists and oceanographers, including graduate and post-graduate students. It can be used as a complement to standard textbooks in geophysical fluid dynamics, dynamical meteorology, and physical oceanography. The book is self-contained and provides, in a concise manner, all necessary prerequisites, except for basic mathematics. It is divided in to three parts. Part I is a geophysical fluid dynamics course in RSW terms and is supplied with problems/exercises which complement demonstrations in the main text. Solutions of the problems can be obtained from the Editorial Office on demand. Part II contains advanced topics and studies of principal dynamical phenomena. Part III considers some modern developments in rotating shallow-water modelling. Each chapter of the book ends by a short résumé of the main points, and bibliographic remarks, for better orientation. A detailed Index serves the same purpose.

The first, and partially the second part are based on the course in geophysical fluid dynamics which is being taught by the author at the Université P. and M. Curie and École Normale Supérieure (Paris). The rest of the book is largely based on the original works of the author with students and collaborators.

Acknowledgements

A considerable part of material of this book is based on my work with friends and collaborators F. Bouchut, S. Medvedev, G. Reznik, and T. Dubos as well as with former masters and PhD students, some of them now colleagues, M. BenJelloul, E. Gouzien, J. Gula, N. Lahaye, J. Lambaerts, J. LeSommer, R. Plougonven, B. Ribstein, M. Rostami, E. Scherer, A. Stegner, and M. Tort, all of which is gratefully acknowledged.

Contents

Part I

Modelling Large-scale Oceanic and Atmospheric Flows: From Primitive to Rotating Shallow-Water Equations and Beyond

1

Introduction

Even a brief look at satellite images of the Earth, like the one shown in the left panel of Figure 1.1, makes it clear how thin is the Earth atmosphere (or at least its part determining weather) with respect to the Earth radius. Indeed, the depth of the whole atmosphere is several tens of kilometres, and its weather-active part, the troposphere, is about 10 kms high in mid-latitudes, becoming higher towards the equator. Alternatively, if the whole of the atmosphere is compressed in a homogeneous layer, it would have thickness of about 10 km, which is negligible compared to the Earth radius of about 6400 km. Nevertheless, as already clear from Figure 1.1, many dynamical processes are going on in this thin fluid film, and a lot of structures of considerable horizontal size are distinguishable in cloud forms (although clouds do not act as ideal dynamical tracers as they can appear and disappear due to thermodynamical processes). For example, a pair of synoptic disturbances, with their characteristic comma-shaped form are visible in the Northern hemisphere in the figure. Such disturbances are large-scale cyclonic vortices produced by instabilities of the upper-tropospheric jet-stream. A meandering tropospheric equatorial jet can be also identified in the cloud pattern. Such jets are at the origin of tropical cyclones. Other examples of dynamical structures distinguishable in clouds will be given throughout this book. The ocean is also a thin film of fluid, compared with the Earth size, with maximal depths of several kilometres. Again, satellite images reveal existence of dynamical structures of large horizontal scale in the ocean, like oceanic vortices visible in the plankton bloom shown in the right panel of Figure 1.1. A typical ratio of thickness of the fluid layer at the scales of the dynamical structures shown in Figure 1.1 to its horizontal extent is very small. A natural desire then arises to exploit the *shallowness* of such structures while trying to understand their dynamics. As usual in physics, everything depends on the scales of interest. We focus here on synoptic-scale motions in the atmosphere, like synoptic disturbances in the left panel of Figure 1.1, having a characteristic horizontal scale of 1000 kms, or meso-scale motions in the ocean, like vortices in the right panel of Figure 1.1, having a characteristic scale of several tens of kilometres. On the contrary, if we consider, for example, an intense small-scale atmospheric vortex, like a tornado, the notion of shallowness would not be appropriate.

Geophysical Fluid Dynamics. Vladimir Zeitlin,
Oxford University Press (2018). © Vladimir Zeitlin. DOI: 10.1093/oso/9780198804338.001.0001

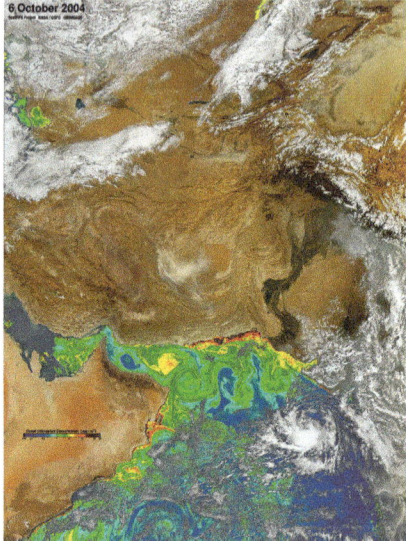

Figure 1.1 *Satellite images of the globe and of vortices in Arabian Sea (false colours). Courtesy NOAA.*

The use of quasi two-dimensional models for understanding synoptic-scale motions in the atmosphere and meso-scale motions in the atmosphere goes back to the early days of meteorology and physical oceanography, and this tradition is perpetuated. Principal dynamical phenomena in both media at these scales were first understood in such models, and the first successful weather prediction was made with the help of a model of this kind.

Unlike other branches of physics where, even now, there is a freedom of imagination for inventing models governing fundamental processes, meteorology and oceanography, in what concerns their dynamical essence, are based on the hydrodynamical equations which were well established more than two centuries ago. These equations constitute the dynamical cores of weather and climate prediction models. The physics of the processes which are to be included in these models, like phase transitions of water or radiative transfer, is also well known and well understood. It is, rather, a necessity to treat simultaneously a number of physical processes at very different space- and time scales, together with difficulties of numerical implementation, which explain why accurate weather prediction is possible only for a couple of days, and why climate models diverge in predicting, for example, humidity and precipitations. The (necessary) complexity of 'all-inclusive' weather and climate prediction models, and a number of parameterisations of small-scale processes, inevitable in constructing numerical implementations, often mask the details of dynamics. Correct reproduction of these details is, however, indispensable for checking consistency of big models themselves. Quasi two-dimensional models, which are the subject of this book, are based on considerable simplifications with respect to the big models using the full 'primitive' dynamical equations. These simplifications are based on systematic approximations of the full equations of motion and elimination of certain spatial and temporal scales. Their justification comes from the quasi-dimensional character of synoptic and meso-scale motions in, respectively, atmosphere and ocean, and they often represent an optimal compromise between complexity and fidelity of representation of dynamical phenomena at these scales. The quasi two-dimensional models are much easier to implement numerically; they often allow an analytical, or semi-analytical, treatment of the problems, and provide a physical intuition, which is precious in analysing observational data or outputs of 'big' models.

In the following, we introduce the primitive equations model and construct a hierarchy of approximations to the primitive equations, based on space and time averaging. We explain the principal properties of models obtained in this way, their dynamical content, the ways to include more elaborate physics like phase transitions, and describe their applications to various dynamical phenomena.

2

Primitive Equations Model

2.1 Preliminaries

Geophysical fluid dynamics (GFD), as applied to oceanic and atmospheric flows, means hydrodynamics in a rotating frame, in spherical geometry (for large scales), and complex domains (coasts, topography/bathymetry), with thermal effects and stratification, and, in general, of multi-phase fluids (air + water vapour + liquid water + ice in the atmosphere, (salty) water + ice in the ocean). By large scales, on which this book is focused, we mean scales in the range from the planetary scale $\sim 10^4$ km to synoptic scales in the atmosphere, $\sim 10^3$ km, and meso-scales in the ocean, $10\text{--}10^2$ km. The atmospheric and oceanic data at these scales reveal two different types of dynamical entities: vortices and waves, cf. Figures 2.1 to 2.3. Note that we do not mean here sound waves (typically small scale, and fast) in the atmosphere, and surface waves (swell) in the ocean, but instead mean internal waves propagating in the interior of the fluid and having considerable wave-lengths (cf. Figure 2.3). We will see later in this chapter that the very existence of these two types of dynamical 'actors' reflects some deep properties of GFD, and one of the fundamental questions in the field, to be abundantly addressed in what follows, is about dynamical coupling, or uncoupling, of waves and vortices.

In this chapter, we will recall the fundamentals of fluid mechanics, introduce the standard approximations, leading to the primitive equations (PE) model for large-scale atmospheric and oceanic motions, and analyse the basic properties of this model.

2.2 A crash course in fluid dynamics

2.2.1 The perfect fluid

Governing equations

The fluid is a continuous mechanical system so its evolution is derived from the local momentum and mass conservation. It is also a thermodynamical system with an equation

Geophysical Fluid Dynamics. Vladimir Zeitlin,
Oxford University Press (2018). © Vladimir Zeitlin. DOI: 10.1093/oso/9780198804338.001.0001

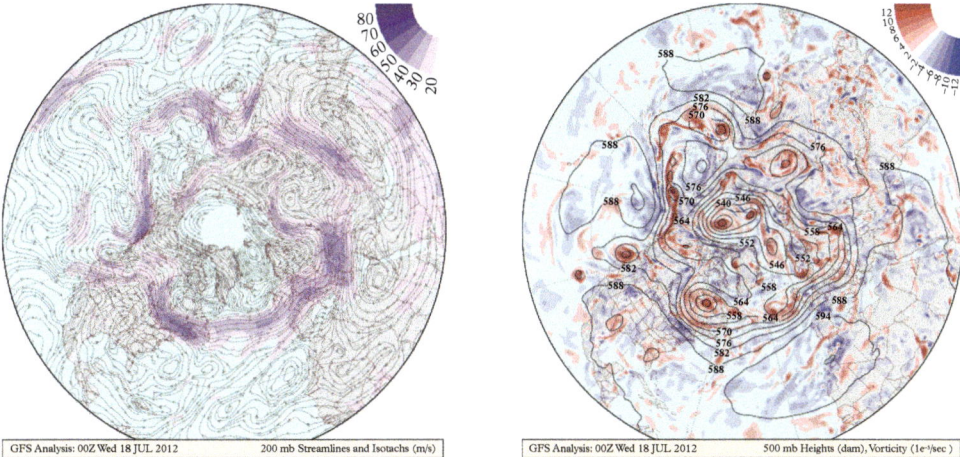

Figure 2.1 *Streamlines of the flow and values of the modulus of velocity (colour) at the 200 mb level (left panel), and relative vorticity (colour) and isolines of geopotential at the 500 mb level (right panel) of the atmosphere in the Northern hemisphere, as follows from the analysis of atmospheric data. Images generated by the Center for Ocean–Land–Atmosphere Studies (COLA) at George Mason University from the data from the forecast models of NOAA's National Center for Environmental Prediction (NCEP).*

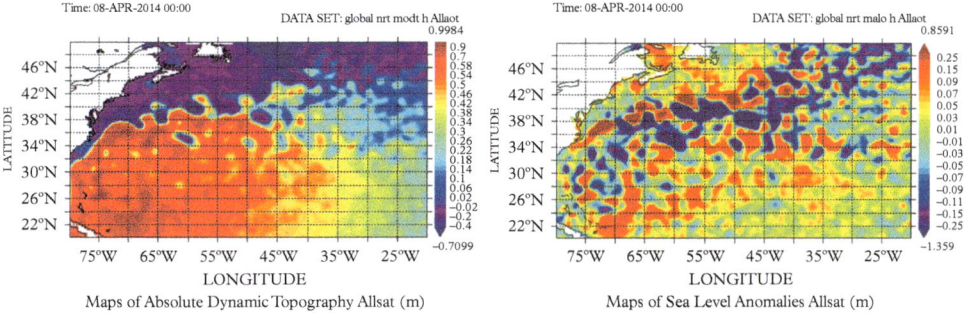

Figure 2.2 *The Gulf Stream (left panel) and related vortices (right panel) as seen in satellite altimetry data. Velocity, to the leading order, follows isopleths of the height anomaly, presented in the figure. The altimeter products were produced and distributed by the Copernicus Marine and Environment Monitoring Service (CMEMS).*

of state. The *Eulerian* description of the perfect fluid in terms of the vector velocity field $\mathbf{v}(\mathbf{x}, t)$, and scalar density and pressure fields $\rho(\mathbf{x}, t)$, $P(\mathbf{x}, t)$, which are defined at each point $\mathbf{x}$ of the volume occupied by the fluid at any time t, is based on the following equations of motion: the Euler equation expressing conservation of momentum,

$$\rho \left(\frac{\partial \mathbf{v}}{\partial t} + \mathbf{v} \cdot \boldsymbol{\nabla} \mathbf{v} \right) = -\boldsymbol{\nabla} P + \mathbf{F}, \tag{2.1}$$

Figure 2.3 *Internal waves in the atmosphere (left) and in the ocean (right), as seen by satellite observations. Part of the coast line (left panel) and an island (right panel) give an idea of spatial scale. Courtesy NASA.*

where $\mathbf{F}$ represents an external forcing, and the continuity equation

$$\frac{\partial \rho}{\partial t} + \mathbf{\nabla} \cdot (\rho \mathbf{v}) = 0. \tag{2.2}$$

These equations are written in a universal vector form. If otherwise not explicitly stated, we assume a Cartesian system of coordinates x, y, z, formed by a triple of orthonormal unit vectors $\hat{\mathbf{x}}$, $\hat{\mathbf{y}}$, $\hat{\mathbf{z}}$, with the radius vector $\mathbf{x} = x\hat{\mathbf{x}} + y\hat{\mathbf{y}} + z\hat{\mathbf{z}}$. The gradient operator $\mathbf{\nabla}$ is then expressed as $\mathbf{\nabla} = \hat{\mathbf{x}}\frac{\partial}{\partial x} + \hat{\mathbf{y}}\frac{\partial}{\partial y} + \hat{\mathbf{z}}\frac{\partial}{\partial z}$, and the velocity field as $\mathbf{v} = \hat{\mathbf{x}}u + \hat{\mathbf{y}}v + \hat{\mathbf{z}}w$, in terms of its components. The dot notation is used for the scalar product of two vectors. A shorthand notations $\frac{\partial(\ldots)}{\partial x} \equiv \partial_x(\ldots) \equiv (\ldots)_x$ etc., will be often used for partial derivatives throughout the book.

To close the system (2.1), (2.2), an equation of state relating pressure P and density ρ is needed. In a general case, the equation of state for a one-phase system (we postpone the treatment of multi-phase systems to Chapter 15) has the form

$$P = P(\rho, s), \tag{2.3}$$

where $s(\mathbf{x}, t)$ is the entropy per unit mass in the fluid. A case of particular interest in fluid dynamics is the *barotropic fluid* (also called homentropic or isentropic), with

$$P = P(\rho) \leftrightarrow s = \text{const}, \tag{2.4}$$

which is sufficient to close the system (2.1) to (2.2). In the case of *baroclinic fluid* (2.3) an additional equation for s is necessary. In the perfect fluid, with no heat exchanges between fluid masses, the entropy is simply advected by the flow:

$$\frac{\partial s}{\partial t} + \mathbf{v} \cdot \mathbf{\nabla} s = 0. \tag{2.5}$$

A particular case of the barotropic fluid is incompressible fluid. Incompressibility means volume conservation, whence

$$\nabla \cdot \mathbf{v} = 0. \tag{2.6}$$

Pressure is no more an independent variable and can be expressed in terms of velocity and its derivatives. In the case of constant density,

$$\rho = \text{const}, \quad \text{and} \quad \nabla \cdot (\mathbf{v} \cdot \nabla \mathbf{v}) = -\frac{1}{\rho} \nabla^2 P. \tag{2.7}$$

In the case of variable density,

$$\frac{\partial \rho}{\partial t} + \mathbf{v} \cdot \nabla \rho = 0, \quad \text{and} \quad \nabla \cdot (\mathbf{v} \cdot \nabla \mathbf{v}) = -\nabla \cdot \left(\frac{\nabla P}{\rho} \right). \tag{2.8}$$

Of course, in order to obtain the expression of pressure in terms of velocity and density, the operators on the right-hand side of (2.7) and (2.8) have to be inverted. *Inversion problems* of this type recurrently arise in applications, as we will see shortly.

Euler–Lagrange duality

Lagrangian description of fluid dynamics treats the fluid as an ensemble of fluid parcels with time-dependent positions $\mathbf{X}(\mathbf{x}, t)$, $\mathbf{X}(\mathbf{x}, 0) = \mathbf{x}$. A key concept is Euler–Lagrange duality which stems from the fact that, due to continuity of the fluid, any given point of the flow domain is, at the same time, a position of some fluid parcel, and that the Eulerian velocity at this point is, at the same time, the Lagrangian velocity of this parcel. This duality allows us to establish the expression of the time derivative along the trajectory of the fluid parcel, the Lagrangian derivative, in Eulerian terms immediately:

$$\frac{d}{dt} = \frac{\partial}{\partial t} + \mathbf{v} \cdot \nabla \mathbf{v}. \tag{2.9}$$

and the equivalence between Newton's second law for the parcel

$$\rho(\mathbf{X}, t) \frac{d^2 \mathbf{X}}{dt^2} = -\nabla_{\mathbf{X}} P(\mathbf{X}, t) + \mathbf{F}, \tag{2.10}$$

where $\nabla_{\mathbf{X}}$ is the gradient with respect to coordinates $\mathbf{X}$, and the Euler equation (2.1). The mass conservation for any fluid element is written in Lagrangian terms as

$$\rho_i(x) d^3 \mathbf{x} = \rho(\mathbf{X}, t) d^3 \mathbf{X}, \leftrightarrow \rho_i(x) = \rho(\mathbf{X}, t) \mathcal{J} \tag{2.11}$$

where ρ_i is the initial distribution of density, and $d^3 \mathbf{x}$ and $d^3 \mathbf{X}$ are initial and current elementary volumes respectively. The Jacobi determinant (Jacobian) in this formula is defined as

$$
\mathcal{J} =
\begin{vmatrix}
\dfrac{\partial X}{\partial x} & \dfrac{\partial X}{\partial y} & \dfrac{\partial X}{\partial z} \\[2mm]
\dfrac{\partial Y}{\partial x} & \dfrac{\partial Y}{\partial y} & \dfrac{\partial Y}{\partial z} \\[2mm]
\dfrac{\partial Z}{\partial x} & \dfrac{\partial Z}{\partial y} & \dfrac{\partial Z}{\partial z}
\end{vmatrix}
\equiv \dfrac{\partial(X, Y, Z)}{\partial(x, y, z)}.
$$

Fluid velocity is the Lagrangian derivative of the parcel position: $\mathbf{v}(\mathbf{X}, t) = \frac{d\mathbf{X}}{dt} \equiv \dot{\mathbf{X}}$, where we introduced the dot notation for Lagrangian time derivatives, to be used throughout the book. The incompressibility in Lagrangian terms means that $\mathcal{J} = 1$. As is easy to check, by taking Lagrangian time derivative of this relation we obtain the Eulerian incompressibility condition of zero velocity divergence. The advection of entropy (2.5) acquires a transparent meaning of conservation of entropy by each fluid parcel $\dot{s} = 0$, in Lagrangian language, the same for the advection of density in incompressible fluid.

Energy and thermodynamics

Let us recall the first principle of thermodynamics for one-phase systems, as applied to reversible processes (this assumption is consistent with the perfect fluid hypothesis):

$$
\delta\epsilon = T\delta s - P\delta v, \tag{2.12}
$$

where ϵ is internal energy per unit mass, $v = \frac{1}{\rho}$ is specific volume, and δ denotes variations of corresponding quantities. (The intensive, in a thermodynamical sense, variables are denoted by capitals.) For the enthalpy per unit mass defined as $h = \epsilon + Pv$ we get

$$
\delta h = T\delta s + v\delta P. \tag{2.13}
$$

The energy density of the fluid is a sum of kinetic and internal energy densities:

$$
e = \frac{\rho \mathbf{v}^2}{2} + \rho\epsilon. \tag{2.14}
$$

Local conservation of energy always holds for the perfect fluid:

$$
\frac{\partial e}{\partial t} + \boldsymbol{\nabla} \cdot \left[\rho\mathbf{v}\left(\frac{\mathbf{v}^2}{2} + h\right) \right] = 0. \tag{2.15}
$$

In the case of barotropic fluid

$$
\delta h = \frac{\delta P}{\rho} \quad\Rightarrow\quad \frac{\boldsymbol{\nabla} P}{\rho} = \boldsymbol{\nabla} h, \tag{2.16}
$$

and we get a gradient on the right-hand side of the momentum equation. Dividing both sides of the Euler equation (2.1) by ρ and taking the curl of both of them makes the right-hand side disappear. This convenient property will be exploited shortly.

Kelvin circulation theorem

The circulation of the fluid velocity around a contour Γ consisting of fluid parcels, and moving with the fluid (the contour, thus, is a Lagrangian object), is defined as

$$\gamma = \int_{\Gamma} \mathbf{v} \cdot d\mathbf{l} = \int_{S_{\Gamma}} (\nabla \wedge \mathbf{v}) \cdot d\mathbf{l}, \tag{2.17}$$

where S_{Γ} is an arbitrary surface bounded by Γ, $d\mathbf{l}$ is an oriented contour element, and Stokes theorem was used in order to transform the contour integral into a surface one. We thus see that circulation is controlled by the curl of velocity, the *vorticity*. The Kelvin theorem, which is one of the most remarkable results in fluid mechanics, states that,

- for barotropic fluids,

$$\frac{d\gamma}{dt} = 0, \tag{2.18}$$

- for baroclinic fluids,

$$\frac{d\gamma}{dt} = -\int_{\Gamma} \frac{\nabla P}{\rho} \cdot d\mathbf{l}. \tag{2.19}$$

Variational principles for the dynamics of perfect fluids

As a conservative mechanical system, dynamics of the perfect fluid can be derived from a variational (Hamilton's) principle. A natural framework for formulating such principle is the Lagrangian description. From the Lagrangian viewpoint, the equations of motion are Newton's equations for the positions of fluid parcels $\mathbf{X}$ labelled by their coordinates $\mathbf{x}$ at the initial moment $t = 0$. An additional change of independent variables $\mathbf{x} \mapsto \mathbf{a}$ to some other set of Lagrangian labels $\mathbf{a}$ is often convenient. The well-known rules of construction of the action principle for a mechanical system of N elements, described in terms of generalised coordinates $\mathbf{X}_i$, $i = 1, 2, \ldots, N$ and corresponding generalised velocities $\dot{X}_i$, say that the Lagrangian L is a difference between the total kinetic energy $K = \sum_{i=1}^{N} \frac{m_i \dot{X}_i^2}{2}$ and the total potential energy U of the system. Once the action $\mathcal{S}$ is known in terms of the Lagrangian,

$$\mathcal{S} = \int dt\, L\left[\mathbf{X}, \dot{X}\right] = \int dt\, \left(K\left[\mathbf{X}, \dot{X}\right] - U\left[\mathbf{X}\right]\right), \tag{2.20}$$

the equations of motion follow from the principle of least action $\delta S = 0$:

$$\frac{d}{dt}\left[\frac{\delta L}{\delta \dot{X}}\right] - \frac{\delta L}{\delta \mathbf{X}} = 0. \tag{2.21}$$

These rules apply as well to the continuous system of fluid parcels, the Lagrangian being expressed via Lagrangian density and summation becoming integration over the domain of the flow. The kinetic energy density of the fluid is $\rho(\mathbf{X}, t)\left(\frac{d\mathbf{X}}{dt}\right)^2$, and the total potential energy consists of internal energy $\mathcal{E}$ plus potential energy due to the presence of gravity. Omitting the latter, $U = \mathcal{E} = \rho\epsilon(v, s)$, where the internal energy density ϵ is considered as a function of volume and entropy of the fluid per unit mass. An element of the volume of the fluid can be measured in terms of differentials of $\mathbf{X}$, and we arrive at the following expression for the Lagrangian:

$$L = \int d^3\mathbf{X}\,\rho(\mathbf{X}, t)\left[\left(\frac{d\mathbf{X}}{dt}\right)^2 - \epsilon(v, s)\right]. \tag{2.22}$$

We should now recall the mass conservation (2.11), which allows us to replace the integration measure in (2.22) by the time-independent one: $\rho_i(\mathbf{x})d^3\mathbf{x}$. At the same time, the volume per unit mass of the fluid v is expressed as

$$v = \frac{1}{\rho} = \frac{1}{\rho_i}\frac{\partial(X, Y, Z)}{\partial(x, y, z)}. \tag{2.23}$$

Together with Lagrangian conservation of the entropy density s,

$$\dot{s} = 0 \Rightarrow s(\mathbf{X}, t) = s_i(\mathbf{x}), \tag{2.24}$$

this gives a closed expression of the Lagrangian in terms of dynamical variables $\mathbf{X}$ and their derivatives. It is, however, technically more convenient to make a further change of independent variables, introducing new Lagrangian labels $\mathbf{a} = (a, b, c)$, such that $\frac{\partial(a,b,c)}{\partial(x,y,z)} = \frac{\rho_i}{\rho_0}$, where ρ_0 is a constant which may be taken to be unity without loss of generality. This allows us to eliminate the initial density from the Lagrangian and get

$$L = \int d^3\mathbf{a}\,\mathcal{L} = \int d^3\mathbf{a}\left(\left(\frac{d\mathbf{X}}{dt}\right)^2 - \epsilon\left(\frac{\partial(X, Y, Z)}{\partial(a, b, c)}, s_i\right)\right). \tag{2.25}$$

The equations of motion of a baroclinic perfect fluid are Euler–Lagrange equations following from (2.25) by the standard calculus of variations, and remembering that, by definition, the pressure is $P = \frac{\partial\mathcal{E}}{\partial v}$.

The incompressible fluid is of special importance in what follows. As was already mentioned, in this case the pressure is not an independent variable. The Lagrangian takes the form

$$L = \int d^3\mathbf{a} \left(\left(\frac{d\mathbf{X}}{dt} \right)^2 - P \left(\frac{\partial(X, Y, Z)}{\partial(a, b, c)} - 1 \right) \right), \tag{2.26}$$

with pressure entering the Lagrangian as a Lagrangian multiplier to the incompressibility constraint.

Note that the *Hamiltonian* description of the fluid may be derived according to the standard rules, by defining a generalised momentum density $\mathcal{P} = \frac{\delta \mathcal{L}}{\delta \dot{X}}$ and the Hamiltonian density

$$\mathcal{H} = \mathcal{P} \cdot \dot{X} - \mathcal{L}. \tag{2.27}$$

The equations of motion then are obtained from the canonical Poisson brackets of $\mathcal{P}$ and X with the full Hamiltonian $H = \int d^3\mathbf{a}\,\mathcal{H}$:

$$\dot{X} = \{\mathcal{P}, \mathcal{H}\}, \quad \dot{\mathcal{P}} = \{X, \mathcal{H}\}. \tag{2.28}$$

The Poisson bracket is defined as $\{A, B\} = \frac{\delta A}{\delta X} \frac{\delta B}{\delta \mathcal{P}} - \frac{\delta B}{\delta X} \frac{\delta A}{\delta \mathcal{P}}$, and obeys the Jacobi identity $\{\{A, B\}, C\} + \{\{B, C\}, A\} + \{\{C, A\}, B\} = 0$. Note that systems with constraints, like (2.26), need special treatment in this context.

2.2.2   Real fluids: incorporating molecular transport

The approximation of perfect fluid neglects completely the molecular transport, taking into account only the macroscopic fluxes of momentum, mass, and internal energy (heat). Relaxing this approximation means including corrections to the macroscopic fluxes due to molecular ones. These latter may be calculated from the flux–gradient relations

$$\mathbf{f}_A = -k_A \nabla A, \tag{2.29}$$

where A stands for any thermodynamical variable, $\mathbf{f}_A$ is the corresponding molecular flux, and k_A is related molecular transport coefficient. These relations express the general thermodynamic principle stating that molecular fluxes tend to restore thermodynamic equilibrium. Adding molecular fluxes in this way results in:

- Viscosity corrections to the Euler equation for the incompressible fluid, giving the Navier–Stokes equation

$$\frac{\partial \mathbf{v}}{\partial t} + \mathbf{v} \cdot \nabla \mathbf{v} = -\frac{\nabla P}{\rho} + \nu \nabla^2 \mathbf{v}, \ \nabla \cdot \mathbf{v} = 0. \tag{2.30}$$

- Diffusivity corrections to the continuity equation

$$\frac{\partial \rho}{\partial t} + \nabla \cdot (\rho \mathbf{v}) = D \nabla^2 \rho. \tag{2.31}$$

- Thermal conductivity corrections to the heat/temperature advection giving the heat equation

$$\frac{\partial T}{\partial t} + \mathbf{v} \cdot \boldsymbol{\nabla} T = \chi \boldsymbol{\nabla}^2 T. \qquad (2.32)$$

Here ν, D, χ are kinematic viscosity, diffusivity, and thermo-conductivity, the molecular transport coefficients for momentum, mass, and energy, respectively, all having dimension $\left[\frac{L^2}{T}\right]$. As is well known, the parameters governing the dissipative effects are not these dimensional quantities, but instead are corresponding non-dimensional numbers constructed with the help of typical velocity and length scales of the flow V, L. Thus the Reynolds number is $Re = VL/\nu$, and the Peclet number is obtained by replacement $\nu \to D$ or $\nu \to \chi$. The ratio of the two is called the Prandtl number. The key observation in the context of large-scale oceanic and atmospheric flow modelling is that kinematic viscosities of seawater and air are very small, of the order of 10^{-6} and 10^{-5} m^2/s, respectively. For large-scale motions discussed earlier, and their typical velocities (of the order of tens of centimetres per second and ten metres per second, respectively), the corresponding Reynolds numbers are huge (the same for Peclet numbers, as Prandtl numbers are of the order unity), and sufficiently far from the boundaries, where boundary layers are inevitable because the distance to the boundary plays the role of L and thus the Reynolds numbers are small, the effects of molecular dissipation can be safely neglected.

2.3   Rotation, sphericity, and tangent plane approximation

2.3.1   Hydrodynamics in the rotating frame with gravity

As is well known, for motions in a rotating frame inertia forces appear in the Newton's second law: the Coriolis force and the centrifugal force. We start from the equation of motion of a free material point in the frame rotating with constant angular velocity $\boldsymbol{\Omega}$

$$m\frac{d\mathbf{v}}{dt} + 2m\boldsymbol{\Omega} \wedge \mathbf{v} + m\boldsymbol{\Omega} \wedge (\boldsymbol{\Omega} \wedge \mathbf{x}) = 0, \qquad (2.33)$$

where the wedge notation is used for the vector product of two vectors, m is the mass of the point, $\mathbf{x}(t)$ its position, and $\mathbf{v} = \dfrac{d\mathbf{x}}{dt}$ its velocity. On this basis, using Lagrangian description of the fluid and Euler–Lagrange duality, writing down the Euler equations in the rotating frame under the influence of gravity is straightforward:

$$\frac{\partial \mathbf{v}}{\partial t} + \mathbf{v} \cdot \boldsymbol{\nabla}\mathbf{v} + 2\boldsymbol{\Omega} \wedge \mathbf{v} = -\frac{\boldsymbol{\nabla} P}{\rho} + \mathbf{g}^*. \qquad (2.34)$$

Here, gravity and centrifugal acceleration are added up to form the effective gravity:

$$\mathbf{g}^* = \mathbf{g} - \boldsymbol{\Omega} \wedge (\boldsymbol{\Omega} \wedge \mathbf{x}). \qquad (2.35)$$

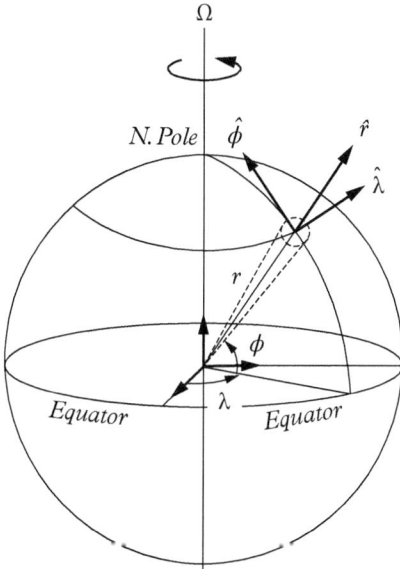

Figure 2.4 *Spherical coordinates.*

The recipe to include rotation of the reference frame into the variational principle of Section 2.2.1 is well known: a vector potential for the angular velocity $\boldsymbol{\Omega}$ should be introduced, 'augmenting' the Lagrangian density $\mathcal{L}\left(X, \dot{X}\right)$:

$$\mathcal{L} \to \mathcal{L} + \mathcal{R} \cdot \dot{X}, \quad \mathcal{R}: \nabla_X \wedge \mathcal{R} = 2\boldsymbol{\Omega}, \tag{2.36}$$

where it is sufficient to choose $\mathcal{R} = \boldsymbol{\Omega} \wedge X$.

A key fact about terrestrial atmosphere and oceans is that they are shallow, as compared with the Earth's radius, as was already mentioned in the Introduction. This means that the variations of effective gravity are negligible across the fluid envelopes of the Earth. Moreover, it is easy to check that correction to gravity due to the centrifugal acceleration is of the order of 1 per cent. It will be neglected in what follows, which constitutes the first step of the so-called *traditional approximation* in GFD. Gravity then is directed towards the centre of the Earth, whose mean surface will be assimilated to the sphere, which is the second step of the traditional approximation. It is worth noting that, in any case, the centrifugal acceleration is *potential*; that is it can be written as a gradient of some function. This allows us to absorb it into pressure in the Euler equations.

2.3.2 Hydrodynamics in spherical coordinates and the 'traditional' approximation in GFD

Although the vector notation which was used until now is coordinate-independent, in view of approximations to be made, and for the sake of physical clarity, it is useful to rewrite the equations of motion in spherical coordinates presented in Fig. 2.4. The Euler and the continuity equations take the following form:

$$\frac{dv_r}{dt} - \underbrace{\frac{v_\lambda^2 + v_\phi^2}{r}} - \underbrace{2\Omega \cos\phi v_\lambda} + g = -\frac{1}{\rho}\partial_r P, \qquad (2.37)$$

$$\frac{dv_\lambda}{dt} + \frac{v_r v_\lambda - v_\phi v_\lambda \tan\phi}{r} + 2\Omega\left(-\sin\phi v_\phi + \underbrace{\cos\phi v_r}\right) = -\frac{1}{\rho r \cos\phi}\partial_\lambda P, \qquad (2.38)$$

$$\frac{dv_\phi}{dt} + \frac{v_r v_\phi + v_\lambda^2 \tan\phi}{r} + 2\Omega \sin\phi v_\lambda = -\frac{1}{\rho r}\partial_\phi P, \qquad (2.39)$$

$$\frac{d\rho}{dt} + \rho\left[\frac{1}{r^2}\frac{\partial(r^2 v_r)}{\partial r} + \frac{1}{r \cos\phi}\left(\frac{\partial(\cos\phi v_\phi)}{\partial\phi} + \frac{\partial v_\lambda}{\partial\lambda}\right)\right] = 0, \qquad (2.40)$$

where subscripts denote the corresponding components of the velocity field, and the Lagrangian derivative in spherical coordinates is

$$\frac{d}{dt} = \frac{\partial}{\partial t} + v_r\partial_r + \frac{v_\phi}{r}\partial_\phi + \frac{v_\lambda}{r \cos\phi}\partial_\lambda.$$

According to the already made approximation, gravity appears only in the equation for vertical motions (2.37). It should be stressed that the gravity acceleration $g \approx 10\,\text{m/s}^2$ is huge with respect to average vertical accelerations observed at large and medium scales in the atmosphere and the oceans, which is consistent with the fact that in order to accelerate large-scale fluid masses at such rate an enormous amount of energy would be needed. As a consequence, atmospheric and oceanic motions at the scales in question are close to the hydrostatic equilibrium, that is the equilibrium between the pressure and gravity forces. Neglecting vertical accelerations of the fluid (the first under-braced term) in (2.37) with respect to the gravity acceleration constitutes the third step in the traditional approximation. The next step consists in neglecting other under-braced terms in the Euler equations which express, respectively, correction to the vertical acceleration due to curvature in spherical coordinates, the vertical component of the Coriolis force, small with respect to g, and the contribution of the vertical velocity into the horizontal components of the Coriolis force. The latter approximation is related to the fact that, due to already mentioned shallowness of the atmosphere and the oceans at the scales of interest, typical vertical velocities are much lower than typical horizontal velocities, on average. The last step in the traditional approximation consists in neglecting the variation of vertical positions of fluid masses as compared to the Earth's radius, that is by replacing $r \to R = \text{const}$ in the denominators of all expressions in the Euler and continuity equations.

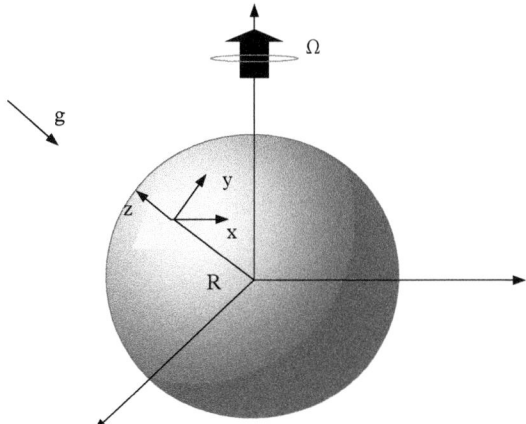

Figure 2.5 *A sketch of the tangent plane and related Cartesian coordinates.*

2.3.3 The tangent plane approximation

Even with all simplifications of the traditional approximation, the equations of motion in spherical coordinates remain complex. A further simplification is based on the observation that synoptic-scale motions in the atmosphere and meso-scale motions in the ocean typically evolve at scales which are large but still much smaller than the Earth's radius. This allows us to neglect the effect of sphericity by passing to the description of the motion on the plane tangent to the Earth's surface. The traditional choice of Cartesian coordinates on the tangent plane in the Northern hemisphere is x for zonal, y for meridional, and z for the vertical direction, as shown in Figure 2.5. The Euler equations on the tangent plane in the traditional approximation thus have the form

$$\frac{\partial \mathbf{v}}{\partial t} + \mathbf{v} \cdot \boldsymbol{\nabla}\mathbf{v} + f\hat{\mathbf{z}} \wedge \mathbf{v} + g\hat{\mathbf{z}} = -\frac{\boldsymbol{\nabla} P}{\rho}, \tag{2.41}$$

where f is the Coriolis parameter, twice the vertical component of the angular velocity $\boldsymbol{\Omega}$. If the meridional variation of this component is neglected we get the f *plane* approximation: $f = const$. If the meridional variation is retained, it is proportional, in the leading order, to the meridional coordinate y on the tangent plane, with a proportionality constant β, whence the β *plane* approximation results: $f = f_0 + \beta y$. The γ *plane* approximation is used for the tangent plane at the pole, with $f = f_0 - \gamma (x^2 + y^2)$.

2.4 Primitive equations in the oceanic and atmospheric context

We are now ready to formulate the model governing the dynamics of large-scale oceanic and atmospheric motions, the primitive equations model. We will consider first the oceanic and the atmospheric cases separately, but then show that under certain approximations, and under judicious change of variables, the equations of the model are, in fact, the same for both.

2.4.1 Oceanic context

A typical density field in the ocean may be decomposed in three parts:

$$\rho(\mathbf{x}, t) = \rho_0 + \rho_s(z) + \sigma(x, y, z; t), \quad \rho_0 = \text{const}, \quad \rho_0 \gg ||\rho_s|| \gg ||\sigma||. \qquad (2.42)$$

Here ρ_s represents the mean stratification, σ is the density perturbation, and there is, roughly, two orders of magnitude difference between the variable and constant parts of the density. The meso-scale motions in the ocean are close to the hydrostatic equilibrium, which we impose:

$$g\rho + \frac{\partial P}{\partial z} = 0. \qquad (2.43)$$

It thus follows from (2.42) that

$$P = P_0(z) + P_s(z) + \pi(x, y, z; t), \qquad (2.44)$$

where each entry in the r.h.s. is related to the corresponding part of density in (2.42) via the hydrostatic balance relation (2.43). To a very good approximation oceanic water is incompressible at the scales of interest:

$$\nabla \cdot \mathbf{v} = 0, \qquad (2.45)$$

whence

$$\frac{\partial \rho}{\partial t} + \mathbf{v} \cdot \nabla \rho = 0, \qquad (2.46)$$

as follows from the continuity equation. This means that ocean motions are barotropic in such approximation, and pressure is not an independent variable, as was explained above in Section 2.2.1.

By dividing velocity field into horizontal and vertical parts $\mathbf{v} = \mathbf{v}_h + \hat{\mathbf{z}}w$, which is suggested by the hydrostatic approximation, introducing the geopotential $\Phi = P/\rho_0 = \Phi_0 + \Phi_s + \phi$, and neglecting the variable part of density in the denominator of the r.h.s. of the Euler equation for horizontal velocity, we get

$$\frac{\partial \mathbf{v}_h}{\partial t} + \mathbf{v} \cdot \nabla \mathbf{v}_h + f\hat{\mathbf{z}} \wedge \mathbf{v}_h = -\nabla_h \phi, \qquad (2.47)$$

where $\nabla_h = \mathbf{x}\partial_x + \mathbf{y}\partial_y$. Equations (2.42) - (2.47) is a simplified version of hydrostatic Primitive Equations (PE) model for the meso-scale motions in the ocean. Dissipative effects are excluded from these equations, as in what follows we are mostly interested in the intrinsic dynamical properties of the system in the limit of very high Reynolds and Peclet numbers. Molecular dissipation may be always restored according to the recipes of Section 2.2.2. Thermodynamics of sea water is also excluded from these equations. An important element is *salinity* S of sea water, which, in the absence of dissipative effects is just advected by the flow $dS/dt = 0$. Yet, together with temperature T, or specific entropy s (which is also simply advected in the absence of heat sources and thermal conductivity) it enters the equation of state of the sea water $\rho = \rho(S, T)$. Empirical formulas are known for this equation, although a linear dependence of density on both S and T provides a reasonable fit. These elements can be easily added to the equations (2.43–2.47), but will be beyond our scope of interest in the following, although they are indispensable to understand the general circulation of the ocean.

PE should be supplied with boundary conditions. In the horizontal directions, the basin boundaries (realistic or idealised) with free-slip (i.e. vanishing of the normal component of velocity) boundary conditions are to be imposed in the absence of dissipation, yet for theoretical and numerical studies doubly periodic boundary conditions are convenient. Obviously, if dissipation is reintroduced, the boundary conditions become no-slip (i.e. vanishing of both normal and tangent components of velocity). As small-scale surface waves are beyond the scope of large-scale modelling, the simplest vertical boundary condition at the surface is the rigid lid, otherwise the free-surface (the surface considered as a material one) boundary condition should be imposed, which is technically more difficult to treat, especially numerically. The material surface condition means that vertical velocity at the surface is equal to a Lagrangian derivative of the position of the surface, expressing the fact that the surface consists of fluid parcels. Similarly, in the presence of bottom topography, the bottom surface should be treated as material one in the absence of dissipation and bottom boundary layer. An external forcing (e.g. wind stress) can be easily included in the horizontal momentum equations, if a forced-dissipative problem is to be considered, with appropriate parameterisations of the related surface and bottom boundary layers.

2.4.2 Atmospheric context

The basic facts about the atmosphere at synoptic scales are that the mean pressure is monotonically decreasing with height and that synoptic motions are close to hydrostatic equilibrium and have small vertical velocities, on average. For the *dry* atmosphere, which will be our subject until Chapter 15, an important characteristic is potential temperature, which is directly related to the entropy. It is conserved by air masses and, thus, simply advected, in the absence of diabatic effects.

Quick reminder of thermodynamics of the dry atmosphere

The equation of state of the dry air considered as an ideal gas is

$$P = \rho RT, \quad c_{p,v} = T \left(\frac{\partial s}{\partial T} \right)_{P,V} = const, \quad c_p - c_v = R. \tag{2.48}$$

The explicit expression for the entropy density in terms of other thermodynamical variables in this case is

$$s = c_p \ln T - R \ln P + const.$$ (2.49)

Entropy is constant in adiabatic processes:

$$s = \text{const} \ \Rightarrow c_p \frac{dT}{T} - R\frac{dP}{P} = 0, \Rightarrow T = T_s \left(\frac{P}{P_s}\right)^{\frac{R}{c_p}}.$$ (2.50)

Potential temperature is defined as a temperature the air parcel would have if adiabatically brought to a reference pressure level P_s:

$$\theta = T\left(\frac{P_s}{P}\right)^{\frac{R}{c_p}}, \quad s = c_p \ln \theta + \text{const.}$$ (2.51)

The variation of geopotential ϕ, considered as a thermodynamical variable, is defined as work against gravity, necessary to lift a unit mass over a distance δz in the vertical

$$\delta\phi = g\,\delta z.$$ (2.52)

Pressure as vertical coordinate

The monotonicity of mean atmospheric pressure in the vertical allows us to use pressure as vertical coordinate. In conjunction with hydrostatics this change of variables leads to substantial simplifications in the primitive equations. If the roles of z and pressure P as independent and dependent variables are inverted, expressing δz in (2.52) in terms of δP with the help of the hydrostatic relation gives

$$\delta\phi = -\frac{\delta P}{\rho} = -\frac{RT}{P}\delta P,$$ (2.53)

and hence

$$\frac{\partial \phi}{\partial P} = -\frac{RT}{P}.$$ (2.54)

With this change of independent and dependent variables, the air density ρ may be eliminated from the right-hand side of the horizontal momentum equation, which thus becomes the gradient of geopotential. Indeed, using the 'triangular' relation for a triple of functionally dependent variables P, x, z:

$$\left(\frac{\partial P}{\partial x}\right)_z \left(\frac{\partial x}{\partial z}\right)_P \left(\frac{\partial z}{\partial P}\right)_x = -1,$$ (2.55)

we get

$$\left(\frac{\partial P}{\partial x}\right)_z = -\left(\frac{\partial P}{\partial z}\right)_x \left(\frac{\partial z}{\partial x}\right)_P = \rho \left(\frac{\partial \phi}{\partial x}\right)_P, \tag{2.56}$$

and similarly for the derivative with respect to y. An even more drastic simplification, which seems counterintuitive at the first glance, is that air becomes effectively incompressible with this choice of vertical coordinate, with non-divergent velocity $\mathbf{\nabla} \cdot \mathbf{v} = 0$. Indeed, with the help of the hydrostatic relation, the mass element in pressure coordinates becomes, up to a constant, the volume element:

$$\rho\, dxdydz = -\frac{1}{g} dxdydP. \tag{2.57}$$

Hence, local mass conservation means local volume conservation in new coordinates. The Lagrangian conservation of mass, leading to the continuity equation in Eulerian terms, thus reads in new coordinates as Lagrangian conservation of volume, leading to incompressibility in Eulerian terms. Evidently, the definition of the vertical velocity changes: $w \to \omega = \frac{dP}{dt}$. The adiabatic equations of motion thus become

$$\begin{cases}
\mathbf{\nabla} \cdot \mathbf{v} = \mathbf{\nabla}_h \cdot \mathbf{v}_h + \partial_p \omega = 0, \\[2ex]
\dfrac{\partial \mathbf{v}_h}{\partial t} + \mathbf{v} \cdot \mathbf{\nabla} \mathbf{v}_h + f\hat{\mathbf{z}} \wedge \mathbf{v}_h = -\mathbf{\nabla}_h \phi, \\[2ex]
\dfrac{\partial \theta}{\partial t} + \mathbf{v} \cdot \mathbf{\nabla} \theta = 0, \\[2ex]
\dfrac{\partial \phi}{\partial P} = -\dfrac{RT}{P} = -\dfrac{R}{P}\left(\dfrac{P}{P_s}\right)^{\frac{R}{c_p}} \theta,
\end{cases} \tag{2.58}$$

where $\mathbf{\nabla}_h$ is understood in the sense of (2.56). An inconvenience of this set of equations is the explicit dependence of the coefficient in the last equation on the new vertical coordinate P. An additional change of vertical coordinate to a certain function of pressure ('pseudo-height'), instead of pressure itself, together with the hypothesis of smallness of the vertical velocity, eliminates this dependence.

'Pseudo-height' coordinate

This new vertical coordinate is defined as follows:

$$\bar{z} = z_0 \left(1 - \left(\frac{P}{P_s}\right)^{\frac{R}{c_p}}\right) \equiv z_0 \left(1 - \left(\frac{P}{P_s}\right)^{\frac{\gamma-1}{\gamma}}\right), \tag{2.59}$$

with the normalisation constant

$$z_0 = \frac{\gamma}{\gamma - 1} \frac{P_s}{g\rho_s} \approx 28\,\text{km}, \tag{2.60}$$

where the typical sea-level values of P_s, ρ_s are taken. The mass element thus becomes

$$\rho\, dx\, dy\, dz = -\frac{1}{g}\, dx\, dy\, dP = r(\bar{z})\, dx\, dy\, d\bar{z}, \tag{2.61}$$

where r is called pseudo-density. It may be easily calculated from the Jacobian of the change of variables $P \to \bar{z}$ and depends only on $\bar{z}$. The Lagrangian conservation of mass in pseudo-height coordinates leads to the continuity equation for the pseudo-density, by an argument similar to that leading to (2.57):

$$r\left(\boldsymbol{\nabla}_h \cdot \mathbf{v}_h + \frac{\partial \bar{w}}{\partial \bar{z}}\right) + \bar{w}\frac{\partial r}{\partial \bar{z}} = 0, \quad \mathbf{v} = (\mathbf{v}_h, \bar{w}), \quad \bar{w} = \dot{\bar{z}}. \tag{2.62}$$

Assuming $\bar{z} \ll z_0$ and $|\bar{w}| \ll 1$, we get

$$\boldsymbol{\nabla}_h \cdot \mathbf{v}_h + \frac{\partial \bar{w}}{\partial \bar{z}} = -\frac{\bar{w}}{r}\frac{\partial r}{\partial \bar{z}} = \frac{\bar{w}}{(\gamma - 1)z_0\left(1 - \frac{\bar{z}}{z_0}\right)}. \tag{2.63}$$

The last term is small at $\bar{z} \ll z_0$, remembering that average vertical velocities of synoptic-scale motions are small. So if it can be neglected, in the absence of forcing and dissipation we arrive at the atmospheric primitive equations in the form

$$\begin{cases} \dfrac{\partial \mathbf{v}_h}{\partial t} + \mathbf{v} \cdot \boldsymbol{\nabla}\mathbf{v}_h + f\hat{\mathbf{z}} \wedge \mathbf{v}_h = -\boldsymbol{\nabla}_h \phi, \\[2mm] -g\dfrac{\theta}{\theta_0} + \dfrac{\partial \phi}{\partial \bar{z}} = 0, \\[2mm] \dfrac{\partial \theta}{\partial t} + \mathbf{v} \cdot \boldsymbol{\nabla}\theta = 0; \quad \boldsymbol{\nabla} \cdot \mathbf{v} = 0, \end{cases} \tag{2.64}$$

which are identical to the oceanic primitive equations with the replacement $\sigma \to -\theta$. If forcing and dissipation are to be included this leads to appearance of external forces and viscous terms in the first, and of heat sources and thermal conductivity in the third equations.

The fact that primitive equations have the same form for the ocean and the atmosphere is of primary importance in GFD because the same notions, methods, approximations, and intuition apply to both.

2.4.3 Remarkable properties of the PE dynamics

Potential vorticity conservation

Probably the most important dynamical fact about PE in the absence of dissipation is existence of a specific Lagrangian invariant, the potential vorticity (PV). To be concrete, let us take the PE in the oceanic context:

$$\begin{cases} \dfrac{\partial \mathbf{v}_h}{\partial t} + \mathbf{v} \cdot \nabla \mathbf{v}_h + f\hat{\mathbf{z}} \wedge \mathbf{v}_h + \nabla_h \phi & = 0, \\[2ex] g\dfrac{\sigma}{\rho_0} + \partial_z \phi & = 0, \\[2ex] \dfrac{\partial \sigma}{\partial t} + \mathbf{v} \cdot \nabla \sigma + w\rho'_s(z) & = 0, \\[2ex] \nabla_h \cdot \mathbf{v}_h + \partial_z w & = 0. \end{cases} \qquad (2.65)$$

We define the relative vorticity for PE:

$$\boldsymbol{\zeta} = -\partial_z v\hat{\mathbf{x}} + \partial_z u\hat{\mathbf{y}} + (\partial_x v - \partial_y u)\hat{\mathbf{z}}. \qquad (2.66)$$

Note that this expression differs from the curl of velocity $\nabla \wedge \mathbf{v}$, which is the standard definition of vorticity, although its divergence is still zero. If a characteristic vertical scale H and a characteristic horizontal scale L are introduced it is easy to see that missing terms in (2.66) are of the order H/L. This is consistent with the hydrostatic hypothesis, and the fact that PE are derived for motions with vertical scales much smaller than the horizontal scales. The corresponding absolute vorticity, which is a sum of thus-defined relative vorticity and the planetary vorticity $\hat{\mathbf{z}}f$, is

$$\boldsymbol{\zeta}_a = \boldsymbol{\zeta} + \hat{\mathbf{z}}f, \quad \nabla \cdot \boldsymbol{\zeta}_a = 0. \qquad (2.67)$$

(It is easy to see that circulation of velocity introduced by overall rotation of the system around z axis with angular velocity $\Omega = f/2$ corresponds, by Stokes theorem, to the vorticity $\hat{\mathbf{z}}f$.)

Taking curl of the momentum equations in (2.65), and using the 'hydrodynamic' identity of vector analysis:

$$\mathbf{v} \cdot \nabla \mathbf{v} = \frac{1}{2}\nabla \mathbf{v}^2 - \mathbf{v} \wedge (\nabla \wedge \mathbf{v}), \qquad (2.68)$$

we arrive at the following equation for $\boldsymbol{\zeta}_a$:

$$\frac{d\boldsymbol{\zeta}_a}{dt} = \boldsymbol{\zeta}_a \cdot \nabla \mathbf{v} + \frac{g}{\rho}\hat{\mathbf{z}} \wedge \nabla\rho. \qquad (2.69)$$

A direct calculation, with the help of the equation of the advection of density and the equation (2.69), shows that the potential vorticity (PV), defined as

$$q = \boldsymbol{\zeta}_a \cdot \nabla\rho, \qquad (2.70)$$

is conserved by each fluid parcel:

$$\frac{dq}{dt} = 0. \tag{2.71}$$

This is a manifestation of the general Ertel theorem in the context of the primitive equations. As follows from the definition of PV, it is non-zero even if there is no motion of the fluid, having the background value $q_0 = f\rho_s'(z)$. That is why it is useful to introduce the notion of PV anomaly, PVA, which is defined as $q - q_0$.

Wave–vortex dichotomy and spectral gap

The second remarkable fact about PE is that they contain two different kinds of dynamical entities: vortices and waves. To see this, it is sufficient to analyse the spectrum of small perturbations over the state of rest—a standard approach in physics to look up the dynamical content of a model. We will take, for simplicity, a state of rest with linear stratification in the f plane, with periodic or decay boundary conditions at spatial infinity, and linearise the equations about it. This means that we consider amplitudes of three components of velocity perturbation u, v, w, geopotential perturbation ϕ, and density perturbation σ over the state of rest with background density ρ_s to be infinitesimal, so that their bilinear combinations can be neglected:

$$\begin{cases} u_t - fv + \phi_x = 0, \\ v_t + fu + \phi_y = 0, \\ \phi_z + \dfrac{g}{\rho_0}\sigma = 0, \\ \sigma_t + w\rho_s' = 0, \\ u_x + v_y + w_z = 0. \end{cases} \tag{2.72}$$

We successively eliminate σ, which leads to

$$\phi_{zt} + wN^2 = 0, \tag{2.73}$$

and then w. We introduced here the Brunt–Väisälä frequency $N^2 = -\frac{g\rho_s'}{\rho_0}$. We obtain in this way

$$\begin{cases} u_t - fv + \phi_x = 0, \\ v_t + fu + \phi_y = 0, \\ u_x + v_y - N^{-2}\phi_{zzt} = 0. \end{cases} \tag{2.74}$$

If the Brunt–Väisälä frequency is constant, this is a system of linear equations with constant coefficients which can be treated by the method of Fourier, where the general solution is represented as a superposition of harmonic waves, each of which has the form

$$(u, v, \phi) = (u_0, v_0, \phi_0)e^{i(\omega t - \mathbf{k} \cdot \mathbf{x})} + c.c. \tag{2.75}$$

Here, ω et $\mathbf{k} = (k_x, k_y, k_z)$ are frequency and wave number, respectively, and $c.c.$ denotes complex conjugation here and in what follows. For each wave we get an algebraic system of homogeneous equations for the amplitudes (u_0, v_0, ϕ_0):

$$\begin{pmatrix} i\omega & -f & -ik_x \\ f & i\omega & -ik_y \\ -ik_x & -ik_y & i\frac{\omega}{N^2}k_z^2 \end{pmatrix} \begin{pmatrix} u_0 \\ v_0 \\ \phi_0 \end{pmatrix} = 0. \tag{2.76}$$

The condition of existence of nontrivial solutions of this system is

$$\det \begin{pmatrix} i\omega & -f & -ik_x \\ f & i\omega & -ik_y \\ -ik_x & -ik_y & i\frac{\omega}{N^2}k_z^2 \end{pmatrix} = 0, \tag{2.77}$$

which gives

$$\omega \left(\omega^2 - \left(N^2 \frac{k_x^2 + k_y^2}{k_z^2} + f^2 \right) \right) = 0. \tag{2.78}$$

The three roots of this equation correspond to two different kinds of solutions:

- Time-independent solutions with $\omega = 0$,
- Propagative waves with dispersion relation:

$$\omega = \pm \sqrt{N^2 \frac{k_x^2 + k_y^2}{k_z^2} + f^2}. \tag{2.79}$$

It is easy to see that the frequency of the wave solutions is bounded from below by f, and thus there is a gap in the spectrum of small perturbations in the model. This gap reflects a difference in physical nature of the solutions. Indeed, by linearising (2.71), one can easily see that for small perturbations it becomes

$$\partial_t \left(v_x - u_y + fN^{-2}\phi_{zz} \right) = 0. \tag{2.80}$$

Exactly the same equation follows by cross-differentiating the first and the second equations in (2.74) and combining the result with the third one. Thus, the stationary solution of the dispersion relation corresponds to the conservation of the linearised PV and is thus a 'vortex' one. On the contrary, from the polarisation relations for the wave solutions it follows that the PV of these latter is zero (see Problem 8). Waves, thus, do not bear PV. Therefore, we see that PE describe two physically different entities, vortices and waves, with the difference defined in a clear-cut way by the PV content of the solutions. The waves in question are internal inertia–gravity waves propagating in the interior of the fluid due to combined effects of inertia (rotation) and gravity, which creates a restoring force once the stratified fluid is displaced with respect to the state of rest. These are the waves shown in Figure 2.3 earlier.

2.4.4   What do we lose by assuming hydrostatics?

It is important to understand the limits of the traditional approximation. The non-traditional contributions to the Coriolis force will be discussed in Chapter 16. We draw the reader's attention here to the limits of hydrostatic approximation itself. Let us consider an incompressible non-stratified fluid with constant density $\rho_0 = 1$. The corresponding Euler equations in the rotating frame without gravity read

$$\partial_t \mathbf{v} + \mathbf{v} \cdot \nabla \mathbf{v} + f\hat{\mathbf{z}} \wedge \mathbf{v} = -\nabla P, \quad \nabla \cdot \mathbf{v} = 0. \tag{2.81}$$

The linearisation of these equations immediately gives

$$\begin{cases} u_t - fv + P_x = 0 \\ v_t + fu + P_y = 0 \\ w_t + P_z = 0, \\ u_x + v_y + w_z = 0 \end{cases} \tag{2.82}$$

leading to wave solutions, the *inertial* (gyroscopic) waves, with dispersion relation

$$\omega^2 = f^2 \frac{k_z^2}{k_x^2 + k_y^2 + k_z^2}. \tag{2.83}$$

As is easy to see, these waves are sub-inertial, and are therefore filling the spectral gap discussed in Section 2.4.3. If we work with the non-hydrostatic version of (2.72), that is we replace

$$g\frac{\rho}{\rho_0} = -\phi_z \rightarrow \frac{dw}{dt} + g\frac{\rho}{\rho_0} = -\phi_z, \tag{2.84}$$

instead of (2.78) we get

$$\omega \left[\omega^2 - \left(N^2 \frac{k^2 + l^2}{k^2 + l^2 + m^2} + f^2 \frac{m^2}{k^2 + l^2 + m^2} \right) \right] = 0, \tag{2.85}$$

and for propagative waves we obtain the dispersion relation.

$$\omega^2 = N^2 \frac{k^2 + l^2}{k^2 + l^2 + m^2} + f^2 \frac{m^2}{k^2 + l^2 + m^2}. \tag{2.86}$$

It is easy to see that in the case of sufficiently strong stratification $N^2 > f^2$

$$f^2 \leq \omega^2 \leq N^2, \tag{2.87}$$

and the spectral gap between zero and the inertial frequency still exists. However, the frequencies of inertia–gravity waves are now bounded from above by N, unlike (2.79), which means that the latter formula is valid only for motions with horizontal scales much larger than vertical ones. On the other hand, for weak stratifications $N < f$, sub-inertial waves exist, with frequencies bounded from below by N.

2.5 Summary, comments, and bibliographic remarks

Thus, starting from standard hydrodynamic equations in the rotating frame with gravity we constructed, on the basis of the observed properties of large-scale atmospheric and oceanic motions, a simplified primitive equations (PE) model. A fact of utmost importance is that the model, in its non-dissipative version, possesses a Lagrangian invariant, PV, which allows us to distinguish between two dynamical entities described by the model: PV-less inertia–gravity waves and PV-bearing vortex motions.

The first acquaintance with the PE model in this chapter will be complemented in Chapter 16 by a more in-depth consideration of the primitive equations on the rotating sphere, where a variational principle and conservation laws for this model will be established.

There are a number of classical books on geophysical fluid dynamics and atmosphere-ocean dynamics (Eckart 1960, Holton 1979, Gill 1982, McWilliams 2006, Pedlosky 1982, Vallis 2006) with excellent introductions into the subject. Majda (2003) and Cullen (2006) give a more mathematically oriented view of GFD. The material of Section 2.2 is standard. It may be found in a concise form in Landau and Lifshits (1975), except for variational principles. For systematic application of these latter in hydrodynamics and geophysical fluid dynamics, see Salmon (1998) and Holm, Marsden, and Ratiu (2002) in more mathematically advanced form. Discussion and applications of primitive equations may be found in all of these books, with special emphasis on the atmosphere in Holton (1979). Details on the equation of state of seawater can be found, for example in Gill (1982). Traditional approximation, as named in Eckart (1960), goes back to Laplace (1799). The simplifying hypotheses of Earth sphericity can be relaxed, with the use of oblate spheroidal coordinates instead of spherical ones, an approach which we do not pursue. Pressure coordinates are standard in meteorology. Although the derivation of primitive equations in pressure coordinates given above is standard, (cf. Holton 1979), it is heuristic as it does not contain a clean treatment of the curvature of the isobaric surfaces. A systematic approach necessitates the use of curvilinear coordinates, which is beyond the scope of this book. The pseudo-height coordinate was introduced in Hoskins and Bretherton (1972). The fact that linear inertia–gravity waves do not carry PV is well known but is rarely stressed in the textbooks.

2.6   Problems

2.1 Derive (2.9).

2.2 Demonstrate that Eqs. (2.2) and (2.11) are equivalent.

2.3 Prove Kelvin theorem by direct calculation.

2.4 Show that Euler–Lagrange equations (2.21) with the Lagrangian (2.26) are equivalent to the Euler equations for the incompressible fluid with constant density.

2.5 Derive Euler and continuity equations in spherical coordinates. Determine the conditions of validity of the tangent plane approximation.

2.6 Derive the vorticity equation (2.69) in the PE model.

2.7 By direct calculation prove Lagrangian conservation of potential vorticity (2.71).

2.8 Find the polarisation relations, that is the expressions of u_0 and v_0 in terms of ϕ_0 in (2.75), which correspond to different roots of (2.78).

2.9 Show that inertia–gravity waves bear no potential vorticity anomaly.

2.10 Find polarisation relations for gyroscopic waves. Compare them with inertia-gravity waves.

3

Simplifying Primitive Equations: Rotating Shallow-Water Models and their Properties

The primitive equation (PE) model can be further simplified while retaining its most essential properties. The simplification is based on the already emphasised disparity of horizontal and vertical scales of synoptic motions in the atmosphere, and meso-scales motions in the ocean, and allows us to obtain a conceptual GFD model, which is the main subject of this book: the rotating shallow-water model (RSW). It permits, at the same time, physically transparent analysis and interpretation and efficient numerical tools for simulations of fundamental dynamical phenomena. We show here how the RSW model(s) can be obtained from PE by vertical averaging. In order to stress the generality of the approach, we apply it to the equations of motion without supposing incompressibility. The same procedure holds for PE for incompressible fluid with obvious simplifications. It can be found, with some modifications, in Chapters 15 and 16.

3.1 Vertical averaging of horizontal momentum and mass conservation equations

We start from the Euler and the continuity equations in the presence of rotation. By combining them, we obtain equations of horizontal motion in the form

$$\begin{cases} \partial_t(\rho u) + \partial_x(\rho u^2) + \partial_y(\rho v u) + \partial_z(\rho w u) - f\rho v = -\partial_x p, \\ \partial_t(\rho v) + \partial_x(\rho u v) + \partial_y(\rho v^2) + \partial_z(\rho w v) + f\rho u = -\partial_y p, \end{cases} \tag{3.1}$$

which express the local conservation of the horizontal momentum of the fluid in the presence of the Coriolis force. We will integrate these equations between two arbitrary material surfaces in the interior of the fluid. The vertical position of each surface is given

Geophysical Fluid Dynamics. Vladimir Zeitlin,
Oxford University Press (2018). © Vladimir Zeitlin. DOI: 10.1093/oso/9780198804338.001.0001

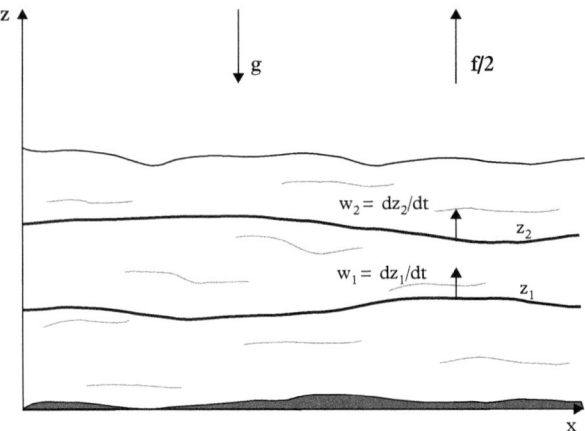

Figure 3.1 *Motion of material surfaces in the fluid.*

by $z = z_{1,2}(x, y, t)$. By definition of material surfaces they move with the local vertical velocity of the fluid, cf. Figure 3.1:

$$w|_{z_i} = \frac{dz_i}{dt} = \partial_t z_i + u \partial_x z_i + v \partial_y z_i, \quad i = 1, 2. \tag{3.2}$$

The Leibniz formula

$$\int_{z_1}^{z_2} dz \partial_x F = \partial_x \int_{z_1}^{z_2} dz F - \partial_x z_2 \, F|_{z_2} + \partial_x z_1 \, F|_{z_1}, \tag{3.3}$$

and its analogues for y and t differentiations, will be used for any function $F(x, y, z, t)$ integrated over the interval with variable limits.

With the help of (3.2) and (3.3) we get

$$\partial_t \int_{z_1}^{z_2} dz \rho u + \partial_x \int_{z_1}^{z_2} dz \rho u^2 + \partial_y \int_{z_1}^{z_2} dz \rho u v$$

$$-f \int_{z_1}^{z_2} dz \rho v = -\partial_x \int_{z_1}^{z_2} dz p - \partial_x z_1 \, p|_{z_1} + \partial_x z_2 \, p|_{z_2}. \tag{3.4}$$

Analogously,

$$\partial_t \int_{z_1}^{z_2} dz \rho v + \partial_x \int_{z_1}^{z_2} dz \rho u v + \partial_y \int_{z_1}^{z_2} dz \rho v^2$$

$$+f \int_{z_1}^{z_2} dz \rho u = -\partial_y \int_{z_1}^{z_2} dz p - \partial_y z_1 \, p|_{z_1} + \partial_y z_2 \, p|_{z_2}. \tag{3.5}$$

We can similarly integrate the continuity equation

$$\partial_t(\rho) + \partial_x(\rho u) + \partial_y(\rho v) + \partial_z(\rho w) = 0, \tag{3.6}$$

and get

$$\partial_t \int_{z_1}^{z_2} dz\rho + \partial_x \int_{z_1}^{z_2} dz\rho u + \partial_y \int_{z_1}^{z_2} dz\rho v = 0. \tag{3.7}$$

We introduce notation μ for the vertically integrated density

$$\mu = \int_{z_1}^{z_2} dz\rho, \tag{3.8}$$

and the mass-weighted vertical average for any quantity F:

$$\langle F \rangle = \frac{1}{\mu} \int_{z_1}^{z_2} dz\rho F. \tag{3.9}$$

Equations (3.4), (3.5), and (3.7) become

$$\partial_t \left(\mu \langle u \rangle \right) + \partial_x \left(\mu \langle u^2 \rangle \right) + \partial_y \left(\mu \langle uv \rangle \right) - f\mu \langle v \rangle =$$
$$- \partial_x \int_{z_1}^{z_2} dzp - \partial_x z_1 \left. p \right|_{z_1} + \partial_x z_2 \left. p \right|_{z_2}, \tag{3.10}$$

$$\partial_t \left(\mu \langle v \rangle \right) + \partial_x \left(\mu \langle uv \rangle \right) + \partial_y \left(\mu \langle v^2 \rangle \right) + f\mu \langle u \rangle =$$
$$- \partial_y \int_{z_1}^{z_2} dzp - \partial_y z_1 \left. p \right|_{z_1} + \partial_y z_2 \left. p \right|_{z_2}, \tag{3.11}$$

$$\partial_t \mu + \partial_x \left(\mu \langle u \rangle \right) + \partial_y \left(\mu \langle v \rangle \right) = 0. \tag{3.12}$$

The pressure at any point inside the fluid layer between z_1 and z_2 can be expressed in terms of pressure at the lower material surface and the vertical position of the point by integrating the hydrostatic balance condition

$$p(x, y, z, t) = -g \int_{z_1}^{z} dz' \rho(x, y, z', t) + \left. p \right|_{z_1}. \tag{3.13}$$

Alternatively, in terms of the pressure at the upper surface,

$$p(x, y, z, t) = g \int_{z}^{z_2} dz' \rho(x, y, z', t) + \left. p \right|_{z_2}. \tag{3.14}$$

Equations (3.10) to (3.12) are exact, the only hypothesis made by now is hydrostatics, if pressure is to be expressed via (3.13), (3.14). Yet these equations are not closed. In order to obtain the equations for averaged velocities a closure hypothesis is necessary, expressing the second moments $\langle u^2 \rangle$, $\langle uv \rangle$, etc., in terms of first moments (the averages themselves), as usual. The simplest one is the mean-field hypothesis, as it is called in physics, expressing higher moments as products of averages. It means negligible variations of all fields in the vertical direction (which is plausible for shallow enough layers), and corresponds to *columnar motion* of the fluid between the chosen material surfaces. In the mean-field approximation,

$$\langle uv \rangle \approx \langle u \rangle \langle v \rangle, \quad \langle u^2 \rangle \approx \langle u \rangle \langle u \rangle, \quad \langle v^2 \rangle \approx \langle v \rangle \langle v \rangle. \tag{3.15}$$

It should be emphasised that corrections to the mean-field approximation can be introduced, if necessary, with the help of 'turbulent' viscosity/diffusivity parameterisations, when differences between the second moments and products of first moments are considered to be proportional to the gradients of corresponding mean fields. In this way terms containing Laplacian of velocity would appear in the averaged momentum equation, and similarly for the averaged continuity equation.

Introducing the vertical average of density $\bar{\rho}$

$$\bar{\rho} = \frac{1}{(z_2 - z_1)} \int_{z_1}^{z_2} dz \rho, \quad \mu = \bar{\rho}(z_2 - z_1), \tag{3.16}$$

we can express the pressure in terms of $\bar{\rho}$:

$$p(x, y, z, t) \approx -g\bar{\rho}(z - z_1) + p|_{z_1}, \tag{3.17}$$

or

$$p(x, y, z, t) \approx +g\bar{\rho}(z_2 - z) + p|_{z_2}, \tag{3.18}$$

supposing that the fluid layer is shallow.

For any pair of material surfaces, from (3.10), (3.11), (3.15), (3.17), with the help of (3.12), (3.16), we get the following momentum and mass equations, respectively:

$$\bar{\rho}(z_2 - z_1)\left(\partial_t \langle \mathbf{v}_h \rangle + \langle \mathbf{v}_h \rangle \cdot \nabla_h \langle \mathbf{v}_h \rangle + f\hat{\mathbf{z}} \wedge \langle \mathbf{v}_h \rangle\right) = \tag{3.19}$$

$$-\nabla_h \left(-g\bar{\rho}\frac{(z_2 - z_1)^2}{2} + (z_2 - z_1)\,p|_{z_1}\right) - \nabla_h z_1\,p|_{z_1} + \nabla_h z_2\,p|_{z_2},$$

$$(\bar{\rho}(z_2 - z_1))_t + \nabla_h \cdot (\bar{\rho}(z_2 - z_1)\langle \mathbf{v}_h \rangle) = 0, \tag{3.20}$$

where $\mathbf{v}_h = (u, v)$. In what follows we will adopt a simplifying hypothesis of constant mean density: $\bar{\rho} = const$. A more general choice, $\bar{\rho} = \bar{\rho}(x, y, t)$, is also possible, leading to the so-called thermal shallow-water equations, which will be the subject of Chapter 14.

The averaged continuity equation (3.20) then takes a simpler form:

$$((z_2 - z_1))_t + \nabla_h \cdot ((z_2 - z_1)\langle \mathbf{v}_h \rangle) = 0, \tag{3.21}$$

expressing local conservation of the volume of a fluid column. All variables in (3.19), (3.21) are functions of horizontal coordinates and time. These equations are *master equations*, which together with hydrostatic relations (3.17), (3.18), allow us to build the multi-layer rotating shallow-water models for any number of layers, with any boundary conditions on top and bottom of a stack of the layers, by applying the following recipe:

- choose N material surfaces $z_1, z_2, \ldots, z_N$,
- write down the master equations for each layer (z_i, z_{i+1}), $i = 1, 2, \ldots, N - 1$,
- define the nature of the upper and lower boundaries z_1 and z_N (free surface, rigid lid/bottom, bottom topography, etc.),
- require continuity of pressure across each interface.

The free-surface boundary condition at z_N, the uppermost material surface, presumes a given (usually constant) pressure at the surface $p|_{z_N} = p_0$, the boundary condition of the interface between the layers corresponds to the continuity of pressure across it, the rigid lid/bottom boundary conditions mean simply that related z_i are constant, and the boundary condition at the bottom topography $b(x, y)$ is $z_1 = b(x, y)$ for the lowermost z_1.

As already mentioned, the master equations (3.19) to ((3.20) allow for generalisations, like 1) non-constant $\bar{\rho} = \bar{\rho}(x, y, t)$ with advection of $\bar{\rho}$ and additional term in the pressure gradient (thermal shallow-water equations), 2) additional terms in the momentum and mass equations of the type $\nabla_h^2 \mathbf{v}_h$, $\nabla_h^2(z_{i+1} - z_i)$ to account for deviations from the mean-field approximation and/or molecular dissipation, and 3) additional fluxes across the interfaces, to be added while expressing w_i in terms of dz_i/dt, in order to model convection and exchanges with boundary layers. These generalisations will be discussed in Part III.

3.2 Archetype models

3.2.1 One-layer RSW model

The classical RSW model results from applying the master equation to a single layer with the flat bottom at $z_1 = 0$ and a free surface under a constant pressure at $z_2 = h$:

$$\begin{cases} \partial_t \mathbf{v} + \mathbf{v} \cdot \nabla \mathbf{v} + f\hat{\mathbf{z}} \wedge \mathbf{v} + g\nabla h = 0, \\ \partial_t h + \nabla \cdot (\mathbf{v}h) = 0. \end{cases} \tag{3.22}$$

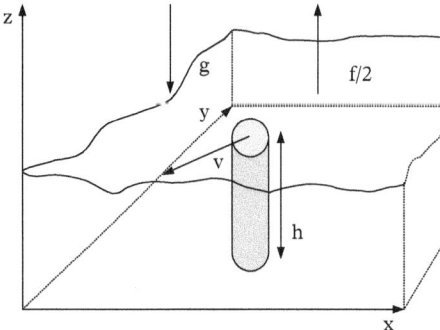

Figure 3.2 *A sketch of the one-layer RSW configuration with a flat bottom.*

The model describes the motion of fluid columns of variable depth h moving with velocity $\mathbf{v} = (u, v)$, cf. Figure 3.2. If the model is to be applied to the case of nontrivial bathymetry, h should be replaced by $h - b(x, y)$ in the second equation. Inspection of equations (3.22) makes immediately clear that, modulo the rotation term, they are equivalent to the equations of two-dimensional gas dynamics for a barotropic gas with density h and the equation of state $P = g\frac{h^2}{2}$. This useful analogy allows us to use, in particular, the powerful numerical methods developed for gas dynamics in GFD.

3.2.2   Two-layer RSW model with a rigid lid

The equations of the model follow from the application of master equations to the fluid between the flat bottom $z_1 = 0$ and the lid $z_3 = H$ with a material surface $z = z_2(x, y, t) \equiv h(x, y, t)$ inside the fluid separating layers of different density, cf. Figure 3.3. Generalisation to nontrivial bottom topography is obtained by the replacement $z_1 \rightarrow b(x, y)$. The equations of motion read

$$\begin{cases} \partial_t \mathbf{v}_1 + \mathbf{v}_1 \cdot \nabla \mathbf{v}_1 + f\hat{\mathbf{z}} \wedge \mathbf{v}_1 = -\dfrac{1}{\rho_1} \nabla \left. p \right|_H, \\[6pt] \partial_t \mathbf{v}_2 + \mathbf{v}_2 \cdot \nabla \mathbf{v}_2 + f\hat{\mathbf{z}} \wedge \mathbf{v}_2 = -\dfrac{1}{\rho_2} \nabla \left. p \right|_H - g\dfrac{\rho_2 - \rho_1}{\rho_2} \nabla h, \\[6pt] \partial_t h + \nabla \cdot (\mathbf{v}_2 h) = 0, \\[6pt] \partial_t (H - h) + \nabla \cdot (\mathbf{v}_1 (H - h)) = 0, \end{cases} \qquad (3.23)$$

where $\mathbf{v}_{1(2)}, \rho_{1(2)}$ are velocity and density in the upper (lower) layer, and we omitted bars over the means. Throughout this book, layer 1 will be always the upper layer in the oceanic context, and the lower layer in the atmospheric context, by reasons explained shortly. Note the appearance of *reduced gravity* $g' = g\frac{\rho_2 - \rho_1}{\rho_2}$ in (3.23).

3.2.3   Two-layer RSW model with a free upper surface

The equations of the model follow from the application of master equations to two layers of fluid of different density: one between the flat bottom $z_1 = 0$ and the interface $z_2 = h_2$,

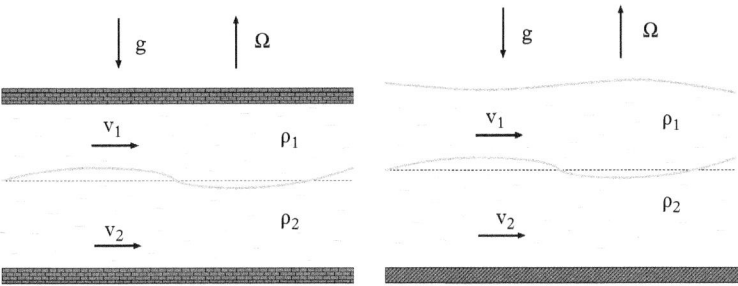

Figure 3.3 *Sketches of the two-layer RSW configurations with a rigid lid (left panel) and free upper surface (right panel).*

and another between the interface and the free surface $z_3 = h_1 + h_2$, cf. Figure 3.3. Generalisation to nontrivial topography is obtained by the replacement $z_1 \to b(x, y)$. The equations of the model read

$$
\begin{cases}
\partial_t \mathbf{v}_1 + \mathbf{v}_1 \cdot \nabla \mathbf{v}_1 + f \hat{\mathbf{z}} \wedge \mathbf{v}_1 = -g \nabla (h_1 + h_2), \\
\partial_t \mathbf{v}_2 + \mathbf{v}_2 \cdot \nabla \mathbf{v}_2 + f \hat{\mathbf{z}} \wedge \mathbf{v}_2 = -g \nabla (r h_1 + h_2), \\
\partial_t h_1 + \nabla \cdot (\mathbf{v}_1 h_1) = 0, \\
\partial_t h_2 + \nabla \cdot (\mathbf{v}_2 h_2) = 0,
\end{cases}
\tag{3.24}
$$

where $r = \frac{\rho_1}{\rho_2} \leq 1$ is the density ratio, and $h_{1,2}$ are the thicknesses of the layers.

For future use and for completeness we also give a variant of this model adapted to the atmospheric context. It is obtained by averaging the primitive equations in pseudo-height pressure coordinates, using the hydrostatic relation between the geopotential and potential temperature, and supposing that the layers have constant mean potential temperatures $\theta_{1,2}$. The layers are upside down, as compared to the oceanic context, in a sense that static stability corresponds to θ increasing with height, and also because it is not the pressure but the geopotential which is constant on the ground. The ground is therefore a 'free surface' for pseudo-height (pressure), while the upper boundary is fixed and isobaric, that is of constant z. The equations of the model are

$$
\begin{cases}
\partial_t \mathbf{v}_1 + (\mathbf{v}_1 \cdot \nabla) \mathbf{v}_1 + f \hat{\mathbf{z}} \wedge \mathbf{v}_1 = -g' \nabla (h_1 + h_2), \\
\partial_t \mathbf{v}_2 + (\mathbf{v}_2 \cdot \nabla) \mathbf{v}_2 + f \hat{\mathbf{z}} \wedge \mathbf{v}_2 = -g' \nabla (h_1 + \alpha h_2), \\
\partial_t h_1 + \nabla \cdot (h_1 \mathbf{v}_1) = 0, \\
\partial_t h_2 + \nabla \cdot (h_2 \mathbf{v}_2) = 0,
\end{cases}
\tag{3.25}
$$

where $\alpha = \frac{\theta_2}{\theta_1} \geq 1$ is the stratification parameter for the atmosphere, and $g' = g \frac{\theta_1}{\theta_0}$. The prime over g will be further omitted. Note that the one-layer RSW model can be, obviously, also reinterpreted in the atmospheric context along the same lines.

It should be emphasised that equations (3.25) are obtained from the primitive equations, while our derivation in section 3.1 was based on the full Euler and continuity equations. It is easy to see that the derivation goes along the same lines for primitive equations, with replacement of mass-weighted vertical averages by geometric averages. We will come back to the derivation of RSW models starting from primitive equations in Chapters 14 and 16.

3.2.4 RSW model on the sphere

The hypothesis of columnar motion, which is the foundation of shallow-water modelling, is independent of the tangent plane approximation. Vertical averaging of the primitive equations leading to RSW models can be straightforwardly performed in spherical geometry. By neglecting the variation of the fluid depth with respect to the planet's radius, that is by replacing $r \to R$ everywhere in the denominators of the equations (2.37) to (2.40) without non-traditional terms, and by vertical averaging between a rigid spherical 'bottom' and a free surface, we obtain, along the same lines as shown earlier, the standard one-layer RSW model on the sphere:

$$
\begin{cases}
\partial_t u + \dfrac{u}{R\cos\phi}\partial_\lambda u + \dfrac{v}{R}\partial_\phi u - \dfrac{uv\tan\phi}{R} - 2\Omega\sin\phi\, v = -\dfrac{g}{R\cos\phi}\partial_\lambda h, \\[2mm]
\partial_t v + \dfrac{u}{R\cos\phi}\partial_\lambda v + \dfrac{v}{R}\partial_\phi v + \dfrac{u^2\tan\phi}{R} + 2\Omega\sin\phi\, u = -\dfrac{g}{R}\partial_\phi h, \\[2mm]
\partial_t h + \dfrac{1}{R\cos\phi}\left(\partial_\lambda(hu) + \partial_\phi(hv\cos\phi)\right) = 0.
\end{cases}
\tag{3.26}
$$

Generalisations of these equations to several layers are straightforward.

3.3 Vortices and waves in rotating shallow-water models

3.3.1 One-layer RSW model

Conservation laws

By construction, equations (3.22) express the local momentum and mass conservation. The energy density of the model

$$
e = h\frac{\mathbf{v}^2}{2} + g\frac{h^2}{2}
\tag{3.27}
$$

is a sum of kinetic and potential energy densities and obeys the conservation law:

$$
\partial_t e + \nabla \cdot \left(\mathbf{v}h\left(\frac{\mathbf{v}^2}{2} + gh\right)\right) = 0.
\tag{3.28}
$$

Thus, the total energy, $E = \int dxdy\, e$, does not change in time for an isolated system. The RSW model inherits from the parent PE model the conservation of potential vorticity (PV), which is built from the relative vorticity $\zeta = v_x - u_y$, Coriolis parameter f, and the fluid depth h:

$$q = \frac{\zeta + f}{h}. \tag{3.29}$$

Here $\zeta + f$ is the absolute vorticity, the vorticity measured by an observer in the non-rotating frame, and f is planetary vorticity, the vorticity due to the overall rotation of the system, as was explained in Chapter 2. Lagrangian conservation of PV,

$$\frac{dq}{dt} = (\partial_t + \mathbf{v} \cdot \nabla)\, q = 0, \tag{3.30}$$

follows by combining the equations of vorticity, which is obtained by cross-differentiation of the equations for the two components of velocity:

$$\frac{d(\zeta + f)}{dt} + (\zeta + f)\nabla \cdot \mathbf{v} = 0, \tag{3.31}$$

and conservation of the volume of the fluid column:

$$\frac{dh}{dt} + h\nabla \cdot \mathbf{v} = 0, \tag{3.32}$$

which give

$$\frac{d}{dt}\frac{\zeta + f}{h} = \frac{1}{h}\frac{d}{dt}(\zeta + f) - \frac{\zeta + f}{h^2}\frac{d}{dt}h = 0. \tag{3.33}$$

It should be emphasised that the PV (3.29) which, at first glance has not much to do with the Ertel PV (2.70) of Chapter 2, is in fact directly related to it. Indeed, there is no explicit density gradient in RSW equations but there is a density jump across the free surface. This jump, $\Delta\rho$, and the thickness of the layer $\Delta z = z_2 - z_1 = h$ can be combined to form a (crudely) discretised vertical derivative of density: $\frac{\Delta\rho}{\Delta z} = \frac{\Delta\rho}{h}$. The shallow-water PV then is a product of absolute vorticity, which has only a vertical component $\zeta + f$ in RSW, and the discretised density gradient, in full accordance with the definition of Ertel PV (2.70). Thus, qualitatively, the dynamics of one-layer RSW may be thought of as a dynamics of an ensemble of fluid columns of variable depth moving in the rotating plane, each preserving its potential vorticity, cf. Figure 3.2.

Conservation of PV is expressed in Eulerian terms as time-independence of any integral of the form

$$C_{\mathcal{F}} = \int dxdy\, h\mathcal{F}(q), \tag{3.34}$$

over the domain of the flow, where $\mathcal{F}$ is an arbitrary function of PV. Such invariants are called Casimir invariants.

Spectrum of small perturbations

For small perturbations about the state of rest with $\mathbf{v} = 0$, $h = H_0 = const$ in the f plane, quadratic terms in the equations of motion of the model can be neglected, as explained in Chapter 2, and the following linearised equations result:

$$\begin{cases} u_t - fv + g\eta_x = 0, \\ v_t + fu + g\eta_y = 0, \\ \eta_t + H_0(u_x + v_y) = 0, \end{cases} \tag{3.35}$$

where u, v are the two components of velocity perturbation, and η is the perturbation of the free surface. The system (3.35) on the f plane has constant coefficients and can be treated by the Fourier method. By looking for solutions in the form of harmonic waves

$$(u, v, \eta) = (u_0, v_0, \eta_0)e^{i(\omega t - \mathbf{k} \cdot \mathbf{x})} + c.c., \tag{3.36}$$

where $\mathbf{k} = (k, l)$ is the two-dimensional wave number, we get a system of homogeneous algebraic equations for the amplitudes (u_0, v_0, η_0). The condition of existence of nontrivial solutions of this algebraic system is

$$\det \begin{pmatrix} i\omega & -f & -igk \\ f & i\omega & -igl \\ -iH_0k & -iH_0l & i\omega \end{pmatrix} = 0, \tag{3.37}$$

which leads to the dispersion relation between the frequency and wave number:

$$\omega \left(\omega^2 - gH_0\mathbf{k}^2 - f^2 \right) = 0. \tag{3.38}$$

This equation has three roots corresponding 1) to stationary solutions with $\omega = 0$, and 2) to propagative waves with the dispersion relation

$$\omega = \pm\sqrt{gH_0\mathbf{k}^2 + f^2}, \quad |\omega| \geq f. \tag{3.39}$$

The most striking property of this dispersion relation is the *spectral gap* between two types of solutions which expresses, as in the parent PE model (cf. Section 2.4.3), the wave–vortex dichotomy, see Figure 3.4. The waves in the model are surface waves produced by the combined effects of inertia (rotation) and gravity, the inertia–gravity waves. In the absence of rotation they become surface gravity waves. It is worth noting that the frequency of the waves has a minimum at $|\mathbf{k}| = 0$ with non-zero value $\omega = f$. This means that there exist nontrivial solutions with infinite wavelength, which are called inertial oscillations, and that very long waves have very small group velocity. As is well known, the energy of the wave packet propagates with the group velocity. Thus, long-wave

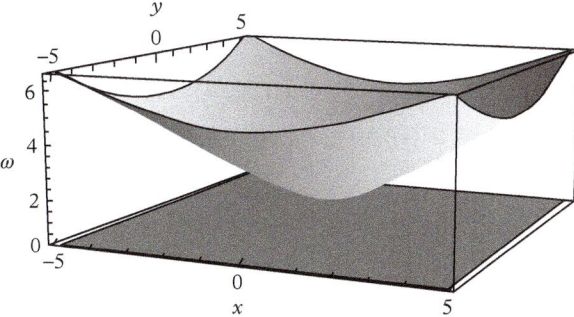

Figure 3.4 *Dispersion relation for inertia–gravity waves in the RSW model. $c = \sqrt{gH_0} = 1, f = 1$, negative values $\omega < 0$ are not shown. Solution $\omega = 0$ is also displayed in order to illustrate the spectral gap.*

perturbations will have a tendency to stay, for a long time, at their initial locations. We will see manifestations of this phenomenon in subsequent chapters. The solution $\omega = 0$ of (3.38) corresponds to the linearised PV conservation equation, as in the full primitive equations in Chapter 2.

It should be stressed that the analysis of the spectrum becomes much more complicated on the beta plane, when the coefficients of the linearised system are y-dependent. We will postpone this analysis until some additional simplifications are made.

Thus, we see that already the simplest vertically averaged model reproduces the characteristic features of the parent PE model: it maintains the wave–vortex dichotomy, and incorporates slow vortex motions related to Lagrangian conservation of PV and having zero frequency in linear approximation, and fast wave motions with frequencies separated from those of vortex motion by a spectral gap. The model also allows for inclusion of bathymetry. These key dynamical elements, and lateral boundaries (which we will study in Chapter 4), are perfectly captured by the artist's eye in Hiroshige's painting in Figure 3.5.

Figure 3.5 *Artist's view of shallow vortex dynamics: vortices, waves, and topography in shallow water in the Hiroshige's triptych 'View of the Whirlpools at Awa', 1857.*

3.3.2   Two-layer RSW model

Conservation laws

The specifics of the equations of motion of the model (3.24) (we will be considering the more general case of the two-layer model with a free surface, a reduction to the rigid-lid model is straightforward) is that while the local conservation of mass is ensured layer-wise, the exchange terms appear in the momentum equations. Indeed, we can rewrite the momentum equations in (3.24) in the form

$$\begin{cases} (\partial_t + \mathbf{v}_1 \cdot \nabla)(\mathbf{v}_1 h_1) + \mathbf{v}_1 h_1 \nabla \cdot \mathbf{v}_1 + f\hat{\mathbf{z}} \wedge (\mathbf{v}_1 h_1) = -g\nabla \dfrac{h_1^2}{2} - gh_1 \nabla h_2, \\[3mm] (\partial_t + \mathbf{v}_2 \cdot \nabla)(\mathbf{v}_2 h_2) + \mathbf{v}_2 h_2 \nabla \cdot \mathbf{v}_2 + f\hat{\mathbf{z}} \wedge (\mathbf{v}_2 h_2) = -g\nabla \dfrac{h_2^2}{2} - rgh_2 \nabla h_1. \end{cases} \tag{3.40}$$

The second term on the right-hand side of each equation takes into account the forcing of one layer by another, and does not allow us to rewrite these equations in a conservative form, even in the absence of the Coriolis force. (Obviously, the left-hand side of each equation can be rewritten as time derivative plus divergence of the respective components of the momentum density.) Only the total momentum density $\rho_1 \mathbf{v}_1 h_1 + \rho_2 \mathbf{v}_2 h_2$ thus obeys a local conservation law.

The energy of the system is a sum of potential and kinetic energies of the layers, $E = E_1 + E_2$, which are defined, up to a constant, as

$$\begin{cases} E_1 = \int dxdy\, \rho_1 \left(\dfrac{h_2}{2}\mathbf{v}_2^2 + gh_1 h_2 + g\dfrac{h_2^2}{2} \right), \\[3mm] E_2 = \int dxdy\, \rho_1 \left(\dfrac{h_1}{2}\mathbf{v}_1^2 + g\dfrac{h_1^2}{2} \right). \end{cases} \tag{3.41}$$

In the absence of forcing and dissipation, the energy is conserved. Following the lines of the demonstration in the one-layer case, it can be easily shown that PV in each layer

$$q_i = \frac{\zeta_i + f}{h_i}, \quad i = 1, 2, \tag{3.42}$$

is a Lagrangian invariant $\frac{dq_i}{dt} = 0$.

Spectrum of small perturbations

Linearising the system (3.24) about the state of rest with unperturbed thicknesses $H_{1,2}$, respectively, we get

$$\begin{cases} \partial_t \mathbf{v}_1 + f\hat{\mathbf{z}} \wedge \mathbf{v}_1 + g\nabla\,(\eta_1 + \eta_2) = 0, \\ \partial_t \eta_1 + H_1 \nabla \cdot \mathbf{v}_1 = 0, \\ \partial_t \mathbf{v}_2 + f\hat{\mathbf{z}} \wedge \mathbf{v}_2 + g\nabla\,(r\eta_1 + \eta_2) = 0, \\ \partial_t \eta_2 + H_2 \nabla \cdot \mathbf{v}_2 = 0. \end{cases} \tag{3.43}$$

In this subsection, for illustrative purposes, we choose the technically simplest configuration with $H_1 = H_2 = \frac{H}{2}$. Equations (3.43) can be simplified by means of decomposition of dependent variables in barotropic (+) and baroclinic (−) components, which reads, in the case $H_1 = H_2$:

$$\mathbf{v}^\pm = \sqrt{r}\mathbf{v}_1 \pm \mathbf{v}_2, \quad \eta^\pm = 2\left(\sqrt{r}\eta_1 \pm \eta_2\right), \tag{3.44}$$

where we introduced an additional factor 2 in the second equation for technical convenience. In terms of the baroclinic and the barotropic components (3.44), the equations (3.43) become

$$\begin{cases} \partial_t \mathbf{v}^+ + f\hat{z} \wedge \mathbf{v}^+ + g\frac{1+\sqrt{r}}{2}\boldsymbol{\nabla}\eta^+ = 0, \\ \partial_t \eta^+ + H\boldsymbol{\nabla} \cdot \mathbf{v}^+ = 0, \end{cases} \tag{3.45}$$

$$\begin{cases} \partial_t \mathbf{v}^- + f\hat{z} \wedge \mathbf{v}^- + g\frac{1-\sqrt{r}}{2}\boldsymbol{\nabla}\eta^- = 0, \\ \partial_t \eta^- + H\boldsymbol{\nabla} \cdot \mathbf{v}^- = 0. \end{cases} \tag{3.46}$$

As follows from (3.45), (3.46), the system is decoupled into two rotating shallow-water subsystems (the barotropic and the baroclinic ones) with different effective gravity accelerations $\frac{1\pm\sqrt{r}}{2}g$, respectively. The wave spectrum of rotating shallow water in the infinite domain is given by (3.39). Hence, for barotropic (upper sign) and baroclinic (lower sign) inertia–gravity waves we get the dispersion relations, cf. Figure 3.6,

$$\omega_\pm^2 = c_\pm^2 \mathbf{k}^2 + f^2, \tag{3.47}$$

where ω is the wave frequency and $\mathbf{k} = (k, l)$ is the wave vector, and we introduced the phase velocities of the baroclinic and barotropic gravity waves:

$$c_\pm = \sqrt{gH\frac{1 \pm \sqrt{r}}{2}}. \tag{3.48}$$

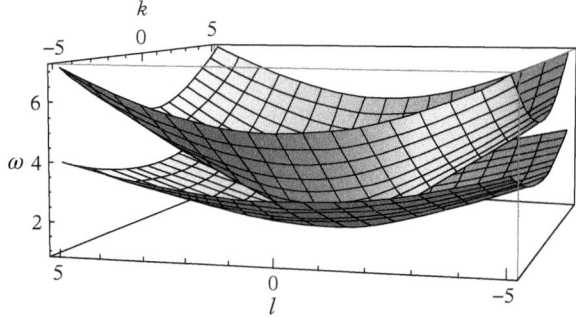

Figure 3.6 *Dispersion relation for inertia–gravity waves in the two-layer RSW model. $c_+ = 1$, $c_- = 0.3$, $f = 1$ (non-positive values $\omega \leq 0$ are not shown).*

The full dispersion relation for the small perturbations also contains two stationary solutions, corresponding to linearised Lagrangian conservation of PV layer-wise, as is easy to see.

Thus, the dynamical content of the two-layer RSW consists, again, of slow vortex motions and fast waves, although the baroclinic effects lead to doubling of each species. An important observation is that baroclinic waves are slower, and barotropic ones are faster, which is a general and model-independent fact.

3.4   Lagrangian approach and variational principles for shallow-water models

3.4.1   Lagrangian formulation of one-layer RSW

RSW in Lagrangian coordinates

The view of the RSW as an ensemble of fluid columns of variable depth moving in the plane can be given a literal meaning if the coordinates of the columns along their trajectories $(X(x,y;t), Y(x,y;t))$, where (x,y) is a position of the column at $t = 0$, are used as dynamical variables. The velocity of the column in these terms is $(\dot{X}, \dot{Y}) = (u(X, Y; t), v(X, Y; t))$. This gives the Lagrangian description of the model along the general lines explained in Section 2.2.1. Let us consider, for simplicity, a configuration with a flat bottom. The conservation of the volume of the column is expressed as

$$h(X, Y; t)\, dX\, dY = h_I(x, y)\, dx\, dy, \tag{3.49}$$

where h_I is the initial distribution of thickness. Therefore, h is not an independent dynamical variable, and is expressed in terms of Lagrangian coordinates as

$$h(X, Y; t) = \frac{h_I(x, y)}{\mathcal{J}} \tag{3.50}$$

where $\mathcal{J} = \frac{\partial(X,Y)}{\partial(x,y)}$ is the Jacobian of the transformation $(x, y) \rightarrow (X, Y)$. These formulae represent a transposition of general Euler–Lagrange duality rules of Section 2.2.1 to the shallow-water context. The momentum equations are written as

$$\begin{cases} \ddot{X} - f\dot{Y} = -g\partial_X h = -\dfrac{g}{\mathcal{J}} \dfrac{\partial(h, Y)}{(x, y)}, \\[2mm] \ddot{Y} + f\dot{X} = -g\partial_Y h = -\dfrac{g}{\mathcal{J}} \dfrac{\partial(X, h)}{(x, y)}, \end{cases} \tag{3.51}$$

where, the relation (3.50) is to be used. Note that by an additional change of variables the initial (continuous) thickness h_I can be 'straightened' $h_I \rightarrow H = $ const, and we will be supposing this latter choice later in this chapter, taking the constant to be equal to one without loss of generality. This fact expresses a liberty of choosing the Lagrangian 'labels' (x, y) at convenience. The Lagrangian equations of motion become especially simple and intuitive if a one-dimensional reduction of the RSW is considered, which will be used in Chapters 7 and 8.

Hamilton's principle for RSW

In the absence of rotation ($f = 0$), the equations (3.51) can be obtained from a physically transparent variational (Hamilton's) principle with a Lagrangian in the standard form of difference between kinetic and potential energy of the fluid, cf. Section 2.2.1. Indeed, the mass of the fluid column is $\rho h\, dX dY$, where ρ is the density of the fluid, which is constant by construction of the model. Thus, the Lagrangian has the form

$$L = \rho \int dxdy\, \mathcal{L} = \rho \int h dX dY \left(\frac{\dot{X}^2 + \dot{Y}^2}{2} - \frac{gh(X, Y)}{2} \right), \tag{3.52}$$

where the integration is performed over the whole domain occupied by the flow. The first term in the integrand is the kinetic energy of a fluid column, and the second is its potential energy. The factor 1/2 in front of the second term is due to the fact that the potential energy is obtained by vertical integration of the potential energy density of fluid particles which constitute the column $\int_0^h dz\, \rho g z$. According to (3.49), $h dX\, dY$ can be replaced by $h_I(x, y)\, dx\, dy$, and the above-mentioned change of Lagrangian labels $h_I dxdy \to dxdy$ eliminates h_I. In this way we get

$$L = \rho \int dxdy \left(\frac{\dot{X}^2 + \dot{Y}^2}{2} - \frac{gh(X, Y)}{2} \right), \tag{3.53}$$

and the variation of the Lagrangian density, taking into account (3.50), gives

$$\begin{aligned}
\delta\mathcal{L} &= \dot{X}\delta\dot{X} + \dot{Y}\delta\dot{Y} + \frac{gh_I}{2\mathcal{J}}\, \delta\mathcal{J} \\
&= \dot{X}\delta\dot{X} + \dot{Y}\delta\dot{Y} + \frac{gh_I}{2\mathcal{J}} \left[\frac{\partial\, (\delta X, Y)}{\partial\, (x, y)} + \frac{\partial\, (X, \delta Y)}{\partial\, (x, y)} \right].
\end{aligned} \tag{3.54}$$

By integrating by parts in time and in x, y, imposing vanishing of the variations at the integration limits, and using (3.50), we arrive at

$$\begin{aligned}
\delta\mathcal{S} &= \rho \int dt \int dxdy\, \delta\mathcal{L} \\
&= \rho \int dt \int dxdy \left[-\ddot{X}\delta X - \ddot{Y}\delta Y - g \left(\frac{\partial\left(\frac{h_I}{2\mathcal{J}^2}, Y \right)}{\partial\, (x, y)} \delta X + \frac{\partial\left(X, \frac{h_I}{2\mathcal{J}^2} \right)}{\partial\, (x, y)} \delta Y \right) \right] \\
&= \rho \int dt \int dxdy \left[\left(-\ddot{X} - g\frac{\partial\, (h, Y)}{\partial\, (X, Y)} \right) \delta X + \left(-\ddot{Y} - \frac{\partial\, (X, h)}{\partial\, (X, Y)} \right) \delta Y \right], \tag{3.55}
\end{aligned}$$

and the condition of vanishing variation of $\mathcal{S}$ at any $\delta X, \delta Y$ gives (3.51) with $f = 0$.

Rotation can be introduced following the recipe of Section 2.3.1 which leads, in the f-plane approximation, to

$$S = \rho \int dt\,dx\,dy\,\mathcal{L} = \rho \int dx\,dy \left(\frac{\dot{X}^2 + \dot{Y}^2}{2} - \frac{gh(X,Y)}{2} - fY\dot{X} \right). \tag{3.56}$$

The passage to the β-plane approximation and introduction of bottom topography in the variational principle are straightforward. We will come back to these subjects in Chapter 16.

Symmetries and conservation laws

The main advantage of the variational formulation of any physical system is that it allows us to obtain its conservation laws from symmetries corresponding to invariance of the action with respect to various transformations of independent coordinates (Noether theorem). For example, the fact that the Lagrangian does not explicitly depend on time is equivalent to the invariance with respect to time shifts. Differentiation of the Lagrangian in time and integration by parts then gives

$$\int dx\,dy \left[\frac{d}{dt} \left(\frac{\delta\mathcal{L}}{\delta\dot{X}}\dot{X} + \frac{\delta\mathcal{L}}{\delta\dot{Y}}\dot{Y} - \mathcal{L} \right) + \frac{\partial}{\partial x} \left(\frac{\delta\mathcal{L}}{\delta X_x}\dot{X} + \frac{\delta\mathcal{L}}{\delta Y_x}\dot{Y} \right) + \frac{\partial}{\partial y} \left(\frac{\delta\mathcal{L}}{\delta X_y}\dot{X} + \frac{\delta\mathcal{L}}{\delta Y_y}\dot{Y} \right) \right], \tag{3.57}$$

which is equivalent to the energy conservation law (3.28). Here and later the notation $X_x = \partial_x X$ etc. is used, for compactness.

This interpretation of the conservation laws allows us to understand the nature of probably the most important of them, the PV conservation, which appears to be a consequence of invariance with respect to relabelling of fluid particles (columns). In order to understand the relation between the symmetry with respect to relabelling and PV, let us first obtain the expression of the latter in Lagrangian terms. The Lagrangian relative vorticity ζ has the following form:

$$\zeta = \frac{\partial\dot{Y}}{\partial X} - \frac{\partial\dot{X}}{\partial Y} = \frac{\partial(\dot{Y},Y)}{\partial(X,Y)} - \frac{\partial(X,\dot{X})}{\partial(X,Y)} = \frac{1}{\mathcal{J}} \left[\frac{\partial(\dot{Y},Y)}{\partial(x,y)} - \frac{\partial(X,\dot{X})}{\partial(x,y)} \right], \tag{3.58}$$

and the Lagrangian expression for PV is given by

$$q = \frac{\zeta + f}{h} = \frac{1}{h_I} \left[\frac{\partial(\dot{Y},Y)}{\partial(x,y)} - \frac{\partial(X,\dot{X})}{\partial(x,y)} + f\mathcal{J} \right]. \tag{3.59}$$

As was already mentioned, h_I can be removed by an additional change of independent variables. Relabelling means such change of independent variables that does not affect physical quantities. A change of variables is characterised by its Jacobian. As follows from (3.50), in order not to modify the thickness distribution, a relabelling

$(x, y) \rightarrow (x', y') = (x + \delta x, y + \delta y)$ should not change $\mathcal{J}$. It is easy to see that this means that

$$\delta x = -\partial_y \chi, \quad \delta y = +\partial_x \chi, \tag{3.60}$$

where $\chi(x, y)$ is an arbitrary function. The velocity of the column should not change under relabelling either, which means that

$$\frac{d}{dt}\left(X(x + \delta x, y + \delta y), Y(x + \delta x, y + \delta y)\right) = \frac{d}{dt}\left(X(x, y), Y(x, y)\right) \tag{3.61}$$

whence we get, for the x velocity,

$$\begin{aligned}
\delta \dot{X} &= \dot{X}(x + \delta x, y + \delta y) - \dot{X}(x, y) \\
&= \dot{X}_x \delta x + \dot{X}_y \delta y = -X_x \frac{d}{dt}\delta x - X_y \frac{d}{dt}\delta y,
\end{aligned} \tag{3.62}$$

and similarly for the y velocity. By definition, h, and hence the potential energy part of the Lagrangian, does not change under relabelling. The variation of the action (3.56) due to relabelling thus consists in variation of the kinetic energy part, and gives

$$\begin{aligned}
\delta \mathcal{S} &= \rho \int dt \int dx dy \left[-\left(\dot{X} - fY\right)\left(X_x \frac{d}{dt}\delta x - X_y \frac{d}{dt}\delta y\right) - \dot{Y}\left(Y_x \frac{d}{dt}\delta x - Y_y \frac{d}{dt}\delta y\right) \right] \\
&= \rho \int dt \int dx dy \left[-\left(\left(\dot{X} - fY\right)X_x + \dot{Y}Y_x\right)\frac{d}{dt}\delta x - \left(\left(\dot{X} - fY\right)X_y + \dot{Y}Y_y\right)\frac{d}{dt}\delta y \right] \\
&= \rho \int dt \int dx dy \left[\frac{d}{dt}\left(\left(\dot{X} - fY\right)X_x + \dot{Y}Y_x\right)\delta x + \frac{d}{dt}\left(\left(\dot{X} - fY\right)X_y + \dot{Y}Y_y\right)\delta y \right],
\end{aligned} \tag{3.63}$$

where an integration by parts in time was used to obtain the last line. Recalling (3.60), integrating by parts once more, and requiring that the variation of the action vanishes for any χ, we get

$$\begin{aligned}
\frac{d}{dt}&\left[\partial_y\left(\left(\dot{X} - fY\right)\partial_x X + \dot{Y}\partial_x Y\right) - \partial_x\left(\dot{X} - fY\right)\left(\partial_y X + \dot{Y}\partial_y Y\right)\right] \\
&= \frac{d}{dt}\left[\frac{\partial\left(\dot{X} - fY, X\right)}{\partial(x, y)} + \frac{\partial\left(\dot{Y}, Y\right)}{\partial(x, y)}\right] = 0.
\end{aligned} \tag{3.64}$$

Comparing this with (3.59), we recognise the conservation of PV (3.30) in Lagrangian form.

3.4.2 Lagrangian formulation of two-layer RSW

Lagrangian description of two-layer RSW with a free surface

Equations (3.24) can be easily rewritten in the Lagrangian form:

$$\ddot{X}_1 + f\hat{z} \wedge \dot{X}_1 = -g\nabla_{X_1}(h_1 + h_2),$$
$$\ddot{X}_2 + f\hat{z} \wedge \dot{X}_2 = -g\nabla_{X_2}(rh_1 + h_2). \tag{3.65}$$

Here $X_{1,2} = (X_{1,2}, Y_{1,2})$ are functions of Lagrangian labels (x, y), as in the one-layer case, and $h_{1,2}$ are not independent variables, as mass conservation layer-wise in (3.24) gives

$$h(X_i, Y_i; t) = h_{I_i}(x, y)\frac{\partial(x, y)}{\partial(X_i, Y_i)}, \quad i = 1, 2. \tag{3.66}$$

The 'foreign' h in the equations (3.65) (e.g. h_2 in the first equation) is considered as a function of 'native' X (correspondingly, X_1 in the first equation, and similarly for the second equation). As in the one-layer case, the initial thicknesses h_{I_i} can be rendered constant by an additional transformation of Lagrangian labels, which will be assumed.

Hamilton's principle for two-layer RSW

The variational principle can be constructed along the same lines as in the one-layer case, by considering the Lagrangian as a difference between kinetic and potential energies layer-wise. Rewriting in Lagrangian variables all terms entering the expression for the energy is straightforward, except for the exchange term which mixes variables belonging to different layers. The full Lagrangian thus has three entries:

$$L = L_1 + L_2 + L_{12}. \tag{3.67}$$

where

$$L_i = \rho_i \int dxdy \left(\frac{\dot{X}_i^2 + \dot{Y}_i^2}{2} - \frac{gh_i(X, Y)}{2}\right), \quad i = 1, 2, \tag{3.68}$$

and (3.66) is understood. As to the exchange term $L_{12} = \rho_1 \int dxdy\, h_1 h_2$, cf. (3.41), it can be checked that it can be written as

$$L_{12} = \rho_1 \int dx_1 dy_1 \int dx_2 dy_2\, \delta\left(X_1(x_1, y_1) - X_2(x_2, y_2)\right), \tag{3.69}$$

where δ denotes the Dirac's delta function of the corresponding argument, with the property

$$\int dxdy\, F(\mathbf{x})\delta(\mathbf{x} - \mathbf{x}') = F(\mathbf{x}'), \quad \mathbf{x} = (x, y).$$

While varying this expression with respect to, for example, X_1, the integration measure $dx_2 dy_2$ in (3.69) should be changed to $dX_2 dY_2$, which leads to appearance of the corresponding Jacobian, and hence h_2, via (3.66). The integration over $dX_2 dY_2$ is then lifted with the delta functions in the integrand of (3.69), which transforms, at the same time, h_2 into a function of X_1, Y_1. It can be directly verified that the variational principle with the Lagrangian (3.67)-(3.69) leads to the two-layer RSW equations (3.65). As in the one-layer case it can be verified that PV conservation layer-wise follows from the invariance of the action with respect to relabelling, also layer-wise.

3.5   Summary, comments, and bibliographic remarks

Thus, we have seen that vertical averaging of the primitive equations between material surfaces results, under hypothesis of columnar motion, in a simplified RSW model for a fluid layer. The model inherits the PV conservation and wave–vortex dichotomy of the parent model. The baroclinic effects can be incorporated in the model in its multi-layer versions, as well as nontrivial topography.

RSW equations as a vertically averaged model of the fluid envelopes of the Earth were, probably, first considered in Jeffries (1925), and were used by classics of geophysical fluid dynamics (Rossby 1938, Obukhov 1949) as a conceptual model of fundamental dynamical phenomena. The derivation of the RSW models above follows Zeitlin (2007). Usually the shallow-water equations are derived from the Euler equations for the incompressible fluid with a free surface under columnar motion hypothesis. It is, however, important to realise that they represent a vertically averaged motions in a more general context.

Formulation of the Hamilton's principle for RSW is most natural and physically transparent in terms of Lagrangian variables. It was introduced in Salmon (1982) both for one- and two-layer models, as well as interpretation of PV conservation in terms of particle relabelling symmetry. The corresponding Hamiltonian dynamics follows directly by applying the general rules recalled in Section 2.2.1. For an alternative formulation see Allen and Holm (1996).

A number of issues raised while building foundations of weather and climate modelling were first addressed and solved in the RSW model, such as the initialisation problem (Baer and Tribbia 1977, Machenhauer 1977). A series of tests, which became compulsory in developing dynamical cores of general circulation models, was elaborated within the RSW model on the sphere (Williamson *et al.* 1992). It should be emphasised that vertical discretisation of the primitive equations, necessary for numerical implementations, de facto leads to appearance of a stack of fluid layers, each obeying the RSW-like dynamics. We will be following a long tradition of using this conceptual model for understanding fundamental dynamical phenomena throughout this book. Thousands of papers based on RSW models exist in the literature and it is impossible to cite them all. Key citations will be given in subsequent chapters, in relation with the corresponding topics.

3.6 Problems

3.1 Derive the equations of the one- and two-layer RSW models from the master equations (3.19), (3.20).

3.2 Derive the equations (3.25) by vertical integration of the primitive equations for the atmosphere in pseudo-height coordinates.

3.3 Derive the RSW equations in cylindrical coordinates.

3.4 Derive the local conservation of energy (3.28) in the RSW model from the equations of motion (3.22).

3.5 Derive the vorticity equation (3.31).

3.6 Demonstrate that the Casimir invariant (3.34) is an integral of motion.

3.7 Find the polarisation relations, that is the expressions of u_0 and v_0 in terms of η_0 in (3.36) which correspond to different roots of the dispersion relation (3.38). Show that the root $\omega = 0$ corresponds to linearised conservation of PV.

3.8 Show that linear inertia–gravity waves have zero PV anomaly $q - \frac{f}{H_0}$.

3.9 By using the barotropic–baroclinic decomposition (3.44), obtain equations (3.45) and (3.46) from (3.43).

3.10 Demonstrate that (3.57) is equivalent to (3.28).

3.11 In the absence of rotation, demonstrate that invariance of the action functional of the one-layer RSW with respect to spatial translations leads to momentum conservation and derive the corresponding momentum conservation equations in terms of Lagrangian variables.

4

Wave Motions in Rotating Shallow Water with Boundaries, Topography, at the Equator, and in Laboratory

In the Chapter 3 we were mostly interested in intrinsic dynamical properties of the RSW models and did not discuss non-trivial boundary conditions. We will see in the present chapter that lateral boundaries and topography, or a passage from the mid-latitude to equatorial tangent plane, lead to the appearance of completely new elements in the dynamics. We will concentrate in this chapter on wave motions; the slow vortex motions will be considered in subsequent chapters. As we will see, the above-mentioned new elements engender new wave species, some of them sub-inertial, yet in the long-wave limit they remain well-separated in frequency from the fast inertia–gravity waves, and thus do not destroy the spectral gap discussed in the previous chapter.

4.1 Introducing lateral boundaries and shelf

In this section we will be considering the effects of the idealised rectilinear 'coast' and associated shelf-like bathymetry. As previously, we will be not including viscous effects, supposing that related boundary layers are sufficiently thin with respect to characteristic span-wise scales of the flow.

4.1.1 Kelvin waves in RSW with an idealised coast

One-layer RSW equations in a half-plane with a rectilinear coast, for example, at $x = 0$, represent the simplest configuration with a lateral boundary. To see how the seemingly innocuous change from the whole plane to a half-plane modifies dynamical content of the model, we linearise the equations of motion, as in Section 3.3:

$$\begin{cases} u_t - fv + g\eta_x = 0, \\ v_t + fu + g\eta_y = 0, \\ \eta_t + H_0(u_x + v_y) = 0. \end{cases} \quad (4.1)$$

Geophysical Fluid Dynamics. Vladimir Zeitlin,
Oxford University Press (2018). © Vladimir Zeitlin. DOI: 10.1093/oso/9780198804338.001.0001

These equations should now be supplied with a non-penetration (free-slip) boundary condition at the boundary, $u|_{x=0} = 0$, which renders the system inhomogeneous in x. Still, the Fourier transformation in y and t is legitimate: $(u, v, \eta) = \int dl\, d\omega\, (\bar{u}_0, \bar{v}_0, \bar{\eta}_0)e^{i(ly - \omega t)} + c.c.$, giving for each Fourier harmonic

$$\begin{cases} -i\omega\bar{u}_0 - f\bar{v}_0 + g\bar{\eta}_0' = 0, \\ -i\omega\bar{v}_0 + f\bar{u}_0 + ilg\bar{\eta}_0 = 0, \\ -i\omega\bar{\eta}_0 + H_0(il\bar{v}_0 + \bar{u}_0') = 0. \end{cases} \tag{4.2}$$

This system can be reduced, under the assumption that $\omega \neq f$, to a single equation for $\bar{\eta}_0$

$$\bar{\eta}_0'' + (\omega^2 - f^2 - gH_0l^2)\bar{\eta}_0 = 0, \tag{4.3}$$

while

$$\bar{u}_0 = i\frac{l\bar{\eta}_0 - \omega\bar{\eta}_0'}{\omega^2 - f^2}, \tag{4.4}$$

and, hence, the boundary condition in terms of $\bar{\eta}_0$ is

$$(fl\bar{\eta}_0 - \omega\bar{\eta}_0')\big|_{x=0} = 0. \tag{4.5}$$

Solutions of (4.3) are now of two types:

- *free* inertia–gravity waves with

$$\omega^2 - f^2 - gH_0l^2 \equiv gH_0k^2 > 0, \Rightarrow \tag{4.6}$$

$$\bar{\eta}_0 \propto e^{\pm ikx}, \omega^2 = f^2 + gH_0(k^2 + l^2). \tag{4.7}$$

- *trapped* at the boundary waves with

$$\omega^2 - f^2 - gH_0l^2 \equiv -\kappa^2 < 0, \tag{4.8}$$

$$\bar{\eta}_0 \propto e^{-\kappa x}, \Rightarrow \kappa > 0. \tag{4.9}$$

The second type of solution is exponentially growing at $x < 0$. This is why it was discarded considering the whole plane, while it is perfectly admissible in the right half-plane. Such waves are called boundary Kelvin waves. They are trapped near the coast, and are dispersionless. Indeed, the boundary condition (4.5) implies that $\kappa = -\sqrt{gH_0}\frac{fl}{\omega}$, whence $\omega^2 - f^2 - gH_0l^2 + gH_0\frac{f^2l^2}{\omega^2} = 0$, and $\omega^2 = gH_0l^2$. As $\kappa > 0$, we get $\omega = -\sqrt{gH_0}l$, and $\eta \propto e^{-\frac{x}{R_d}}$, where $R_d = \frac{\sqrt{gH_0}}{f}$ is the *Rossby deformation radius*. It is constructed of global parameters and represents an intrinsic scale of the RSW model, which we will encounter

many times throughout the book. Due to their dispersionless character, any packet of
Kelvin waves:

$$(u, v, \eta) = (0, K(y + \sqrt{gH_0}t), -K(y + \sqrt{gH_0}t))e^{-\frac{x}{R_d}}, \qquad (4.10)$$

where K is an arbitrary function, is a solution of linearised RSW equations. From the
previous calculations it is clear that the scaling where time is measured in units of f^{-1},
and distances in units of R_d, is natural and will be used from now on.

Kelvin waves are, therefore, travelling along the boundary, leaving it on their right (it is
easy to check this rule for any orientation of the boundary). The normal to the boundary
component of the velocity is absent in the Kelvin-wave solution, and the along-boundary
velocity and height anomaly are opposite. Needless to say that the same treatment may be
repeated in the two-layer model, with doubling of waves of each species, cf. Figure 4.1.
As is clear from their dispersion relation, Kelvin waves fill the gap between fast inertia–
gravity waves and slow vortex motions (one can easily check that a nontrivial solution of
the system (4.1) with $\omega = 0$ still exists, corresponding to the conservation of potential
vorticity (PV) in the linear approximation). Thus, dynamical content and properties of
the system with lateral boundaries are changed with respect to the unbounded case. Yet,
Kelvin waves, which can have arbitrarily low frequencies, do not bear PV, as can be
checked by a direct calculation (cf. Problem 4.1 below). In this sense they resemble fast
inertia–gravity waves.

The 'standard' inertia–gravity waves are also affected by the boundary, although their
dispersion relation does not change with respect to the case of unbounded domain. Free
propagation of inertia–gravity waves in any direction now is modified by the boundary,

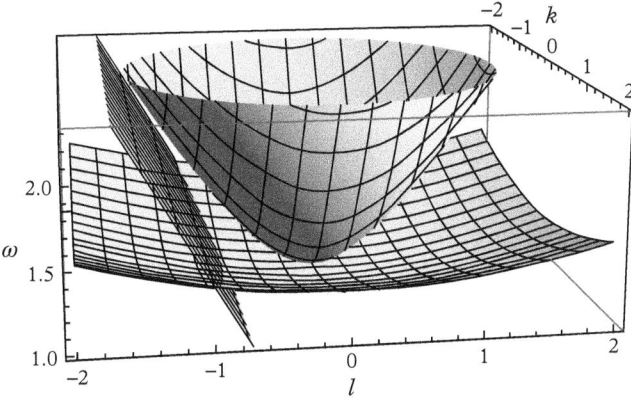

Figure 4.1 *Dispersion relation for inertia–gravity and Kelvin
waves in the two-layer RSW model. To avoid overcharging the
figure, baroclinic Kelvin waves are not shown. Upper surface:
barotropic inertia–gravity waves; Lower surface: baroclinic
inertia-gravity waves; Inclined plane: barotropic Kelvin waves.*

where they experience reflection. A general solution of the wave equation for these waves should be therefore considered as a combination of incident and reflected waves, $(u, v, \eta) = (u_i, v_i, \eta_i) + (u_r, v_r, \eta_r)$. Under the scaling introduced earlier,

$$
\begin{cases}
(u_i, v_i, \eta_i) = A_i \left(\dfrac{-k\omega + il}{\omega^2 - 1}, \dfrac{l\omega + ik}{\omega^2 - 1}, 1 \right) e^{i(-kx+ly-\omega t)} + \text{c.c.}, \\[3mm]
(u_r, v_r, \eta_r) = A_r \left(\dfrac{k\omega + il}{\omega^2 - 1}, \dfrac{l\omega - ik}{\omega^2 - 1}, 1 \right) e^{i(kx+ly-\omega t)} + \text{c.c.}.
\end{cases}
\tag{4.11}
$$

The lateral boundary condition leads to a relation between the amplitudes of incident and reflected waves:

$$
u_i + u_r |_{x=0} = 0, \; \Rightarrow A_r = A_i \frac{k\omega - il}{k\omega + il}, \; \omega^2 = 1 + k^2 + l^2,
\tag{4.12}
$$

which expresses Snell's law in this context.

4.1.2   Waves in RSW with idealised coast and a shelf

A shelf means variable depth of the fluid in the state of rest near the coast, cf. Figure 4.2. So we rewrite the one-layer RSW equation by dividing the full fluid depth in two parts: the non-perturbed depth $H(x, y)$, and a perturbation of the free surface $\eta(x, y, t)$:

$$
\begin{cases}
u_t + uu_x + vu_y - fv + g\eta_x = 0, \\
v_t + uv_x + vv_y + fu + g\eta_y = 0, \\
\eta_t + [[(H(x,y) + \eta)u]_x + [(H(x,y) + \eta)v]_y = 0,
\end{cases}
\tag{4.13}
$$

and consider, for simplicity, a one-dimensional topography $H = H(x)$. With the above scaling, the linearised non-dimensional shallow-water equations in the presence of topography/bathymetry are

$$
\begin{cases}
u_t - v + \eta_x = 0, \\
v_t + u + \eta_y = 0, \\
\eta_t + (Hu)_x + (Hv)_y = 0.
\end{cases}
\tag{4.14}
$$

We work with the straight coast introduced in Section 4.1.1, and suppose that H depends only of the perpendicular to the coast variable x. We will be distinguishing the cases of abrupt and gentle shelves, with rapid variation of H close to the boundary $x = 0$ in the first case, and variations of the order unity in the second.

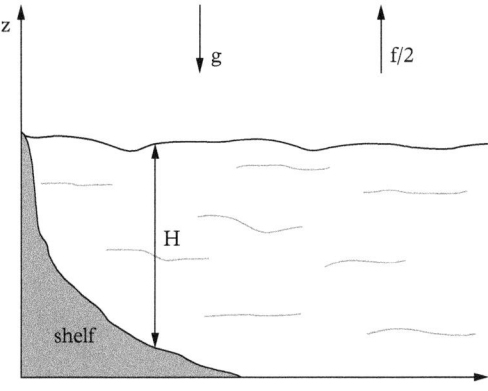

Figure 4.2 *Schematic representation of the shallow-water model with a shelf.*

Abrupt shelf

The typical horizontal scale of the shelf in this case is $L << R_d$. We thus have a small parameter $\frac{L}{R_d} = \epsilon$. The non-dimensional H, and its boundary conditions are

$$H = H\left(\frac{x}{\epsilon}\right), \quad H|_{x=0} = 0, \quad H|_{x=\infty} = 1.$$

Looking for wave solutions $(u, v, \eta) = (\bar{u}_0, \bar{v}_0, \bar{\eta}_0)e^{i(ly-\omega t)} + \text{c.c.}$, we get, in non-dimensional form, with the prime denoting x-derivative

$$\begin{cases} -i\omega\bar{u}_0 - \bar{v}_0 + \bar{\eta}_0' = 0, \\ -i\omega\bar{v}_0 + \bar{u}_0 + il\bar{\eta}_0 = 0, \\ -i\omega\bar{\eta}_0 + ilH\bar{v}_0 + (H\bar{u}_0)' = 0, \end{cases} \tag{4.15}$$

which can be reduced to a single equation:

$$\left(H\bar{\eta}_0'\right)' + (\omega^2 - 1 - l^2 H - \frac{l}{\omega}H')\bar{\eta}_0 = 0. \tag{4.16}$$

We have a small parameter ϵ at our disposal, so we can use multi-scale asymptotic expansions, which consist in representing all variables in series in ϵ, introducing multiple time- and/or space scales, and solving the problem order by order in ϵ. We will see several applications of this technique throughout the book. We thus consider two regions in x, the open ocean with a flat bottom, as the variations of H cease rapidly out of the coast, and the coastal domain, where these variations can not be neglected.

In the open ocean the wave equation (4.16) becomes

$$\bar{\eta}_0'' + (\omega^2 - 1 - l^2)\bar{\eta}_0 = 0. \tag{4.17}$$

It obviously has free-wave solutions, incident and reflected by the coast, but we are interested in the trapped-near-the-coast solutions of the form $\bar{\eta}_0^{(h)} = Ae^{-\kappa x}$, with $\kappa > 0$, and $\kappa^2 = l^2 + 1 - \omega^2$. We will be supposing that $\kappa = \kappa_0 + \epsilon \kappa_1 + \ldots$, and $\omega = \omega_0 + \epsilon \omega_1 + \ldots$.

In the coastal domain,

$$\frac{1}{\epsilon^2} \left(H(\xi) \bar{\eta}_0^{(c)}(\xi)' \right)' + \left(\omega^2 - 1 - l^2 H(\xi) - \frac{1}{\epsilon} \frac{l}{\omega} H'(\xi) \right) \bar{\eta}_0^{(c)} = 0, \qquad (4.18)$$

where we introduced a 'stretched' independent variable $\xi = \dfrac{x}{\epsilon}$, and we are looking for a solution in the form

$$\bar{\eta}_0^{(c)}(\xi) = \bar{\eta}^{(0)}(\xi) + \epsilon \bar{\eta}^{(1)}(\xi) + \ldots. \qquad (4.19)$$

We get the following equations for $\bar{\eta}^{(0)(n)}$, $n = 0, 1$:

$$\left(H(\xi) \bar{\eta}^{(0)}(\xi)' \right)' = 0, \qquad (4.20)$$

$$\left(H(\xi) \bar{\eta}^{(1)}(\xi)' \right)' - \frac{l}{\omega_0} H'(\xi)) \bar{\eta}^{(0)}(\xi) = 0. \qquad (4.21)$$

Integrating (4.20) we get

$$H(\xi) \bar{\eta}^{(0)}(\xi)' = C = \text{const}. \qquad (4.22)$$

If $C \neq 0$, a singularity at $x = 0$ is inevitable, and hence $\bar{\eta}^{(0)} = \text{const}$. At $x = \epsilon \xi$, $\bar{\eta}_0^{(h)}$ is given by the series

$$\bar{\eta}_0^{(h)} = A \left(1 - \kappa_0 \epsilon \xi + \frac{1}{2} \kappa_0^2 (\epsilon \xi)^2 - \epsilon^2 \kappa_1 \xi + \ldots \right). \qquad (4.23)$$

Matching it to $\bar{\eta}_0^{(c)}$ gives, to the lowest order in ϵ, that $\bar{\eta}^{(0)} = A$. Integrating (4.21) we thus obtain

$$\left(H(\xi) \bar{\eta}^{(1)}(\xi)' \right) - \frac{l}{\omega_0} H'(\xi)) A = C_1 = \text{const}. \qquad (4.24)$$

Regularity of the solution for $\bar{u}_0, \bar{v}_0$ necessitates $C_1 = 0$. Hence,

$$\bar{\eta}^{(1)} = \frac{l}{\omega_0} A\xi + C_2. \qquad (4.25)$$

Next-order matching of $\bar{\eta}_0^{(c)}(\xi) = \bar{\eta}^{(0)} + \epsilon \bar{\eta}^{(1)}$ to $\bar{\eta}_0^{(h)}$ at $x = \epsilon \xi$ gives $\frac{l}{\omega_0} = -\kappa_0$, and $C_2 = 0$. Since $\kappa^2 = l^2 + 1 - \omega^2$, $\omega^2 \neq 1$, it follows that $\kappa_0 = 1$. We have therefore proved that, to the lowest order in the non-dimensional width of the shelf, the trapped wave over the abrupt shelf is just the Kelvin wave of Section 4.1.1. Going further to the next orders is not straightforward and necessitates taking into account logarithmic corrections to the asymptotic expansion in powers of ϵ.

Shelf with a gentle slope

We now consider a shelf with characteristic variations of a scale comparable with the deformation radius. The linearised RSW equations with a shelf can be still reduced to a single equation, which we write in dimensional form:

$$\left(gH\bar{\eta}_0'\right)' + \left(\omega^2 - f^2 - gHl^2 - \frac{fl}{\omega}gH'\right)\bar{\eta}_0 = 0. \tag{4.26}$$

For a general shelf profile $H(x)$ this equation can not be solved analytically. There exist, however, some profiles where all calculations can be done by hand. A well-known example is the Ball's model with an exponential profile of H:

$$H(x) = H_0(1 - e^{-ax}). \tag{4.27}$$

A convenient scaling in the case of linearised equations over a gentle shelf is the following: typical width of the shelf L as length scale, $L/\sqrt{gH_0}$ as time scale, typical velocity scale U for velocity, and $U\sqrt{H/g}$ as the scale of free-surface elevation. It will be used later with f denoting non-dimensional f henceforth.

If we are looking for trapped-near-the-coast solutions, the changes of dependent and independent variables $x \to s = e^{-ax}$, $\bar{\eta}_0 \to s^p\bar{\eta}_0$, where p is defined by

$$\omega^2 - f^2 - l^2 = -p^2 < 0, \tag{4.28}$$

cast the equation (4.26) to the hypergeometric form

$$s(1 - s)\tilde{\eta}_0'' + [\gamma - (\alpha + \beta + 1)]\,\tilde{\eta}_0' - \alpha\beta\tilde{\eta}_0 = 0, \tag{4.29}$$

where

$$\gamma = 2p + 1, \quad \alpha = p + \frac{1}{2} - \sqrt{l^2 - \frac{fl}{\omega} + \frac{1}{4}}, \quad \beta = p + \frac{1}{2} + \sqrt{l^2 - \frac{fl}{\omega} + \frac{1}{4}}. \tag{4.30}$$

A regular at $x = 0$ and decaying at $x \to \infty$ solution of the hypergeometric equation exists if parameter α is a negative integer, $\alpha = -n$, $n = 0, 1, \ldots$. In this case,

$$\bar{\eta}_0 = s^p F(-n, \beta, \gamma, s), \quad n = 0, 1, \ldots, \tag{4.31}$$

where

$$F(-n, \beta, \gamma, s) = \sum_{m=0}^{n} \frac{(-n)_m(\beta)_m}{(\gamma)_m m!} s^m, \quad (a)_m := a(a+1)\ldots(a+m-1), \tag{4.32}$$

is a truncated hypergeometric series. From the condition $\alpha = -n$ we get the dispersion relation for trapped waves:

$$p + \frac{1}{2} + n = \sqrt{l^2 - \frac{fl}{\omega} + \frac{1}{4}}, \ n = 0, 1, \ldots . \tag{4.33}$$

The solution for propagating (incident and reflected) Poincaré waves is obtained by the replacement $p \to ik$ everywhere in the above-displayed formulas. The solution is then given in terms of hypergeometric functions

$$\bar{\eta}_0 = A \left[e^{-ikx} F(\alpha^*, \beta^*, \gamma^*, s) - r e^{ikx} F(\alpha, \beta, \gamma^*, s) \right], \tag{4.34}$$

where A is the amplitude of the wave, $*$ means complex conjugation, and r is the reflection coefficient:

$$r = \frac{\Gamma(\alpha)\Gamma(\beta)\Gamma(\alpha^* + \beta^*)}{\Gamma(\alpha^*)\Gamma(\beta^*)\Gamma(\alpha + \beta)}, \ \Gamma - \text{gamma function.} \tag{4.35}$$

The dispersion relation (4.33) is represented in Figure 4.3. Although it was obtained in the Ball's model, analysis of equation (4.26) with arbitrary depth profile using the general theory of ordinary differential equations shows that, qualitatively, the dispersion diagram is always of this form for regular shelf profiles with a single zero at the coast. Namely, there are always:

- A single Kelvin wave with unidirectional propagation (mode with $n = 0$ in the left half-plane of Figure 4.3),

- A discrete spectrum of sub-inertial unidirectional waves with $\omega < f$ (shelf waves, only the lowest mode with $n = 1$ is displayed in the figure). Both shelf and Kelvin waves propagate leftwards, looking at the coast,

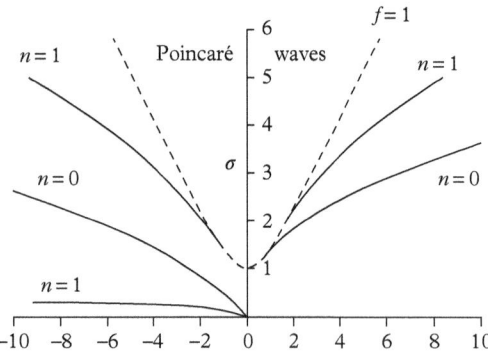

Figure 4.3 *Dispersion diagram in the Ball's model. Only the modes with $n = 0, 1$ are displayed.*

- A discrete spectrum of supra-inertial waves with $\omega > f$ (edge waves), which may propagate in both directions along the coast, only the lowest modes with $n = 0$ and $n = 1$ are displayed in the figure,
- A continuous spectrum of incident/reflected inertia–gravity (Poincaré) waves which is filling the domain above the dashed curve in Figure 4.3.

We thus see that in the presence of a shelf three new classes of coastal trapped waves appear: shelf waves, edge waves, and topographic waves, while Kelvin waves acquire small dispersion.

4.2   Waves over topography/bathymetry far from lateral boundaries

4.2.1   Topographic waves

Wave spectrum in the presence of escarpment

As in Section 4.1.2, we will see that topography can trap waves and thus modify the spectrum of small perturbations in the model. Let us consider a characteristic case of an escarpment: a region of localised non-zero gradient of H. As an example, we will take an escarpment with a linear slope represented in Figure 4.4. Linearisation leads again to the equation (4.26), although the boundary conditions are different. Far from the escarpment in the x direction the depth is constant, albeit not the same at the left- and

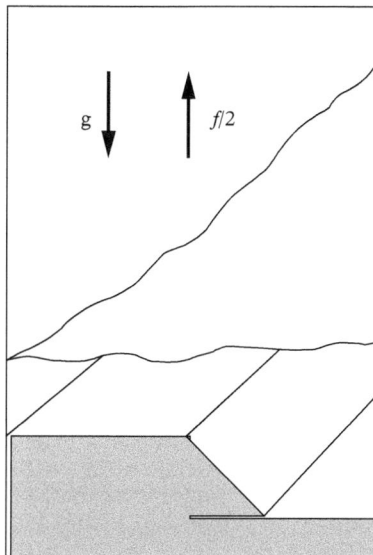

Figure 4.4 *A sketch of the rotating shallow water over a linear escarpment.*

right-hand sides: $H = H_\pm$ = const. Hence the left and right asymptotics of the fluid height $\bar\eta_{0\pm}$ obey the equations

$$gH_\pm\bar\eta_{0\pm}'' + (\omega^2 - f^2 - gH_\pm l^2)\bar\eta_{0\pm} = 0. \tag{4.36}$$

At each side of the escarpment solutions of (4.36) are of two kinds, depending on the signs of expressions $p_\pm^2 = \omega^2 - f^2 - l^2 gH_\pm$. In the case of positive $p_\pm^2$ we have harmonic in x solution, that is a wave propagating towards or out of escarpment. On the contrary, in the case of negative $p_\pm^2$, we have exponentially growing or decaying solution, which is legitimate, once a decay out of escarpment is guaranteed. Obviously, in order to get a solution over the whole x axis, these asymptotic solutions should be matched with the inner solution in the region of escarpment. Equation (4.26) can be rarely solved analytically on the whole x axis. For topography with discontinuous derivatives, as in Figure 4.4, matching at the discontinuities should provide continuity of pressure (i.e. h), and continuity of the transverse momentum (i.e. uh). It can be shown that these conditions are equivalent to the continuity of $\bar\eta_0(x)$ and its derivative $\bar\eta_0'(x)$. It is clear that solutions of the first kind correspond to inertia–gravity waves propagating in any direction in the $x - y$ plane. Such solutions always exist. Solutions of the second kind, if they exist, would correspond to waves trapped by the escarpment and propagating along it. In the case of escarpment of Figure 4.4, situated between $x = \pm L$ with the slope which may be taken to be unity, without loss of generality, the non-dimensional equation (4.26), under the same scaling as in Section 4.1.2 becomes

$$\left((H_m - x)\bar\eta_0'\right)' + \left(\omega^2 - f^2 - l^2(H_m - x) + \frac{fl}{\omega}\right)\bar\eta_0 = 0, \tag{4.37}$$

where $H_m = \frac{H_+ + H_-}{2H_0}$ is non-dimensional mean depth. This is an equation which can be transformed to a confluent hypergeometric form, and explicitly solved in terms of confluent hypergeometric functions M and U. The general solution of (4.37) is

$$\bar\eta_0(x) = C_1 U\left(-\frac{-fl - f^2\omega - l\omega + \omega^3}{2l\omega}, 1, 4l - 2lx\right)$$

$$+ C_2 M\left(\frac{-fl - f^2\omega - l\omega + \omega^3}{2l\omega}, 1, 4l - 2lx\right), \tag{4.38}$$

where $C_{1,2}$ = const. As already stated, this solution should be matched to the solutions $\bar\eta_0(x) = C_\pm e^{\mp\sqrt{-p_\pm^2}}$ decaying at $\pm\infty$, respectively. Continuity of $\bar\eta_0$ and $\bar\eta_0'$ at $x = \pm 1$ gives a system of four homogeneous linear algebraic equations for the constants $C_\pm$, $C_{1,2}$, and its solvability condition leads to the dispersion relation $\omega = \omega(l)$. There is a series of dispersion curves, two of them being represented in Figure 4.5. The obtained wave solutions can propagate only in the negative direction along the escarpment, that is leaving the shallower region on their right. Multiple curves with diminishing eigenfrequencies

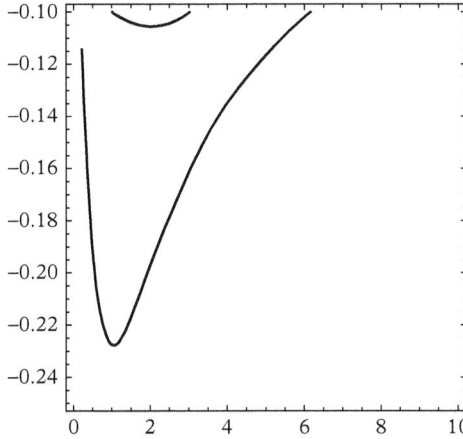

Figure 4.5 *Dispersion diagram $\omega = \omega(l)$ for topographic waves trapped by a linear escarpment. Only two lowest modes with, respectively, zero and one nodes in the x direction across the escarpment are shown. Negative ω for positive l means propagation in negative direction in y.*

correspond to the eigenfunctions with increasing number n of nodes across the escarpment, according to the general theory of the Sturm–Liouville problems, of which we have an example here. The phase portrait of the ground mode represents a sequence of alternating positive and negative thickness (pressure) anomalies localised at the escarpment, with the velocity field following the isobars. We will see examples of such modes in Chapter 10. We do not present the construction of propagating inertia–gravity wave solutions, which goes along the same lines, but with positive $p_\pm^2$. This procedure gives scattering and reflection of the inertia–gravity waves by the escarpment, according to the general theory of wave scattering. It should be emphasised that eigenfrequencies of the inertia–gravity waves are always much higher than eigenfrequencies of the trapped topographic waves.

Thus, localised changes in bathymetry lead to the appearance of trapped waves propagating across the gradient of the bottom topography. These are topographic Rossby waves, and by their properties (unidirectional propagation and the character of the dispersion curve) they resemble other types of Rossby waves, which we well see later in this and the next chapter. This resemblance is not accidental, as variations of bathymetry produce gradients of potential vorticity, which are at the origin of Rossby waves propagating across them. We should stress that although we obtained the earlier results with a specific bathymetry profile, they are general and are qualitatively the same for any monotonous profile of escarpment.

4.2.2 Mountain (lee) waves in RSW

Up to now we were studying the question how boundaries and/or topography modify the spectrum of small perturbations over the state of rest. We will now consider a situation where topography is localised: a configuration with a single mountain, and consider perturbations (mountain waves) created by this mountain in a mean flow. This is the first time (other examples will follow) that we treat waves in the presence of the mean flow, and the latter will be chosen to be the simplest possible, that is it will be horizontally uniform.

The one-layer RSW equations on the f plane in the presence of the mean zonal flow $-U\hat{x}$ and topography, after a proper non-dimensionalisation, read

$$u_t + (u - F)u_x + vu_y - (Bu)^{-\frac{1}{2}}v + h_x = 0,$$

$$v_t + (u - F)v_x + vv_y + (Bu)^{-\frac{1}{2}}u + h_y = 0, \qquad (4.39)$$

$$h_t + ((u - F)(h - Mb))_x + (v(h - Mb))_y = 0.$$

Here h is position of the free surface, $b(x, y)$ is topography, and we introduced typical scales: L in the horizontal directions, H (non-perturbed thickness) in the vertical, $\sqrt{gH}$ for velocity, and $T = L/\sqrt{gH}$ for time. This scaling is convenient as it reduces the number of parameters to three essential ones: Froude (or Mach) number of the mean flow $F = \frac{U}{\sqrt{gH}}$, non-dimensional mountain height $M = \frac{b_{max}}{H}$, and Burger number: $Bu = \frac{gH}{f^2L^2}$.

We are considering the linear limit of these equations by assuming that topography is of weak amplitude $M \ll 1$; that is we are excluding situations where the mountain 'sticks out' of the fluid layer forming an island. After linearisation, standard manipulations lead to a single wave equation for η, the perturbation of h with respect to H, with a stationary source produced by topography:

$$\left((\partial_t - F\partial_x)^2 - \nabla^2 + Bu^{-1}\right)\eta = M(F^2\partial_x^2 + Bu^{-1})b. \qquad (4.40)$$

We are looking for time-independent solutions, wave patterns created by the flow passing over the mountain. They are obtained by inversion of the stationary version of (4.40) with the help of the Green's function

$$\eta(x, y) = M \int dx' dy' (F^2\partial_x^2 + (Bu)^{-1})b(x - x', y - y')G(x', y'), \qquad (4.41)$$

where, by definition,

$$\left((-F\partial_x)^2 - \nabla^2 + (Bu)^{-1}\right)G(x - x', y - y') = \delta(x - x')\delta(y - y'), \qquad (4.42)$$

and $\delta(x)$ denotes Dirac's delta function which obeys the relations

$$\int dx\, F(x)\delta(x - x') = F(x'), \qquad \int dx\, \delta(x) = 1.$$

Let us start from the simplest stationary one-dimensional case with no dependence on t and y (a stationary flow over a straight ridge in two dimensions). The equation (4.40) becomes

$$\left((F^2 - 1)\partial_{xx}^2 + (Bu)^{-1}\right)\eta = M(F^2\partial_x^2 + (Bu)^{-1})b. \qquad (4.43)$$

Excluding a special case $F = 1$ where the solution is straightforward, we can divide both sides by $F^2 - 1$ and get a differential operator $\partial_{xx}^2 + \frac{1}{\mathcal{U}}$ on the left-hand side, where $\mathcal{U} = Bu(F^2 - 1)$. The equation for the Green's function then is

$$G''(x - x') + \frac{1}{\mathcal{U}}G(x - x') = \delta(x - x'). \tag{4.44}$$

Applying Fourier transformation, we get

$$G(x - x') = \int_{-\infty}^{+\infty} dk \, \frac{e^{i(k(x-x'))}}{k^2 - \frac{1}{\mathcal{U}}}, \tag{4.45}$$

where the Fourier representation of the delta function: $\delta(x) = \int_{-\infty}^{+\infty} dk \, e^{ikx}$ is used. In *subcritical regime*, $F < 1$ and $\mathcal{U} < 0$, the calculation of the integral is straightforward and gives

$$G(x - x') = \pi\sqrt{-\mathcal{U}} \, e^{-\frac{|x-x'|}{\sqrt{-\mathcal{U}}}}. \tag{4.46}$$

In *supercritical regime*, $F > 1$ and $\mathcal{U} > 0$, the integrand in (4.45) is singular, and the integral should be calculated by the method of residues. The boundary condition of upstream decay of perturbations dictates shifting the singularity to the upper half-plane of complex k, leading to

$$G(x - x') = \begin{cases} \pi\sqrt{\mathcal{U}} \sin\frac{(x-x')}{\sqrt{\mathcal{U}}}, & x - x' > 0 \\ 0, & x - x' < 0. \end{cases} \tag{4.47}$$

Physically speaking, the Green's function gives the stationary wave produced by a point mountain as seen in the pressure field. The above results mean that a mountain in a subcritical flow does not generate waves, while in a supercritical flow a standing wave-field of a special form is produced downstream.

In a two-dimensional case, the Green's function can be calculated along the same lines. More details on Green's functions for stationary wave problems can be found below in Chapter 5. By using Fourier transformation and method of residues, the following expression is obtained:

$$G(x, y) = \begin{cases} \frac{1}{2\sqrt{F^2-1}}\mathcal{J}_0\left(\frac{Bu^{-\frac{1}{2}}\sqrt{x^2-(F^2-1)y^2}}{\sqrt{F^2-1}}\right), & x > 0, \, x^2 < (F^2-1)y^2, \\ 0, \text{ otherwise,} \end{cases} \tag{4.48}$$

where we supposed that the flow is arriving towards the mountain from the left. $\mathcal{J}_0$ here and below denotes the Bessel function of the first kind. The isolines (isobars) of this Green's function are shown in Figure 4.6, together with the mountain waves, as seen in the inversion layer in the atmosphere, which can be modelled with the help of RSW.

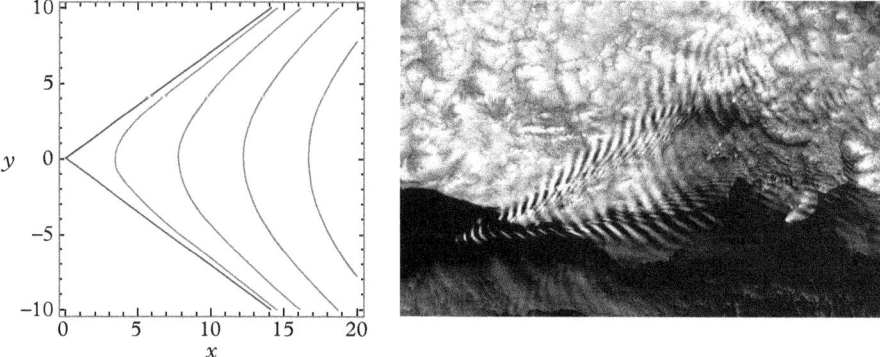

Figure 4.6 *Mountain waves behind a localised obstacle (lee waves) as calculated in linearised RSW model (left panel, isobars shown), and as seen in a satellite image (right panel, courtesy NASA) of clouds on top of the atmospheric inversion layer.*

4.3   Waves in outcropping flows

In multi-layer models the interfaces between the layers may touch each other, or touch the bottom or the free surface, thus leading to a termination of one of the layers (vanishing thickness). Such out- or in cropping of an interface represents a particular case of lateral boundary which is, however, not fixed but moving with the fluid. Configurations of this kind are appropriate in the oceanic context for modelling density currents resulting from, for example, river outflows at the surface, or outpouring of saltier water at the bottom. In the two-layer model, if the lower layer is very deep, such configuration can be modelled by a single terminating shallow-water layer, cf. Figure 4.7.

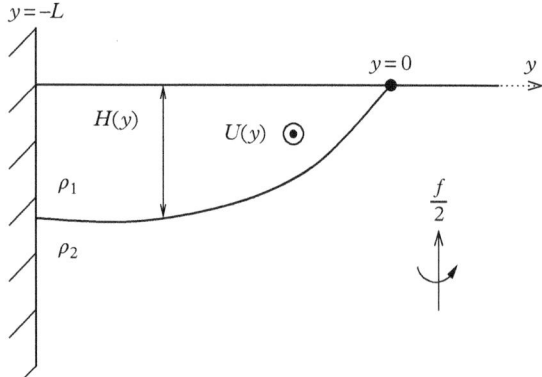

Figure 4.7 *Schematic representation of an outcropping coastal current.*

An outcropping necessarily means a nontrivial profile of the layer thickness h in a steady state and, hence, via the geostrophic balance, a non-zero mean velocity, see below. So a steady state with outcropping is necessarily a current, and this is the second time that we are led to consider waves on the background of the mean current. However, contrary to Section 4.2, the latter will be necessarily non-uniform. In order to limit the horizontal extension of the current, we will bound it by an idealised coast at some distance from, and parallel to, the outcropping. Our basis state will thus be a straight coastal current. The outcropping line is a density front, as density makes a jump while crossing it, cf. Figure 4.7.

Equations of the rotating shallow-water model in the f plane

$$\begin{cases} u_t + uu_x + vu_y - fv = -g'h_x, \\ v_t + uv_x + vv_y + fu = -g'h_y, \\ h_t + (hu)_x + (hv)_y = 0, \end{cases} \tag{4.49}$$

still hold, the only difference with the configuration with the coast considered earlier is an additional boundary condition at the outcropping. Here we put the x axis along the coast and the outcropping line, and introduced the reduced gravity g' originating from the underlying two-layer model. The prime will be omitted in what follows.

The basic state is assumed to be steady, and thus is given by the fields $u = U(y)$, $v = 0$, and $h = H(y)$, such that

$$U(y) = -\frac{g}{f}H_y(y) \tag{4.50}$$

which, as is easy to see, provides an exact stationary solution of the equations (4.49). This solution is an example of *geostrophic balance*, which will be considered in the Chapter 5.

We linearise equations (4.49) about this steady state. The linearised equations, where $u(x, y, t)$, $v(x, y, t)$, and $h(x, y, t)$ now denote the perturbations to the basic state fields, are

$$\begin{cases} u_t + Uu_x + vU_y - fv = -g\,h_x, \\ v_t + Uv_x + fu = -g\,h_y, \\ h_t + Uh_x = -(Hu_x + (Hv)_y). \end{cases} \tag{4.51}$$

We non-dimensionalise the problem by introducing the time scale f^{-1}, the horizontal scale L, which is the unperturbed width of the current, the velocity scale fL, and the vertical scale $(fL)^2/g'$. We will use from now on non-dimensional variables without changing notation. The following non-dimensional equations result:

$$\begin{cases} u_t + U u_x + v U_y - v = -h_x, \\ v_t + U v_x + u = -h_y, \\ h_t + U h_x = -(H u_x + (Hv)_y). \end{cases} \tag{4.52}$$

We impose the free-slip boundary condition at the coast: $v(-1) = 0$. The outcropping line consists of fluid parcels: it is a material line. The boundary conditions at the outcropping therefore are

$$H(y) + h(x, y, t)|_{y=Y_0} = 0, \quad \frac{dY_0}{dt} = v\Big|_{y=Y_0}. \tag{4.53}$$

Here $y = 0$ is the location of the free streamline of the balanced flow, $Y_0(x, t)$ is the position of the perturbed free streamline, and $\frac{d}{dt}$ is the Lagrangian derivative. It is easy to check that the linearised boundary conditions are equivalent to the following relation between the perturbation of the position of the free streamline and the value of the height perturbation:

$$Y_0 = -\frac{h}{H_y}\Big|_{y=0}, \tag{4.54}$$

and to the last equation in (4.49) evaluated at $y = 0$. Hence, the only constraint to impose on the solutions of (4.52) is regularity at $y = 0$.

The PV of the mean flow, in non-dimensional terms, is

$$Q(y) = \frac{1 - U_y}{H(y)}, \tag{4.55}$$

with the geostrophic equilibrium constraint $U(y) = -H_y(y)$. The thickness profile in the basis state is therefore given by the solution of the following order differential equation (ODE):

$$H_{yy}(y) - Q(y)H(y) + 1 = 0, \quad H(0) = 0, \; H_y(0) = -U_0, \tag{4.56}$$

where $U(0) = U_0$ is the mean flow velocity at the outcropping. In this section we consider mean flows with constant PV. A technical advantage of such configuration is that if solutions of (4.52) are sought in the form of waves with wave number k propagating along the coast with the phase velocity c, that is $(u, v, h) = (\bar{u}(y), \bar{v}(y), \bar{h}(y))e^{ik(x-ct)} + c.c.$, then the corresponding wave equation does not have singularity, which is otherwise the case, at *critical levels* y_c such that $U(y_c) - c = 0$. Incidentally, observations show that PV in coastal currents is often well mixed.

Let us demonstrate the absence of critical layer singularities in the simplest case of zero-PV flow. It follows from $Q = 0$ that $U_y - 1 = 0$, and hence the two last terms on the left-hands side of the first equation in (4.52) cancel. After the Fourier transformation in x and t, the latter equations become

$$\begin{cases} (U-c)\bar{u} + \bar{h} = 0, \\ ik(U-c)\bar{v} + \bar{u} + \bar{h}_y = 0, \\ ik(U-c)\bar{h} + ikH\bar{u} + (H\bar{v})_y = 0, \end{cases} \qquad (4.57)$$

and by eliminating all variables in favour of $\bar{u}$ we get

$$\left(H\bar{u}_y\right)_y - k^2\left[H - (U-c)^2\right]\bar{u} = 0. \qquad (4.58)$$

This equation does not have a singularity at $U-c = 0$ due to positiveness of H (except at the outcropping, where it does not pose a problem in view of the boundary conditions). In the case of constant but non-zero PV, the proof goes along the same lines but is technically more involved. Both cases are illustrations of Lin's theorem stating that a neutral wave mode, a mode with real c, with a critical level can exist in the flow only if the gradient of PV vanishes at the critical level.

In flows with zero PV $Q = 0$, and

$$H(y) = -U_0 y - \frac{y^2}{2}, \quad U(y) = U_0 + y, \qquad (4.59)$$

where U_0 is an arbitrary constant. Flows with constant non-zero PV, $Q(y) = Q_0 \neq 0$ have thickness and velocity which are given by solutions of (4.56):

$$\begin{cases} H(y) = \frac{1}{Q_0}[1 - U_0\sqrt{Q_0}\,sinh(\sqrt{Q_0}y) - cosh(\sqrt{Q_0}y)], \\ U(y) = U_0\,cosh(\sqrt{Q_0}y) + \frac{1}{\sqrt{Q_0}}\,sinh(\sqrt{Q_0}y). \end{cases} \qquad (4.60)$$

Examples of such profiles are presented in Figure 4.8. Due to the nontrivial profiles of thickness and velocity in the steady state, the linearised problem cannot be reduced to an algebraic system by application of the Fourier transformation to (4.52). Separating y and x, t variables, and looking for solutions in the form of waves with wave number k and phase velocity c along the coast/outcropping, an eigenvalue problem for c arises in a form of a system of ordinary differential equations in y. As shown earlier, in the case of zero PV this system can be reduced to a rather simple ODE (4.58) for one of the variables. In the case of constant non-zero PV, this kind of equation becomes rather involved, and it is simpler to solve the original system of three ordinary differential equations directly. Apart from rare special cases, it cannot be solved analytically and necessitates a numerical treatment. In Figure 4.9 we present the dispersion relation between phase velocity c and wave number k for the waves in the flow with constant PV obtained by the pseudospectral collocation method. It should be emphasised that complex eigenvalues of c can appear at some values of parameters, leading to *instability*, which will be the subject of Chapter 10. Here we focus on the stable flows only. As follows from Figure 4.9, we

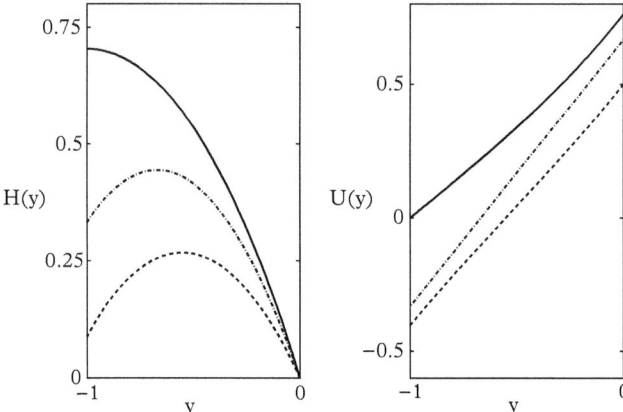

Figure 4.8 *Examples of the basic state thicknesses (left panel) and velocities (right panel) for constant PV flows with $U_0 = -\sinh(-1)/ \cosh(-1)$ (thick line), $U_0 = 1/2$ (dotted line), and a zero PV flow (dashed-dotted line).*

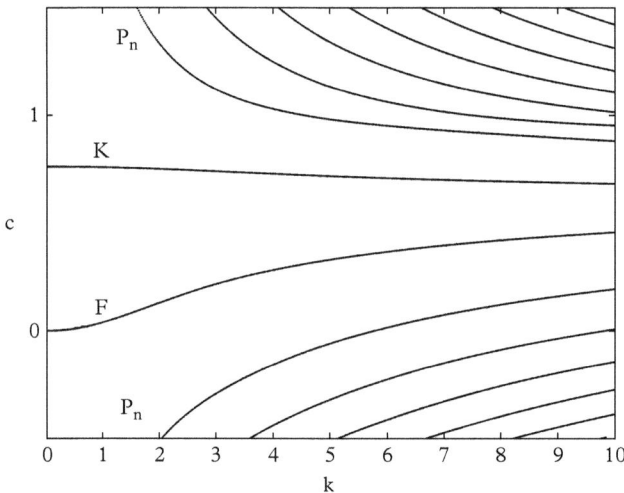

Figure 4.9 *Dispersion diagram for waves in the flow with $Q_0 = 1$. K is coastal Kelvin wave, F is frontal wave, P_n is Poincaré (inertia–gravity) wave, n is number of nodes of the mode in the span-wise direction.*

recover the usual spectrum which comprises 1) inertia–gravity waves, which are often called Poincaré waves in the bounded domains, with two senses of propagation; as we have a boundary-value problem in y, the spectrum of these waves consists of a series of eigenmodes $P_n(y)$ with n denoting the number of nodes in y direction, 2) a dispersionless Kelvin wave (K), and 3) a new type of wave (F) which is a wave produced by

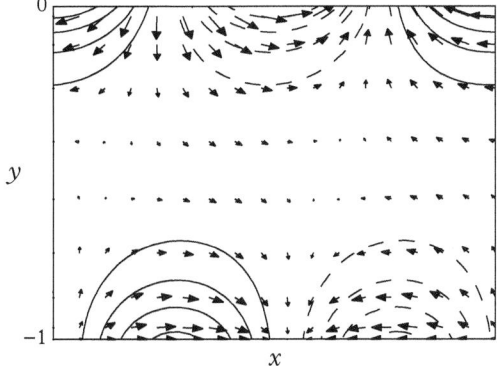

Figure 4.10 *Pressure (contours) and velocity (arrows) anomalies of Kelvin (bottom) and frontal (top) waves propagating over a uniform PV flow flow with $Q_0 = 1$.*

the displacement of the outcropping (frontal) line. At low current velocities this wave is geostrophically balanced; velocity follows isobars and the wave, therefore, is a close relative of Rossby waves which we will be studying later. It is produced, as these latter, by a gradient of PV originating from the change from a constant value at one side of the outcropping to zero at the other side. Figure 4.10, where a snapshot of Kelvin and frontal waves simultaneously propagating on the background of the coastal current, gives an idea of associated pressure and velocity anomalies.

4.4 Equatorial waves

4.4.1 Equatorial waves in one-layer model

RSW at the equator

If the tangent plane is situated at the equator, the vector of angular velocity of the planet is parallel to the tangent plane, which strongly influences dynamics. Let us consider the one-layer RSW model on the equatorial beta-plane. The equations of motion read

$$\begin{cases} \partial_t \mathbf{v} + \mathbf{v} \cdot \nabla \mathbf{v} + \beta y \hat{\mathbf{z}} \wedge \mathbf{v} + g \nabla h = 0, \\ \partial_t h + \nabla \cdot (\mathbf{v} h) = 0. \end{cases} \tag{4.61}$$

If decay boundary conditions at $y \to \pm \infty$ are imposed, the dynamics is confined in the equatorial region.

Scaling

As the constant part of the Coriolis parameter, f_0, is zero at the equator, the scaling which was previously used for the f plane should be modified. Thus, the former definition of the deformation radius, which is of primary importance in RSW dynamics, is not applicable and is changed to $R_e = \left(\frac{\sqrt{gH}}{\beta} \right)^{\frac{1}{2}}$, the equatorial deformation radius. We will be using the following characteristic scales:

- spatial scale L: equatorial deformation radius, $L \sim R_e$,
- time scale: $T \sim (\beta L)^{-1}$, it also changes with respect to mid-latitude tangent plane, and necessarily includes the spatial scale, as there is no more intrinsic time scale in the model,
- velocity scale $U \sim \frac{g\Delta H}{\beta L^2}$, where ΔH is the typical value of the deviation of h with respect to its non-perturbed value H.

With such scaling, the equatorial Rossby number coincides with the nonlinearity parameter λ, which measures relative deviations of the free surface:

$$Ro = \epsilon = \frac{U}{\beta L^2} = \frac{\Delta H}{H} = \lambda. \tag{4.62}$$

The non-dimensional equations of motion, thus, are

$$\begin{cases} \partial_t \mathbf{v} + \epsilon \mathbf{v} \cdot \nabla \mathbf{v} + y\hat{\mathbf{z}} \wedge \mathbf{v} + \nabla h = 0, \\ \partial_t h + \nabla \cdot \mathbf{v} + \epsilon \nabla \cdot (\mathbf{v}h) = 0. \end{cases} \tag{4.63}$$

Wave spectrum

As usual, to determine the dynamical content of the model we proceed by formal linearisation of (4.63):

$$\begin{cases} u_t - yv + h_x = 0, \\ v_t + yu + h_y = 0, \\ h_t + u_x + v_y = 0. \end{cases} \tag{4.64}$$

This system of equations has an explicit dependence on the meridional coordinate in its coefficients. It thus does not admit a direct Fourier approach, and needs special treatment. The following change of variables,

$$f = \frac{1}{2}(u + h); \quad g = \frac{1}{2}(u - h), \tag{4.65}$$

allows us to simplify the system (4.64):

$$\begin{cases} f_t + f_x + \frac{1}{2}(v_y - yv) = 0, \\ g_t - g_x - \frac{1}{2}(v_y + yv) = 0, \\ v_t + y(f + g) + (f - g)_y = 0, \end{cases} \tag{4.66}$$

and makes apparent the characteristic operators $\hat{L}_\pm = \partial_y \pm y$. As is known, for example from quantum mechanics, there is a special set of orthonormal functions $\phi_n(y)$, $-\infty < y < +\infty$, upon which $\hat{L}_\pm$ act in a particularly simple way:

$$\phi_n' + y\phi_n = \sqrt{2n}\phi_{n-1}, \quad \phi_n' - y\phi_n = -\sqrt{(2n+1)}\phi_{n+1}. \tag{4.67}$$

These functions are Gauss–Hermite (or parabolic cylinder) functions,

$$\phi_n(y) = \frac{H_n(y)e^{-\frac{y^2}{2}}}{\sqrt{2^n n! \sqrt{\pi}}}, \tag{4.68}$$

where H_n are Hermite polynomials:

$$H_0 = 1, \quad H_1 = 2y, \quad H_2 = 4y^2 - 2, \quad \dots \tag{4.69}$$

The functions $\phi_n(y)$ are solutions of the following ordinary differential equation:

$$\phi_n''(y) + (2n + 1 - y^2)\phi_n(y) = 0, \tag{4.70}$$

with decay boundary conditions at infinity. Solutions of linearised equations of the equatorial RSW are conveniently expressed in the basis of Gauss–Hermite functions.

It is easy to see that there exist a particular solution of the system (4.66) with $v \equiv 0$. From the first two equations we obtain in this case:

$$f_t + f_x = 0, \ g_t - g_x = 0, \ \Rightarrow f = F(x - t, y), \ g = G(x + t, y), \tag{4.71}$$

and the third equation gives

$$y(f + g) + (f - g)_y = 0, \ \Rightarrow F \propto e^{-\frac{y^2}{2}}, \ G \propto e^{+\frac{y^2}{2}}. \tag{4.72}$$

The decay boundary conditions at $y \pm \infty$ impose $G \equiv 0$, therefore the solution is

$$u = F_0(x - t)e^{-\frac{y^2}{2}}; \ h = F_0(x - t)e^{-\frac{y^2}{2}}; \ v = 0. \tag{4.73}$$

In this form the solution is reminiscent of the coastal Kelvin wave and represents the equatorial Kelvin wave. As its coastal counterpart, the equatorial Kelvin wave has a unique sense of propagation eastwards, and is non-dispersive. Distributions of zonal velocity and thickness perturbations in a harmonic equatorial Kelvin wave are shown in Figure 4.11. A characteristic feature of the equatorial Kelvin wave is the absence of meridional velocity, which is the analogue of absence of across-coast component of velocity in coastal Kelvin waves.

Another particular solution of (4.66) exists, with $g = 0, f \neq 0, v \neq 0$. From (4.66) we obtain in this case

$$\begin{cases} f_t + f_x + \frac{1}{2}(v_y - yv) = 0, \\ v_y + yv = 0, \\ v_t + yf + f_y = 0. \end{cases} \tag{4.74}$$

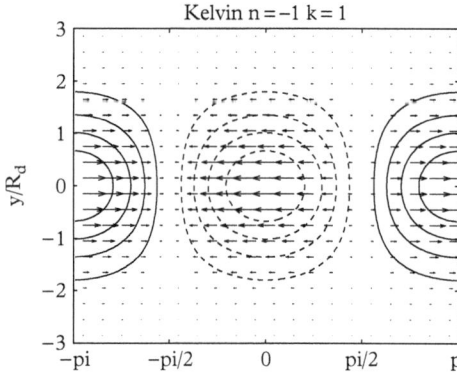

Figure 4.11 *Pressure (contours) and velocity (arrows) distribution in the equatorial Kelvin wave with zonal wave number k = 1.*

Proceeding by separation of variables,

$$v = v_0(x, t)\,\phi_0(y), \quad f = F_1(x, t)\,\phi_1(y), \tag{4.75}$$

and using (4.67) we obtain a pair of equations with constant coefficients for unknown functions $F_1(x, t)$, $v_0(x, t)$:

$$F_{1_t} + F_{1_x} - \frac{1}{\sqrt{2}}v_0 = 0, \quad v_{0_t} + \sqrt{2}F_1 = 0. \tag{4.76}$$

Looking for wave solutions $\propto e^{i(\omega t - kx)}$ we get the dispersion relation

$$\omega = \frac{k}{2} \pm \sqrt{\frac{k^2}{4} + 1}. \tag{4.77}$$

Corresponding eastward (upper sign) and westward (lower sign) propagating waves are called Yanai or mixed Rossby-gravity waves in the, respectively, atmospheric and oceanic contexts. Distributions of thickness and velocity perturbations in such waves are shown in Figure 4.12. The meaning of the latter denomination will become clear from the dispersion diagram below. A typical feature of Yanai waves is non-zero meridional velocity at the equator.

In order to find a general solution of the equatorial wave problem we proceed by elimination of u and h (or f and g) in favour of v, which gives

$$\partial_t \left(\nabla^2 v - y^2 v - \partial_{tt} v \right) + \partial_x v = 0. \tag{4.78}$$

Expansion of v in a series in ϕ_n: $v = \sum_n v_n(x, t)\phi_n(y)$ and the use of (4.70) lead to the following equation for $v_n(x, t)$:

$$\partial_t \left[\partial_{xx}^2 v_n - (2n + 1)v_n - \partial_{tt}^2 v_n \right] + \partial_x v_n = 0. \tag{4.79}$$

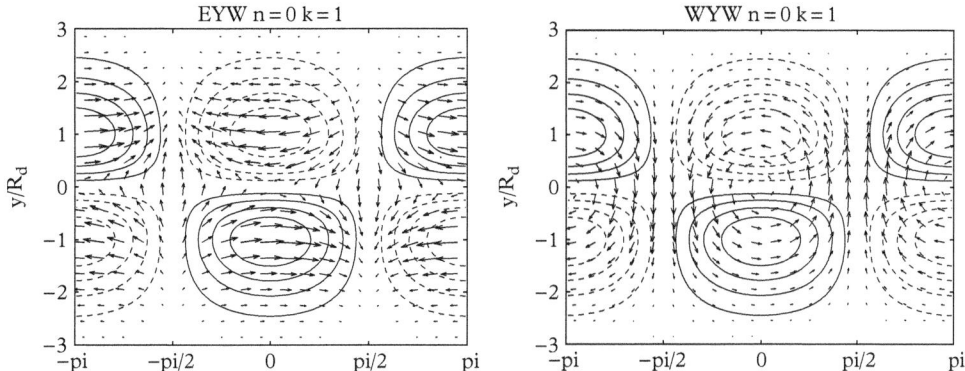

Figure 4.12 *Pressure (contours) and velocity (arrows) distribution in the equatorial eastward- (left panel) and westward- (right panel) propagating Yanai waves with zonal wave number $k = 1$.*

By Fourier-transformation $\tilde{v}_n(k, t) = \int dx e^{-ikx} v_n(x, t) + c.c.$ we obtain

$$\partial_{ttt}^3 \tilde{v}_n + (k^2 + 2n + 1)\partial_t \tilde{v}_n - ik\tilde{v}_n - 0, \tag{4.80}$$

which has the general solution

$$\tilde{v}_n = v_{n_1}(k)e^{-i\omega_{n_1} t} + v_{n_2}(k)e^{-i\omega_{n_2} t} + v_{n_3}(k)e^{-i\omega_{n_3} t}, \tag{4.81}$$

where ω_{n_α}, $\alpha = 1, 2, 3$ are three roots of the dispersion relation:

$$\omega_{n_\alpha}^3 - (k^2 + 2n + 1)\omega_{n_\alpha} - k = 0. \tag{4.82}$$

Here, the lowest value of ω at given k and n corresponds to a wave with unidirectional westward propagation. Its phase portrait, that is the mutual orientation of velocity vectors and isobars, shows that these latter are aligned, cf. Figure 4.13. This is an equatorial Rossby wave. Two other roots of (4.82) correspond to equatorial inertia–gravity waves, with two possible directions of propagation, cf. Figure 4.14. It should be emphasised that n gives a number of nodes in the meridional direction of the meridional component of velocity v. By construction, the dispersion relation (4.82) is applicable only for $n \geq 1$. The Kelvin and Yanai waves, obtained earlier as particular solutions, fall out of this general formula. It is, however, important to notice that

- a formal application of the dispersion relation at $n = -1$ gives

$$\omega^3 - (k^2 - 1)\omega - k = 0, \Rightarrow (\omega - k)(\omega^2 + \omega k + 1) = 0, \tag{4.83}$$

and if the *positive* root $\omega = k$ is retained, the Kelvin-wave solution is recovered,
- a formal application of the dispersion relation at $n = 0$ gives

$$\omega^3 - (k^2 + 1)\omega - k = 0, \Rightarrow (\omega + k)(\omega^2 - \omega k - 1) = 0; \tag{4.84}$$

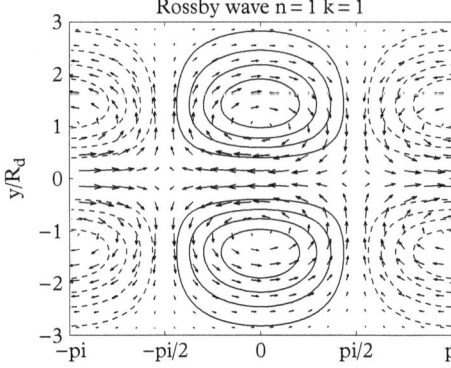

Figure 4.13 *Pressure (contours) and velocity (arrows) distribution in the equatorial Rossby wave with zonal wave number $k = 1$.*

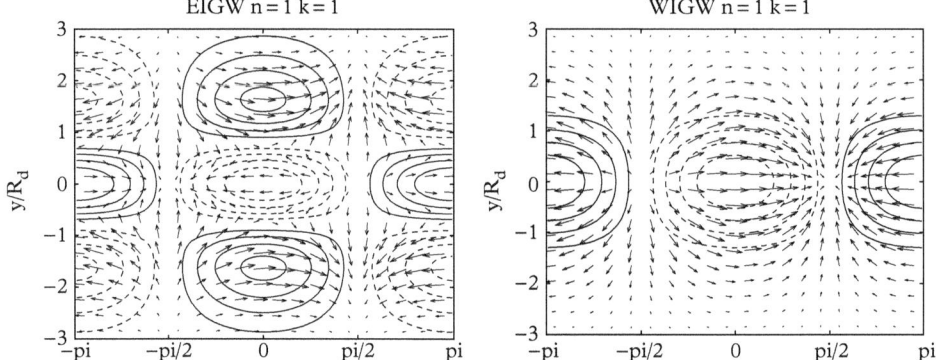

Figure 4.14 *Pressure (contours) and velocity (arrows) distribution in the equatorial eastward- (left panel) and westward- (right panel) propagating inertia–gravity waves with zonal wave number $k = 1$.*

and if *positive* roots of $\omega^2 - \omega k - 1 = 0$ are retained, Yanai-wave solutions are recovered.

Finally, it should be emphasised that the expression $\omega_n = -\frac{k}{2n+1+k^2}$ gives an excellent approximation, with precision $\sim 2\%$, for the Rossby-wave branch of the dispersion diagram.

Inspection of the dispersion diagram for equatorial waves presented in Figure 4.15 shows that inertia–gravity waves have dispersion relations similar to their f plane counterparts, with the role of meridional wave number played by the index of the corresponding Gauss–Hermite function. Rossby waves are propagating only in the westward direction, having much lower phase velocities than inertia–gravity waves. Yet, the gap

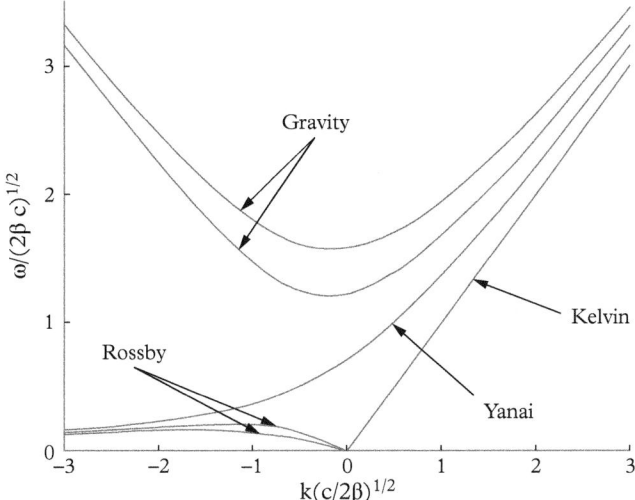

Figure 4.15 *Dispersion diagram for equatorial waves in the one-layer RSW. Only two lowest meridional modes for Rossby and inertia–gravity waves are shown.*

between inertia–gravity and Rossby waves is filled by 1) eastward-propagating dispersionless Kelvin waves, like in the f plane in the presence of a coast, and 2) by Yanai waves which, if propagating westward, have dispersion close to that of Rossby waves and, if propagating westward, have dispersion close to that of inertia–gravity waves, which explains their alternative denomination, mixed Rossby-gravity waves. Still, in the long-wave part of the wave spectrum, there is a well-pronounced gap between inertia–gravity waves and the rest of the spectrum. Another important difference with the f plane is that the minimal frequency of inertia–gravity waves correspond to small, but non-zero zonal wave numbers. The equatorial beta plane, thus, exhibits a wave guide along the equator, with a number of different zonally propagating wave species, which are confined in the meridional direction, and divided into fast and slow groups.

Let us emphasise that equatorial waves in the full PE model have the same horizontal structure as that predicted by the RSW theory presented earlier, which is confirmed by observational data. We present in Figure 4.16 a comparison of the horizontal structure of an equatorial inertia–gravity waves as predicted above with observations. Figure 4.17 shows a satellite image of a twin cyclone over the equatorial Pacific, visible in cloud structure and associated with an equatorial Rossby wave with symmetric, with respect to equator, pressure anomaly, cf. Figure 4.13.

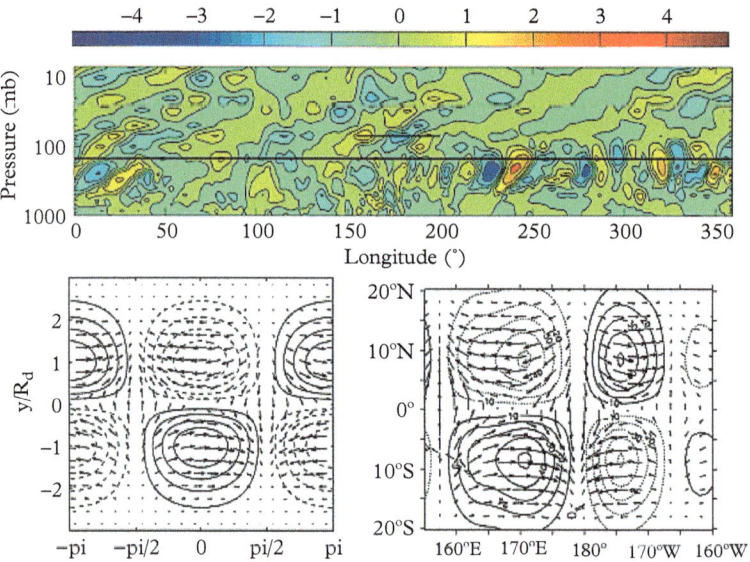

Figure 4.16 *Top panel: vertical section along the equator of the eastward-propagating anomaly of meridional velocity, with the horizontal black line showing average position of the tropopause, as follows from ERA40 dataset. Short black line: horizontal section at 70 hPa presented in the lower right plot. Bottom left panel: horizontal distribution of wind and geopotential anomaly in EIGW, as predicted by shallow-water theory. Bottom right panel: horizontal section of the observed fields.*

Figure 4.17 *Symmetric with respect to an equator twin depression visible in the cloud cover in a satellite image and associated with an equatorial Rossby wave. Courtesy NOAA.*

4.4.2 Waves in two-layer RSW with a rigid lid on the equatorial beta plane

Equations of motion

The equations of the model are standard two-layer model ones (3.23) with a replacement $f \to \beta y$, and we will be working in the often used in oceanography limit $\rho_2 \to \rho_1$, so that density difference enters only through the reduced gravity g':

$$
\begin{cases}
\partial_t \mathbf{v}_i + \mathbf{v}_i \cdot \nabla \mathbf{v}_i + \beta y \hat{\mathbf{z}} \wedge \mathbf{v}_i + \dfrac{1}{\rho_i} \nabla \pi_i = 0 \,, \ i = 1, 2; \\[2mm]
\partial_t h_i + \nabla \cdot (h_i \mathbf{v}_i) = 0 \,, \ i = 1, 2; \\[2mm]
\pi_2 = \pi_1 + \rho_2 g' h_2 \,, \quad g' = g \dfrac{\rho_2 - \rho_1}{\rho_1} \,, \quad h_1 + h_2 = H.
\end{cases}
\tag{4.85}
$$

Scaling and baroclinic–barotropic decomposition

Similarly to the one-layer case, we are using

- Baroclinic equatorial deformation radius: $L \sim \left(\dfrac{\sqrt{g'H}}{\beta} \right)^{\frac{1}{2}}$ as spatial scale,

- $T \sim (\beta L)^{-1}$ as time scale,

- $U \sim \dfrac{g' \Delta H}{\beta L^2}$ as velocity scale, where ΔH is typical deviation of the interface with respect to H

It is convenient to decompose the velocity in barotropic and baroclinic components which are defined in this case as follows:

$$
\mathbf{v}_{bt} = \frac{h_1 \mathbf{v}_1 + h_2 \mathbf{v}_2}{H}, \quad \mathbf{v}_{bc} = \mathbf{v}_2 - \mathbf{v}_1.
\tag{4.86}
$$

The barotropic velocity is subject to incompressibility constraint:

$$
\nabla \cdot (h_1 \mathbf{v}_1 + h_2 \mathbf{v}_2) = H \nabla \cdot \mathbf{v}_{bt} = 0,
\tag{4.87}
$$

which follows from the rigid-lid constraint $h_1 + h_2 = $ const and continuity equations. This allows us to express the barotropic velocity via the barotropic stream-function ψ:

$$
\mathbf{v}_{bt} = \hat{\mathbf{z}} \wedge \nabla \psi.
\tag{4.88}
$$

We rewrite the equations of motion in terms of ψ, $\mathbf{v}_{bc} \equiv \mathbf{v} = (u, v)$, $h_2 \equiv h$:

$$
\nabla^2 \psi_t + \psi_x = \epsilon \left[-\mathcal{J}(\psi, \nabla^2 \psi) - s(\partial_{xx} - \partial_{yy}) \left[(1 + \epsilon q h)(uv) \right] \right.
$$
$$
\left. + s \partial_{xy} \left(u^2 - v^2 \right) \right],
\tag{4.89}
$$

$$\mathbf{v}_t + \nabla h + y\hat{\mathbf{z}} \wedge \mathbf{v} = \epsilon \left[-\mathcal{J}(\psi, \mathbf{v}) + \mathbf{v} \cdot \nabla(\hat{\mathbf{z}} \wedge \nabla \psi) - q\mathbf{v} \cdot \mathbf{v} \right.$$
$$\left. + \epsilon s \left(2h\mathbf{v} \cdot \nabla \mathbf{v} + \mathbf{v}\mathbf{v} \cdot \nabla h \right) \right], \tag{4.90}$$

$$h_t + \nabla \cdot \mathbf{v} = \epsilon \left[-\mathcal{J}(\psi, h) + \epsilon s \nabla \cdot \left(h^2 \mathbf{v} \right) - q\nabla \cdot (\mathbf{v}h) \right], \tag{4.91}$$

where ϵ is the Rossby number, as usual, and we introduced parameters related to the unperturbed thickness ratio: $q = (H - 2H_2)/H$ and $s = H_1 H_2/H^2$.

Wave spectrum

In order to find the wave spectrum of the model we consider the linearised version of (4.89) to (4.91):

$$\begin{cases} \nabla^2 \psi_t + \psi_x = 0, \\ \mathbf{v}_t + \nabla h + y\hat{\mathbf{z}} \wedge \mathbf{v} = 0, \\ h_t + \nabla \cdot \mathbf{v} = 0. \end{cases} \tag{4.92}$$

We see immediately that for the baroclinic component the equations are exactly the same as in the one-layer case, the only difference residing in scaling, where the reduced gravity g' is used instead of 'normal' gravity g. For the barotropic component we get an equation, which is a linearised version of the vorticity equation on the β- plane, to be considered in Chapter 5. We already know the wave solution for the first, while the second can be easily treated by the Fourier method. Thus, the wave spectrum of the model consists of

- Baroclinic waves trapped in the equatorial wave guide

$$(u, v, h) = (iU_n(y), V_n(y), iH_n(y)) \, A e^{i(kx - \omega_n)t} + c.c., \tag{4.93}$$

  with dispersion relation

$$\omega_n^3 - (k^2 + 2n + 1)\omega_n - k = 0; \quad n = -1, 0, 1, 2, \dots, \tag{4.94}$$

  that is equatorial Kelvin, Yanai, Rossby, and inertia–gravity waves,
- Barotropic 'free' Rossby waves, as they are vorticity waves, and do not 'feel' the equatorial wave guide

$$\psi_0 = A_\psi e^{i(kx - \omega t + ly)} + c.c., \tag{4.95}$$

  with dispersion relation

$$\sigma = -k/(k^2 + l^2). \tag{4.96}$$

We thus deduce that the equatorial wave guide in the model confines baroclinic waves, but is transparent for barotropic waves. We will come back to this peculiar property in Chapter 12.

4.5 Waves in rotating annulus

A long tradition in geophysical fluid dynamics is laboratory modelling in rotating tanks. Annular channels between a pair of cylinders are natural to use in such experiments and, indeed, this geometry is frequently used, for example to study baroclinic instability. We will address in detail the instabilities of flows in rotating annuli in Chapter 11, while here we make a first acquaintance with wave-spectrum of such flows in the simplest one-layer configuration.

4.5.1 RSW in cylindrical geometry

We consider the one-layer RSW model in the plane rotating with angular velocity Ω, so the Coriolis parameter is $f = 2\Omega$, and use polar coordinates (r, θ). The fluid evolves in an annulus with boundaries situated at $r = r_1$ and $r = r_2 > r_1$. The equations of the model then are

$$\begin{cases} \dfrac{d}{dt}u - \left(f + \dfrac{v}{r}\right)v - r\Omega^2 = -g\partial_r h\,, \\[2mm] \dfrac{d}{dt}v + \left(f + \dfrac{v}{r}\right)u = -g\dfrac{\partial_\theta h}{r}\,, \\[2mm] \dfrac{d}{dt}h + h\mathbf{\nabla}\cdot\mathbf{v} = 0. \end{cases} \qquad (4.97)$$

Here h is the depth of the layer, (u, v) are radial and azimuthal components of velocity, respectively, $\frac{d}{dt} = \partial_t + u\partial_r + \frac{v}{r}\partial_\theta$ is Lagrangian derivative in polar coordinates. The boundary conditions are free-slip: $u = 0$ at $r = r_1, r_2$.

It should be emphasised that in the equations (4.97) the centrifugal acceleration $r\Omega^2$ is not included in the pressure term, as is traditionally done in the primitive equations, cf. Chapter 2. Equilibrium between the centrifugal force and pressure force gives a steady solution of (4.97), corresponding to solid rotation of the fluid layer. Obviously, the radial profile of thickness $H(r)$ should be parabolic in this case. We can, nevertheless, consider a more general situation of axisymmetric stationary solution with non-zero azimuthal velocity $V(r)$. As is easy to see, in order to be a stationary solution $V(r)$ and $H(r)$ should obey the following equation:

$$fV + \frac{V^2}{r} + r\Omega^2 = g\partial_r H. \qquad (4.98)$$

This is a so-called cyclo-geostrophic equilibrium, that is equilibrium between the centrifugal and Coriolis forces, on the one hand, and the pressure force, on the other hand. We will postpone a discussion of general cyclo-geostrophic equilibria until Chapter 11 and limit ourselves here by solid rotation states with parabolic profiles of $H(r)$. Hence $V(r)$ should be proportional to r. If the proportionality constant is zero, we recover the solid rotation, if it is not we have a solid super-rotation, which can be realised in experiments, see Chapter 11.

Equations (4.97), linearised about a state of cyclo-geostrophic equilibrium, with the same notation for the perturbations as for the full fields, are

$$
\begin{cases}
\partial_t u + \dfrac{V}{r}\partial_\theta u - fv - 2\dfrac{Vv}{r} = -g\partial_r h \, , \\[2mm]
\partial_t v + u\partial_r V + \dfrac{V}{r}\partial_\theta v + fu + \dfrac{Vu}{r} = -g\dfrac{\partial_\theta h}{r} \, , \\[2mm]
\partial_t h + \dfrac{1}{r}(rHu)_r + \dfrac{1}{r}H\partial_\theta v + \dfrac{V}{r}\partial_\theta h = 0 \, .
\end{cases}
\tag{4.99}
$$

By introducing the time scale $f^{-1} = (2\Omega)^{-1}$, the horizontal scale $r_0 = r_2 - r_1$, the vertical scale $H_0 = H(r_1)$ and the velocity scale $V_0 = r_0\Omega$, we obtain the following non-dimensional equations, again without change of notation:

$$
\begin{cases}
\partial_t u + \dfrac{V}{r}\partial_\theta u - v - 2\dfrac{Vv}{r} = -Bu\,\partial_r h \, , \\[2mm]
\partial_t v + u\partial_r V + \dfrac{V}{r}\partial_\theta v + u + \dfrac{Vu}{r} = -Bu\dfrac{\partial_\theta h}{r} \, , \\[2mm]
\partial_t h + \dfrac{1}{r}(rHu)_r + \dfrac{1}{r}H\partial_\theta v + \dfrac{V}{r}\partial_\theta h = 0 \, ,
\end{cases}
\tag{4.100}
$$

where we introduced the Burger number $Bu = (R_d/r_0)^2$, and the deformation radius $R_d = (gH_0)^{\frac{1}{2}}/(2\Omega)$. Normal modes are introduced in the standard way:

$$
(u(r,\theta), v(r,\theta), h(r,\theta)) = (\tilde{u}(r), \tilde{v}(r), \tilde{h}(r))\, exp\,[ik(\theta - ct)] + c.c.,
$$

where k is the azimuthal wave number, which should be integer, and c is the azimuthal phase velocity. Omitting tildes, we thus get the following problem for eigenvalues c and eigenfunctions u, v, h:

$$
\begin{cases}
k\left(\dfrac{V}{r} - c\right)u - \left(1 + 2\dfrac{V}{r}\right)v = -Bu\,h_r \, , \\[2mm]
-\left(1 + \dfrac{V}{r} + V_r\right)u + k\left(\dfrac{V}{r} - c\right)v = -kBu\dfrac{h}{r} \, , \\[2mm]
-\dfrac{(rHu)_r}{r} + k\dfrac{H}{r}v + k\left(\dfrac{V}{r} - c\right)h = 0 \, ,
\end{cases}
\tag{4.101}
$$

where here and below we denote the r derivative by the corresponding subscript, if it does not lead to confusion.

4.5.2   Analytic solution of the eigenvalue problem

For parabolic $H(r)$ the eigenvalue problem (4.101) can be solved analytically. Indeed, by eliminating u and v in favour of h,

$$u = \frac{Bu\, h_r k(\frac{V}{r} - c) + \frac{Bu}{r} kh(1 + 2\frac{V}{r})}{(1 + 2\frac{V}{r})(1 + \frac{V}{r} + V_r) - k^2(\frac{V}{r} - c)^2},$$

$$v = \frac{k(\frac{V}{r} - c)\frac{Bu}{r} kh + (1 + \frac{V}{r} + V_r)Bu\, h_r}{(1 + 2\frac{V}{r})(1 + \frac{V}{r} + V_r) - k^2(\frac{V}{r} - c)^2},$$
(4.102)

where we suppose that $(1 + 2\frac{V}{r})(1 + \frac{V}{r} + V_r) - k^2(\frac{V}{r} - c)^2 \neq 0$, we obtain the following ordinary differential equation for h:

$$(rH(\tfrac{V}{r} - c)h')' - (V' - \tfrac{V}{r})Hh' +$$

$$\left[[H(1 + 2\tfrac{V}{r})]' - k^2\tfrac{H}{r}(\tfrac{V}{r} - c) - \tfrac{r}{Bu}(\tfrac{V}{r} - c)((1 + 2\tfrac{V}{r})(1 + \tfrac{V}{r} + V_r)) - k^2(\tfrac{V}{r} - c)^2)\right] h = 0,$$
(4.103)

where prime denotes ordinary derivative with respect to r. Solid-body (super)rotations of the fluid in cyclo-geostrophic equilibrium are given by a one-parameter family of solutions of (4.98):

$$V(r) = \alpha r, \quad H(r) = \frac{(1 + \alpha)^2}{8Bu} r^2,$$
(4.104)

and we get from (4.103)

$$(r^3 h')' + \left[2\frac{1 + 2\alpha}{\alpha - c} - k^2 - \frac{((1 + 2\alpha)^2) - k^2(\alpha - c)^2)}{2\beta Bu}\right] rh = 0,$$
(4.105)

where we introduced the shorthand notation $\beta = \frac{(1+\alpha)^2}{8Bu}$, not to be confused with beta in the Coriolis parametre. For the sake of compactness, we also introduce the notation:

$$A = \left[2\frac{1 + 2\alpha}{\alpha - c} - k^2 - \frac{((1 + 2\alpha)^2) - k^2(\alpha - c)^2)}{2\beta Bu}\right].$$
(4.106)

General solution of (4.105) is easy to find:

$$h(r) = C_1 r^{\alpha_+} + C_2 r^{\alpha_-}, \quad \alpha_{\pm} = -1 \pm \sqrt{1 - A}.$$
(4.107)

Solutions of the eigenproblem (4.101) must satisfy the boundary conditions $u(r_1) = u(r_2) = 0$ which gives, using (4.102),

$$\left[(\alpha - c)h_r + \frac{(1 + 2\alpha)h(r)}{r}\right]\Bigg|_{r=r_1, r_2} = 0.$$
(4.108)

From (4.107) and (4.108) we get the following system of linear algebraic equations for $C_{1,2}$:

$$\begin{cases} [\alpha_+(\alpha-c)+(1+2\alpha)] \, C_1 r_1^{\alpha+-1} + [\alpha_-(\alpha-c)+(1+2\alpha)] \, C_2 r_1^{\alpha--1} = 0 \,, \\ [\alpha_+(\alpha-c)+(1+2\alpha)] \, C_1 r_2^{\alpha+-1} + [\alpha_-(\alpha-c)+(1+2\alpha)] \, C_2 r_2^{\alpha--1} = 0 \,, \end{cases} \tag{4.109}$$

and the condition of existence of nontrivial solutions:

$$[\alpha_+(\alpha-c)+(1+2\alpha)] \, [\alpha_-(\alpha-c)+(1+2\alpha)] \left[r_1^{\alpha+-1} r_2^{\alpha--1} - r_1^{\alpha--1} r_2^{\alpha+-1} \right] = 0. \tag{4.110}$$

Two different solutions for A thus arise:

$$A = 2\frac{1+2\alpha}{\alpha-c} - \left(\frac{1+2\alpha}{\alpha-c}\right)^2 , \tag{4.111}$$

and

$$A = 1 + \left(\frac{n\pi}{\log(\frac{r_1}{r_2})}\right)^2 , \quad n = 0, 1, 2, \ldots. \tag{4.112}$$

Kelvin modes

The first solution (4.111) combined with (4.106) gives a bi-quadratic equation for the phase speed c:

$$(\alpha-c)^4 - \left(\beta Bu + \left(\frac{1+2\alpha}{k}\right)^2\right)(\alpha-c)^2 + \beta Bu \left(\frac{1+2\alpha}{k}\right)^2 = 0 \tag{4.113}$$

which has the roots

$$c = \alpha \pm \sqrt{\beta Bu}, \tag{4.114}$$

$$c = \alpha \pm \frac{1+2\alpha}{k} \tag{4.115}$$

The solutions (4.114) are non-dispersive and correspond to the eigenmodes with $u \equiv 0$. They, thus, describe two Kelvin modes propagating along the inner and outer boundaries, respectively. We show in Figure 4.18 the phase portraits of Kelvin modes with $k = 2$, that is, the isopleths of the thickness

$$h(r) = C_1 r^3 \text{ or } h(r) = C_2 r^{-3}, \tag{4.116}$$

and the related velocity fields, as follows from (4.102) and (4.104). The characteristic structure of the Kelvin waves, discussed in Section 4.1.1, is clearly recognisable in the

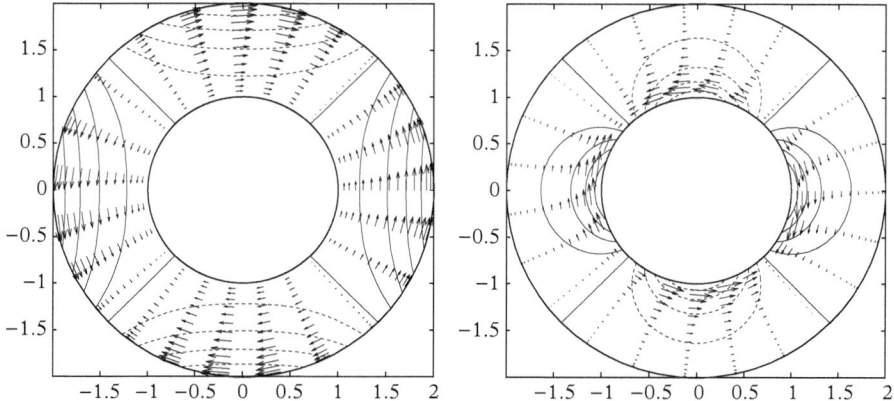

Figure 4.18 *Pressure and velocity fields for Kelvin modes with wave number $k = 2$ propagating along the outer (left panel) and the inner (right panel) walls. These modes correspond to curves (b) and (d), respectively, in Fig. 4.20.*

figure. The second pair of roots (4.115) correspond to the degenerate case $(1 + 2\frac{V}{r})$ $(1 + \frac{V}{r} + V_r) - k^2(\frac{V}{r} - c)^2 = 0$, which was discarded in the previous subsection. As can be seen from (4.101), or directly from (4.108), they do not correspond to any nontrivial eigenfunction.

Rossby and Poincaré modes

The second solution (4.112), when combined with (4.106), gives a third-order equation for the phase speed c at each value of $n = 1, 2, 3, \ldots$:

$$\frac{k^2}{\beta Bu}(\alpha - c)^3 - \left[k^2 + 1 + \left(\frac{n\pi}{\log(\frac{r_1}{r_2})}\right) + \frac{1 + 2\alpha}{\beta Bu}\right](\alpha - c) + 2(1 + 2\alpha) = 0. \quad (4.117)$$

For each n, we thus have a set of three different solutions, corresponding to three roots of (4.117) each given by:

$$h(r) = C_1 r^{i\frac{n\pi}{\log(\frac{r_1}{r_2})} - 1} + C_2 r^{-i\frac{n\pi}{\log(\frac{r_1}{r_2})} - 1} \quad (4.118)$$

where

$$\frac{C_1}{C_2} = -\frac{(\alpha - c)\alpha_- + (1 + 2\alpha)}{(\alpha - c)\alpha_- + (1 + 2\alpha)} \frac{r_1^{\alpha_-}}{r_1^{\alpha_+}}. \quad (4.119)$$

(It should be noted that the case $n = 0$, and hence $A = 1$, is degenerate: the related height field may be obtained by calculating the limit, and leads to the logarithmic in r solution.) The phase portraits of these solutions at $k = 2$ are shown in figure 4.19. The left panel of the figure displays flow velocity turning around pressure, with velocity vectors aligned

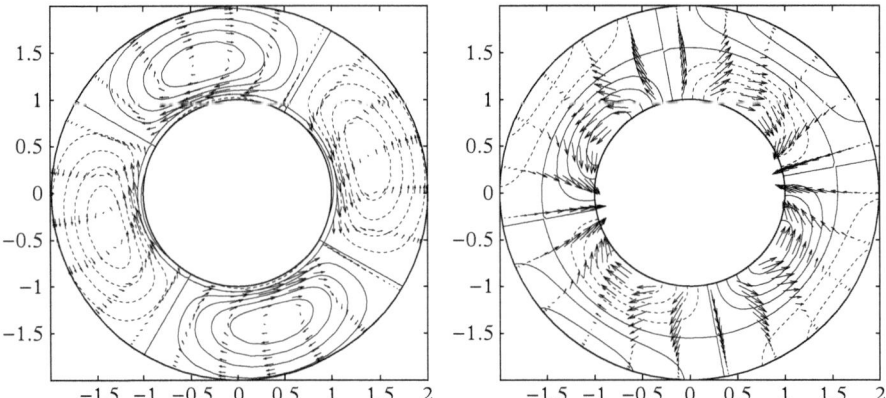

Figure 4.19 *Pressure and velocity fields for n = 1 mode of the Rossby wave (left panel) and the Poincaré wave (right panel) with wave number k = 2. These modes correspond to (c) and (a), respectively, in Fig. 4.20 below.*

with the isobars. We have seen a very similar phase portrait for the equatorial Rossby wave in Section 4.5.1, and will see in Chapter 5 that this is the graphical expression of the *gestrophic balance*. This wave is the Rossby wave in the annulus, and it is also a close relative of the waves trapped by the escarpment, which were considered in Section 4.2. The second and third solutions have the structure displayed in the right panel of the figure (the difference between the two is that they propagate in opposite directions), which is totally different. These waves are inertia–gravity waves, or Poincaré waves.

The dispersion diagram in terms of phase velocity $c = \omega/k = c(k)$ for thus obtained eigen solutions of the problem (4.101) is presented in Figure 4.20. It confirms

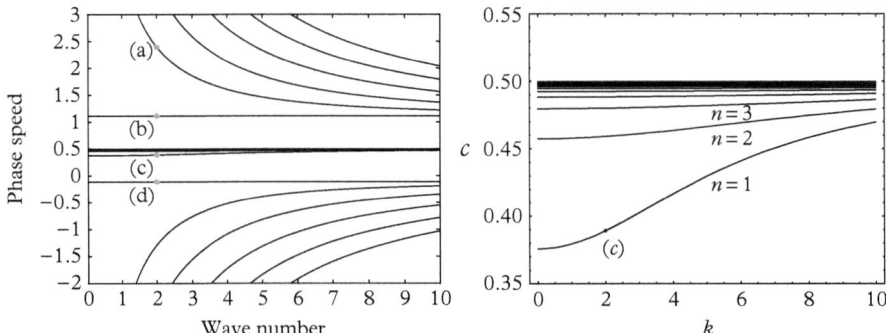

Figure 4.20 *Left panel: dispersion diagram c = c(k) for the solutions of equations (4.117) and (4.113) with α = 1/2. (a) Poincaré modes (cf. right panel of Fig. 4.19), (b) and (d) Kelvin modes (cf. Fig. 4.18), (c) Rossby modes (cf. left panel of Fig. 4.19). The spectrum of k is discrete, but for better visualisation c(k) is presented in a form of continuous curves; only the integer values of k correspond to realisable solutions. Right panel: zoom of the left panel showing slow Rossby modes with different n.*

the identification of the waves, as the dispersion curves $c(k)$ of Poincaré waves (e.g. (a)), and Kelvin waves (e.g. (b)), if converted to $\omega(k) = kc(k)$ form, give familiar dispersion curves of inertia–gravity and Kelvin waves, respectively, cf. Figure 4.1. The Rossby-wave part of the dispersion diagram on the left panel of the figure resembles, qualitatively and quantitatively, the corresponding parts of dispersion diagrams for waves over escarpment, Figure 4.5, and equatorial waves, Figure 4.13, which explains the common name.

The above-described analytic results provide an excellent way to benchmark the numerical pseudo-spectral collocation method for solving linear spectral problems, which was first applied in Section 4.3, and will be systematically used in the following. The comparison of numerical and analytic results shows absolute coincidence already at low numerical resolutions.

4.6   Summary, comments, and bibliographic remarks

The most striking consequence of the presence of lateral boundaries and/or topography, and of the specificity of the equatorial tangent plane approximation, is the appearance of wave guides and corresponding trapped modes, which considerably enriches the wave spectrum of the system. Laboratory flows in rotating annuli are wave guides *par excellence*. We will see in subsequent chapters numerous examples of particular dynamics of fluid motions involving wave-guide modes. An important property of Kelvin waves appearing in all flow geometries considered earlier, is that they are dispersionless and, thus, breaking, forming Kelvin fronts, as we will see later.

On the other hand, we see the appearance of a new class of Rossby waves: over escarpment, at the equator, and in the rotating annulus. They are slow and unidirectional, with velocity, roughly, following the isobars. The physical origin of Rossby waves in different set-ups is also the same: the variations of background potential vorticity which are induced either by variable thickness in the rest state, or varying planetary vorticity, or background velocity shear. All Rossby waves have phase velocities much lower than those of other wave species. Although the Rossby waves 'live' inside the spectral gap described in the previous Chapter, their slowness still allows for time-scale separation and, thus, preserves slow–fast motion dichotomy in the model. Two types of waves in the above-described configurations bridge Rossby and inertia–gravity wave parts of the spectrum. They are Yanai waves at the equator, and frontal waves at the outcropping. Both are sometimes called mixed Rossby-gravity waves by this reason, and both resemble inertia–gravity waves in the high-frequency, and Rossby waves in the low-frequency limits. Nevertheless, there still is a significant frequency gap between these waves and inertia–gravity waves in the long-wave limit. This fact will be exploited in the next chapter.

The presentation of Section 4.1.1 is standard. The calculation of abrupt shelf asymptotics in Section 4.1.2 follows unpublished work by G. Reznik. Necessary facts on hypergeometric functions can be found in Abramowitz and Stegun (1964). General theory of coastal waves for arbitrary shelf profile was developed in Huthnance (1975).

The illustration of this general theory in the case of Ball's model (Ball 1967) follows Reznik and Zeitlin (2011). General theory of waves over escarpment was given in Longuet-Higgins (1968a,b). Section 4.3 gives an illustration of this theory in the simplest configuration. Examples of wave spectra in many different idealised configurations of shelves and escarpments can be found in LeBlond and Mysak (1978). Mountain waves are treated in numerous papers and books. Flow over an obstacle is a classical topic in hydrodynamics, including hydraulics, that is the shallow-water framework. Extension of such studies to the rotating fluid allows for conceptual understanding of general problematic of mountain waves. Expression for the corresponding Green's function can be found, for example, in Esler, Rump, and Johnson (2009). The method of residues can be found in standard textbooks such as Morse and Feshbach (1953), as well the Sturm–Liouville theory evoked earlier. We are following Gula and Zeitlin (2010) while introducing waves in flows with outcropping in Section 4.3. The pseudo-spectral collocation method, which was used for the first time in this chapter for finding wave solutions in flows with outcropping, and in the rotating annulus, and which will be systematically used throughout subsequent chapters, consists in the following algorithm: 1) application of Fourier transformations in time and in the spatial direction, say x, which is 'neutral', in a sense that there is no dependence on the corresponding coordinate in the equations, 2) discretisation of the system of ordinary differential equations resulting in the remaining direction, say y, and the use of finite-difference operators instead of d/dy, which gives a system of algebraic equations, 3) numerical solution of thus obtained algebraic eigenproblem for eigenfrequencies and eigenmodes. In order to avoid the Runge phenomenon, the Chebyshev collocation points are used for discretisation and, correspondingly, Chebyshev differential operators for derivatives (Trefethen 2000). For example, in Section 4.5 a complete basis of Chebyshev polynomials was used to obtain a discrete equivalent of the system which is achieved by evaluating (4.101) on a discrete set of N collocation points (typically a rather low resolution $N = 50$ to 100 is sufficient) and using the Chebyshev differentiation matrix to discretise the spatial derivatives. The eigenvalues and eigenvectors of the resulting operator are computed with the help of Matlab© routine 'eig'. The occurrence of spurious eigenvalues is common in such discretisation procedures, especially in the presence of critical levels. Persistence of the obtained eigenvalues is checked by recomputing the spectrum with increasing N, and verifying convergence.

The material of Section 4.4.1 is standard and can be found in many articles and books (e.g. Gill 1982). The new variables f and g were introduced in Gill (1980). Approximate form of the dispersion relation for equatorial Rossby waves was given in Boyd (1980). Presentation of waves in two-layer equatorial model in Section 4.4.2 follows Reznik and Zeitlin (2007). Section 4.5 is based on Gula, Zeitlin, and Plougonven (2009) Analytic solution of RSW equations linearised about a stationary state with a parabolic profile of thickness was given in Killworth (1983), where this problem was considered for a parabolic lens.

Figure 4.3 is reproduced from Reznik and Zeitlin (2011), Figures 4.7, 4.8, 4.9, and 4.10 are reproduced from Gula and Zeitlin (2010), Figures 4.18, 4.19, and 4.20 are reproduced from Gula, Zeitlin, and Plougonven (2009), with permissions. Figures 4.11 to

4.15 are reproduced from Bouchut, LeSommer, and Zeitlin (2005) with the permission of AIP publishing. Figure 4.16 is taken from LeSommer, Teitelbaum, and Zeitlin (2006).

4.7 Problems

4.1 Demonstrate that coastal Kelvin waves bear no PV anomaly.

4.2 Show that there exist two Kelvin-wave solutions in the two-layer RSW with free surface. Determine their nature.

4.3 Find dispersion and polarisation relations for the Kelvin-wave solution in the Ball's model. Compare them with corresponding solutions for an abrupt shelf.

4.4 Find the wave solutions and their dispersion relation for a step-like escarpment, with a profile in a form of Heaviside function (topographic Kelvin wave).

4.5 Obtain the expressions for the Green's function (4.46), (4.47), (4.48).

4.6 Demonstrate (4.60).

4.7 Obtain (4.78).

4.8 Derive polarisation relations for all types of equatorial waves.

4.9 Derive the equations (4.89) to (4.91).

4.10 Using one-layer RSW equations in polar coordinates, with outcropping at some $r = R$, 1) demonstrate that an axisymmetric lens with parabolic profile of thickness and linear profile of azimuthal velocity is an exact solution, 2) linearise RSW equations about this solution and find the spectrum of small perturbations.

5

Getting Rid of Fast Waves: Slow Dynamics

The essential property of the parent primitive equation (PE) and simplified vertically averaged rotating shallow-water (RSW) models is wave–vortex dichotomy, with substantial difference in characteristic time scales of fast inertia–gravity waves and of slow, typically vortex, motions. As we have seen in Chapter 4, Rossby waves, which are in fact vorticity waves, also belong to slow motions. In view of the practical importance of predicting the latter (e.g. in the meteorological context, where large-scale Rossby waves and associated vortices are synonymous to synoptic perturbations), and also of the difficulties to resolve numerically substantially different scales simultaneously, it is desirable to get rid of fast waves in simulations. One could therefore think about further simplification of the RSW models by fast-time averaging. We will pursue such an approach in Chapter 8. In this chapter we will use a more traditional method of direct filtering of inertia–gravity waves by a choice of scaling adapted to vortex motions. In this way we will obtain the classical barotropic and baroclinic quasi-geostrophic (QG) models. These models were historically the first to be extensively applied in geophysical fluid dynamics both for practical (e.g. the first successful weather forecast) and theoretical (e.g. understanding the baroclinic instability) purposes.

5.1 General properties of the horizontal motion. Geostrophic equilibrium

We first recall the part of primitive equations corresponding to the horizontal motion:

$$\frac{\partial \mathbf{v}_h}{\partial t} + \mathbf{v} \cdot \nabla \mathbf{v}_h + f \hat{z} \wedge \mathbf{v}_h = -\nabla_h \Phi. \tag{5.1}$$

Here the subscript h refers to horizontal components, and Φ is the total geopotential comprising a background and a perturbation parts $\Phi = \bar{\Phi} + \phi$. We consider the fluid motion on the tangent beta plane $f = f_0(1 + \beta y)$ and introduce the corresponding geopotential heights $\bar{\Phi} = gH$, $\phi = gh$. The following scaling is natural to use for

Geophysical Fluid Dynamics. Vladimir Zeitlin,
Oxford University Press (2018). © Vladimir Zeitlin. DOI: 10.1093/oso/9780198804338.001.0001

large-scale vortex motions, which are, intuitively, the objects with a single horizontal scale, a single velocity scale, and a time scale defined by vortex turnover time. Hence we choose a scale V for horizontal velocity $\mathbf{v}_h = (u, v)$, the scale W for vertical velocity w, with $W \ll V$, a scale L for horizontal coordinates, and the scale $H \ll L$ for the verical one, and finally a scale $T \sim L/V$ for time. The intrinsic spatial scale is the deformation (Rossby) radius:

$$R_d = \frac{\sqrt{gH}}{f_0}. \tag{5.2}$$

Therefore, the large-scale horizontal motions can be characterised by the following non-dimensional parameters:

- Rossby number: $Ro = \frac{V}{f_0 L}$,
- Burger number: $Bu = \frac{R_d^2}{L^2}$,
- Nonlinearity parameter measuring relative deviation of the isopycnals from their rest positions. $\lambda - \Delta II/II$,
- Non-dimensional gradient of f: $\tilde{\beta} \sim \beta L$

The key parameter in geophysical fluid dynamics is the Rossby number. If we recall that the lower bound for the inertia gravity wave frequencies is f_0 and, thus, the typical time scale for these waves is $\mathcal{O}(f_0^{-1})$, the Rossby number can be interpreted as a ratio of wave to vortex time scales and is, thus, typically small, as fast wave and slow vortex motions are separated by the spectral gap, as discussed in Chapter 3. As is easy to see from (5.1), Ro measures the relative strength of Lagrangian acceleration. For negligible Ro the Coriolis acceleration on the left-hand side of (5.1) is balanced by the acceleration due to the pressure force on the right-hand side leading to the *geostrophic equilibrium*, that is the equilibrium between the Coriolis and pressure forces:

$$f\hat{\mathbf{z}} \wedge \mathbf{v}_g = -\nabla_h \Phi, \tag{5.3}$$

which defines the *geostrophic wind* $\mathbf{v}_g$. If beta-effect is neglected, f is replaced by f_0. As follows from observations, synoptic vortices in the atmosphere and meso-scale vortices in the ocean are close to the geostrophic equilibrium. Graphically, the geostrophic balance correspond to the wind blowing along the isobars, leaving the higher-pressures on the right (in the Northern hemisphere), which is easy to identify in meteorological maps.

It should be stressed that if geostrophic balance (5.3) is combined with hydrostatic balance, for example in the atmospheric context

$$g\frac{\theta}{\theta_0} = \partial_z \Phi, \tag{5.4}$$

the *thermal wind* relations between the horizontal gradient of temperature and vertical shear of the geostrophic wind follow:

$$f_0 \hat{\mathbf{z}} \wedge \partial_z \mathbf{v}_g = -\frac{g}{\theta_0} \nabla_h \theta, \tag{5.5}$$

which is an extremely useful diagnostic tool for flows with small Rossby numbers, permitting by itself to explain many qualitative features of the observed large-scale flows.

5.2   Slow dynamics in a one-layer model

Using the scaling adopted for horizontal motions in Section 5.1 we can write down a non-dimensional version of the RSW equations

$$\begin{cases} Ro\,(\partial_t \mathbf{v} + \mathbf{v} \cdot \nabla \mathbf{v}) + (1 + \tilde{\beta}y)\hat{\mathbf{z}} \wedge \mathbf{v} = -\dfrac{\lambda Bu}{Ro}\nabla\eta\,, \\ \lambda\partial_t\eta + \nabla \cdot (\mathbf{v}(1 + \lambda\eta)) = 0\,, \end{cases} \tag{5.6}$$

where we introduced a non-dimensional deviation η of the free surface from its rest value: $h = H(1 + \lambda\eta)$. As was already said, under this scaling, in order to be close to the geostrophic equilibrium the motion should have a small Rossby number. At the same time, the combination of parameters $\frac{\lambda Bu}{Ro}$ should be of the order 1, otherwise the balance is impossible. This last condition ties together three of the four parameters of the model. As to the fourth, the non-dimensional β, at the scale of typical large-scale oceanic and atmospheric vortices it is small, of the order of Ro or less. The just described conditions do not uniquely prescribe a dynamical regime. Examples of possible regimes close to geostrophy $Ro \equiv \epsilon \ll 1$ are:

- quasi-geostrophic (QG), corresponding to weak nonlinearity and scales of the order of the deformation radius:

$$\lambda \sim Ro, \; \tilde{\beta} \sim Ro \Rightarrow Bu \sim 1, \Rightarrow L \sim R_d. \tag{5.7}$$

- frontal geostrophic (FG), also called semi-geostrophic and corresponding to strong nonlinearity and scales much larger than the deformation radius:

$$\lambda \sim 1, \; \tilde{\beta} \sim Ro \Rightarrow Bu \sim Ro, \Rightarrow L \gg R_d. \tag{5.8}$$

5.2.1   Derivation of the QG equations

Let us write down the non-dimensional RSW equations in the QG regime (5.7):

$$\begin{cases} \epsilon\,(\partial_t \mathbf{v} + \mathbf{v} \cdot \nabla \mathbf{v}) + (1 + \epsilon y)\hat{\mathbf{z}} \wedge \mathbf{v} = -\nabla\eta\,, \\ \epsilon\partial_t\eta + \nabla \cdot (\mathbf{v}(1 + \epsilon\eta)) = 0\,. \end{cases} \tag{5.9}$$

The velocity field will be sought in a form of asymptotic series in Ro:

$$\mathbf{v} = \mathbf{v}^{(0)} + \epsilon \mathbf{v}^{(1)} + \epsilon^2 \mathbf{v}^{(2)} + \dots . \tag{5.10}$$

It is not necessary to expand η; in this way velocity will be *slaved* to pressure.

At the order ϵ^0 we obtain from the momentum equation in (5.6) the geostrophic equilibrium in non-dimensional terms:

$$u^{(0)} = -\partial_y \eta, \quad v^{(0)} = \partial_x \eta \quad \Rightarrow \quad \partial_x u^{(0)} + \partial_y v^{(0)} = 0. \tag{5.11}$$

At the same time, the second equation in (5.6) also leads, to the leading order, to the condition of zero divergence of velocity. The scaling assumptions are, thus, self-consistent. The zero-order advective derivative is

$$\frac{d^{(0)}}{dt} \dots = \partial_t \dots + u^{(0)} \partial_x \dots + v^{(0)} \partial_y \dots \equiv \partial_t \dots + J(\eta, \dots), \tag{5.12}$$

and we recall the definition of two-dimensional Jacobians:

$$J(A, B) \equiv \partial_x A \partial_y B - \partial_y A \partial_x B. \tag{5.13}$$

At the order ϵ^1 we get the first ageostrophic correction to the velocity field:

$$u^{(1)} = -\frac{d^{(0)}}{dt} v^{(0)} - yu^{(0)}, \quad v^{(1)} = \frac{d^{(0)}}{dt} u^{(0)} - yv^{(0)}. \tag{5.14}$$

Unlike the geostrophic wind, the ageostrophic correction has non-zero divergence

$$\partial_x u^{(1)} + \partial_y v^{(1)} = -\frac{d^{(0)}}{dt} \nabla^2 \eta - v^{(0)}, \tag{5.15}$$

and after injecting this expression into (5.9) we obtain the evolution equation for the height (pressure) anomaly:

$$\frac{d^{(0)}}{dt} \left(\nabla^2 \eta - \eta \right) + \partial_x \eta = 0 \leftrightarrow \frac{d^{(0)}}{dt} \left(\nabla^2 \eta - \eta + y \right) = 0. \tag{5.16}$$

If dimensions are restored, this equation reads:

$$\frac{d^{(0)}}{dt} \left(\frac{f_0^2}{gH_0} \left(\frac{gh}{f_0} \right) - \nabla^2 \left(\frac{gh}{f_0} \right) - f_0 (1 + \beta y) \right) = 0. \tag{5.17}$$

The equation (5.16), or (5.17), is the celebrated quasi-geostrophic (QG) equation. It has the physical meaning of Lagrangian conservation of PV in the QG approximation, that is advection of PV calculated in the lowest-order approximation in Ro (advection by geostrophic wind), and can be alternatively obtained by directly applying the QG scaling in the equation of Lagrangian conservation of PV in the one-layer RSW, see Problem 5.1. It should be stressed that inertia–gravity waves are filtered out by the QG scaling, which is consistent with the just-mentioned physical meaning of the QG dynamics and the fact that inertia–gravity waves do not bear PV anomaly(see Problem 2.7).

5.2.2   Rossby waves and vortex dynamics: β plane vs f plane

Let us rewrite (5.16) using (5.12):

$$\partial_t \mathbf{\nabla}^2 \eta - \partial_t \eta + \mathcal{J}(\eta, \mathbf{\nabla}^2 \eta) + \partial_x \eta = 0.$$

A formal linearisation gives

$$\partial_t \eta - \partial_t \mathbf{\nabla}^2 \eta - \partial_x \eta = 0. \tag{5.18}$$

The wave solutions $\eta \propto e^{i(kx+ly-\omega t)}$ of (5.18) exist, with the dispersion relation

$$\omega = -\frac{k}{k^2 + l^2 + 1}. \tag{5.19}$$

If dimensions are restored, this gives

$$\omega = -\beta \, \frac{k}{k^2 + l^2 + R_d^{-2}}. \tag{5.20}$$

These solutions are Rossby waves. They are vorticity waves arising from the 'elasticity' of the iso-PV contours. We have already made a first acquaintance with these waves in Chapter 4. The gradient of the background potential vorticity—the planetary vorticity gradient β—is at the origin of the Rossby wave on the β plane. Yet, as we have seen in Chapter 4, other sources of variation of background PV also lead to appearance of Rossby waves. Unlike inertia–gravity waves, Rossby waves are strongly dispersive, cf. Figure 5.1. The dispersion curves of various types of Rossby waves of Chapter 4 are similar to cross-sections of Figure 5.1 at a fixed meridional wave number l, and the approximate expression of the dispersion law for equatorial Rossby waves given in Section 4.4 confirms this fact. At a fixed meridional wave number l the frequency of the Rossby wave is not monotonous as a function of the zonal wave number k. This means that the sign of the zonal component of the group velocity of Rossby waves changes with changing wavelength. This, in turn, means that short Rossby waves carry energy eastwards, while the long ones westwards. It should be also emphasised that dispersion of long Rossby waves is weak, the dispersion curve $\omega(k)$ being close to the straight line near the origin, at fixed l. This fact will lead to important consequences which will be analysed later.

Passing from the β- to f-plane approximation removes the planetary vorticity gradients and, thus, the *raison d'être* of Rossby waves. The non-dimensional QG equation on the f plane reads:

$$\partial_t \mathbf{\nabla}^2 \eta - \partial_t \eta + \mathcal{J}(\eta, \mathbf{\nabla}^2 \eta) = 0, \tag{5.21}$$

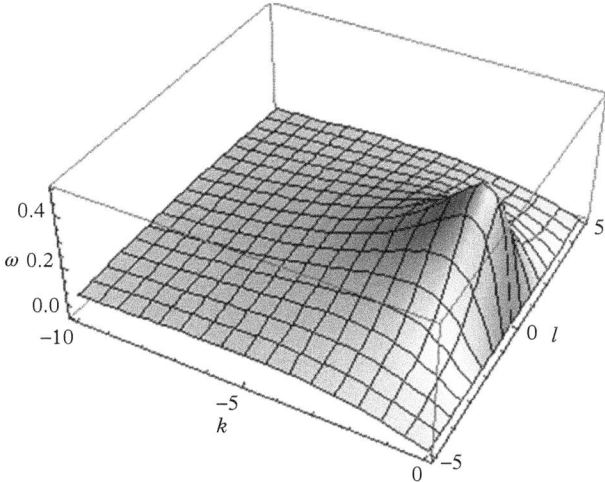

Figure 5.1 *Dispersion relation (5.19) for Rossby waves in one-layer QG model. Negative values $\omega < 0$ are not shown.*

and is equivalent to the two-dimensional (2D) Euler equations for an incompressible fluid, written in terms of stream function, with a stream-function η and modified stream-function - vorticity relation. While this relation is $\zeta = \nabla^2 \eta$ in a 2D Euler system, it becomes here

$$\zeta = \nabla^2 \eta - \eta. \tag{5.22}$$

Obviously, the linearisation of (5.21) gives only stationary solutions, which is consistent with the absence of inertia–gravity waves in the model, and with existence of a zero-frequency root in the dispersion relation (3.38) of the parent RSW model. In dimensional terms $\zeta = \nabla^2 \eta - \frac{1}{R_d^2}\eta$, and the standard 2D Euler dynamics is recovered in the limit $R_d \to \infty$. We will come back to the relation between QG and 2D Euler dynamics in Chapter 6, dedicated to vortex dynamics.

5.2.3 QG dynamics in the presence of topography. Mountain Rossby waves

In the presence of topography/bathymetry b, the Lagrangian conservation of PV in one-layer RSW model takes the form

$$\frac{d}{dt}\left(\frac{\zeta + f}{h - b}\right) = 0. \tag{5.23}$$

For small-amplitude topography $|b| = \mathcal{O}(Ro)$ applying the QG approximation in the β plane to this equations leads to the following non-dimensional equation:

$$\nabla^2 \eta_t - \eta_t + \eta_x + \mathcal{J}(\eta, \nabla^2 \eta + b) = 0, \tag{5.24}$$

where we use the same notation for non-dimensional b. As previously in Section 4.2.2, in order to understand the influence of topography upon the QG dynamics we will consider perturbations to a mean zonal flow with velocity U which are induced by a localised topography, and look for stationary solutions. As the perturbations to the mean flow are supposed to be small, we could proceed by straightforward linearisation. Yet, the specific 'Jacobian' nonlinearity of (5.24) allows us to obtain a general nonlinear solution in this set-up. Indeed, in the case of time-independent η, equation (5.24) becomes

$$\mathcal{J}(\eta, \nabla^2 \eta + b + y) = 0, \tag{5.25}$$

and its general solution is given by

$$\nabla^2 \eta + b + y = \mathcal{F}(\eta), \tag{5.26}$$

where $\mathcal{F}$ is an arbitrary function. For $\eta = -Uy + \psi$ we get

$$\nabla^2 \psi + b + y = \mathcal{F}(\psi - Uy). \tag{5.27}$$

We look for waves generated by a flow encountering a localised topography, therefore far upstream, that is at $x \to +\infty$ for $U < 0$, and $x \to -\infty$ for $U > 0$, the perturbation ψ vanishes and (5.27) becomes

$$y = \mathcal{F}(-Uy), \tag{5.28}$$

which allows us immediately to determine the unknown function $\mathcal{F}(\eta) = -\frac{\eta}{U}$. We thus get the following equation for ψ:

$$U\nabla^2 \psi + \psi = -Ub. \tag{5.29}$$

In order to understand the nature of solutions and to develop physical intuition, let us consider the simplest case of a meridionally uniform topography (a ridge) $b = b(x)$. In this case, the perturbation to the zonal mean flow ψ is also meridionally uniform: $\psi = \psi(x)$. The equation (5.29) becomes

$$\psi''(x) + \frac{1}{U}\psi(x) = -b(x), \tag{5.30}$$

where prime denotes the x differentiation. The solution is given by the inversion of the operator on the left-hand side, that is its Green's function:

$$\psi(x) = -\int_{-\infty}^{+\infty} dx' \, G(x - x')b(x'), \tag{5.31}$$

where

$$G''(x - x') + \frac{1}{U}G(x - x') = \delta(x - x').$$

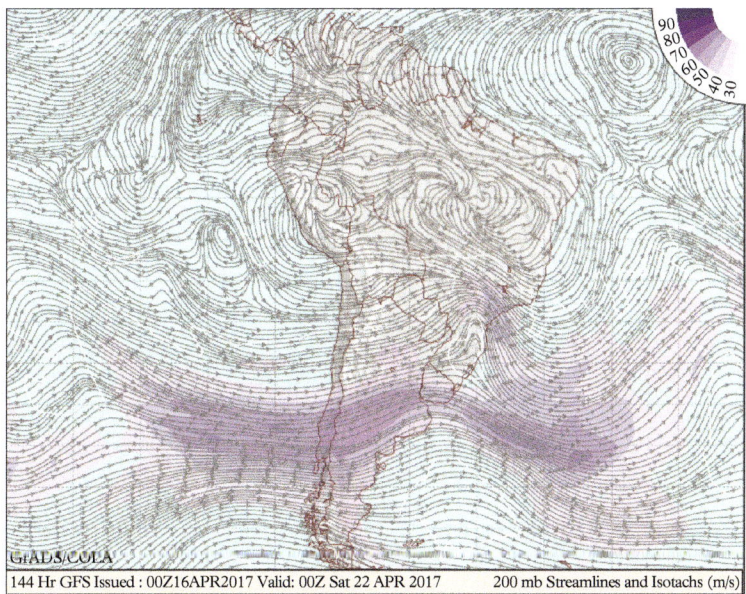

144 Hr GFS Issued : 00Z16APR2017 Valid: 00Z Sat 22 APR 2017   200 mb Streamlines and Isotachs (m/s)

Figure 5.2 *Rossby wave produced by upper tropospheric jet over the Andes Cordillera mountain ridge, as seen in a forecast map. Colour: absolute value of wind velocity, arrows: streamlines. Images generated by the Center for Ocean–Land–Atmosphere Studies (COLA) at George Mason University from the data from the forecast models of NOAA's National Center for Environmental Prediction (NCEP).*

This equation was already solved in Chapter 4, and we can use the results with the replacement $\mathcal{U} \to U$. Thus, in the case of an easterly flow $U < 0$, no waves are produced by the ridge, while in the case of a westerly flow, stationary oscillations arise behind the ridge. Physically speaking, the Green's function is the stream function of the perturbation behind an infinitely thin ridge. It is clear without further calculation that the stream-wise oscillations in westerly flow will result behind any localised ridge, which can be confirmed by direct calculations with simplified ridge profiles. These oscillations are stationary Rossby waves produced by topography, and there are numerous examples of such Rossby wave in the data, cf. Figure 5.2

We now return to the full two-dimensional equation (5.29). Solution of this equation is given by the Green's function $G(x - x', y - y')$:

$$U\nabla^2 G + G = \delta(x - x')\delta(y - y'). \tag{5.32}$$

In order to get the expression for $G(x - x', y - y')$ we proceed by making Fourier transformation $G(x - x', y - y') = \int dk dl\, e^{i(k(x-x')+l(y-y'))} G(k, l)$, which leads to

$$\left[-U(k^2 + l^2) + 1\right] G(k, l) = 1. \tag{5.33}$$

Thus,

$$G(\mathbf{x} - \mathbf{x}') = \int_{-\infty}^{+\infty} dk dl \, \frac{e^{i(k(x-x')+l(y-y'))}}{-U(k^2 + l^2) + 1}. \tag{5.34}$$

For $U < 0$ the integrand in (5.34) is non-singular, and the integration can be carried out straightforwardly, by using polar coordinates in Fourier space:

$$
\begin{aligned}
G(x - x', y - y') &= \int_0^{+\infty} |\mathbf{k}| d|\mathbf{k}| \int_0^{2\pi} d\theta \, \frac{e^{i|\mathbf{k}||\mathbf{x}-\mathbf{x}'|\cos\theta}}{-U\mathbf{k}^2 + 1} \\
&= \int_0^{+\infty} \frac{|\mathbf{k}| d|\mathbf{k}|}{-U\mathbf{k}^2 + 1} \int_0^{2\pi} d\theta \, \cos\left(|\mathbf{k}||\mathbf{x}-\mathbf{x}'|\cos\theta\right) \\
&= 2\pi \int_0^{+\infty} \frac{|\mathbf{k}| d|\mathbf{k}|}{-U\mathbf{k}^2 + 1} \mathcal{J}_0\left(|\mathbf{k}||\mathbf{x}-\mathbf{x}'|\right) = \frac{2\pi}{|U|} K_0\left(\frac{|\mathbf{x}-\mathbf{x}'|}{\sqrt{|U|}}\right).
\end{aligned}
\tag{5.35}
$$

Here $\mathcal{J}_0$ is the Bessel function and K_0 is modified Bessel (Macdonald) function. For $U > 0$, the integrand of the last integral in (5.35) is singular. Calculating it in the principal value sense gives

$$G(x - x', y - y') = -\frac{\pi}{2U} Y_0\left(\frac{|\mathbf{x}-\mathbf{x}'|}{\sqrt{U}}\right), \tag{5.36}$$

where Y_0 is Bessel function of the second kind (Neumann function). We recall that $\mathcal{J}_0(r)$ and $Y_0(r)$ are oscillating and slowly decaying with r, while $K_0(r)$ is exponentially decaying.

As follows from the behaviour of $K_0(r)$, a localised topography (a mountain) will produce exponentially decaying non-oscillating perturbation downstream in the easterly flow. As to the westerly flow, the Green's function (5.36) gives weakly decaying oscillations both downstream and upstream, which is inconsistent with the boundary conditions we used for derivation of (5.29), as the decay of the Neumann function $\sim |\mathbf{x}-\mathbf{x}'|^{-\frac{1}{2}}$ is not sufficiently fast. Thus, this solution should be corrected by a solution of the homogeneous problem 'killing' the oscillations far upstream. The correction can not be found in closed form and is expressed as a series of Bessel functions $\sum_{n=1}^{\infty} \frac{1}{2n-1} \mathcal{J}_{2n-1}\left(\frac{|\mathbf{x}-\mathbf{x}'|}{\sqrt{U}}\right) \cos(2n-1)\phi$, where ϕ is the polar angle on the $x-y$ plane. Example of a flow obtained in this way behind a circular mountain is given in Figure 5.3. The oscillations are Rossby waves generated by the PV produced by topography, and propagating due to the beta effect. Note that we did not limit ourselves by linear waves of infinitesimal amplitude above, yet the problem was reduced to a linear equation. So, the amplitude of the solution is arbitrary, which may lead to appearance of closed streamlines, cf. Fig. 5.3. By construction, the solution is valid only outside of the zones delimited by closed streamlines, and some additional boundary conditions are needed to extrapolate the solution inside.

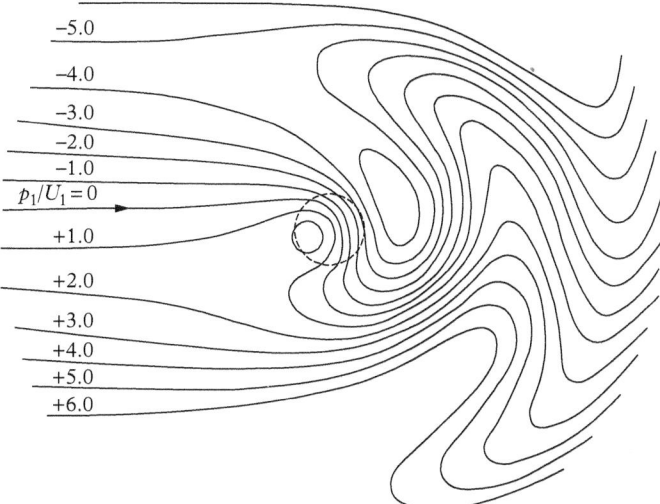

-5.0
-4.0
-3.0
-2.0
-1.0
$p_1/U_1 = 0$
+1.0
+2.0
+3.0
+4.0
+5.0
+6.0

Figure 5.3 *Streamlines corresponding to a stationary Rossby wave produced by an axially symmetric mountain indicated by a dashed circle.*

5.2.4   Frontal geostrophic dynamics

We already mentioned in Section 5.2 that QG regime is not unique at small Rossby numbers. As an example we sketch below the derivation of the model for slow vortex dynamics in the frontal geostrophic (FG) regime, which is characterised by a strong nonlinearity, cf. (5.8). The non-dimensional equations of motion in the case of strong deviations of the interface read:

$$\begin{cases} \epsilon \left(\partial_t \mathbf{v} + \mathbf{v} \cdot \nabla \mathbf{v} \right) + (1 + \epsilon y)\hat{\mathbf{z}} \wedge \mathbf{v} = -\nabla h \,, \\ \partial_t h + \nabla \cdot (\mathbf{v} h) = 0 \,. \end{cases} \tag{5.37}$$

It does not make sense to separate the mean and the perturbation in the height field at $\lambda = \mathcal{O}(1)$, and thus the mass conservation equation has no small parameter. The momentum equation has exactly the same form as in the QG regime, modulo the replacement $\eta \to h$. This means that, while we still get the geostrophic wind equations (5.11) for velocity in the leading order in ϵ, the mass equation implies $\partial_t h \equiv 0$, as the leading-order velocity is non-divergent, so a slower time scale should be chosen in the FG regime: $\partial_t \to \epsilon \partial_\tau$. This leads, in the leading order in ϵ, to the following evolution equation for h:

$$\partial_\tau h - \mathcal{J}(h, \nabla^2 h)h - \mathcal{J}\left(h, \frac{(\nabla h)^2}{2} \right) - h \partial_x h = 0. \tag{5.38}$$

5.3 Slow dynamics in the two-layer model with a rigid lid

Although the one-layer (quasi-)barotropic QG equation provides the simplest model for slow vortex motions, it misses an essential element, the baroclinicity, that is variations of the geostrophic stream function in the vertical. A passage to the multi-, at least two-layer, model is needed to include the baroclinic effects. To get a baroclinic extension of the QG model, we start with the two-layer RSW with a rigid lid.

5.3.1 Derivation of the QG equations

Parameters

Together with universal for horizontal motions Rossby number $Ro = \frac{V}{f_0 L}$, non-dimensional typical deviation of the interface λ, and non-dimensional gradient of the Coriolis parameter $\tilde{\beta}$, new specific parameters appear in the two-layer model: the aspect ratio $d = \frac{H_1}{H_2} = \frac{\bar{h}_1}{\bar{h}_2}$, where $\bar{h}_{1,2} = \frac{H_{1,2}}{H}$ are non-dimensional thicknesses of the layers, while the total thickness is $H = H_1 + H_2$, and the stratification parameter $s = \frac{\rho_2 - \rho_1}{\rho_2} \equiv 1 - r$, where r is the density ratio. Along with the barotropic deformation radius $R_{0_d} = \frac{\sqrt{gH}}{f_0}$, its baroclinic counterpart arises, $R_d = \frac{\sqrt{sgH}}{f_0}$, and the related baroclinic Burger number, $Bu = \frac{R_d^2}{L^2}$. The reduced gravity $g' = gs$ naturally arises in these formulae.

Characteristic scales

Baroclinic deformation radius is the natural horizontal length scale $L \sim R_d$, and the total depth H was already used earlier as a characteristic vertical scale. Vortex turnover time $T \sim L/V$, where V is a typical horizontal velocity scale, which will be chosen the same in both layers, is a typical time scale. Pressures in the layers are scaled as $P_i \sim \rho_i V L f_0$.

Derivation of the QG equations

The non-dimensional equations of the two-layer RSW with a rigid lid take the form

$$
\begin{cases}
\epsilon \dfrac{d_i}{dt} \mathbf{v}_i + (1 + \tilde{\beta}y)\hat{z} \wedge \mathbf{v}_i = -\boldsymbol{\nabla}\pi_i, \quad i = 1, 2, \\[2mm]
-\lambda \dfrac{d_1}{dt}\eta + (\bar{h}_1 - \lambda\eta)\boldsymbol{\nabla} \cdot \mathbf{v}_1 = 0, \\[2mm]
\lambda \dfrac{d_2}{dt}\eta + (\bar{h}_2 + \lambda\eta)\boldsymbol{\nabla} \cdot \mathbf{v}_2 = 0, \\[2mm]
\pi_2 - r\pi_1 = \dfrac{\lambda Bu}{\epsilon}\eta.
\end{cases}
\tag{5.39}
$$

Here $\frac{d_i}{dt} = \partial_t + \mathbf{v}_i \cdot \boldsymbol{\nabla}$, as usual, and η is the non-dimensional deviation of the interface from its equilibrium position. As in the one-layer case, the QG regime is defined by the following choice of parameters:

$$\lambda \sim \tilde{\beta} \sim \epsilon \ll 1, \quad \Rightarrow \quad L \sim R_d \leftrightarrow Bu \sim 1, \tag{5.40}$$

although it should be not forgotten that the Burger number is baroclinic now. Asymptotic expansions in ϵ are the same as in the one-layer case:

$$
\begin{aligned}
u_i &= u_i^{(0)} - \epsilon \left[\partial_t v_i^{(0)} + \mathcal{J}(\pi_i, v_i^{(0)}) + y u_i^{(0)} \right] + \dots \\
v_i &= v_i^{(0)} + \epsilon \left[\partial_t u_i^{(0)} + \mathcal{J}(\pi_i, u_i^{(0)}) - y v_i^{(0)} \right] + \dots,
\end{aligned}
\tag{5.41}
$$

with the geostrophic wind layer-wise in the leading order:

$$u_i^{(0)} = -\partial_y \pi_i, \quad v_i^{(0)} = \partial_x \pi_i. \tag{5.42}$$

As in the one-layer case, we are not expanding pressure variables, which means that velocity field is slaved to the pressure field.

Divergence of velocity in each layer arises in the next order:

$$\partial_x u_i^{(1)} + \partial_y v_i^{(1)} = -\left[\partial_t \boldsymbol{\nabla}^2 \pi_i + \mathcal{J}(\pi_i, \boldsymbol{\nabla}^2 \pi_i) + \partial_x \pi \right], \tag{5.43}$$

giving, after substitution in the η equations in (5.39), the following pair of equations:

$$\partial_t \eta + \mathcal{J}(\pi_i, \eta) - (-1)^i \bar{h}_i \left[\partial_t \boldsymbol{\nabla}^2 \pi_i + \mathcal{J}(\pi_i, \boldsymbol{\nabla}^2 \pi_i) + \partial_x \pi \right] = 0, \quad i = 1, 2. \tag{5.44}$$

By expressing the interface deviation η in terms of $\pi_{1,2}$ via the last equation in (5.39), we get a closed system of two equations for pressures in the layers.

$$\frac{d_i^{(0)}}{dt} \left[\nabla^2 \pi_i - (-1)^i F_i (\pi_2 - r \pi_1) + y \right] = 0, \quad i = 1, 2, \tag{5.45}$$

where we switched to the frequently used notation $\bar{h}_i^{-1} = F_i$. Here the Lagrangian derivatives

$$\frac{d_i^{(0)}}{dt} (\dots) = \partial_t (\dots) + \mathcal{J}(\pi_i, \dots), \quad i = 1, 2 \tag{5.46}$$

represent advection by the geostrophic wind layer-wise. It should be emphasised that on the f plane (5.45) becomes a system of coupled 2D Euler equations with modified stream-function–vorticity relations layer-wise.

5.3.2   Rossby waves in the two-layer QG model

Let us take, for simplicity, a frequently used (in oceanography) limit of weak stratification $r \to 1$, and formally linearise the equations (8.68). We thus get

$$
\begin{cases}
\partial_t \left[\nabla^2 \pi_1 + F_1 (\pi_2 - \pi_1) \right] + \partial_x \pi_1 = 0, \\
\partial_t \left[\nabla^2 \pi_2 - F_2 (\pi_2 - \pi_1) \right] + \partial_x \pi_2 = 0.
\end{cases}
\tag{5.47}
$$

Looking for wave solutions $\pi_i = A_i e^{i(\mathbf{k} \cdot \mathbf{x} - \omega t)} + c.c.$ we get the following condition of existence of nontrivial solutions of the resulting algebraic equations for the wave amplitudes A_i:

$$
\det \begin{pmatrix}
\omega(\mathbf{k}^2 + F_1) + k_x & -\omega F_1 \\
-\omega F_2 & \omega(\mathbf{k}^2 + F_2) + k_x
\end{pmatrix} = 0.
\tag{5.48}
$$

The dispersion relation follows:

$$
\omega = -\frac{k_x}{2\mathbf{k}^2(\mathbf{k}^2 + F_1 + F_2)} \left[(2\mathbf{k}^2 + F_1 + F_2)(F_1 + F_2) \right].
\tag{5.49}
$$

Its two solutions correspond to (see Problem 6 at the end of this chapter):

- a faster barotropic mode: $\omega_{bt} = -\frac{k_x}{\mathbf{k}^2}$,
- a slower baroclinic mode: $\omega_{bc} = -\frac{k_x}{(\mathbf{k}^2 + F_1 + F_2)}$.

As in the one-layer case, these waves appear due to the β - effect. As was already emphasised in Chapter 3, all wave species are doubled in the two-layer case, with appearance of one barotropic and one baroclinic mode of each kind, which is what we observe here.

5.3.3   Baroclinic instability: first acquaintance

The Phillips model

The classical model of the baroclinic instability is the Phillips model, which is based on the two-layer QG model considered in the limit $s \to 0$, and hence $\eta \to \pi_2 - \pi_1$:

$$
\frac{d_i^{(0)}}{dt} \left[\nabla^2 \pi_i - (-1)^i F_i \eta + y \right] = 0, \ i = 1, 2.
\tag{5.50}
$$

An exact solution of these equations is a constant zonal flow $u_i = U_i = \mathrm{const}$ in geostrophic equilibrium: $U_i = -\partial_y \pi_i$, $i = 1, 2$. If $U_1 \neq U_2$ the flow has a vertical shear corresponding to the *inclined interface*: $\eta = \pi_2 - \pi_1 = (U_1 - U_2) y$. Hence, there is available potential energy in such a flow. We already encountered a situation where the background thickness of the flow varied with one of the coordinates in Chapter 4. The overall linear dependence on y here, together with constancy of velocity, allows for a simple analytic treatment.

Wave solutions and linear instability

Linearisation of the QG equations about the background flow

$$\pi_i = -U_i y + \phi_i, \quad ||\phi|| \ll 1,$$

gives:

$$(\partial_t + U_i \partial_x) \left[\nabla^2 \phi_i - (-1)^i F_i(\phi_2 - \phi_1)\right] + \left[1 - (-1)^i F_i(U_1 - U_2)\right] \partial_x \phi_i = 0. \qquad (5.51)$$

Looking for the wave solutions $\phi_i = A_i e^{i(\mathbf{k} \cdot \mathbf{x} - \omega t)} + c.c.$, and using the notation: $c = \omega/k_x$, $U_1 - U_2 = \Delta U$, we get the system of algebraic equations for the wave amplitudes $A_{1,2}$:

$$\begin{cases} A_1 \left[(c - U_1)(\mathbf{k}^2 + F_1) + 1 + F_1(U_1 - U_2)\right] - A_2(c - U_1)F_1 = 0, \\ -A_1(c - U_2)F_2 + A_2 \left[(c - U_2)(\mathbf{k}^2 + F_2) + 1 - F_2(U_1 - U_2)\right] = 0. \end{cases} \qquad (5.52)$$

The following dispersion relation for nontrivial solutions results:

$$c = U_2 + \frac{1}{2\mathbf{k}^2(\mathbf{k}^2 + F_1 + F_2)} \left[\left(\Delta U \mathbf{k}^2(\mathbf{k}^2 + 2F_2) - \mathbf{k}^2(2\mathbf{k}^2 + F_1 + F_2)\right)\right.$$

$$\left. \pm \left[(F_1 + F_2)^2 + 2\Delta U \mathbf{k}^4(F_1 - F_2) - \mathbf{k}^4 (\Delta U)^2 (4F_1 F_2 - \mathbf{k}^4)\right]^{\frac{1}{2}}\right]. \qquad (5.53)$$

An inspection of this formula shows that for sufficiently strong shears ΔU, and sufficiently small $|\mathbf{k}|$ one of the solutions for c, and hence for the frequency ω, has a positive imaginary part. Therefore, the amplitude of the corresponding linear wave is growing exponentially. Such situation corresponds to a *linear instability*. Of course, the linearisation quickly stops to be valid if this happens, and fully nonlinear treatment is necessary to follow the evolution of the fluid. A special Chapter 10 is devoted to studies of instabilities of jets, of which we see here the first example. This one, however, is of primary importance, as it is the *baroclinic instability*, playing a special role in the atmosphere and oceans. The instability is, indeed, baroclinic, as it is easy to show that the corresponding eigenmode is baroclinic, in the sense explained in the Section 5.3.2.

Baroclinic instability in the *f* plane

Baroclinic instability exists also on the f plane. As we have just seen, the instability is related to Rossby waves propagating in the background flow. As already mentioned, the physical origin of Rossby waves is the presence of the gradient of the background potential vorticity. It is the gradient of the planetary vorticity on the β plane which plays this role for 'ordinary' Rossby waves in the one-layer model considered earlier. In the absence of planetary vorticity gradient on the f plane, this role is played by the slope of the interface between the layers. Indeed, constant gradients of PV appear in each layer due to this slope, having opposite signs. This means that Rossby waves propagate in opposite directions in upper and lower layers. As we will see in Chapter 10, this is a necessary condition of instability. The dispersion relation for the resulting waves is

$$c = \frac{1}{2(\mathbf{k}^2 + F_1 + F_2)} \left[U_1(\mathbf{k}^2 + 2F_2) + U_2(\mathbf{k}^2 + 2F_1) \pm \left[(\Delta U)^2 \left(\mathbf{k}^4 - 4F_1F_2 \right) \right]^{\frac{1}{2}} \right].$$
(5.54)

As follows from this formula, unlike the β-plane case, there exists a range of unstable long-wave (small $\mathbf{k}$) modes for *any* shear. There are two important conclusions to draw from this result, which remain valid beyond the shallow-water model: 1) the baroclinic instability is a long-wave one, 2) β effect plays a stabilising role. We will come back to the analysis of the baroclinic instability, and other instabilities, in the full two-layer RSW in Chapter 10.

5.3.4   Frontal geostrophic regimes

As in the one-layer case, the constraint of smallness of the interface deviations can be relaxed, leading to the frontal geostrophy (FG). To analyse this regime we will be using the same scaling as in Section 5.3.1, with a slight difference that pressure in each layer will be scaled with its own velocity: $P_i \sim \rho_i V_i L f_0$. While the velocity scales are of the same order at comparable thicknesses of the layers when the aspect ratio is of the order one $d = \mathcal{O}(1)$, they are different at significantly different thicknesses, and small d. If one thinks about slow motions in the ocean, an obvious interpretation of the two-layer model is in terms of the thermocline and the deep ocean, which corresponds to the latter case. Although the corresponding FG equations can also be derived in the first case, they seem to be less physically relevant, and we will not present them, limiting ourselves by a configuration with a thin upper layer in which we choose $d = \mathcal{O}(\epsilon)$, and $V_2 \sim \epsilon V_1$. It it easy to see that, as in the one-layer case, at small Rossby numbers and large isopycnal deviations, dynamical consistency requires the baroclinic Burger number to be small. It should be emphasised that the characteristic vertical scale in this context, defining in particular the Burger number, is the thickness of the upper layer. We will not give the details of calculations, which follow the same lines as in the one-layer case, with additions concerning the lower-layer variables. For simplicity, we will limit ourselves by the f plane approximation. The procedure results in a pair of coupled equations for the barotropic component of pressure $P = h_1\pi_1 + h_2\pi_2$, where $h_{1,2}$ are non-dimensional thicknesses of the layers, and for the interface displacement η, which is proportional to the baroclinic component of pressure.

$$\partial_\tau \eta + \mathcal{J}(P, \eta) + \mathcal{J}\left(\eta, (1-\eta)\nabla^2\eta - \frac{(\nabla\eta)^2}{2} \right) = 0.$$
(5.55)

$$\partial_\tau \nabla^2 P + \mathcal{J}(P, \nabla^2 P) + \mathcal{J}\left(\eta, (1-\eta)\nabla^2\eta + \frac{(\nabla\eta)^2}{2} \right) = 0.$$
(5.56)

Both equations express conservation of PV, respectively, barotropic and baroclinic. Apart from the advection by the barotropic component of the geostrophic wind, given by the second term in (5.55), this equation coincides with (5.38) if one remembers that with the chosen scaling the non-dimensional thickness of the upper layer is $h_1 = 1 - \eta$. For a very deep lower layer, barotropic velocity coincides with the lower-layer velocity. Hence, for motionless lower layer we recover the one-layer FG equation, which is a consistency check. Note also that by adding the equations (5.55) and (5.56), we obtain that the combination $\nabla^2 P - \eta$ is simply advected by the barotropic velocity.

5.4   Slow dynamics in two-layer model with a free surface

Although the rigid-lid two-layer QG model allows us to introduce baroclinic effects and understand the baroclinic instability, its parent model, the two-layer RSW with a rigid lid is not well-suited for numerical implementations, as will be clear from Chapter 7. Therefore, to compare the results obtained within the QG model to those of the parent RSW model, we need a free-surface version of the two-layer QG model. It is derived here. The FG versions of the two-layer model with a free surface can be obtained along the same lines as previously and are not presented.

5.4.1   Equations of motion, parameters, and scaling

The equations of the model can be rewritten in the same form as in the rigid-lid case:

$$\begin{cases} \partial_t \mathbf{v}_i + (\mathbf{v}_i \cdot \nabla)\mathbf{v}_i + f\hat{z} \wedge \mathbf{v}_i = -\dfrac{\nabla \pi_i}{\rho_i}, \\[2ex] \partial_t h_i + \nabla(\mathbf{v}_i h_i) = 0, \quad i = 1, 2 \end{cases} \tag{5.57}$$

with the dynamical boundary conditions at the interface and at the free surface:

$$\begin{cases} \pi_2 - \pi_1 = g(\rho_2 - \rho_1)h_2, \\[2ex] \pi_1 = g\rho_1(h_1 + h_2). \end{cases} \tag{5.58}$$

We define the interface displacement $\eta_2 = h_2 - H_2$, where H_2 is the thickness of the layer two at rest, and the free surface displacement $\eta_1 = h_1 + h_2 - H$, where $H = H_1 + H_2$ is the total thickness at rest.

We use the same scaling as in the rigid-lid configuration, supposing that characteristic amplitudes of interface and free-surface displacements are of the same order:

$$\begin{cases} u = Vu^*, \quad (x, y) = L(x^*, y^*), \quad t = \dfrac{L}{V}t^*, \\[2ex] \pi_i = \rho_i VLf\pi_i^*, \\[2ex] h_1 = H(F_1^{-1} + \lambda(\eta_1^* - \eta_2^*)), \\[2ex] h_2 = H(F_2^{-1} + \lambda\eta_2^*). \end{cases} \tag{5.59}$$

Here, V, L, and λ are typical scales for velocity, length, and non-dimensional surface and interface displacements, and $F_i = H/H_i \equiv \bar{h}_i^{-1}$, as earlier. Asterisks indicate non-dimensional variables and will be omitted. We choose the QG scaling with $\epsilon \sim \lambda \sim \tilde{\beta}$ and $L \sim R_d$.

5.4.2   QG equations

We will not repeat the asymptotic expansion procedure for the non-dimensional equations which follows the same lines as before. To the leading order the non-dimensional pressures π_i play the role of quasi-geostrophic stream functions for the geostrophic velocities in respective layers. In the next approximation divergence of the velocity field becomes non-zero, and we obtain the QG equations:

$$\begin{cases} \dfrac{d_1^{(0)}}{dt} \left[\nabla^2 \pi_1 + \dfrac{F_1}{1-r}(\pi_2 - \pi_1) + y \right] = 0, \\[4mm] \dfrac{d_2^{(0)}}{dt} \left[\nabla^2 \pi_1 - \dfrac{F_2}{1-r}(\pi_2 - r\pi_1) + y \right] = 0, \end{cases} \tag{5.60}$$

where $\dfrac{d_i^{(0)}}{dt}$ are the same as earlier, cf. (5.46). Note that these equations differ from the ones with a rigid lid, cf. (8.68) by the absence of the density ratio r in the equation for the upper layer (other appearances of this parameter may be removed by rescaling either the parameter F_1 or the length scale). We do not repeat the linearisation and derivation of dispersion relation for baroclinic and barotropic Rossby waves, which can be done along the same lines as in the preceding section.

5.5   Large-scale slow dynamics in the presence of wave guides

5.5.1   A reminder on multi-scale asymptotic expansions

In this section we encounter for the second time the technique of multiple-scale asymptotic expansions, which was already briefly introduced in Section 4.1.2, and will play a central role in Chapter 8. Let us recall its main ideas. Up to now in this chapter we were expanding all variables in asymptotic series in the small parameter ϵ, the Rossby number, after having chosen a scaling for independent variables. A particular feature of the scaling which we were using earlier is that there was always both a single spatial and a single time scale. In the situation where there are multiple scales, they are necessarily of different orders of magnitude, and should be ranged with the help of some small parameter δ as L_1, $L_2 = \delta L_1, \ldots$, and similarly t_1, $t_2 = \delta t_1, \ldots$. A relation between small parameters δ and ϵ should be fixed, too. Derivatives with respect to independent variables become

$\partial_t = \partial_{t_1} + \delta \partial_{t_2} + \ldots$, and similar for the spatial variable(s). The asymptotic expansions of dependent variables in series in powers of the largest of the small parameters, together with expansions of all derivatives, as above, are injected in the equations of motion, which are considered order by order in this small parameter. The procedure results in a series of linear problems. The lowest-order problem is homogeneous, and the higher-order problems are inhomogeneous, with forcing terms determined by the previous-order results. These linear problems can be easily solved, unless there exist obstacles to the solvability, the so-called secular, or resonant, terms in the forcing. Such terms, if left unattended, would lead to resonant growth of the higher-order solutions. This growth would invalidate the perturbation theory, where higher-order terms have to be small corrections to the lower-order ones. The dependence on higher-order, slower times results, precisely, from eliminating these resonances, and thus rendering the whole system solvable, and the perturbation theory consistent. We will see a number of applications of this general scheme in what follows.

5.5.2 Slow motions in the presence of a lateral boundary

As already mentioned, the presence of a lateral boundary, even in its simplest form, without underlying shelf, fills the spectral gap between vortex motions and inertia–gravity waves, because of the presence of non-dispersive Kelvin waves. Yet, even a simple inspection of the dispersion diagram for one-layer rotating shallow water in a half-plane presented in Figure 5.4 shows that for small along-shore wave numbers the frequencies of inertia gravity and Kelvin wave are well separated. This gives us a hint that slow and fast motions still could be dynamically split at large along-shore scales. We will scale the equations of motion in a way to take advantage of this observation.

We introduce a small parameter

$$\delta = \frac{L_x}{L_y} \tag{5.61}$$

which takes into account the disparity of the along- and across-shore scales of motion, and suppose that the across-shore velocity u is also small (it is identically zero for

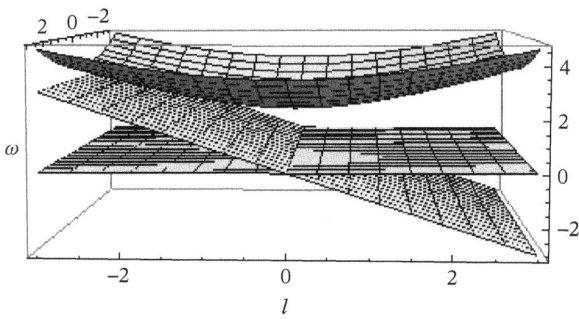

Figure 5.4 *Dispersion relation for internal–gravity and Kelvin waves. Upper surface: inertia–gravity waves; inclined plane: barotropic Kelvin waves. Solution $\omega = 0$ corresponding to vortex motions is represented by horizontal plane. l is along-shore wave number.*

coastal Kelvin waves): $u \to \delta u$. We will also choose the first slow time scale $t_1 = \delta t$ and will introduce, if necessary, even slower time scales $t_2, t_3 \ldots$. We then get the non-dimensional equations of 1- layer RSW with a coast:

$$\begin{cases} \delta^2 u_{t_1} + \delta^2 \epsilon (uu_x + vu_y) - v = -h_x, \\ \delta v_{t_1} + \epsilon (uv_x + vv_y) + \delta u = -\delta h_y, \\ h_{t_1} + u_x + v_y + \epsilon \left[(hv)_y + (hu)_x \right] = 0, \end{cases} \tag{5.62}$$

which are subject to the boundary condition $u|_{x=0} = 0$. Before proceeding further, a relation between the two small parameters, the Rossby number ϵ and the aspect ratio δ, should be fixed and we choose $\epsilon \sim \delta$.

To the lowest order in δ the system is linear:

$$\begin{cases} -v^{(0)} + h_x^{(0)} = 0, \\ v_{t_1}^{(0)} + u^{(0)} + h_y^{(0)} = 0, \\ h_{t_1}^{(0)} + u_x^{(0)} + v_y^{(0)} = 0. \end{cases} \tag{5.63}$$

By injecting the expression for $v^{(0)}$ from the first equation into the remaining two, and by combining them, the system (5.63) is reduced to a single equation:

$$\left(h_{xx}^{(0)} - h^{(0)} \right)_{t_1} = 0, \tag{5.64}$$

which is the leading-order expression of the conservation of quasi-geostrophic potential vorticity, cf. (5.21), for the motions with small δ. However, (5.64) is satisfied identically by $h^{(0)} \propto e^{\pm x}$. Only the minus sign in the exponential is compatible with the boundary condition, which leads to the Kelvin-wave solution considered in Section 4.1:

$$u = 0, \quad h^{(0)} = K(y + t_1)e^{-x}, \quad v^{(0)} = -K(y + t_1)e^{-x}. \tag{5.65}$$

Apart from this particular solution, $h^{(0)}$ is t_1- independent, as follows from (5.64), which, in turn, means that velocity is stationary and is given by the geostrophic balance. Therefore, to this order, the long-wave motion consists of slow Kelvin waves and even slower geostrophic motions, which is in agreement with what is seen in the long-wave (close to the origin) portion of Figure 5.4.

To the next order

$$\begin{cases} -v^{(1)} + h_x^{(1)} = 0, \\ v_{t_1}^{(1)} + u^{(1)} + h_y^{(1)} = -\left(v_{t_2}^{(0)} + u^{(0)} v_x^{(0)} + v^{(0)} v_y^{(0)} \right) \equiv \mathcal{R}_v, \\ h_{t_1}^{(1)} + u_x^{(1)} + v_y^{(1)} = -\left(h_{t_2}^{(0)} + (u^{(0)} h^{(0)})_x + (v^{(0)} h^{(0)})_y \right) \equiv \mathcal{R}_h, \end{cases} \tag{5.66}$$

where we introduced a dependence on the next slow time t_2. The QG part of the flow, unlike the Kelvin-wave part, has non-zero across-shore velocity $u^{(1)}$. We can eliminate it by combining the two last equations in (5.66). We then use the remaining equation to get an equation for $h^{(1)}$:

$$\left(h_{xx}^{(1)} - h^{(1)}\right)_{t_1} = -\mathcal{R}_h + \mathcal{R}_{v_x}. \tag{5.67}$$

In order to avoid secular growth of the first-order solution, the t_1- average of the right-hand side of this equation should vanish. A straightforward calculation gives

$$\left(h_{xx}^{(0)} - h^{(0)}\right)_{t_2} + \mathcal{J}\left(h^{(0)}, h_{xx}^{(0)} - h^{(0)}\right) = 0. \tag{5.68}$$

This is the standard QG equation in the f plane in the limit $\frac{L_x}{L_y} \to 0$, cf (5.21).

In the Kelvin-wave part of the flow $u \equiv 0$. The first-order equations (5.66) can be combined in this case to give

$$\left(h_x^{(1)} - h^{(1)}\right)_{t_1} - \left(h_x^{(1)} - h^{(1)}\right)_y = -\mathcal{R}_h + \mathcal{R}_v. \tag{5.69}$$

If $h^{(1)}$ belongs to the Kelvin-wave part of solution depending only on $y + t_1$, then the condition of absence of secular growth of the Kelvin wave is

$$\int_0^\infty dx\, e^{-x}\, (\mathcal{R}_h - \mathcal{R}_v) = 0. \tag{5.70}$$

Indeed, by integration by parts and using the first equation in (5.66) we get

$$\int_0^\infty dx\, \left[\left(v^{(1)} - h^{(1)}\right)_{t_1} - \left(v^{(1)} - h^{(1)}\right)_y\right] = -h^{(1)}(0, y, t_1)_{t_1} + h^{(1)}(0, y, t_1)_y, \tag{5.71}$$

and the last expression is zero for the Kelvin wave. The constraint (5.70) leads to the following evolution equation for the Kelvin wave:

$$h_{t_2}^{(0)} + \frac{3}{4} h^{(0)} h_y^{(0)} = 0. \tag{5.72}$$

Due to the fact that Kelvin waves carry no PV anomaly, they do not contribute to the evolution of the QG component of the flow, which obeys (5.68). In turn, the QG component has zero contribution into the evolution of the Kelvin wave, as is clear from (5.72). Thus, the two parts of the slow motion, QG vortices and Kelvin waves, are dynamically split and non-interacting, at least up to second-order corrections. As we will see in Chapter 7, the simple-wave equation (5.72) leads to wave breaking in finite (although slow) time. This process is illustrated by Figure 5.5. Breaking means that inclusion of dissipative effects, which are usually neglected throughout this book, becomes inevitable

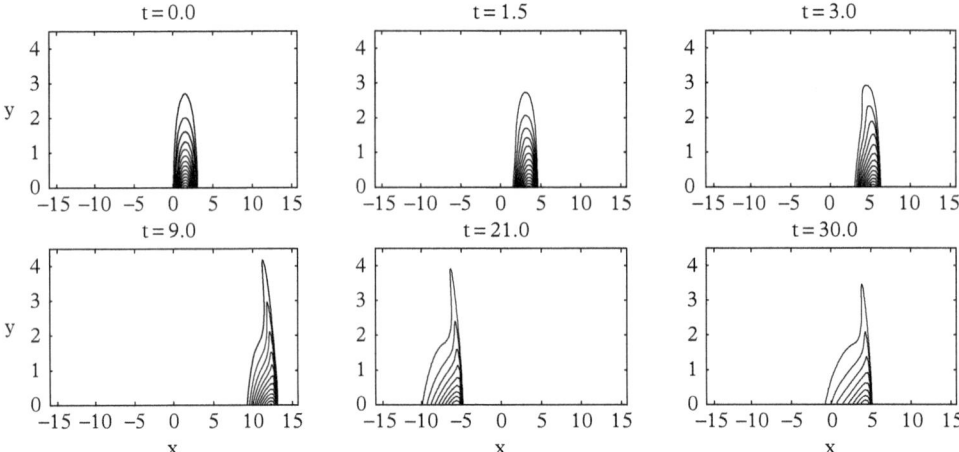

Figure 5.5 *Breaking of a localised packet of coastal Kelvin waves with positive thickness anomaly propagating along the straight boundary at the x axis. Time is measured in units of f^{-1}, and distances in units of R_d. Shown: isolines of thickness anomaly.*

beyond the breaking point. Dissipation does produce PV, if included into the PV evolution equation, and thus breaking will recouple the two components of the flow, although weakly, if dissipation is small.

Let us mention that although the standard quasi-geostrophic dynamics is still valid in the presence of a lateral boundary, the behaviour of vortices which it describes is modified. The boundary conditions of vanishing normal component of the velocity means that the boundary is a streamline of the geostrophic stream function. This means, in turn, that the boundary acts as a mirror. For example, a monopolar vortex near the boundary, due to the presence of its mirror image, behaves like a dipole and moves along the boundary. We will see more of dynamics of vortices in the presence of a boundary in subsequent chapters.

5.5.3   Slow motions over escarpment

As we have already seen in Section 5.2.3, for topography of weak amplitude $|b|_{\max} \sim Ro \ll 1$ the QG approximation leads to the non-dimensional equation (5.24). The term η_x in this equation is absent if the f-plane approximation is used. It is obvious that a linear profile of topography $b(x) = b_0 + b_1 x$ would mimic the beta effect, and lead to appearance of Rossby waves propagating across the gradient of topography, the topographic Rossby waves.

The case of order-1 topography may be treated similarly to the previous subsection. We consider escarpment-type topography which, according to what was just said, produces a wave guide for topographic Rossby waves. The topography b will have constant asymptotics $b_{\pm}$ at $x \to \pm\infty$, and a transition zone, supposed to be monotonous in the vicinity of $x = 0$ (otherwise extra spatial scales are to be considered). We introduce disparity of scales across (x direction) and along (y- direction) the escarpment, which is measured by a small parameter δ, as well as corresponding velocity-scale disparity, as in Section 5.5.2. We suppose that typical non-dimensional amplitude of the thickness anomaly is of the order of the Rossby number ϵ, and consider that motions are slow, and evolve in slow time scale $t_1 = \delta t$. The non-dimensional equations with this scaling take the form

$$\begin{cases} \delta^2 u_{t_1} + \delta^2 \epsilon (u u_x + v u_y) - v = -h_x, \\ \delta v_{t_1} + \epsilon (u v_x + v v_y) + \delta u = -\delta h_y, \\ h_{t_1} + u_x + v_y + [(1 - b + \epsilon h) v]_y + [(1 - b + \epsilon h) u]_x = 0. \end{cases} \tag{5.73}$$

We choose, as in Section 5.5.2, $\epsilon \sim \delta$, and get in the lowest-order approximation:

$$\begin{cases} -v^{(0)} + h_x^{(0)} = 0, \\ v_{t_1}^{(0)} + u^{(0)} + h_y^{(0)} = 0, \\ h_{t_1}^{(0)} + [(1 - b) u^{(0)}]_x + [(1 - b) v^{(0)}]_y = 0. \end{cases} \tag{5.74}$$

This system of linear equations can be straightforwardly reduced to a single one:

$$\left[h^{(0)} - ((1 - b) h_x^{(0)})_x \right]_{t_1} - \left[(1 - b) h_y^{(0)} \right]_x + \left[(1 - b) h_x^{(0)} \right]_y = 0. \tag{5.75}$$

At both sides, far from the escarpment where $b = b_\pm = $ const, this equation becomes, respectively,

$$\left(\frac{1}{1 - b_\pm} h^{(0)} - h_{xx}^{(0)} \right)_{t_1} = 0, \tag{5.76}$$

which, like (5.64), expresses the conservation of quasi-geostrophic PV, with different deformation radii defined by different depths of the fluid far left and far right from the escarpment. Equation (5.76) means that the solution at each side of the escarpment, to this order, is either stationary, or is exponentially decaying far right and far left, with the exponents $\mp\sqrt{1 - b_\pm}$, respectively. Solutions of the first kind are standard geostrophic motions whose evolution takes place at an even slower time scale $t_2 = \delta^2 t$. Solutions of the second kind represent the outer tails of topographic waves propagating along the escarpment, and trapped in its vicinity. Indeed, solutions of the full equations (5.75) can be sought in the form of propagating waves $h^{(0)} = \hat{h}^{(0)}(x) e^{i(\omega t_1 - ly)} + c.c.$, which leads to the ordinary differential equation for $\hat{h}^{(0)}$:

$$\left(1 - \frac{b'(x)}{c} \right) \hat{h}^{(0)} - \left((1 - b(x)) \hat{h}_x^{(0)} \right)_x = 0, \tag{5.77}$$

where $c = \frac{\omega}{l}$ is the phase velocity of the wave. It is clear, prior to any detailed investigation, that if this eigenproblem has decaying at $x = \pm\infty$ solutions, which can be shown to always be the case, the corresponding eigenvalues are some constants depending on the profile of escarpment. Thus the dispersion 'curves' in the $\omega - l$ plane are just straight lines. These are the long-wave (small l) asymptotics of the solutions found in Section 4.3, as can be explicitly shown in the case of the linear escarpment, when (5.77) becomes a confluent hypergeometric equation.

Therefore, the slow motions in the presence of escarpment are long trapped topographic waves propagating along the escarpment, and even slower quasi-geostrophic motions. Although we did not go further to study nonlinear corrections, it can be done following the same lines as in Section 5.5.4 below. We should emphasise that the dispersion of the long topographic waves being weak, it appears in the next order, together with nonlinearity. The two effects combined lead to a similar result as that obtained below for equatorial Rossby waves with small zonal wave numbers.

5.5.4   Slow motions at the equator

We have, thus, seen that dynamical separation of fast and slow variables, first described in rotating shallow water on the infinite f plane is still operational in the presence of coastal and/or topographical wave guides, in spite of the presence of slow wave-guide modes. The equatorial dynamics looks more complicated, due to proliferation of wave species. The spectral gap between low-frequency Rossby wave and high-frequency inertia–gravity waves at the equator is filled by non-dispersive Kelvin waves, on the one hand, and by mixed Rossby-gravity (Yanai) waves on the other hand. It turns out, however that in the long-wave limit there is still a separation of scales of different kinds of waves, as can be inferred from Figure 5.6.

We will be working with the one-layer RSW in the equatorial beta-plane approximation, and using the equatorial scaling of Section 4.4.1. As in the case of the coastal and topographic wave guides, we are considering motions with the scale disparity along and across the equatorial wave guide, which is measured by the parameter δ:

$$\delta = \frac{L_y}{L_x} \ll 1, \tag{5.78}$$

the smallness of δ being equivalent to the long-wave approximation at the equator. Correspondingly, we will suppose that meridional velocities of such motions are small: $v \to \delta v$, and that motions are slow: $t \to t_1 = \delta t$. We will subsequently introduce slower times $t_2, t_3, \ldots$. The non-dimensional equations of motion of the model are:

$$\begin{cases} u_{t_1} + \epsilon(uu_x + vu_y) - yv = -h_x, \\ \delta^2 v_{t_1} + \delta^2\epsilon(uv_x + vv_y) + yu = -h_y, \\ h_{t_1} + u_x + v_y + \epsilon\big((hv)_y + (hu)_x\big) = 0, \end{cases} \tag{5.79}$$

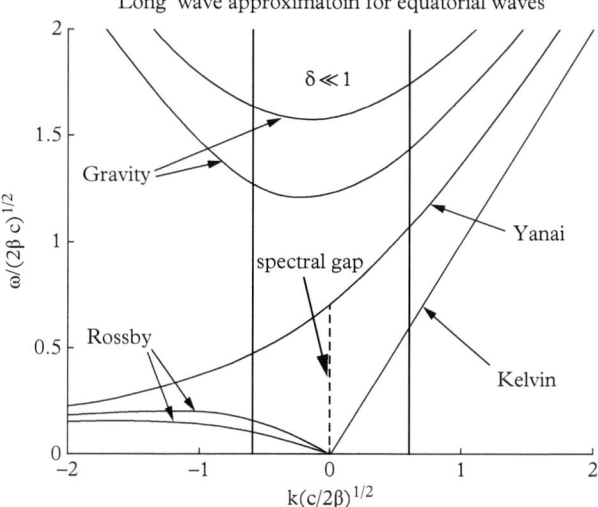

Figure 5.6 *The long-wave region of the dispersion diagram for equatorial waves in the one-layer RSW.*

and decay boundary conditions at $y \to \pm\infty$ are understood. Both parameters ϵ and δ are supposed to be small, and their ratio should be fixed. We choose $\epsilon \sim \delta^2$, the reasons of this choice will be explained shortly.

Linear approximation: long Rossby and Kelvin waves

To the lowest order in δ we get the linear equations:

$$
\begin{cases}
u^{(0)}_{t_1} - y v^{(0)} + h^{(0)}_x = 0, \\
y u^{(0)} + h^{(0)}_y = 0, \\
h^{(0)}_{t_1} + u^{(0)}_x + v^{(0)}_y = 0.
\end{cases}
\tag{5.80}
$$

Similarly to the coastal wave-guide case of Section 5.5.2, this system possesses two kinds of wave solutions, one with $v^{(0)} \equiv 0$ and one with $v^{(0)} \neq 0$. We start with the second.

By differentiating the first two equations in (5.80) in time, and eliminating $h^{(0)}$ with help of the third equation we get

$$
\begin{cases}
u^{(0)}_{t_1 t_1} - u^{(0)}_{xx} = y v^{(0)}_{t_1} + v^{(0)}_{xy}, \\
y u^{(0)}_{t_1} - u^{(0)}_{xy} = v^{(0)}_{yy},
\end{cases}
\tag{5.81}
$$

and by applying the operator $\partial^2_{t_1 t_1} - \partial^2_{xx}$ to the second equation in (5.81) we eliminate $u^{(0)}$. We thus get an equation for $v^{(0)}$:

$$\left(y^2 v^{(0)} - v^{(0)}_{yy}\right)_{t_1 t_1} - v^{(0)}_{xt_1} = 0, \tag{5.82}$$

which has two types of solutions: time-independent and propagative ones. As follows from (5.80), stationary solutions obey the geostrophic equilibrium relations on the equatorial beta plane, and have zero divergence. Due to the explicit appearance of y, the two conditions are compatible only for geostrophically equilibrated mean zonal flows:

$$u^{(0)}_M = u^{(0)}(y), \quad v^{(0)}_M = 0, \quad h^{(0)}_M = h^{(0)}(y), \quad yu^{(0)} + h^{(0)} = 0. \tag{5.83}$$

It is easy to check that such solutions are also solutions of the full nonlinear system (5.79).

The form of (5.82) suggests looking for propagating solutions in the form

$$v^{(0)}_R = \sum_{n=1}^{\infty} v^{(0)}_{nR} = \sum_{n=1}^{\infty} V_n(x + c_n t_1)\phi_n(y), \tag{5.84}$$

where $\phi_n(y)$ are Gauss–Hermite functions. Summation starting at $n = 1$ means that Yanai wave with $v \propto \phi_0(y)$ is excluded. Substitution of (5.84) into (5.82) immediately gives the eigenvalues

$$c_n = -\frac{1}{2n + 1}. \tag{5.85}$$

We recover, therefore, the long-wave asymptotics of the Rossby part of the equatorial wave spectrum, as the eigenfrequencies $-\frac{k}{2n+1}$ are limits at $k \to 0$ of the approximate expression for eigenfrequencies of equatorial Rossby waves $-\frac{k}{2n+1+k^2}$ introduced in Section 4.4.1. The solutions are, thus, long equatorial Rossby waves propagating westwards with constant phase velocities, which are given by the slopes of the corresponding dispersion curves of Figure 5.6 at the origin, and depend on the meridional structure of the modes. The zonal velocity and the thickness are obtained from the forced solutions of the inhomogeneous equations

$$\begin{cases} u^{(0)}_{t_1} + h^{(0)}_x = yv^{(0)}, \\ h^{(0)}_{t_1} + u^{(0)}_x = -v^{(0)}_y, \end{cases} \tag{5.86}$$

and are given by

$$\begin{cases} u^{(0)}_{nR} = \dfrac{1}{2}\left(\dfrac{\sqrt{2(n+1)}}{1+c_n}V_n(x+c_nt_1)\phi_{n+1}(y) + \dfrac{\sqrt{2n}}{1-c_n}V_n(x+c_nt_1)\phi_{n-1}(y)\right), \\[4mm] h^{(0)}_{nR} = \dfrac{1}{2}\left(\dfrac{\sqrt{2(n+1)}}{1+c_n}V_n(x+c_nt_1)\phi_{n+1}(y) - \dfrac{\sqrt{2n}}{1-c_n}V_n(x+c_nt_1)\phi_{n-1}(y)\right), \end{cases} \tag{5.87}$$

where $\mathcal{V}_n$ denotes the primitive of V_n: $\mathcal{V}_n = \int^x V_n(x')\, dx'$. General solution of the linearised system is a linear combination of such modes with $n = 1, 2, \ldots$, without $n = 0$, because there is no Yanai wave by construction.

The above-described procedure is not applicable if $v^{(0)} \equiv 0$. Solutions of (5.80) in this case obey the equations:

$$\begin{cases} u_{t_1}^{(0)} + h_x^{(0)} = 0, \\ yu^{(0)} + h_y^{(0)} = 0, \\ h_{t_1}^{(0)} + u_x^{(0)} = 0. \end{cases} \tag{5.88}$$

These solutions are eastward-propagating equatorial Kelvin waves of the form

$$(u_K^{(0)}, v_K^{(0)}, h_K^{(0)}) = (K(x-t),\, 0,\, K(x-t))\, \phi_0(y). \tag{5.89}$$

The general solution of the linearised problem is, therefore, given by a combination of a zonal flow, and Kelvin and Rossby waves:

$$(u^{(0)}, v^{(0)}, h^{(0)}) = (u_M^{(0)}, 0, h_M^{(0)}) + (u_K^{(0)}, 0, h_K^{(0)}) + \sum_{n=1}^{\infty} (u_{n_R}^{(0)}, v_{n_R}^{(0)}, h_{n_R}^{(0)}), \tag{5.90}$$

where the corresponding contributions are given by (5.83), (5.90), and (5.87).

Nonlinear slow dynamics

It is easy to see that the next-order approximation becomes trivial if we suppose that there is no t_2 dependence in the solution, and we can safely pass to the next order δ^2. To this order we have

$$\begin{cases} u_{t_1}^{(2)} - yv^{(0)} + h_x^{(2)} = -\left(u_{t_3}^{(0)} + u^{(0)} u_x^{(0)} + v^{(0)} u_y^{(0)} \right) \equiv -\mathcal{R}_u^{(0)}, \\ yu^{(2)} + h_y^{(2)} = -v_{t_1}^{(0)} \equiv -\mathcal{R}_v^{(0)}, \\ h_{t_1}^{(2)} + u_x^{(2)} + v_y^{(2)} = -\left(h_{t_3}^{(0)} + (h^{(0)} u^{(0)})_x + (h^{(0)} v^{(0)})_y \right) \equiv -\mathcal{R}_h^{(0)}, \end{cases} \tag{5.91}$$

where we introduced dependence on the next slow time t_3, which means that in the formulae (5.84) and (5.87) for $(u_R^{(0)}, v_R^{(0)}, h_R^{(0)})$ the wave amplitude is supposed to be t_3-dependent: $V_n = V_n(x + c_n t_1; t_3)$. Under hypothesis that $v^{(2)} \neq 0$, by the same manipulations as in the linear approximation, this system of equations can be reduced to a single inhomogeneous equation for $v^{(2)}$:

$$\left(y^2 v^{(2)} - v_{yy}^{(2)} \right)_{t_1 t_1} - v_{t_1 x}^{(2)} = \left(\partial_{tt}^2 - \partial_{xx}^2 \right) \left(-\mathcal{R}_{v_{t_1}}^{(0)} + \mathcal{R}_{h_y}^{(0)} \right) + \left(y\partial_t - \partial_{xy}^2 \right) \left(\mathcal{R}_{h_x}^{(0)} - \mathcal{R}_{u_{t_1}}^{(0)} \right), \tag{5.92}$$

where $\mathcal{R}^{(0)}_{u,v,h}$ are defined in (5.91). In order to have bounded in space and time solutions for $v^{(2)}$, the right-hand side of this equation should verify the solvability conditions. Without dwelling on technical details, let us analyse the structure of different terms qualitatively. The derivatives on the right-hand side of (5.92) are of two kinds. While y derivatives (as well as multiplication by y) change the degree n of the Gauss–Hermite functions $\phi_n(y)$, the x and t_1 derivatives act on the amplitude function $V_n(x + c_n t_1; t_3)$ and, up to a constant, are derivatives of this function with respect to its first argument. There are linear and nonlinear in V_n terms on the right-hand side of (5.92). Among the linear terms there are ones with and without t_3 derivative. The terms with t_3 derivative are present in $\mathcal{R}^{(0)}_h$ and $\mathcal{R}^{(0)}_u$. As follows from the lower-order calculations, $h^{(0)}$ and $u^{(0)}$ are expressed in terms of the primitive of V_n, cf. (5.87). A single linear term without t_3 derivative originates from $\mathcal{R}^{(0)}_v$, in fact from $v^{(0)}_{t_1}$. It contains three derivatives with respect to $x + c_n t_1$, and four derivatives, if everything is expressed in terms of the primitive $\mathcal{V}_n$. It is easy to understand that this term gives a dispersive correction to the constant phase velocity of long equatorial Rossby waves. Finally, nonlinear terms contain products of V_n with different n and their derivative with respect to $x + c_n t_1$. To decide on the resonant character of these terms, initial conditions and boundary conditions in x for the problem should be specified. If initial conditions are such that $\mathcal{V}_n$ in (5.92) are localised functions, as in the adjustment problem treated in Chapter 8, the contributions from products with different n vanish upon integration in $x - t$, as wave packets with different n move with different velocities and do not sufficiently overlap. Hence, the only resonant contributions in this case come from self-interactions, and we get decoupled evolution equations for each n. Otherwise (e.g. in the case of periodic waves), there exist resonant triads with zonal wave numbers and frequencies verifying the resonance conditions $k_1 + k_2 = k_3$, $\omega_1 + \omega_2 = \omega_3$, and evolution equations for different modes are coupled. Dynamics of such resonant wave interactions will be addressed in Chapter 12.

In the case of single-wave resonances, it may be checked that the solvability condition leads to the following evolution equation:

$$\left(V_{n_{t_3}} + \alpha_n V_{n_{xxx}} + \beta_n V_n V_{n_x} \right)_x = 0, \tag{5.93}$$

where α_n, β_n are coefficients defined by the meridional structure of the mode. As we assumed that the function $\mathcal{V}_n$ is spatially localised, this equation can be safely integrated once, leading to the celebrated Korteweg–deVries (KdV) equation. The most striking property of solutions of the KdV equation is that any localised initial perturbation becomes a series of solitary waves (solitons) moving with velocities determined by their amplitudes, solitons of lesser amplitude moving slower. Figure 5.7, which is obtained by numerical simulations with the full RSW model on the equatorial beta plane (and not with the slow-motion reduction!), shows that localised long-wave perturbation indeed follow this pattern.

Above, we analysed only the Rossby-wave part of the general solution of the linearized problem (5.90). It is easy to check that Kelvin-wave part of the linear solution does not contribute to the right-hand side of (5.92) due to the fact that $\mathcal{R}^{(0)}_v$ vanishes for this part

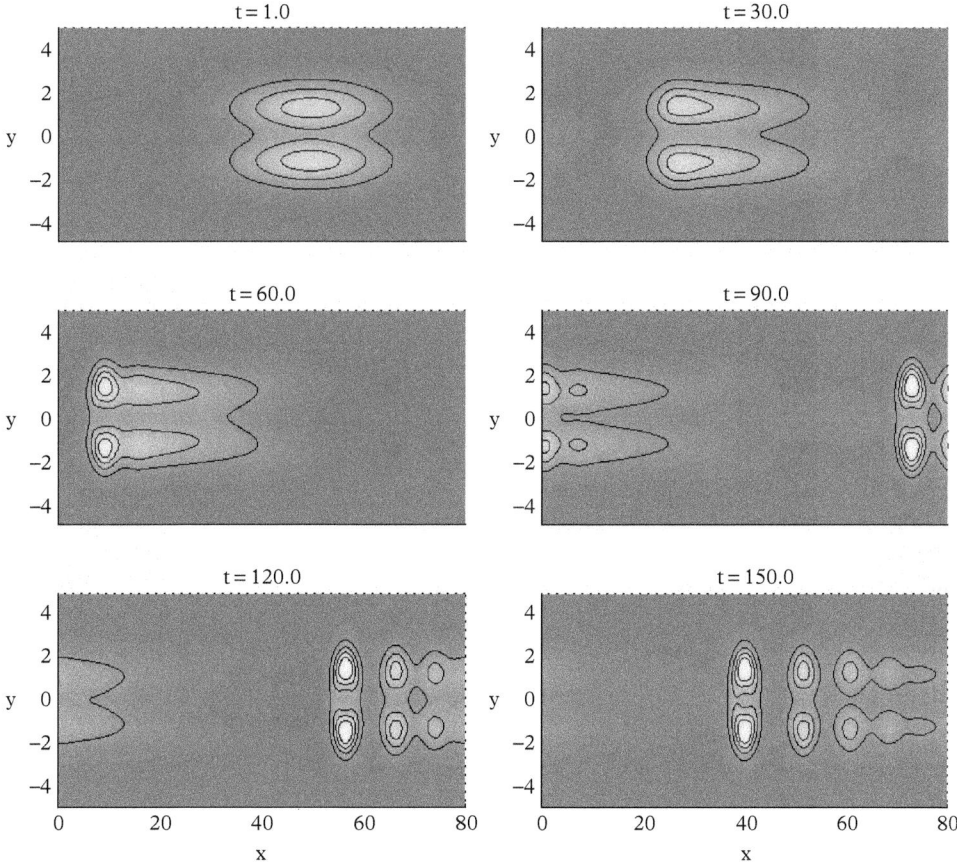

Figure 5.7 *Evolution of a long-wave large-amplitude localised initial perturbation with a meridional structure corresponding to the equatorial Rossby wave. Formation of a soliton tail is clearly seen. Levels of grey: thickness (pressure). Time is measured in $(\beta R_e)^{-1}$, and distances in R_e.*

and that it is annihilated by the action of the operator $\partial_{tt}^2 - \partial_{xx}^2$. Nonlinear evolution of the Kelvin part $K(x - t, t_3)$ of the linear solution can be obtained following similar lines, but after a simpler algebra:

$$K_{t_3} + \gamma K K_x = 0, \qquad (5.94)$$

where $\gamma = \frac{3}{2} \int_{-\infty}^{+\infty} dy\, \phi_0^3(y)$. By the same, as above, reason of propagation in the opposite direction, the Rossby-wave part does not contribute to this equation, and thus Rossby and Kelvin-wave components of the slow motion completely decouple, each obeying its proper evolution equation. It should be stressed again that evolution of the Kelvin

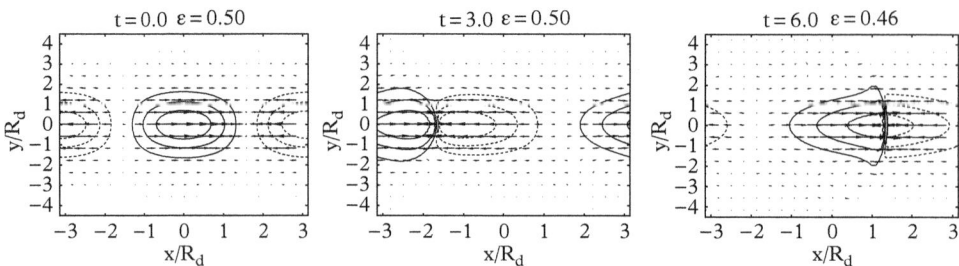

Figure 5.8 *Breaking of an equatorial Kelvin wave . Snapshots of thickness anomaly at* $t = 0, 3, 6 \, (\beta R_e)^{-1}$. *Solid: positive, dashed: negative anomaly.*

component is governed by the simple-wave equation leading to finite-time breaking and formation of so-called Kelvin fronts, which is illustrated in Figure 5.8.

It is obvious that the inviscid theory breaks down at this stage (we will come back to the problem of breaking and front formation in Chapter 7).

Finally, let us comment on the choice of the relation between small parameters ϵ and δ which was made earlier:

- If $\delta^2 \gg \epsilon$, it easy to see that the dynamics up to higher orders is linear. As it is the term $v_{t_1}^{(0)}$ which provides dispersion in the system, it means that dispersion overcomes nonlinearity in this case, and any initial packet of Rossby waves will be dispersed.

- If $\delta^2 \ll \epsilon$ dispersion appears at higher order than nonlinearity, which thus overcomes. This means breaking not only of the Kelvin-wave but also of the Rossby-wave component of slow motion.

- If $\delta^2 \sim \epsilon$, nonlinearity and dispersion compensate each other, as we have just seen, and solitary waves may form from the initial perturbation in the Rossby-wave part of the spectrum

As to the long Kelvin waves, they always break because they are defined by the condition $v^{(0)} = 0$ and are, thus, non-dispersive whatever the relation between δ and ϵ.

5.6   Summary, comments, and bibliographic remarks

Thus, we have seen in this chapter that by an appropriate scaling, corresponding to an intuitive idea of vortex motions, and under hypothesis that the vortex turn–over times are much smaller than that of the planet, which means small Rossby numbers, we obtain the simplified quasi-geostrophic (QG) equations of motion, where inertia–gravity waves are filtered out and do not manifest themselves. In the absence of boundaries, topography, and β effect, the filtering of fast waves is possible because of the presence of a gap between the inertia–gravity wave spectrum and spectrum of vortex motions in the RSW models. QG equations can be obtained along the same lines in the framework of both one- and two-layer RSW models, allowing us in the latter case to include baroclinic effects. The QG equations express conservation of potential vorticity in the leading order under the QG scaling, where advection and vorticity are defined by the geostrophic wind, and deviations of free surface and/or interfaces between layers are small. If these latter are not small, another dynamical regime, frontal geostrophic dynamics, arises.

In the one-layer case, the QG equations on the f plane are similar to the 2D Euler equations in terms of stream function and vorticity, with a modification (screening) of the Green's function due to the finite deformation radius. In the β-plane approximation, QG equations describe nonlinear Rossby waves. Rossby waves also appear in the presence of the geostrophically balanced mean flow. Counter-propagating Rossby waves in the two-layer QG model lead to the baroclinic instability, which exists both in the f-plane and β-plane approximations, the β effect playing a stabilising role.

Coastal, topographic, and equatorial wave guides allow for wave-guide modes filling the above-mentioned spectral gap. Nevertheless, for the wave guides in the f plane we have shown that low-frequency, long-wave wave-guide modes are not interacting with slow vortex motions which obey the QG dynamics. These results can be straightforwardly extended to two-layer models, although we did not presented them, with doubling of all modes in baroclinic and barotropic sub-species.

The use of QG models revolutionised dynamical meteorology and oceanography in the middle of the twentieth century. QG models and applications can be found in thousands of papers on atmosphere and ocean dynamics. The one-layer (barotropic) QG equation was used for the first in the history successful numerical weather prediction (see Lynch 2006) and references therein. Although boundary layers are out of the scope of the present book, it is worth noting that Ekman boundary layers, arising due to combined action of viscosity and rotation, produce vorticity sink or source (depending on the sign of voricity), and that the forced dissipative barotropic QG equation in non-dimensional form reads

$$\frac{d_{QG}(\zeta + y)}{dt} = -D\zeta + \frac{1}{Re}\nabla^2\zeta + F, \tag{5.95}$$

where $\zeta = -\eta + \nabla^2\eta$, $\frac{d_{QG}\zeta}{dt}\cdots = \partial_t\cdots + \mathcal{J}(\eta, \dots)$ represents advection by the geostrophic wind, D is Ekman friction coefficient, F is any forcing, and molecular dissipation $\propto Re^{-1}$ is included for completeness.

The FG models are less known, but are also used for modelling ocean motions, for example (Cushman-Roisin, Sutyrin, and Tang 1992).

Above, the derivation of the one-layer QG and FG models followed Stegner and Zeitlin (1995). Discussion of the Rossby field produced by localised mountains is close to (McCartney 1976). Phillips model (Phillips 1954) provided understanding of the baroclinic instability which is driving synoptic-scale motions in the atmosphere and meso-scale motions in the ocean. A large part of the book Pedlosky (1982) is devoted to the baroclinic instability in layered models, where many useful details can be found.

The derivation of the two-layer rigid-lid models followed Benilov and Reznik (1996), Stegner and Zeitlin (1996), and that of the free-surface two-layer QG model followed Lahaye and Zeitlin (2012). Derivation of slow-motion dynamics at the coast and at the equator is adapted from the papers on geostrophic adjustment in respective configurations (Reznik and Grimshaw (2002), LeSommer, Reznik, and Zeitlin (2004)). Solitons of the long equatorial Rossby wave were first found in Boyd (1980).

Figure 5.3 is reproduced from McCartney (1976), with permissions, Figures 5.6 and 5.7 are reproduced from LeSommer, Reznik, and Zeitlin (2004), with permissions. Figure 5.8 is reproduced from Bouchut, LeSommer, and Zeitlin (2005) with the permission of AIP publishing.

5.7   Problems

5.1 By applying the QG scaling, derive the QG equation in one-layer RSW directly from the PV conservation equation.

5.2 Derive and analyse expressions for phase and group velocity of Rossby wave in one-layer RSW.

5.3 By applying the FG scaling, derive the FG equation (5.38) directly from the PV conservation.

5.4 Linearise equation (5.38) and find wave solutions. What is the physical meaning of the result?

5.5 Find the polarisation relations for the solutions of (5.47) and demonstrate that two solutions represent baroclinic and barotropic modes

5.6 Find the most unstable mode of the baroclinic instability on the f - plane and its growth rate.

5.7 Derive equations (5.55) and (5.56).

5.8 Derive equations (5.68) and (5.72), as indicated in the main text.

5.9 Consider one-layer RSW with a rigid lid and escarpment-type topography ('long-wave approximation'). Analyse wave solutions.

5.10 Show that equation (5.93) possesses localised steady-propagating solutions (solitons).

6

Vortex Dynamics on the f and beta Plane and Wave Radiation by Vortices

6.1 Two-dimensional vortex dynamics

As was shown in Section 5.2.2 of Chapter 5, the one-layer quasi-geostrophic (QG) equation is equivalent to the dynamics of two-dimensional ideal incompressible constant-density fluid with modified relation between the stream function and vorticity. The interest to the two-dimensional fluid dynamics is partially explained by this fact, although it was, and is, a classical area of research in hydrodynamics. We will briefly recall the main facts on 2D vortex dynamics. As explained in Section 5.2.2, the canonical relation $\zeta = \nabla^2 \psi$ between the stream-function ψ and (vertical component of) vorticity ζ should be changed to $\zeta = \nabla^2 \psi - \frac{1}{R_d^2} \psi$, in order to reinterpret the statements established for 2D Euler equations in QG terms.

6.1.1 2D Euler equations in stream-function–vorticity variables

Two-dimensional Euler equations for an incompressible fluid with constant unit density read

$$\partial_t \mathbf{v} + \mathbf{v} \cdot \nabla \mathbf{v} = -\nabla P, \quad \nabla \cdot \mathbf{v} = 0. \tag{6.1}$$

Note that if Coriolis terms with constant f were included in these equations, they would drop out, due to incompressibility, in the ensuing vorticity equation. The stream-function ψ is introduced in order to solve the incompressibility constraint:

$$\mathbf{v} = \hat{\mathbf{z}} \wedge \nabla \psi \leftrightarrow u = -\partial_y \psi, \, v = \partial_x \psi \Rightarrow \nabla \cdot \mathbf{v} = 0, \tag{6.2}$$

and the equations (6.1) are equivalent to

$$\partial_t \nabla^2 \psi + \mathcal{J}(\psi, \nabla^2 \psi) = 0. \tag{6.3}$$

The vertical component of vorticity, by definition, is

$$\zeta = \hat{\mathbf{z}} \cdot (\nabla \wedge \mathbf{v}) = \nabla^2 \psi. \tag{6.4}$$

Geophysical Fluid Dynamics. Vladimir Zeitlin,
Oxford University Press (2018). © Vladimir Zeitlin. DOI: 10.1093/oso/9780198804338.001.0001

Thus, (6.3) expresses Lagrangian conservation of vorticity:

$$\frac{d\zeta}{dt} = \partial_t \zeta + \mathcal{J}(\psi, \zeta) = 0, \tag{6.5}$$

which means that the motion of two-dimensional incompressible fluid is totally determined by the evolution of vorticity. Therefore, the 2D Euler dynamics is, in fact, vortex dynamics. In order to recover velocity from a given vorticity distribution, the *inversion problem* should be solved to determine stream function from the relation $\nabla^2 \psi = \zeta$. This is a classical Poisson problem and is solved with the help of *Green's function*:

$$\psi(\mathbf{x}, t) = \int_{\mathcal{D}} \mathcal{G}(\mathbf{x}, \mathbf{x}') \zeta(\mathbf{x}', t) d\mathbf{x}', \quad \nabla^2 \mathcal{G} = \delta(\mathbf{x} - \mathbf{x}'), \tag{6.6}$$

where $\delta(\mathbf{x})$ is Dirac's delta function. Green's function, thus, gives the stream function of a positive point vortex. If the flow is considered in the whole plane, without boundaries, then

$$\mathcal{G}(\mathbf{x}, \mathbf{x}') = \frac{1}{2\pi} \log |\mathbf{x} - \mathbf{x}'|. \tag{6.7}$$

In QG dynamics the deformation radius R_d plays the role of a *screening radius* for the vorticity. Indeed, to obtain the distribution of the geostrophic stream-function η corresponding to a localised vorticity distribution, the inversion of the operator on the right-hand side of (5.22) is needed. The inverse operator is the Green's function obeying the following equation:

$$\left(\nabla^2 - \frac{1}{R_d^2} \right) \mathcal{G}(\mathbf{x}, \mathbf{x}') = \delta(\mathbf{x} - \mathbf{x}'), \tag{6.8}$$

where we reconstituted dimensions, for clarity. The solution of (6.8) in an unbounded plane is given, up to a sign, by the modified Bessel function $K_0 \left(\frac{|\mathbf{x} - \mathbf{x}'|}{R_d} \right)$. This function, which we have seen in previous chapters has a logarithmic asymptotics at $|\mathbf{x} - \mathbf{x}'| \ll R_d$ and exponentially decaying asymptotic at $|\mathbf{x} - \mathbf{x}'| \gg R_d$. In the limit $R_d \to \infty$, $-K_0$ becomes $\frac{1}{2\pi} \log |\mathbf{x} - \mathbf{x}'|$, and coincides with (6.7).

The kinetic energy of the 2D fluid is

$$E = \frac{1}{2} \int_{\mathcal{D}} \mathbf{v}^2 \, d\mathbf{x} = \frac{1}{2} \int_{\mathcal{D}} (\nabla \psi)^2 \, d\mathbf{x} = \text{const} - \frac{1}{2} \int_{\mathcal{D}} \zeta \psi \, d\mathbf{x}$$

$$= \text{const} - \frac{1}{2} \int_{\mathcal{D}} \int_{\mathcal{D}} d\mathbf{x} d\mathbf{x}' \zeta(\mathbf{x}) \mathcal{G}(\mathbf{x}, \mathbf{x}') \zeta(\mathbf{x}'), \tag{6.9}$$

where integration by parts and the Green's formula (i.e. the Stokes theorem in the plane)

$$\oint_{\Gamma} (P dx + Q dy) = \int_{\mathcal{D}_\Gamma} (\partial_x Q - \partial_y P) \, dx dy \tag{6.10}$$

were used.

It is important to realise that Lagrangian conservation of vorticity (6.5) means that there exist an infinity of conservation laws in Eulerian terms. They are called Casimir invariants, or simply Casimirs, which we already encountered in Chapter 3. Indeed, for any function $\mathcal{F}$

$$\int_{\mathcal{D}} \mathcal{F}(\zeta) \, d\mathbf{x} = \text{const.} \tag{6.11}$$

Thus, for example, the integral over the domain of the flow of any power of vorticity is conserved. In particular, the choice $\mathcal{F}(\zeta) = \zeta$ gives conservation of the mean vorticity, and the choice $\mathcal{F}(\zeta) = \zeta^2$ gives conservation of *enstrophy*. It should be stressed that $\mathcal{F}(\zeta)$ is not necessarily a continuous function. Taking it to be the characteristic function of the area between two vorticity isolines $\zeta = \zeta_-$ and $\zeta = \zeta_+$: $\mathcal{F}(\zeta) = 1$, if $\zeta_- < \zeta < \zeta_+$, and zero otherwise, we get from (6.11) the conservation of this area in time. Analogously, it can be shown that the topology of the vorticity field is preserved; that is the number of maxima, minima, and saddle points, as well as peak values of vorticity. The presence of these conservation laws means that the motion of the fluid is severely constrained, which allows for development of specific methods of analysis and of numerical simulations for the 2D vortex dynamics. With the change of Green's function explained earlier, they are transposable to the QG dynamics in the f-plane approximation. The best known are point-vortex dynamics and contour dynamics. They are based on the fact that arbitrary distribution of vorticity may be approximated by a superposition of point vortices with different intensities or, respectively, by a stack of superimposed patches of constant vorticity. We briefly sketch both below, as well as another less-known discretisation method exploiting specifics of 2D vortex dynamics. However, we first reformulate the 2D Euler equations in Lagrangian form, which turns out to be useful in applications.

6.1.2 Lagrangian formulation of 2D hydrodynamics of a perfect fluid

Lagrangian equations of an incompressible non-rotating fluid of constant unit density in the plane read (cf. (2.10) in Chapter 2),

$$\begin{cases} \ddot{X} = -\dfrac{\partial(P, Y)}{(x, y)} \,, \\[2mm] \ddot{Y} = -\dfrac{\partial(X, P)}{(x, y)} \,, \\[2mm] \dfrac{\partial(X, Y)}{\partial(x, y)} = 1, \end{cases} \tag{6.12}$$

where the dot, as usual, denotes Lagrangian time-derivative d/dt. The pressure P can be eliminated by cross-differentiation, giving

$$\frac{\partial\left(X, \ddot{X}\right)}{\partial\left(x, y\right)} - \frac{\partial\left(\ddot{Y}, Y\right)}{\partial\left(x, y\right)} = 0, \tag{6.13}$$

which is an expression of Lagrangian conservation of the relative vorticity ζ:

$$\frac{d}{dt}\zeta = 0, \quad \zeta = \left[\frac{\partial\left(X, \dot{X}\right)}{\partial\left(x, y\right)} - \frac{\partial\left(\dot{Y}, Y\right)}{\partial\left(x, y\right)}\right] = \frac{\partial\dot{X}}{\partial Y} - \frac{\partial\dot{Y}}{\partial X}, \tag{6.14}$$

in agreement with (6.5). As we will see on several occasions shortly, it is convenient to describe the motion of a two-dimensional fluid in terms of complex coordinates. We can, therefore, introduce complex Lagrangian coordinates, and complex Lagrangian labels:

$$Z = X + iY, \ Z^* = X - iY, \quad \xi = x + iy, \ \xi^* = x - iy. \tag{6.15}$$

In complex notation, (6.13) is equivalent to

$$Re\left(\frac{\partial\left(\ddot{Z}, Z^*\right)}{\partial\left(\xi, \xi^*\right)}\right) = 0. \tag{6.16}$$

where we used the fact that $\frac{\partial\left(\xi, \xi^*\right)}{\partial\left(x, y\right)} = -2i$.

6.1.3   Dynamics of point vortices

Vorticity of a system of N point vortices with current positions $\mathbf{x}_i(t)$ and intensities κ_i is

$$\zeta_N(\mathbf{x}) = \sum_{i=1}^{N}\kappa_i\delta(\mathbf{x} - \mathbf{x}_i). \tag{6.17}$$

If the domain of the flow is an unbounded plane, the corresponding stream function is

$$\psi_N(\mathbf{x}) = \frac{1}{2\pi}\int d\mathbf{x}'\,\log|\mathbf{x} - \mathbf{x}'|\sum_{i=1}^{N}\kappa_i\delta(\mathbf{x}' - \mathbf{x}_i) = \frac{1}{2\pi}\sum_{i=1}^{N}\kappa_i\log|\mathbf{x} - \mathbf{x}_i|, \tag{6.18}$$

which expresses the superposition principle following from the linearity of the relation between vorticity and stream function. The evolution of the vortices consists in advection of each of them by the velocity field created by all others:

$$\dot{\mathbf{x}}_i = \hat{\mathbf{z}}\wedge\nabla\psi_N\big|_{\mathbf{x}=\mathbf{x}_i} = \frac{1}{2\pi}\sum_{j=1}^{N}\kappa_i\frac{\hat{\mathbf{z}}\wedge\mathbf{x}_{ij}}{\mathbf{x}_{ij}^2}, \ i = 1, 2, \ldots, N. \tag{6.19}$$

where $\mathbf{x}_{ij} = \mathbf{x}_i - \mathbf{x}_j$. The intensities of the vortices do not change: $\dot{\kappa} = 0$, as follows from the Kelvin circulation theorem. Thus, the nonlinear partial differential equation for the stream function is replaced by a system of ordinary differential equations, if the vorticity field is approximated by a distribution of point vortices, with obvious computational advantages.

The following quantities are conserved for an ensemble of N point vortices: the energy:

$$E_N = \frac{1}{2} \int \zeta_N(\mathbf{x}) \psi_N(\mathbf{x}) \, d\mathbf{x} = \frac{1}{4\pi} \sum_{i \neq j}^{N} \kappa_i \kappa_j \log |\mathbf{x}_{ij}|, \qquad (6.20)$$

the centroid of vorticity:

$$\mathbf{R}_N = \frac{\int \mathbf{x} \zeta_N(\mathbf{x}) \, d\mathbf{x}}{\int \zeta_N(\mathbf{x}) \, d\mathbf{x}} = \frac{\sum_{i=1}^{N} \mathbf{x}_i \kappa_i}{\sum_{i=1}^{N} \kappa_i}, \qquad (6.21)$$

and the angular momentum:

$$M_N = \frac{1}{2} \int \zeta_N(\mathbf{x}) \mathbf{x}^2 \, d\mathbf{x} = \frac{1}{2} \sum_{i=1}^{N} \kappa_i \mathbf{x}_i^2. \qquad (6.22)$$

A number of simple exact solutions of the point-vortex dynamics can be easily obtained (see Problems 6.4 to 6.6), among them a particular vortex dipole solution, which is given by the following stream function, if its axis is oriented in the y direction:

$$\psi = -\frac{\kappa}{2\pi} \left(\log \sqrt{(y - Vt)^2 + \left(x - \frac{d}{2} \right)^2} - \log \sqrt{(y - Vt)^2 + \left(x + \frac{d}{2} \right)^2} \right), \qquad (6.23)$$

where $V = -\frac{\kappa}{2\pi d}$ The dipole in this configuration is moving in the y direction with the velocity V, and without change of form. The zonal velocity at the dipole's axis is zero:

$$u|_{x=0} = \frac{\kappa}{2\pi} \left[\frac{y - Vt}{(y - Vt)^2 + \left(x - \frac{d}{2} \right)^2} - \frac{y - Vt}{(y - Vt)^2 + \left(x + \frac{d}{2} \right)^2} \right] \equiv 0. \qquad (6.24)$$

6.1.4 Contour dynamics

For a vortex patch of a constant vorticity ζ_0 bounded by a contour Γ, each point on the boundary of the patch, which is a Lagrangian particle, moves according to

$$\dot{\mathbf{x}}_\Gamma = \mathbf{v}_\Gamma = \hat{\mathbf{z}} \wedge \nabla \psi|_{\mathbf{x}_\Gamma}, \qquad (6.25)$$

where

$$\psi = \frac{\zeta_0}{2\pi} \int_{S_\Gamma} d\mathbf{x}' \log |\mathbf{x} - \mathbf{x}'|. \tag{6.26}$$

Hence

$$\dot{x}_\Gamma = -\frac{\zeta_0}{2\pi} \int_{S_\Gamma} d\mathbf{x}' \, \partial_y \log |\mathbf{x} - \mathbf{x}'| = \frac{\zeta_0}{2\pi} \int_{S_\Gamma} d\mathbf{x}' \, \partial_{y'} \log |\mathbf{x}_\Gamma - \mathbf{x}'|,$$

$$= -\frac{\zeta_0}{2\pi} \oint_\Gamma dx' \log |\mathbf{x}_\Gamma - \mathbf{x}'|, \tag{6.27}$$

$$\dot{y}_\Gamma = \frac{\zeta_0}{2\pi} \int_{S_\Gamma} d\mathbf{x}' \, \partial_x \log |\mathbf{x} - \mathbf{x}'| = -\frac{\zeta_0}{2\pi} \int_{S_\Gamma} d\mathbf{x}' \, \partial_{x'} \log |\mathbf{x}_\Gamma - \mathbf{x}'|,$$

$$= -\frac{\zeta_0}{2\pi} \oint_\Gamma dy' \log |\mathbf{x}_\Gamma - \mathbf{x}'|, \tag{6.28}$$

where the Green's formula has been used. In vector notation we have

$$\dot{\mathbf{x}}_\Gamma = -\frac{\zeta_0}{2\pi} \oint_\Gamma d\mathbf{x}' \log |\mathbf{x} - \mathbf{x}'|. \tag{6.29}$$

If several vortex patches with constant vorticities ζ_j are superimposed, then for points on each of their boundaries we get, by superposition principle,

$$\dot{\mathbf{x}}_{\Gamma_i} = -\frac{1}{2\pi} \sum_j^N \zeta_j \oint_{\Gamma_j} d\mathbf{x}' \log |\mathbf{x}_{\Gamma_i} - \mathbf{x}'_j|. \tag{6.30}$$

In numerical implementations the integrals along the contours in (6.30) should be discretised and, as the lengths of the contours, unlike their areas, are not conserved, this discretisation should take into account ubiquitous elongation of the contours and formation of vortex filaments emanating from the vortex. This leads to a necessity to adapt the number of discretisation points and so-called contour surgery algorithms.

A particular example of constant vorticity patch which rotates without change of form at constant angular velocity, like a vortex pair is the classical Kirchhoff vortex, cf. Problem 6.7.

6.1.5   Structure (Casimir)-preserving discretisations of vorticity equation in Fourier space

As mentioned earlier, vortex dynamics in the plane possesses an infinite number of invariants of motion. If point-vortex or contour-dynamics discretisations of continuous vorticity distributions are applied, only part of these integrals of motion is preserved.

Most often, the discretisation of the vorticity equation in numerical simulations is performed in the Fourier space, by using doubly periodic boundary conditions, Fourier-transforming the stream function and vorticity and truncating the resulting Fourier series. Calculations are then performed either entirely in Fourier space (spectral methods) or partially in Fourier, and partially in physical space (pseudo-spectral methods). It is easy to see that such direct truncation in Fourier space does not preserve Casimir invariants. Indeed, let us suppose doubly periodic boundary conditions, renormalise the distances in a way that spatial period is 2π in both x and y directions, and make the Fourier transformation of the vorticity field $\zeta(x, y, t)$

$$\zeta(x, y) = \sum_{m_1, m_2} \zeta_{m_1 m_2} e^{i(m_1 x + m_2 y)} \equiv \sum_{\mathbf{m}} \zeta_{\mathbf{m}} e^{i \mathbf{m} \cdot \mathbf{x}}. \tag{6.31}$$

Then the vorticity equation takes the form

$$\dot{\zeta}_{\mathbf{m}} = - \sum_{\mathbf{k}=(-\infty, -\infty)}^{(+\infty, +\infty)} k^{-2} \mathbf{m} \times \mathbf{k} \, \zeta_{\mathbf{m} + \mathbf{k}} \, \zeta_{-\mathbf{k}} \tag{6.32}$$

where we introduced notation $\mathbf{a} \times \mathbf{b}$ for $\hat{\mathbf{z}} \cdot (\mathbf{a} \wedge \mathbf{b})$.

Let us consider smooth distributions of vorticity. Then the fundamental system of Casimirs is given by powers of vorticity integrated over the domain, cf (6.11). In Fourier space, this gives

$$\mathcal{C}_{\mathcal{N}} = \sum_{\mathbf{n}_1, \mathbf{n}_2, \ldots, \mathbf{n}_{\mathcal{N}}} \zeta_{\mathbf{n}_1}, \zeta_{\mathbf{n}_2} \cdots \zeta_{\mathbf{n}_{\mathcal{N}}}, \quad \sum_{i=1}^{\mathcal{N}} \mathbf{n}_i = 0, \; \mathcal{N} = 1, 2, \ldots, \infty. \tag{6.33}$$

It is straightforward to check that the Casimirs (6.33) are conserved if $\zeta_{\mathbf{m}}$ obey (6.32).

Consider now a truncation of (6.32) at wave numbers $(\pm N, \pm N)$, which is typical for numerical implementations. This gives a $(2N-1)^2$-dimensional dynamical system in

$$\dot{\zeta}_{\mathbf{m}} = - \sum_{\mathbf{k}=(-N, -N)}^{(+N, +N)} k^{-2} \mathbf{m} \times \mathbf{k} \, \zeta_{\mathbf{m} + \mathbf{k}} \, \zeta_{-\mathbf{k}}. \tag{6.34}$$

It is straightforward to check that no Casimir, except the enstrophy, C_2, is conserved anymore, and thus the Fourier truncations change the very nature of the system. However, an issue from this seemingly dead-end situation exists. It consists in modifying the coefficients of interaction between the Fourier modes in (6.32) in such a way that in the limit $N \to \infty$ the original coefficients are reproduced. Let us denote by $\mathcal{B}$ the box in Fourier-space $\left[\left(-\frac{N-1}{2}, -\frac{N-1}{2}\right), \left(\frac{N-1}{2}, \frac{N-1}{2}\right)\right]$, where N is supposed to be odd. The following dynamical system for $N^2 - 1$ modes,

$$\dot{\zeta}_{\mathbf{m}} = -\sum_{\mathcal{B}} \mathbf{k}^{-2} \frac{N}{2\pi} \sin\left(\frac{2\pi}{N}\mathbf{m} \times \mathbf{k}\right) \zeta_{\mathbf{m+k}}|_{modN} \; \zeta_{-\mathbf{k}}, \tag{6.35}$$

where summation on the right-hand side is understood modulo N, in order to always stay in the box $\mathcal{B}$. Equation (6.35) is equivalent to the Fourier-transformed vorticity equation (6.32) in the limit $N \to \infty$, and conserves N Casimirs which, again in the limit $N \to \infty$, reproduce the Casimirs of (6.32):

$$C_2 = \sum_{\mathcal{B}} \zeta_{\mathbf{k}} \zeta_{-\mathbf{k}}, \tag{6.36}$$

$$C_3 = \sum_{\mathcal{B}} \frac{N}{2\pi} \cos\left(\frac{2\pi}{N}\mathbf{m} \times \mathbf{k}\right) \zeta_{\mathbf{k}} \zeta_{\mathbf{l}} \zeta_{-\mathbf{k-l}}, \tag{6.37}$$

$$\dots$$

$$C_M = \sum_{\mathcal{I}^M} \zeta_{\mathbf{i}_1} \dots \zeta_{\mathbf{i}_M} \cos\left(\frac{4\pi}{N} A(\mathbf{i}_1, \dots, \mathbf{i}_M)\right), \tag{6.38}$$

where $M \leq 2N$, and $A(\mathbf{i}_1, \dots, \mathbf{i}_M)$ is the area spanned by the index vectors

$$A(\mathbf{i}_1, \dots, \mathbf{i}_M) = \frac{1}{2} \left(\mathbf{i}_2 \times \mathbf{i}_1 + \mathbf{i}_3 \times (\mathbf{i}_1 + \mathbf{i}_2) + \dots + \mathbf{i}_M \times (\mathbf{i}_1 + \dots + \mathbf{i}_{M-1})\right).$$

The energy of this system is the same as in (6.32):

$$H = \frac{1}{2} \sum_{\mathcal{B}} \mathbf{k}^{-2} \zeta_{\mathbf{k}} \, \zeta_{-\mathbf{k}} \tag{6.39}$$

6.2 Quasi-geostrophic modons in the β- and f-plane approximations

6.2.1 Influence of the beta effect upon a monopolar vortex

In Section 6.1 we considered vortex dynamics in the f-plane approximation, or without rotation at all, which results in the same vorticity equation. It is clear that even a simple single-vortex solution is not valid any more if the beta effect is switched on. The following qualitative considerations illustrated by Figure 6.1 show that a monopolar vortex generates a Rossby-wave 'tail' on the beta plane. Indeed, the term ψ_x appearing due to the β effect in the QG equation, which we reproduce in dimensionless form

$$\nabla^2 \psi_t - \psi_t + \mathcal{J}(\psi, \nabla^2 \psi) + \psi_x = 0, \tag{6.40}$$

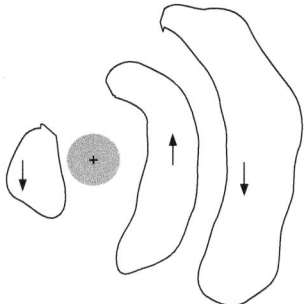

Figure 6.1 *Initial relative vorticity of a monopolar vortex, and secondary vorticity zones produced due to the beta effect. Meridional velocity induced by subsequent vorticity zones is indicated by arrows.*

is the meridional geostrophic velocity v. In an initially axisymmetric vortex, the non linear term in (6.40) vanishes and the only term changing vorticity is the beta term. Cyclonic circulation of initial vortex with positive vorticity has positive values of v on its right, and negative on its left. According to (6.40), they serve, respectively, as sources of negative vorticity anomaly on the right, and positive on the left. The same simple consideration can be repeated for the secondary negative vorticity zone on the right, which creates, in its turn, a zone of positive vorticity anomaly on its own right, and so on. Thus, a sequence of vorticity anomalies of alternating signs appears on the right of the vortex, which is nothing else than a Rossby-wave packet. Hence, the point-vortex solution is not valid on the β plane (some approximate solution can still be constructed, cf. Section 6.5). Yet, a steady moving vortex dipole solution can still be found, cf. Problem 6.8.

6.2.2   Constructing QG modon solutions: one-layer case

We have seen in Section 6.1 that a pair of point vortices of equal intensities and opposite sign is a special vortex configuration which can travel far from its original location, transporting vorticity anomaly, and other quantities. A distributed-vorticity analogue of this solution, the Lamb's dipole, is well known in hydrodynamics. In QG dynamics such solutions are called modons, and exist in both β- and f-plane approximations. They are direct generalisations of Lamb's dipole solutions, and are constructed as follows.

Consider steady translating solutions of (6.40): $\psi(x, y, t) = \psi(x - Ut, y)$. In this case (6.40) becomes a condition that the Jacobian of a pair of functions vanishes:

$$\mathcal{J}(\psi + Uy, \nabla^2\psi + (1 + U)y) = 0, \tag{6.41}$$

which means that these functions are functionally dependent:

$$\nabla^2\psi + (1 + U)y = F(\psi + Uy), \tag{6.42}$$

where F is an arbitrary function. The physical meaning of this equation is that potential vorticity (which coincides with absolute vorticity in this case) is constant along the streamlines in the reference frame moving with the zonal velocity U (co-moving frame).

It should be stressed that F is not necessarily the same over the whole (x, y) plane. If domains with different F exist in the flow, than matching across their boundaries should be made, ensuring continuity of pressure and velocity. We consider spatially localised solutions with $\psi \to 0$ at $r = \sqrt{x^2 + y^2} \to \infty$, and will be looking for solutions decaying out of a circle $r = a$. Then, in the 'external' domain $r > a$ it is sufficient to take F to be a linear function, The condition of decay at infinity allows us immediately to determine its form from (6.42):

$$F(\psi + Uy) = p^2(\psi + Uy), \quad p^2 = \frac{1 + U}{U} > 0. \tag{6.43}$$

The simplest choice of F in the interior domain $r < a$ in this case is to take it also to be linear:

$$\begin{cases} \nabla^2 \psi = p^2(\psi + Uy) - (1 + U)y, & r > a, \\ \nabla^2 \psi = -k^2(\psi + Uy) - (1 + U)y, & r < a, \end{cases} \tag{6.44}$$

Solutions of these equations inside and outside the circular boundary are natural to seek in polar coordinates $x = r \cos\theta$, $y = r \sin\theta$, which straightforwardly gives

$$\begin{cases} \psi = BK_1(pr)\sin\theta, & r > a, \\ \psi = \left[A\mathcal{J}_1(kr) - \dfrac{r}{k^2}(1 + U + Uk^2) \right]\sin\theta, & r < a, \end{cases} \tag{6.45}$$

where $\mathcal{J}_1$ is the oscillating Bessel function of order 1, and K_1 is decaying modified Bessel function of order 1. A and B are constants to be determined from the conditions of matching at $r = a$. The matching uses kinematic and dynamic boundary conditions, which are equivalent to vanishing of the co-moving stream function at the boundary, and continuity of its derivative:

$$\begin{aligned} \psi + Uy|_{a-} &= \psi + Uy|_{a+} = 0, \\ \partial_r \psi|_{a-} &= \partial_r \psi|_{a+}. \end{aligned} \tag{6.46}$$

The two first conditions allow us to determine A and B: $A = \frac{a(1+U)}{k^2 \mathcal{J}_1(ka)}$, $B = -\frac{Ua}{K_1(pa)}$, and the third condition gives k in terms of other parameters:

$$\frac{\mathcal{J}_1'(ka)}{\mathcal{J}_1(ka)} = \frac{1}{ka}\left(1 + \frac{k^2}{p^2}\right) - \frac{k}{p}\frac{K_1'(pa)}{K_1(pa)}. \tag{6.47}$$

This is a transcendental equation, which can be solved numerically. For any (a, p) there exists an infinite series of solutions for k. As it is easy to see, the value of k is related to a number of nodes of the solution (6.45) in meridional direction The minimal value of k gives a dipolar structure. The above construction works also in the f-plane approximation, with straightforward changes in A, B, p, and k. We display the f-plane modon in Figure 6.2. A characteristic feature of the geostrophic modon is its perfect

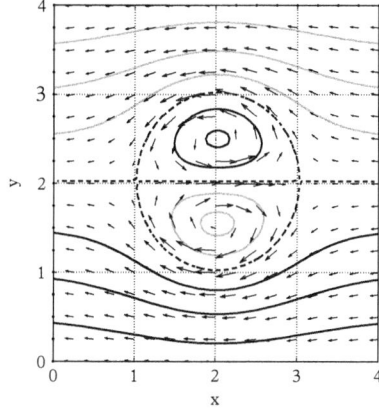

Figure 6.2 *Geostrophic velocity (arrows) and streamlines of ψ + Uy of the QG modon in the f plane.*

cyclone–anticyclone symmetry. The importance of modon solutions is, in particular, in their specific transport properties, as the fluid inside the separatrix is moving with the modon.

6.2.3   Constructing QG modon solutions: two-layer case

General framework

For technical simplicity we consider the two-layer QG model in the f-plane approximation and, for generality and subsequent numerical experiments with full RSW model, we work with the free-surface version of two-layer QG equations discussed in Chapter 5. Let us recall the equations of the model, cf Section 5.4, which we rewrite in terms of geostrophic stream functions (pressures) Ψ_i:

$$\begin{cases} \dfrac{d_1^{(0)}}{dt}\left[\nabla^2\Psi_1 + \dfrac{F_1}{1-s}(\Psi_2 - \Psi_1)\right] = 0, \\[3mm] \dfrac{d_2^{(0)}}{dt}\left[\nabla^2\Psi_2 - \dfrac{dF_1}{1-s}(\Psi_2 - s\Psi_1)\right] = 0. \end{cases} \tag{6.48}$$

Here we use notation s for density ratio $\rho_1/\rho_2 < 1$, to avoid confusion with the radius r, $d = H_1/H_2$ is the depth ratio, $F_i = H_0/H_i$ whence $F_2 = dF_1$, and $\frac{d_i^{(0)}}{dt} = \partial_t + \mathcal{J}(\Psi_i, \cdots)$ is the quasi-geostrophic advective derivative. As in Section 6.1, we are looking for a solution steady moving at a constant speed c, which we choose to be along the x axis, without loss of generality. The stream function depends on $x-ct$ and y only, and the QG equations take the form

$$\begin{cases} \mathcal{J}\left(\Psi_1 + cy, \nabla^2\Psi_1 + \dfrac{F_1}{1-s}(\Psi_2 - \Psi_1)\right) = 0, \\[3mm] \mathcal{J}\left(\Psi_2 + cy, \nabla^2\Psi_2 - \dfrac{dF_1}{1-s}(\Psi_2 - s\Psi_1)\right) = 0, \end{cases} \tag{6.49}$$

which means that the co-moving stream functions $\Psi_i + cy$ and the quasi-geostrophic potential vorticity (PV) of the layers are functionally related, that is

$$
\begin{cases}
\nabla^2 \Psi_1 + \dfrac{F_1}{1-s}(\Psi_2 - \Psi_1) = G_1(\Psi_1 + cy), \\[2mm]
\nabla^2 \Psi_2 - \dfrac{dF_1}{1-s}(\Psi_2 - s\Psi_1) = G_2(\Psi_2 + cy),
\end{cases}
\tag{6.50}
$$

where $G_i(z)$ are arbitrary functions. In the following we make an assumption that they are linear, as in the one-layer case. Again, as in the one-layer case, we look for solutions inside and outside a circular separatrix of radius a, which gives the size of the modon to be constructed. The functions G in the interior and exterior of the separatrix, respectively G^{int}, and G^{ext}, will be not the same, but will obey the matching conditions guaranteeng continuity of pressure and velocity.

Solution in the exterior domain

We are looking for a localised solution, that is $\Psi_i \to 0$ when $s \to \infty$. It then follows from (6.50) that $G_i^{ext}(z) = 0$. We transform the system to the normal form by introducing new variables $T_j^{ext} = \Psi_1^{ext} + \beta_j \Psi_2^{ext}$, where β_j are constants to be determined:

$$
\nabla^2 T_j - \frac{q_j^2}{a^2} T_j = 0, \quad j = 1, 2.
\tag{6.51}
$$

Here,

$$
\frac{q_j^2}{a^2} = -\frac{F_1}{\beta_j}\frac{1 - \beta_j d}{1 - s} = F_1 \frac{1 - s\beta_j d}{1 - s}, \quad j = 1, 2,
\tag{6.52}
$$

which is an algebraic system for β_j at each j:

$$
\begin{pmatrix}
\dfrac{q_j^2}{a^2} - \dfrac{F_1 d}{1-s} & \dfrac{F_1}{1-s} \\[3mm]
\dfrac{F_1 sd}{1-r} & \dfrac{q_j^2}{a^2} - \dfrac{F_1}{1-s}
\end{pmatrix}
\begin{pmatrix}
\beta_j \\[2mm]
1
\end{pmatrix}
= 0.
\tag{6.53}
$$

To have nontrivial solutions for β, the determinant of the matrix on the left-hand side of (6.53) should be zero, which gives

$$
\beta_j = \frac{1}{2ds}\left(1 - d \mp \sqrt{(1-d)^2 + 4ds}\right), \quad j = 1, 2.
\tag{6.54}
$$

As $0 \leq (d, s) \leq 1$, q_j and β_j are real. Furthermore, they only depend on d and s, and not on Rossby nor Burger numbers. A relation between β_1 and β_2 follows from (6.52):

$$(\beta_1 - \beta_2)ds = \frac{1}{\beta_1} - \frac{1}{\beta_2}, \Rightarrow \beta_1\beta_2 = -\frac{1}{ds}. \tag{6.55}$$

We look for a dipolar solution decaying at $r \to \infty$ and, after separation of variables in polar coordinates (r, θ) and retaining the first angular mode, we get the exterior solution

$$T_j = A_j K_1\left(\frac{q_j}{a}r\right)\sin\theta, \tag{6.56}$$

where K_1 is a modified Bessel function of first order, and the constants A_j will be determined by matching with the interior solution.

Solution in the interior domain

We choose $G_i^{int}(x)$ to be linear, as in the one-layer case,

$$G_i^{int}(x) = -S_i x + Q_i \tag{6.57}$$

where S_i and Q_i are constants. We write the interior equations in normal form in terms of the variables $T_j^{int} = \Psi_1^{int} + \alpha_j\Psi_2^{int}$:

$$\nabla^2 T_j + \frac{k_j^2}{a^2}T_j = -(S_1 + \alpha_j S_2)\,c\,r\sin\theta + Q_1 + \alpha_j Q_2, \quad j = 1, 2, \tag{6.58}$$

where

$$\frac{k_j^2}{a^2}T_j = F_1\frac{1 - \alpha_j d}{1 - s}\Psi_2 - F_1\frac{1 - s\alpha_j d}{1 - s}\Psi_1 + S_1\Psi_1 + \alpha_j S_2\Psi_2, \quad j = 1, 2, \tag{6.59}$$

and the same relation as (6.55) follows by the same reasoning: $\alpha_1\alpha_2 = -\frac{1}{ds}$. Analogously, (6.59) can be written in a matrix form. The determinant of the matrix should be zero to have nontrivial solutions, which gives

$$\alpha_j = \frac{1 - s}{F_1 ds}\left[\frac{S_2 - S_1}{2} + \frac{F_1(1 - d)}{2(1 - s)} \pm \frac{1}{2}\sqrt{\left(\frac{F_1(1 - d)}{1 - s} + S_2 - S_1\right)^2 + 4ds\left(\frac{F_1}{1 - d}\right)^2}\right]. \tag{6.60}$$

As to the signs of the coefficients k_j^2, they can be negative, in which case, instead of oscillating ordinary Bessel functions, exponential modified Bessel functions appear in solutions of (6.58). To avoid a singularity at $r = 0$, we have to retain only modified Bessel functions of the first kind $I_n(r)$ in the interior domain in this case. Stream functions are combinations of T_j, so it is not forbidden to have one of the two k_j^2 to be negative. Yet, k_1^2 and k_2^2 can not be both negative because the matching conditions for the first derivative would be not possible to satisfy, except for the trivial case of the flow at rest. Analysis of solutions for $k_{1,2}^2$ in the (S_1, S_2) plane shows that solutions with non-negative k_j^2 always exist. Thus, the interior solution is given by

$$T_j = \frac{Q_1 + \alpha_j Q_2}{k_j^2/a^2} + \left[B_j \mathcal{J}_1 \left(\frac{k_j}{a} r \right) - \left(\frac{S_1 + \alpha_j S_2}{k_j^2/a^2} \right) cr \right] \sin\theta, \quad j = 1, 2 \qquad (6.61)$$

with yet unknown constant B_j, to be determined from matching conditions. If k_2 is imaginary one should replace $B_2 \mathcal{J}_1 \left(\frac{k_2}{a} r \right)$ by $B_2 I_1 \left(\left| \frac{k_2}{a} \right| r \right)$, with purely imaginary B_2.

Matching conditions

At this stage, eight constants remain unknown: A_j, B_j, S_j, and Q_j, $j = 1, 2$. The stream functions must satisfy the conditions of continuity of pressure (stream function) and velocity, which are the same as in the one-layer case layer-wise:

$$\Psi_i^{\text{int}}\big|_{r=a} = \Psi_i^{\text{ext}}\big|_{r=a}, \quad i = 1, 2.$$

$$\frac{\partial \Psi_i^{\text{int}}}{\partial r}\bigg|_{r=a} = \frac{\partial \Psi_i^{\text{ext}}}{\partial r}\bigg|_{r=a}, \quad i = 1, 2. \qquad (6.62)$$

The continuity of radial velocities is ensured by the continuity of the stream functions themselves, and their dependence on θ. In terms of T_j in the interior domain (6.62) they gives

$$T_j^{\text{int}}\big|_{r=a} = \Psi_1^{\text{ext}}\big|_{r=a} + \alpha_j \Psi_2^{\text{ext}}\big|_{r=a},$$

$$\frac{\partial T_j^{\text{int}}}{\partial r}\bigg|_{r=a} = \frac{\partial \Psi_1^{\text{ext}}}{\partial r}\bigg|_{r=a} + \alpha_j \frac{\partial \Psi_2^{\text{ext}}}{\partial r}\bigg|_{r=a}, \quad j = 1, 2. \qquad (6.63)$$

From the first relation and (6.61) it follows immediately that $Q_1 + \alpha_j Q_2 = 0$, because the exterior stream function has no θ-independent component. The matching conditions for the stream functions and their radial derivatives give, with the help of the recursion formulae for Bessel functions,

$$B_j \mathcal{J}_1(k_j) - \frac{S_1 + \alpha_j S_2}{k_j^2/a^2} ca = \frac{(\beta_2 - \alpha_j) A_1 K_1(q_1) + (\alpha_j - \beta_1) A_2 K_1(q_2)}{\beta_2 - \beta_1}, \qquad (6.64)$$

$$B_j k_j \mathcal{J}_2(k_j) = \frac{(\beta_2 - \alpha_j) A_1 q_1 K_2(q_1) + (\alpha_j - \beta_1) A_2 q_2 K_2(q_2)}{\beta_2 - \beta_1}, \quad j = 1, 2. \qquad (6.65)$$

When k_2 is purely imaginary, the replacement $B_2 K_2 \mathcal{J}_2(k_2) \rightarrow i B_2 I_2(|k_2|)$ should be made.

In the two-layer case the circle $r = a$ can be a streamline in both layers, or only in one. So we can consider the two cases: 1) $r = a$ is a streamline in both layers: S_j and $Q_j \neq 0$, 2) $r = a$ is a streamline only in the first (upper) layer: S_2 and $Q_2 = 0$. The two corresponding solutions will be called *quasi-barotropic* and *baroclinic* modons, as we will see shortly that their velocity fields are mostly so.

Quasi-barotropic modon: $G_2(\Psi_2 + cy) \neq 0$

In this case, the following relations for the exterior stream functions hold:

$$\Psi_1|_{r=a} + ca\sin\theta = 0, \quad \Rightarrow \quad \frac{\beta_2 A_1 K_1(q_1) - \beta_1 A_2 K_1(q_2)}{\beta_2 - \beta_1} + ca = 0. \tag{6.66}$$

$$\Psi_2|_{r=a} + ca\sin\theta = 0, \quad \Rightarrow \quad \frac{A_2 K_1(q_2) - A_1 K_1(q_1)}{\beta_2 - \beta_1} + ca = 0. \tag{6.67}$$

giving the values of Aj : $A_1 = -\frac{ca(1+\beta_1)}{K_1(q_1)}$, $A_2 = -\frac{ca(1+\beta_2)}{K_1(q_2)}$. Hence the exterior stream functions are fully determined, and their amplitudes are proportional to the ratio $Ro/Bu = ca$. Then, the constants k_j and α_j are expressed in terms of (S_1, S_2) by (6.59) to (6.60) and we only have to solve the set of equations (6.64) to (6.65) which is nonlinear in S_j but linear in B_j. By writing it in the matrix form

$$\mathcal{M}_j \begin{pmatrix} B_j \\ ca \end{pmatrix} = 0, \quad j = 1, 2, \tag{6.68}$$

where

$$\mathcal{M}_j = \begin{pmatrix} \mathcal{J}_1(k_j) & \dfrac{S_1 + \alpha_j S_2}{k_j^2/a^2} + \dfrac{(\beta_2 - \alpha_j)A_1 K_1(q_1) + (\alpha_j - \beta_1)A_2 K_1(q_2)}{\beta_2 - \beta_1} \\ k_j\mathcal{J}_2(k_j) & \dfrac{(\beta_2 - \alpha_j)A_1 q_1 K_2(q_1) + (\alpha_j - \beta_1)A_2 q_2 K_2(q_2)}{\beta_2 - \beta_1} \end{pmatrix}, \tag{6.69}$$

we require $\det(\mathcal{M}) = 0$, and solve this equation numerically using a standard iterative Newton algorithm.

Baroclinic modon: $G_2(\Psi_2 + cy) = 0$

In this case, we have $S_2 = Q_2 = 0$. In the first layer, the condition that the stream function must be constant at $r = a$, and the relation (6.66) still hold. From (6.59) and (6.64) to (6.65) we can derive two other equations for S_1, A_1, and A_2:

$$k_j \mathcal{J}_2(k_j) \left[\left(1 + \alpha_j + \frac{\alpha_j d F_1}{k_j^2/a^2} \right) ca + \frac{(\beta_2 - \alpha_j)A_1 K_1(q_1) + (\alpha_j - \beta_1)A_2 K_1(q_2)}{\beta_2 - \beta_1} \right]$$

$$- \mathcal{J}_1(k_j) \frac{(\beta_2 - \alpha_j)A_1 q_1 K_2(q_1) + (\alpha_j - \beta_1)A_2 q_2 K_2(q_2)}{\beta_2 - \beta_1} = 0, \quad j = 1, 2 \qquad (6.70)$$

We can then write (6.66) and (6.70) in a matrix form:

$$\mathcal{M} \cdot \begin{pmatrix} A_1 \\ A_2 \\ ca(\beta_2 - \beta_1) \end{pmatrix} = 0, \qquad (6.71)$$

where the matrix elements $\mathcal{M}_{a,b}$ are given by

$$\mathcal{M}_{1,1} = (\alpha_1 - \beta_2)q_1 K_2(q_1)\mathcal{J}_1(k_1) + (\beta_2 - \alpha_1)K_1(q_1)k_1\mathcal{J}_2(k_1),$$

$$\mathcal{M}_{1,2} = (\beta_1 - \alpha_1)q_2 K_2(q_2)\mathcal{J}_1(k_1) + (\alpha_1 - \beta_1)K_1(q_2)k_1\mathcal{J}_2(k_1),$$

$$\mathcal{M}_{1,3} = \left(1 + \alpha_1 + \frac{\alpha_1 d F_1}{k_1^2/a^2} \right) k_1\mathcal{J}_2(k_1),$$

$$\mathcal{M}_{2,1} = (\alpha_2 - \beta_2)q_1 K_2(q_1)\mathcal{J}_1(k_2) + (\beta_2 - \alpha_2)K_1(q_1)k_2\mathcal{J}_2(k_2),$$

$$\mathcal{M}_{2,2} = (\beta_1 - \alpha_2)q_2 K_2(q_2)\mathcal{J}_1(k_2) + (\alpha_2 - \beta_1)K_1(q_2)k_2\mathcal{J}_2(k_2),$$

$$\mathcal{M}_{2,3} = \left(1 + \alpha_2 + \frac{\alpha_2 d F_1}{k_2^2/a^2} \right) k_2\mathcal{J}_2(k_2),$$

$$\mathcal{M}_{3,1} = \beta_2 K_1(q_1), \quad \mathcal{M}_{3,2} = -\beta_1 K_1(q_2), \quad \mathcal{M}_{3,3} = 1.$$

and do not contain the ratio $Ro/Bu = ca$. Solving $\det(\mathcal{M}) = 0$ for S_1, computing A_1, A_2 from the linear system (6.71), and deducing B_j from (6.64) or (6.65), we get a solution of the full problem.

The pressure fields and associated velocities for both barotropic and baroclinic modons are given in Figure 6.3. We see that velocity fields are practically the same in both layers for the quasi-barotropic modon, whereas for the baroclinic modon the velocity in the lower layer is much weaker, in agreement with zero PV in the second layer.

To finish the discussion of QG modons, we should stress that these steady-translating solutions give a standing dipole in a uniform flow by a change of reference frame (co-moving frame). Although the nice symmetric theoretical solutions are difficult to observe, by the simple reason that straight uniform flows are hard to find in nature, qualitatively similar configurations frequently happen in the atmosphere, leading to persisting blockings of the mean atmospheric circulation. An example of such blocking which stayed over Europe for several weeks in 2011 is presented in Figure 6.4.

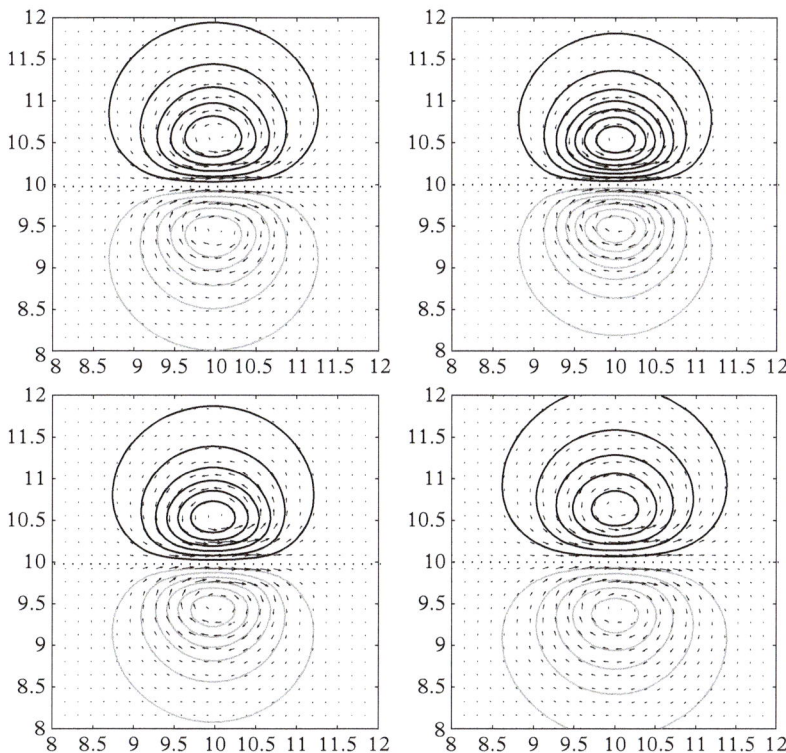

Figure 6.3 *Pressure and velocity fields in the upper (top panel) and lower (bottom panel) layers for the quasi-barotropic (left column) and baroclinic (right column) QG modon solutions of the two-layer free-surface QG model with s = 0.8, and d = 0.85. Black: cyclones, grey: anticyclones. Contour interval is ±0.01 in both layers for the barotropic modon and the upper layer of the baroclinic one, and ±0.006 for the lower layer of the baroclinic modon.*

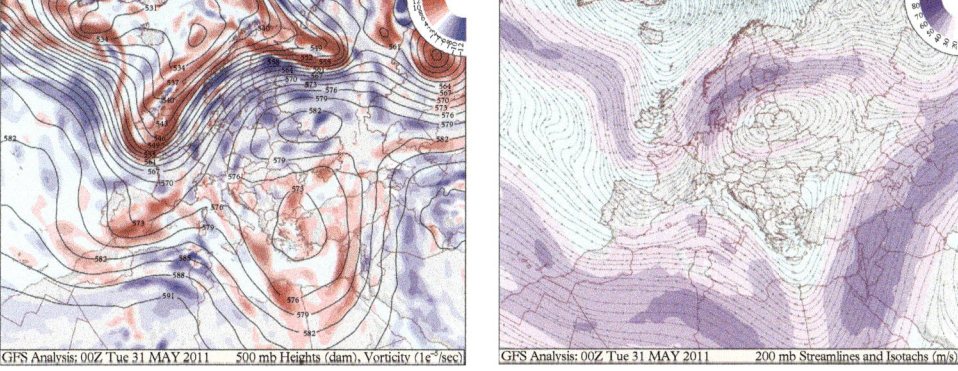

Figure 6.4 *Blocking configuration observed in Spring 2011 over Europe, as follows from the analysis of atmospheric data. Left panel: vorticity (colours: red - cyclonic, blue - anticyclonic) and geopotential height (contours) at the isobaric surface 500mb. Right panel: velocity (colours) and streamlines at the isobaric surface 200mb. Images generated by the Center for Ocean–Land–Atmosphere Studies (COLA) at George Mason University from the data from the forecast models of NOAA's National Center for Environmental Prediction (NCEP).*

6.3 A crash course in 2D turbulence

Systems of point vortices discussed in Section 6.1.3 provide nice examples of nontrivial dynamical systems with a finite number of degrees of freedom. For small number of vortices particular exact solutions can be easily found, cf. the problems at the end of the chapter. Thus, a system of three vortices of arbitrary intensities turns to be completely integrable. However, a four-vortex system is chaotic in the strict mathematical sense. This fact by itself gives an idea why, in general, evolution of arbitrary initial vorticity distribution exhibits *turbulent* motion; that is intuitively speaking, a complex and unpredictable behaviour with a wide range of time and spatial scales involved, although *laminar* regular solutions usually also exist. According to (quasi-) two-dimensional character of the models treated in this book, we focus on 2D turbulence. The standard set-ups in studying numerically 2D turbulence are either a free decay one, starting form some random initial conditions, or a forced-dissipative one, where the 2D Navier–Stokes equation is forced in some way. 2D turbulence differs from the classical three-dimensional (3D) turbulence mainly by the fact that, in the limit of vanishing dissipation, the governing 2D Euler equations possess an infinite number of integrals of motion, including a second positive-definite quadratic invariant, the enstrophy, in addition to energy, while there is none in 3D Euler equations (a unique Casimir, helicity is quadratic, but not positive-definite in 3D). Two-dimensional turbulence is a subject of active research for years, and one of its *raisons d'être* is importance for large-scale geophysical flows, which are quasi two-dimensional. Without going into details, we will sketch the standard ideas allowing us to understand rough properties of turbulent vortex motions.

6.3.1 Reminder on statistical description of turbulence

Solutions of the Navier–Stokes equation corresponding to the turbulent behaviour are represented as a sum of the mean flow and turbulent fluctuations (or pulsations):

$$\mathbf{v} = \mathbf{U} + \mathbf{u},$$

with mean velocity $\mathbf{U}$, and turbulent fluctuations $\mathbf{u}$. Turbulent velocity is considered to be in this context a random process, and splitting of the velocity into the mean and fluctuations is achieved through averaging $\mathbf{U} = \langle \mathbf{v} \rangle$, $\langle \mathbf{u} \rangle = 0$. Averaging $\langle \ldots \rangle$ can be either over ensemble of realisations $i = 1, 2, \ldots$,

$$\langle \mathbf{v}(\mathbf{x}, t) \rangle_E = \lim_{N \to \infty} \frac{1}{N} \sum_{i=1}^{N} \mathbf{v}^{(i)}(\mathbf{x}, t), \tag{6.72}$$

or over time,

$$\langle \mathbf{v}(\mathbf{x}, t) \rangle_T = \lim_{T \to \infty} \frac{1}{T} \int_0^{\infty} d\tau \, \mathbf{v}(\mathbf{x}, t + \tau). \tag{6.73}$$

Usually the ergodic hypothesis $\langle \ldots \rangle_E = \langle \ldots \rangle_T$ is made, neglecting possible differences between the two averages. These latter have following obvious properties:

$$\langle\langle \ldots \rangle\rangle = \langle \ldots \rangle, \quad \langle \text{const} \ldots \rangle = \text{const} \langle \ldots \rangle, \quad \langle A \langle B \rangle \rangle = \langle A \rangle \langle B \rangle, \quad \langle A + B \rangle = \langle A \rangle + \langle B \rangle.$$

Averaging the Navier–Stokes and divergence equations leads to

$$\partial_t \mathbf{U} + \mathbf{U} \cdot \nabla \mathbf{U} + \langle \mathbf{u} \cdot \nabla \mathbf{u} \rangle = -\nabla P + \nu \nabla^2 \mathbf{U}. \tag{6.74}$$

$$\nabla \cdot \mathbf{U} = \nabla \cdot \mathbf{u} = 0, \tag{6.75}$$

and introducing the *Reynolds stress* tensor, $\tau_{ki} = -\langle u_k u_i \rangle$, $\tau_{ki} = \tau_{ik}$, we get the *Reynolds equations* for the mean velocity and its turbulent fluctuations (in tensor notation with $i = 1, .., N$, where N is the spatial dimension of the problem, and summation over repeated indices is understood):

$$\partial_t U_i + U_k \partial_k U_i = \partial_k \tau_{ki} - \partial_i P + \nu \nabla^2 U_i. \tag{6.76}$$

$$\partial_t u_i + U_k \partial_k u_i + u_k \partial_k U_i + u_k \partial_k u_i = +\partial_k \tau_{ki} - \partial_i p + \nu \nabla^2 u_i. \tag{6.77}$$

An important quantity in turbulence is the mean kinetic energy of turbulent fluctuations $e_t = \left\langle \frac{\mathbf{u}^2}{2} \right\rangle$, which obeys the equation following from multiplying (6.77) by u_i and averaging:

$$\partial_t e_t + U_k \partial_k e_t = \tau_{ik} \partial_k U_i - \partial_k \left(\frac{1}{2} \langle u_i u_i u_k \rangle + \langle u_k p \rangle \right) + \nu \nabla^2 \langle u_k u_k \rangle - 2\nu \left\langle (\partial_k u_i)^2 \right\rangle \tag{6.78}$$

As is obvious from the Reynolds equations, they are not closed, and writing down equations for the third moments $\langle u_i u_j u_k \rangle$ appearing in the equation for the second ones, cf (6.78), leads to appearance of the fourth moments, etc. Therefore, some ad hoc hypothesis expressing higher moments in terms of lower ones, a *turbulence closure*, is needed. There is a number of plausible hypotheses, the oldest and best known being the *turbulent viscosity* one expressing the second moments in terms of gradients of the mean field. Such hypotheses are usually sufficient for practical applications, the best-known case is the boundary layer theory. However, theoretically they are not always satisfactory. This is why direct numerical simulations became the most widespread tool in theoretical studies of fully developed turbulence. Such studies, in the context of 2D turbulence, are most often performed in simple periodic domains either with the help of direct truncations in the Fourier space discussed above in Section 6.1.5, the use of Casimir-preserving discretisations being rather rare, or with a contour dynamics method of Section 6.1.4.

6.3.2   Developed turbulence: energy and enstrophy cascades

Turbulence classics: energy cascade, inertial interval,
and Kolmogorow–Obukhov laws

The key idea in the theory of developed 3D turbulence is that of cascade of energy. Solving (6.78), which is essentially non linear at high Reynolds numbers, being virtually impossible, an heuristic picture due to Richardson is that the energy, which is injected in the system by forcing at some spatial scale, can be dissipated only at small scales, where the effective Reynolds number becomes of the order unity, and that in the interval of scales (the inertial interval) between the (large) input scale and (small) dissipation scale the energy cascades towards the small scales without losses. The advantage of such a picture is that it is determined by a single parameter: energy dissipation rate per unit mass ϵ which has the dimension L^2/T^3. (We use the standard notation ϵ here, not to be confused with Rossby number elsewhere in the book.) In the simplest situation when this parameter is constant, simple dimensional estimates allow us to establish dependence of the energy of turbulent pulsations e_t on their scale λ in the inertial interval, the famous Kolmogorov–Obukhov law,

$$e_t(\lambda) \sim (\epsilon\lambda)^{\frac{2}{3}}, \tag{6.79}$$

and the corresponding expression for the energy contained in Fourier modes of turbulent velocity with $k > k_\lambda \sim \frac{1}{\lambda}$:

$$e_t(\lambda) \sim \int_{1/\lambda}^{\infty} E(k)dk, \quad E(k) \sim \epsilon^{\frac{2}{3}}k^{-\frac{5}{3}}. \tag{6.80}$$

The famous −5/3 law is confirmed by many measurements and numerical simulations. This does not, however, mean that such a simple theory bearing the name of Kolmogorov, is universally valid. Indeed, its obvious caveat is the constancy of ϵ, as there is no reason for this quantity to be uniform in space in time, as was noticed by Landau already in early days of Kolmogorov's theory. If ϵ were constant, the same dimensional argument, if applied to higher moments of velocity, would give power-law spectra for all of them with exponents proportional to 1/3. Yet, deviations from such simple power laws are systematically reported both in measurements and numerical simulations reflecting the *intermittency* of turbulence.

Cascades in 2D turbulence

Obviously, similar to the previous subsection reasoning can be applied to 2D turbulence, with the same −5/3 law for the spectral density of energy. Yet, another quadratic in velocity invariant, the enstrophy, exists in 2D Euler equations, as we have seen in Section 6.1.1. The representations of energy and enstrophy through their spectral densities are

$$E = \int_0^{\infty} dk\, E(k), \quad \Omega = \int_0^{\infty} dk\, k^2 E(k),$$

and the cascade argument can be used for enstrophy as well, with enstrophy dissipation rate per unit mass χ replacing ϵ. By scaling arguments, it leads to -1 power law (Batchelor's law) for $\Omega(k)$, and -3 law for the related energy spectrum. A simple argument, known as the Fjortoft's theorem, indicates that it is the cascade of enstrophy which is directed towards the small scales, while the energy cascade is directed towards larger scales in two dimensions. Let us write the unforced 2D Navier–Stokes equation in terms of stream-function ψ, by adding a viscous term in (6.3):

$$\partial_t \nabla^2 \psi + \mathcal{J}(\psi, \nabla^2 \psi) = \nu \nabla^4 \psi. \tag{6.81}$$

In the absence of dissipation $\nu \to 0$ energy $E = \int d\mathbf{x} \, (\nabla \psi(\mathbf{x}))^2$ and enstrophy $\Omega = \int d\mathbf{x} \, (\nabla^2 \psi(\mathbf{x}))^2$ are conserved, as we have already seen. In the Fourier space, $\psi(\mathbf{x}, t) = \int d\mathbf{k} \, \psi(\mathbf{k}, t) e^{-i\mathbf{k}\cdot\mathbf{x}}$, and (6.81) in the absence of dissipation takes the form, cf. (6.32)

$$\mathbf{k}^2 \, \partial_t \psi(\mathbf{k}, t) + \int \int d\mathbf{m} d\mathbf{l} \, \delta(\mathbf{k} - \mathbf{m} - \mathbf{l}) \mathbf{m}^2 \, (\mathbf{m} \times \mathbf{l}) \psi(\mathbf{m}, t) \psi(\mathbf{l}, t) = 0. \tag{6.82}$$

As follows from these equations the interaction among different Fourier modes is *triadic*, by triples of modes with wave numbers $\mathbf{k}, \mathbf{m}, \mathbf{l}$ obeying the condition $\mathbf{k} - \mathbf{m} - \mathbf{l} = 0$. If we limit ourselves by a single triad with wave numbers $\mathbf{k}_1, \mathbf{k}_2, \mathbf{k}_3$, the energy and the enstrophy of the triad are conserved,

$$E = E_1 + E_2 + E_3 = \text{const}, \quad \Omega = k_1^2 E_1 + k_2^2 E_2 + k_3^2 E_3 = \text{const},$$

and it is easy to show (cf. Problem 6.9) that if the median mode loses energy, it goes to the lower wave number mode, and if it loses enstrophy, it goes to the higher wave number mode, in order to respect both conservation laws. Thus, with a forcing at some wave number, a direct (i.e. towards smaller scales, and higher wave numbers) cascade of enstrophy and an inverse (i.e. towards larger scales, and smaller wave numbers) cascade of energy are expected. In the language of vorticity, the enstrophy cascade observed in numerical simulations of 2D decaying (i.e. unforced) turbulence corresponds to formation of thin vorticity filaments, and the energy cascade corresponds to merging of vortices and formation of vortices of greater scale, limited only by the size of the computational domain. We will see examples of both processes in Chapter 9.

6.4   When vortices emit waves: Lighthill radiation

In this section we show that simple vortex systems, if embedded in a system supporting gravity waves, produce continuous wave emission, which is called Lighthill radiation. We work in the technically simplest framework of a point-vortex configuration, and omit rotation, initially.

6.4.1   2D hydrodynamics and vortex-pair solution in complex notation

We have seen in Section 6.1 that it can be advantageous to use complex variables for two-dimensional flows, and we will work with complex coordinates: $z = x + iy$, $z^* = x - iy$. Complex velocity $V = u - iv$, is introduced here in a way that is traditional for potential flows. We will be considering point-vortex systems where the flow is *potential* everywhere, except for locations of vortices, which are of measure zero. Physically speaking, the point vortices represent localised vortices in a situation where the scales of interest are much larger than proper scales of the vortices. The flow, thus, is described by a *complex potential* $\phi(z)$, such that $V = \partial_z\phi$. For example, a point vortex of intensity κ at location z_0 has a complex velocity and potential given by

$$V = \partial_z\phi_1, \quad \phi_1 = -i\frac{\kappa}{2\pi}\log(z - z_0), \tag{6.83}$$

where the minus sign in the last expression is introduced to be consistent with the Green function introduced above. Following the superposition principle, the complex potential for a vortex pair is

$$\phi_2 = -i\frac{\kappa_1}{2\pi}\log(z - a_1 e^{i\Omega t}) - i\frac{\kappa_2}{2\pi}\log(z - a_2 e^{i\Omega t}), \tag{6.84}$$

where $\Omega = \frac{\kappa_1+\kappa_2}{2\pi a^2}$, a is the distance between the vortices, the origin of the coordinate system is placed at the centroid of vorticity, and $a_1 = a\frac{\kappa_2}{\kappa_1+\kappa_2}$, $a_2 = -a\frac{\kappa_1}{\kappa_1+\kappa_2}$. In the following, we will need the asymptotics of solution at $r \to \infty$. For this we use the polar form of the complex coordinate $z = re^{i\theta}$ and perform *multipolar expansion* in inverse powers of r:

$$\phi_2|_{r\to\infty} = -i\frac{\kappa_1 + \kappa_2}{2\pi}\log\left(re^{i\theta}\right) + \frac{i}{4\pi}\frac{\kappa_1\kappa_2}{\kappa_1 + \kappa_2}\left(\frac{a}{r}\right)^2 e^{2i(\Omega t-\theta)} + \dots. \tag{6.85}$$

It should be emphasised that a dipolar term $\propto e^{i(\Omega t-\theta)}$ is absent, as it is proportional to $a_1\kappa_1 + a_2\kappa_2$, and this combination is zero because of the position of the origin.

6.4.2   Gravity waves in cylindrical geometry

In the absence of rotation, the linearised shallow-water equations can be reduced to a single one for the thickness anomaly h:

$$\partial_{tt}^2 h - gH_0\nabla^2 h = 0. \tag{6.86}$$

Energy balance to the leading order in perturbations over the state of rest with thickness H_0 is obtained from (3.28) by linearisation and the use of linearised mass conservation:

$$\partial_t \left(H_0 \frac{\mathbf{v}^2}{2} + g \frac{h^2}{2} \right) = -g H_0 \nabla \cdot (\mathbf{v}h) . \tag{6.87}$$

If we suppose a wave source at the origin, the total energy flux over a circle far from the source is

$$I_E = g H_0 \oint d\mathbf{l} \cdot \hat{\mathbf{z}} \wedge \mathbf{v}h. \tag{6.88}$$

Emission of waves by a localised source is natural to treat in polar coordinates. Let us recall that in polar coordinates the Laplacian entering the wave equations has the form $\frac{1}{r} \frac{\partial}{\partial r} \left(r \frac{\partial}{\partial r} \right) + \frac{1}{r^2} \frac{\partial^2}{\partial \theta^2}$. Solution of the wave equation (6.86) is sought in the form of superposition of azimuthal harmonics with integer azimuthal wave numbers n:

$$h(r, \theta, t) = \sum_n h_n(r) e^{in(\Omega t - \theta)}, \tag{6.89}$$

which allows us to transform the wave equation (6.86) into an ordinary second-order differential eqution in r. A change of independent variable: $r \to \rho = \frac{n\Omega}{\sqrt{gH_0}} r$ further transforms this equation into the classical Bessel equation:

$$h_n'' + \frac{1}{\rho} h_n' + \left(1 - \frac{n^2}{\rho^2} \right) h_n = 0. \tag{6.90}$$

Solution of (6.90) corresponding to wave emission is a combination of fundamental solutions verifying *radiation boundary condition* at $r \to \infty$. As is known, the Hankel function $H_n^{(2)} = \mathcal{J}_n - i Y_n$, where $\mathcal{J}_n$, Y_n are ordinary Bessel functions of order n, verifies the radiation condition of outward energy flux far from the source. The asymptotics of the Hankel function are

$$H_n^{(2)}(\rho)\big|_{\rho \to 0} \to \frac{i}{\pi} n! \left(\frac{\rho}{2} \right)^{-n}, \quad H_n^{(2)}(\rho)\big|_{\rho \to \infty} \to \sqrt{\frac{2}{\pi \rho}} e^{\left(-i \left(\rho - n \frac{\pi}{2} - \frac{\pi}{4} \right) \right)}. \tag{6.91}$$

6.4.3 Lighthill radiation

The main idea

The idea allowing us to associate emission of gravity waves to the rotational motion of the vortex pair is as follows. The motion of the vortex pair has a definite frequency Ω. Hence, the emitted wave field, if any, should have the same frequency. The vortex-pair solution of Section 6.4.1 can be considered as an approximate solution of shallow-water equations if deformations of the free surface have a characteristic length much larger than the distance between the vortices. In the context of wave radiation this means that the characteristic wave length of gravity waves $\lambda = \frac{\sqrt{gH_0}}{\Omega}$ is much larger than a. The ratio of the two gives a characteristic Froude (or Mach) number: $M = \frac{a}{\lambda} = \frac{a\Omega}{\sqrt{gH_0}}$. If $M \to 0$, then the *near field*, if the vortex pair is placed at the origin, can be considered to be *incompressible*, and *far field* to be waves. If the near field and the far field can be matched respecting dynamical boundary condition of *continuity of pressure*, then a wave field is linked to the motion of the vortex pair, and, therefore, the pair radiates waves.

Matching vortex and wave fields

Using the Bernoulli equation for potential flows, the pressure can be expressed via the time derivative of the velocity potential. Hence, the matching condition of the near wave field and the far vortex field is

$$gh^{(waves)}\big|_{r\to 0} = -\partial_t \phi_2^{(vortex)}\big|_{r\to\infty}. \tag{6.92}$$

Due to the structure of the vortex field, uniquely the $n = 2$ azimuthal component is allowed in the wave field. Calculating the time-derivative of (6.85) and using (6.91), we get:

$$gh^{(waves)} = -\frac{i}{8\pi^3 c_0^2} \frac{\kappa_1 \kappa_2 (\kappa_1 + \kappa_2)^2}{(2\pi)^2 a^2} H_2^{(2)}\left(\frac{2\Omega}{c_0} r\right) e^{2i(\Omega t - \theta)}, \tag{6.93}$$

where we introduced the notation $c_0 = \sqrt{gH_0}$ for the phase velocity of gravity waves. Thus, the vortex field and the wave field are matched, and the radiation of gravity waves by the vortex pair is proved. The asymptotics at infinity of the emitted wave field is

$$gh^{(waves)}\big|_{r\to\infty} \propto \frac{\kappa_1 \kappa_2 |\kappa_1 + \kappa_2|^{\frac{3}{2}}}{c_0^{\frac{3}{2}} a^3 r^{\frac{1}{2}}} e^{2i\left(\Omega t - \frac{\Omega r}{c_0} - \theta - \frac{3}{8}\pi\right)}, \tag{6.94}$$

and the wave field is quadrupolar (in polar angle).

6.4.4   Back-reaction of wave radiation

The wave emission provides a sink of energy, which influences the vortex system. This phenomenon is called in physics the back-reaction of wave radiation. The total energy flux of the waves (6.94) emitted by the vortex pair is

$$I_E = gH_0 \oint d\mathbf{l} \cdot \hat{\mathbf{z}} \wedge \mathbf{v}h = H_0 \oint dl\, hv_r, \tag{6.95}$$

where v_r is the radial component of velocity. As $v_r \propto \frac{gh}{c_0}$,

$$I_E \propto \frac{(\kappa_1 \kappa_2)^2 |\kappa_1 + \kappa_2|^3}{c_0^4 a^6} \propto M^4. \tag{6.96}$$

The emission is thus weak. Nevertheless, it is continuous, and its effect upon the vortices is accumulating. Let us write down the energy balance of the vortex pair in the presence of the radiation sink. The energy of the pair, cf. (6.20), is $E_{pair} \propto \kappa_1\kappa_2 \log a$. As $\kappa_{1,2}$, H_0, and c_0 are constant, the only parameter which can change due to the energy sink is the distance a between the vortices:

$$\dot{E} = -I_E \Rightarrow -\frac{\text{const}}{a}\dot{a} = -\frac{\text{const}}{a^6} \Rightarrow a^5\dot{a} = \text{const} \propto \kappa_1\kappa_2|\kappa_1 + \kappa_2|^3. \tag{6.97}$$

The resulting ordinary differential equation is easy to integrate, which results in the following power law:

$$a(t) = a(0)\left(1 + \frac{t}{\tau}\right)^{\frac{1}{6}}, \quad \tau \propto \left(\kappa_1\kappa_2|\kappa_1 + \kappa_2|^3\right)^{-1}. \tag{6.98}$$

Different cases are therefore possible depending on the signs of vortices: 1) $\kappa_1\kappa_2 > 0$, giving *increase* of a in time, and 2) $\kappa_1\kappa_2 < 0$, giving *decrease* of a in time. Remarkably, the latter case leads to a *vortex collapse*, that is vortices merging in finite time τ, and thus qualitatively changing their configuration. Thus, in spite of weak intensity, the Lighthill radiation can lead to qualitative changes in vortex systems.

6.4.5 Lighthill radiation in the presence of rotation

In the demonstration of the Lighthill radiation earlier, we completely ignored the effects of rotation. Yet, we know that rotation leads to the appearance of the screening radius R_d in the system. This new scale interferes with the reasoning based on separation of the system into far and near fields with respect to the source, which we used. To understand how the presence of the deformation radius influences the radiation, let us recall that the wave equation for inertia–gravity waves reads in this case, with the same notation as in (6.86):

$$-\frac{1}{gH_0}\frac{\partial^2 h}{\partial t^2} - \frac{1}{R_d^2}h + \nabla^2 h = 0, \tag{6.99}$$

It is easy to check that in order to have the same form of this equation as (6.90), after having introduced the decomposition in azimuthal harmonics (6.89), the following change of variables is needed:

$$\rho = \frac{n\Omega}{c_0}r\sqrt{1 - Ro_n^{-2}}; \quad Ro_n = \frac{n\Omega}{f}, \tag{6.100}$$

where Ro_n is the equivalent Rossby number of the mode n. It coincides with the Rossby number of the vortex pair $Ro_2 = \frac{2\Omega}{f}$. for the mode $n = 2$, which can be matched, along the same lines as earlier, with the far vortex field. An immediate consequence of (6.100) is that propagative waves, and thus the Lighthill radiation, are possible only at $Ro_n > 1$. Moreover, to avoid exponential decay of the vortex field due to screening beyond R_d, the matching should take place at a distance r such that $a \ll r \ll R_d$, and therefore the Burger number should be large: $Bu \gg 1$. Together with the condition $M \ll 1$, this gives $Bu \gg Ro_2^2$. The overall conclusion following from these considerations is that Lighthill radiation of inertia–gravity waves by vortices is possible only at high Rossby and Burger numbers.

6.5   Summary, comments, and bibliographic remarks

We thus have reviewed the methods of (quasi-) two-dimensional vortex dynamics, which is the essence of slow motions in quasi-geostrophic approximation. Vortex dynamics is also the core of two-dimensional turbulence, relevant in the context of quasi-geostrophic dynamics, with its specific power laws and double energy and enstrophy cascades. We have also seen that vortices can emit waves. Vortices are intrinsically coupled to Rossby waves which are vorticity waves, if gradients of the background PV are present, although specific dipolar vortex structures, the modons, are exempt from leaving a Rossby-wave tail while propagating. We will see in Chapter 9 that modon solutions exist also in full RSW, and have some surprising properties. Vortices can emit inertia–gravity waves, if considered in the full system containing fast motions, and if the intrinsic frequency of vortex motions matches that of waves, which is possibly only at high enough Rossby numbers. The last phenomenon shows the limits of dynamical separation of fast and slow motions in geophysical fluid dynamics, which is thus restrained to small Rossby-number flows.

   The material on 2D vortex dynamics is standard and can be found in many books and textbooks such as Saffman (1992). Point-vortex dynamics is a classical topic in hydro-dynamics, and is also covered in many books. For a review of conformal dynamics, see and related numerical methods see Dritschel (1997). The method of constructing the classical Kirhhoff vortex solution proposed in Problem 6.7 is based on Abrashkin and Yakubovich (2006). This method, which consists, in fact, in the mapping of the vorticity patch onto the unit circle, suggests a variant of contour dynamics using con-formal mappings, which is called conformal dynamics, see Legras and Zeitlin (1992). Structure-preserving truncations of vortex dynamics in Fourier space were proposed in Zeitlin (1991a) on the basis of geometric interpretation of ideal fluid flows. Dynamics of point vortices on the β plane was analysed in Reznik (1992). Lamb's dipole solution can be found in Batchelor (1967), for example. Construction of the QG modon follows the classical paper by Larichev and Reznik (1976). Modons in the two-layer QG with a rigid lid were built in Flierl *et al.* (1980), Their construction in the two-layer QG with a free surface follows Lahaye and Zeitlin (2012), where a detailed study of dependence on stratification and aspect ratio parameters can be found. Turbulence in general, and 2D turbulence in particuar, is an immense topic and we only recalled the most basic notions. The Kolmogorov theory of turbulence, which is, of course, much deeper than the sketch presented here, can be found in the turbulence encyclopaedia by Monin and Yaglom (2007). For an introduction, including 2D turbulence , see Davidson (2004). The cascade of enstrophy in 2D turbulence was discovered by Batchelor (1969). Light-hill radiation by a vortex pair and its back-reaction, with possible vortex collapse, were found by Gryanik (1983), and our presentation follows this paper. Similar analysis for the extended Kirchhoff vortex was performed in Zeitlin (1991b), where the construction of the Kirchhof vortex can be also found. These ideas were further developed in Ford (1994). Analysis of Lighthill radiation in the presence of rotation follows Zeitlin (2008). For radiation boundary conditions in wave equation see Morse (1953).

Figure 6.2 is reproduced from Ribstein, Gula, and Zeitlin (2010) with the permission of AIP publishing. Figure 6.3 is reproduced from Lahaye and Zeitlin (2012), with permission.

6.6 Problems

6.1 Obtain Green's function (6.7) by applying Gauss theorem in two dimensions to a point-vortex configuration.

6.2 Demonstrate (6.11).

6.3 Demonstrate the conservation laws for a system of N vortices.

6.4 Find two-vortex solutions of (6.19), with vortices of intensities κ_1 and κ_2. Show that if $\kappa_1 \neq -\kappa_2$ the vortices rotate about their center of vorticity, which position is defined by $\mathbf{x}_c = \frac{\kappa_1 \mathbf{x}_1 + \kappa_2 \mathbf{x}_2}{\kappa_1 + \kappa_2}$, and if $\kappa_1 = -\kappa_2$, they form a dipole moving along a straight line.

6.5 Demonstrate that three vortices of equal intensity forming an equilateral triangle rotate as a rigid body with constant angular velocity.

6.6 Demonstrate that a point vortex near the lateral boundary moves with a constant velocity along this latter.

6.7 Prove that (6.16) has a solution

$$Z = \alpha \xi e^{i\omega t} + \beta \xi^*,$$

where α and β are arbitrary complex numbers, and ω is a real number. Calculate vorticity of this flow with the help of (6.14), and show that it is constant. Show that if ξ lies inside the unit circle $|\xi| = 1$, the boundary of the domain in Z- space is an ellipse with semi-axes $|\alpha| \pm |\beta|$. Show that the flow

$$\begin{cases} Z = \alpha \xi e^{i\omega t} + \beta \xi^{-1}, \\ V = i\omega \alpha e^{i\omega t} \xi^{*-1}, \end{cases}$$

defined in the domain outside the unit circle $|\xi| = 1$ on the plane of complex ξ is potential, and can be matched to the vortex flow above.

The whole construction gives an elliptic patch of constant vorticity rotating with constant angular velocity, the *Kirchhoff vortex*.

6.8 Demonstrate that a pair of point vortices of opposite intensity (a vortex dipole) which moves with constant velocity U in zonal direction is a solution of the Euler equations for the incompressible fluid on the β plane. Find the spectrum of possible values of U and analyse the resulting velocity field.

6.9 Check the statement of Fjortoft's theorem with a triad $\mathbf{k}_1, a\mathbf{k}_1, b\mathbf{k}_1, b > a$.

7

Rotating Shallow-Water Models as Quasilinear Hyperbolic Systems, and Related Numerical Methods

In this chapter we will give the mathematical background necessary to understand the behaviour of the rotating shallow-water (RSW) model and the algorithms used in numerical simulations, which we have already applied to illustrate Kelvin wave breaking in Chapter 6, and which we will use abundantly throughout the book. It should be stressed that numerical methods for full shallow-water equations are different with respect to those discussed earlier for vortex dynamics. The origin of this difference is in underlying physics: shallow-water models include waves, and wave systems are described by *hyperbolic partial differential equations*, which need specific treatment. In particular, such systems have *weak solutions*, shocks in the first place. As we have already seen, they appear as the result of *wave breaking*, which is a typical nonlinear process. The gas-dynamics analogy explained in Chapter 3 allows us to profit from the huge corpus of related numerical developments. However, the specifics of geophysical field dynamics (GFD) features—rotation, topography, and baroclinicity—pose specific problems in numerical implementations. We will present the main ideas allowing us to resolve, or at least circumvent, these problems. We will not pursue a mathematically rigorous formulation and will remain, rather, on the heuristic level, with an accent on physics behind mathematics. As is often the case, it is easier to understand the mathematics of the models by reducing the number of their spatial dimensions (although lower-dimensional results are not always directly transposable to higher dimensions). That is why we focus on one-dimensional versions of RSW models. As is easy to understand, the models cannot be purely one dimensional, as rotation mixes up two components of velocity via the Coriolis force. So we consider '1.5'-dimensional models, in a sense that, although all dynamical variables will not depend on one of the spatial coordinates, let us say y (which makes no difference with x in the f-plane approximation, mostly used throughout this chapter), the velocity component in this direction will be maintained.

Geophysical Fluid Dynamics. Vladimir Zeitlin,
Oxford University Press (2018). © Vladimir Zeitlin. DOI: 10.1093/oso/9780198804338.001.0001

7.1 One-layer model

7.1.1 1.5-dimensional one-layer RSW model

If all y dependence of all dynamical variables is removed, the equations of the one-layer RSW take the form

$$\begin{cases} \partial_t u + u\partial_x u - fv + g\partial_x h = 0, \\ \partial_t v + u\partial_x v + fu = 0, \\ \partial_t h + u\partial_x h + h\partial_x u = 0. \end{cases} \tag{7.1}$$

They possess two Lagrangian invariants: the one-dimensional reduction of PV

$$q = (\partial_x v + f)/h, \quad \partial_t q + u\partial_x q = 0 \tag{7.2}$$

and the so called geostrophic momentum

$$M = v + fx, \quad \partial_t M + u\partial_x M = 0, \tag{7.3}$$

where x is understood in the Lagrangian sense. Linearisation of (7.1) about a state of rest in the f-plane approximation leads to the standard spectrum of inertia–gravity waves propagating leftwards or rightwards in the x direction

$$\omega = \pm(c_0^2 k^2 + f^2)^{\frac{1}{2}}, \quad c_0 = \sqrt{gH_0}. \tag{7.4}$$

The geostrophic equilibrium

$$fv = g\partial_x h, \quad u = 0 \tag{7.5}$$

becomes an exact solution of the equations of motion in this configuration.

7.1.2 Lagrangian approach to the 1.5-dimensional model

As we will see shortly, it is advantageous to use the Lagrangian description of the model, following the lines adopted in Section 3.4.1. This description uses coordinates along the trajectories of fluid 'parcels' on the x axis $X(x, t)$, where x is a position of the parcel at $t = 0$, as dynamical variables. The velocity of the parcel is $\dot{X} = u(X, t)$, and we adopt the notation $X' = \frac{\partial X}{\partial x}$ from now on. The momentum equations take the form

$$\begin{cases} \ddot{X} - fv + g\partial_X h = 0, \\ \dot{v} + f\dot{X} = 0, \end{cases} \tag{7.6}$$

where v is considered as a function of x and t. The mass conservation equation becomes

$$h(X, t) \, dX = h_I(x) \, dx, \quad \Rightarrow \quad h(X, t) = h_I(x) \partial_X x. \tag{7.7}$$

The Lagrangian equations of motion (7.6) and (7.7) can be reduced to a single partial differential equation. The equation for v is readily integrated in time giving

$$v(x, t) + fX(x, t) = M(x). \tag{7.8}$$

The function $M(x)$ can be determined from initial conditions

$$M(x) = fx + v_I(x). \tag{7.9}$$

Chain differentiation gives

$$\partial_X h = \partial_X \left(h_I(x) \partial_x X \right) = h'_I \left(X' \right)^{-2} - h_I(x) X'' \left(X' \right)^{-3}, \tag{7.10}$$

and a closed equation for X results

$$\ddot{X} + f^2 X + gh'_I \left(X' \right)^{-2} + \frac{gh_I}{2} \left[\left(X' \right)^{-2} \right]' = fM. \tag{7.11}$$

This equation may be rewritten in terms of deviations of the parcels from their initial positions: $\phi(x, t) = X(x, t) - x$:

$$\ddot{\phi} + f^2 \phi + gh'_I \left(\frac{1}{(1 + \phi')^2} \right) + \frac{gh_I}{2} \left(\frac{1}{(1 + \phi')^2} \right)' = fv_I, \tag{7.12}$$

which is to be solved with initial conditions

$$\phi(x, 0) = 0, \quad \dot{\phi}(x, 0) = u_I(x),$$

where u_I is the initial velocity in the x direction.

Utility of the Lagrangian description may be anticipated from the fact that short inertia–gravity waves are dispersionless, cf. (7.4), and thus are expected to break and form shocks at large enough amplitudes, like ordinary sound waves (which they, indeed, are, according to the acoustic analogy which was already stressed in Chapter 3). Lagrangian approach allows us to get a simple and intuitively clear idea of breaking. The archetype model of breaking is the so-called simple-wave equation,

$$u_t + \epsilon u u_x + u_x = 0, \tag{7.13}$$

where we introduced a parameter ϵ (the non-dimensional wave amplitude), which controls nonlinearity. Solutions of the linearised equation (7.13) are wave packets propagating without change of form along the x axis: $u(x, t) = U(x-t)$. By choosing a reference frame moving with the wave and introducing dependence on the slow time $T = \epsilon t$, we cast (7.13) in the form

$$U_T + UU_x = 0, \tag{7.14}$$

and with the help of the Lagrangian coordinate X, such that $U = \dot{X}$, the equation (7.14) can be trivially integrated:

$$\ddot{X} = 0, \Rightarrow \dot{X} = U_I(x), \Rightarrow X(x, T) = x + U_I(x)t. \tag{7.15}$$

Here U_I is the initial distribution of U. Therefore, for any x_1, x_2, such that $x_2 > x_1$ and $U_I(x_2) < U_I(x_1)$, the trajectories of Lagrangian particles intersect in finite time, which means that different Lagrangian particles arrive at the same location, which in turn means that the spatial derivative of the velocity field at this location is infinite, and the wave profile thus steepens and breaks down. We will see shortly how this scenario is realised in the RSW model.

7.1.3 Quasilinear and hyperbolic systems

A system of first-order partial differential equations for dependent variables $\vec{V} = (V_1, V_2, ..., V_N)$ of the form

$$\partial_t V_i(x, t) + \sum_{j=1}^{N} M_{ij}\left(\vec{V}\right) \partial_x V_j(x, t) = R_i\left(\vec{V}\right), \; i = 1, 2, ..., N \tag{7.16}$$

is called quasilinear. Let $\vec{l}^{(\alpha)}$ be the left eigenvectors, and $\xi^{(\alpha)}$ the corresponding left eigenvalues of the matrix M, $\alpha = 1, 2, ...$:

$$\vec{l}^{(\alpha)} \cdot M = \xi^{(\alpha)} \vec{l}^{(\alpha)}. \tag{7.17}$$

Then

$$\vec{l}^{(\alpha)} \cdot \left(\partial_t \vec{V} + M \cdot \partial_x \vec{V}\right) = \vec{l}^{(\alpha)} \cdot \left(\partial_t \vec{V} + \xi^{(\alpha)} \partial_x \vec{V}\right). \tag{7.18}$$

If we introduce *characteristic directions* defining corresponding characteristic curves, or simply characteristics,

$$\frac{dx}{dt} = \xi^{(\alpha)}, \tag{7.19}$$

and advection of the field $\vec{V}$ along a characteristic

$$\dot{\vec{V}} \equiv \frac{d\vec{V}}{dt} = \left(\partial_t + \xi^{(\alpha)} \partial_x\right) \vec{V}, \tag{7.20}$$

the original equations (7.16) can be rewritten in the characteristic form

$$\vec{l}^{(\alpha)} \cdot \dot{\vec{V}} = \vec{l}^{(\alpha)} \cdot \vec{R} \tag{7.21}$$

and, thus, become a system of ordinary differential equations, which obviously provides a great simplification, both from conceptual and computational points of view.

A system (7.16) is called *hyperbolic* if M has N eigenvalues $\xi^{(\alpha)}$, which are all real and different. A further simplification for hyperbolic systems can be achieved if $\vec{l}^{(\alpha)}$ are constant. Then the system can be rewritten in terms of Riemann variables (which become invariants if $\vec{R} = 0$):

$$r^{(\alpha)} = \vec{l}^{(\alpha)} \cdot \vec{V}, \quad \frac{dr^{(\alpha)}}{dt} = \vec{l}^{(\alpha)} \cdot \vec{R}. \tag{7.22}$$

In this terminology, shocks correspond to intersections of characteristics, like in the simple-wave example earlier, which means appearance of a singularity in the derivatives of Riemann invariants. It should be emphasised that finding Riemann invariants in the systems with more than three variables is not guaranteed.

7.1.4 Wave breaking in non-rotating and rotating one-layer RSW

Let us see how the above-described notions apply to the one-dimensional shallow-water model. We start with the model without rotation, which is equivalent to two-dimensional acoustics of a barotropic gas. The non-dimensional equations of the model (the second component of velocity decouples in the absence of rotation) can be rewritten in quasilinear form

$$\partial_t \begin{pmatrix} u \\ h \end{pmatrix} + \begin{pmatrix} u & 1 \\ h & u \end{pmatrix} \partial_x \begin{pmatrix} u \\ h \end{pmatrix} = 0. \tag{7.23}$$

The left eigenvectors and eigenvalues are

$$\vec{l}^{\pm} = (\pm\sqrt{h}, 1), \quad \xi^{\pm} = u \pm \sqrt{h}, \tag{7.24}$$

and the Riemann invariants are

$$r^{\pm} = u \pm 2\sqrt{h}, \quad \text{with} \quad \frac{dr^{\pm}}{dt^{\pm}} = 0, \quad \frac{d}{dt^{\pm}} \equiv \partial_t + \xi^{\pm}\partial_x. \tag{7.25}$$

Breaking and shock formation in the model are diagnosed with the help of the derivatives of the Riemann invariants $D^{\pm} \equiv \partial_x r^{\pm}$ obeying the following evolution equations:

$$\frac{dD^{\pm}}{dt^{\pm}} + \partial_x \xi^{\pm} D^{\pm} = 0, \quad \xi^{\pm} = \frac{3}{4}r^{\pm} + \frac{1}{4}r^{\mp}, \tag{7.26}$$

whence

$$\frac{dD^{\pm}}{dt^{\pm}} + \frac{3}{4}\left(D^{\pm}\right)^2 + \frac{1}{4}D^{\pm}D^{\mp}. \tag{7.27}$$

Let us suppose that one of $D^{\pm}$ is identically zero. Then a Riccati equation along the characteristic arises for the remaining D:

$$\frac{dD}{dt} + \frac{3}{4}(D)^2 = 0, \tag{7.28}$$

with a solution $D \propto (D_I^{-1} + \frac{3}{4}t)^{-1}$ developing a singularity, and hence wave breaking, in finite time if initial D is negative.

If we try a similar approach to 1.5-dimensional RSW equations (7.1) we immediately run into the difficulty of finding Riemann invariants, as the number of dependent variables is three. We will circumvent it by using Lagrangian description. For this we will proceed by a simplification of the system (7.6) and (7.7) with the help of an additional change of independent variables (Lagrangian labels) $x = x(a)$, which 'straightens' the initial elevation profile $h_I(x)$. We thus get the following relation between $\mathcal{J}$, the Jacobian of Lagrangian mapping, and h (we suppose that h is everywhere positive, that is thre is no 'drying'): $\mathcal{J} = \partial X/\partial a = H/h(X,t)$, where we introduced the uniform mean height H. Applying the chain differentiation rule, we get $g\,\partial_X h = \partial_a P$, where $P = gH/(2\mathcal{J}^2)$ is the so-called Lagrangian pressure. Non-dimensional Lagrangian equations of motion for dynamical variables as functions of time and Lagrangian labels a, are

$$\begin{cases} \dot{u} - v + \partial_a P = 0, \\ \dot{v} + u = 0, \\ \dot{\mathcal{J}} - \partial_a u = 0, \end{cases} \tag{7.29}$$

where time is measured in units of f^{-1}, distances in units of $R_d = \sqrt{gH}/f$, and velocities in units of $\sqrt{gH}$. Although, formally, there are three equations, the equation for v may be immediately integrated with the help of Lagrangian conservation of the geostrophic momentum, or that of PV. Hence, we will be considering the first and the third equations in (7.29), where v is not considered as an independent variable, and should be determined from the relation

$$\partial_a v = q(a) - \mathcal{J} \tag{7.30}$$

in terms of PV and $\mathcal{J}$. These equations form an inhomogeneous quasilinear system:

$$\begin{pmatrix} \dot{u} \\ \dot{\mathcal{J}} \end{pmatrix} + \begin{pmatrix} 0 & -\mathcal{J}^{-3} \\ -1 & 0 \end{pmatrix} \partial_a \begin{pmatrix} u \\ \mathcal{J} \end{pmatrix} = \begin{pmatrix} v \\ 0 \end{pmatrix}. \tag{7.31}$$

The left eigenvectors of the advection matrix on the left-hand side are $\left(1, \pm \mathcal{J}^{-\frac{3}{2}}\right)$, and corresponding eigenvalues are $\mu_{\pm} = \pm \mathcal{J}^{-\frac{3}{2}}$. The Riemann invariants are $r_{\pm} = u \pm 2\mathcal{J}^{-\frac{1}{2}}$, and obey the following equations:

$$\dot{r}_{\pm} + \mu_{\pm} \partial_a r_{\pm} = v. \tag{7.32}$$

Expressions of original variables in terms of $r_{\pm}$ are given by

$$u = \frac{1}{2}(r_+ + r_-), \quad \mathcal{J} = \frac{16}{(r_+ - r_-)^2} > 0, \quad \mu_{\pm} = \pm \left(\frac{r_+ - r_-}{4}\right)^3. \tag{7.33}$$

After some algebra we obtain the following equations for the derivatives of the Riemann invariants $D_{\pm} = \partial_a r_{\pm}$:

$$\dot{D}_{\pm} + \mu_{\pm} \partial_a D_{\pm} + \frac{\partial \mu_{\pm}}{\partial r_+} D_+ D_{\pm} + \frac{\partial \mu_{\pm}}{\partial r_-} D_- D_{\pm} = \partial_a v = q(a) - \mathcal{J}. \tag{7.34}$$

These equations can be rewritten, using the derivatives along the characteristics $\frac{dD}{dt_{\pm}} = \dot{D} + \mu_{\pm} \partial_a D$, as

$$\frac{dD_{\pm}}{dt_{\pm}} + \frac{\partial \mu_{\pm}}{\partial r_+} D_+ D_{\pm} + \frac{\partial \mu_{\pm}}{\partial r_-} D_- D_{\pm} = q(\alpha) - \mathcal{J}. \tag{7.35}$$

Wave breaking and shock formation correspond to the loss of smoothness by the Riemann invariants in finite time, that is to $D_{\pm}$ becoming infinite in finite time. The equations (7.35) are the generalised Riccati equations. Qualitative analysis of such equations gives the following conditions of appearance, or absence of shocks:

1. If initial relative vorticity $q - \mathcal{J} = \partial_a v$ is sufficiently negative, breaking always takes place whatever the initial conditions,
2. If relative vorticity is positive, as well as the derivatives of the Riemann invariants at the initial moment, breaking never takes place.

Thus, rotation, in spite of the fact that it introduces dispersion of inertia–gravity waves at the long-wave end of the spectrum, does not prevent wave breaking in rotating shallow water. Note that non-dispersive Kelvin waves appearing in coastal (with abrupt shelf) and equatorial wave guides are strictly dispersionless and, thus, always break. Examples of Kelvin- wave breaking were already presented in Figures 5.5 and 5.8.

7.1.5 Hydraulic theory applied to rotating shallow water

On the one hand, the onset of wave breaking formally means that the shallow-water model reached the limits of its validity. On the other hand, solution with jump discontinuities are meaningful both from physical and mathematical points of view.

Physically, they represent hydraulic jumps. Mathematically, they represent weak solutions of the equations of motion. Obviously, from the physical point of view shocks cannot be treated correctly without introduction of dissipation, but at least in the one-layer case (see the discussion of the two-layer case following), such solutions can be obtained from the dissipative ones in the limit of vanishing viscosity. The shocks, which are discontinuities in the inviscid solutions, may be considered, in the context of large-scale oceanic and atmospheric flows, as proxies of zones of enhanced dissipation and mixing, appearing at the locations of strong velocity gradients in the flow.

In order to define weak solutions of a system of partial differential equations, these latter should be rewritten in a form of conservation laws. Such a form expresses conservation of mass and momentum (modulo Coriolis force) in the RSW model. The conservative form of the one-layer 1.5D RSW equations is

$$\begin{cases} (hu)_t + (hu^2 + \frac{1}{2}gh^2)_x - fhv = 0, \\ (hv)_t + (huv)_x + fhu = 0, \\ h_t + (hu)_x = 0. \end{cases} \tag{7.36}$$

Weak solutions are defined by requiring conservation of mass and momentum across a discontinuity. This conservation is expressed in the form of Rankine–Hugoniot (RH) jump conditions. The standard discontinuity calculus holds for shallow water in view of its equivalence to two-dimensional gas dynamics, and the RH conditions have the form

$$\begin{cases} -\mathcal{U}\,[hu] + [hu^2 + gh^2/2] = 0, \\ -\mathcal{U}\,[hv] + [huv] = 0, \\ -\mathcal{U}\,[h] + [hu] = 0, \end{cases} \tag{7.37}$$

where $\mathcal{U}$ is the speed of the discontinuity and $[A]$ is the jump of any quantity A across the discontinuity, following a fluid parcel: $[A] = A_{front} - A_{rear}$. These conditions do not depend on the Coriolis parameter f and express, respectively, the conservation of momentum across and along the discontinuity, and mass conservation across the discontinuity. Yet, physically relevant solutions are not fully specified, unless the dissipation of the total energy density e across the discontinuity is ensured:

$$-\mathcal{U}\,[e] + \left[(e + gh^2/2)u\right] \le 0, \quad e = \frac{h}{2}(gh + u^2 + v^2), \tag{7.38}$$

which is an analogue of the gas dynamics entropy condition.

From the RH and entropy conditions it follows that the rate of the energy dissipation in every material volume $V(t)$ which contains a discontinuity depends only on the amplitude of the jump

$$\frac{d}{dt}\int_{V(t)} e\,dx = \frac{g}{4}\,[h]^2\,[u]. \tag{7.39}$$

An important question is that of the influence of wave breaking upon the potential vorticity distribution. Indeed, we have seen that wave–vortex dichotomy is based on Lagrangian conservation of PV. As was mentioned in Chapter 5, this conservation is not valid anymore in the presence of dissipation. The dissipation, implicitly contained in the weak solutions can, thus, affect the PV distribution. However, in the 1.5D reduction of the RSW we are considering, this is not possible. Indeed, if we take the continuity equation $[(\mathcal{U} - u)h] = 0$ in the absence of contact discontinuities, that is for $[v] = 0$, we immediately obtain $(\mathcal{U} - u)h[q] = 0$. Hence $[q] = 0$ in one dimension.

The situation is, however, different for curved shocks in the full two-dimensional model, which were presented in Figures 5.5 and 5.8. The jump of PV across the shock in two dimensions can be calculated as follows. Consider a shock propagating through a material volume $V(t)$ in a local reference frame, with normal and tangential unit vectors $(\mathbf{n}, \mathbf{s})$ with respect to the shock. This frame is moving at the local speed of the shock $\mathbf{n}\mathcal{U}_n$. The velocity and Bernoulli function, an important quantity in this context, in the moving frame are

$$\bar{\mathbf{u}} = \begin{pmatrix} \bar{u}^{(n)} \\ \bar{u}^{(s)} \end{pmatrix}_{loc} = \begin{pmatrix} u^{(n)} - \mathcal{U}_n \\ u^{(s)} \end{pmatrix}_{loc} \qquad \bar{B} = gh + \bar{\mathbf{u}}^2/2. \qquad (7.40)$$

The RH conditions become

$$\begin{cases} -\mathcal{U}_n [h] + [hu^{(n)}] = 0, \\ -\mathcal{U}_n [hu^{(n)}] + [hu^{(n)2} + \frac{1}{2}gh^2] = 0, \\ -\mathcal{U}_n [hu^{(s)}] + [hu^{(n)} u^{(s)}] = 0. \end{cases} \qquad (7.41)$$

The jump in $\bar{B}$ is obtained from the first and the second RH conditions (7.41):

$$[\bar{B}] = -\frac{g}{4} \frac{[h]^3}{h_f h_r}, \qquad (7.42)$$

where the subscripts refer to front and rear, respectively. The PV jump across the shock is obtained from the along-front momentum equation:

$$h\bar{u}^{(n)} [q] = -\partial_s [\bar{B}]. \qquad (7.43)$$

Therefore, for moving shocks of any shape,

$$[q] = \frac{g}{4h\bar{u}^{(n)}} \partial_s \frac{[h]^3}{h_f h_r}. \qquad (7.44)$$

The rate of change of the amount of vorticity contained in any material volume V is

$$\frac{d}{dt} \int_V hq \, dV = -\int_{(S)} h\bar{u} [q] \, dS = [\bar{B}]_{front} - [\bar{B}]_{rear}. \qquad (7.45)$$

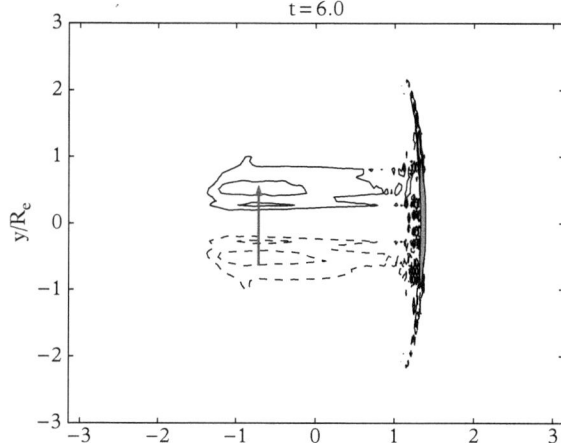

Figure 7.1 *PV production behind a breaking equatorial Kelvin wave with positive thickness anomaly. Positive (negative) values of PV continuous (dashed) contours. Arrow indicates the PV flux due to breaking.*

Figure 7.1 illustrates the PV production and redistribution by a breaking equatorial Kelvin wave, as follows from the simulations with the numerical scheme explained below. Note the formation of a *PV dipole* behind the Kelvin front.

7.1.6 A brief description of finite-volume numerical methods for one-layer RSW

The one-layer RSW model is thus a hyperbolic system. In the absence of rotation and topography it is just a two-dimensional gas dynamics with a specific equation of state. Powerful numerical methods developed in gas dynamics, and in particular finite-volume methods, are therefore transposable into RSW. Rotation, however, represents a stiff source and needs special care, as well as topography. As we have already seen in Chapters 2 and 4 and will see again in Chapter 8, rotation leads to appearance of a spectral gap. In the 1.5-dimensional case treated earlier, slow motion sin geostrophic equilibrium become 'infinitely slow', in other words steady. In order to preserve this fundamental property of the system, numerical schemes should preserve the slowness of the balanced motions (steady states in the 1.5D case), that is they should be 'well-balanced'. Guaranteeing this in the presence of topography and rotation is a nontrivial numerical task. Another requirement in the presence of shocks is that these latter should be dissipative, that is entropy should be increasing across them, and numerical schemes should respect this fundamental property ('entropy-satisfying' schemes). In hydraulics, it is the energy which should decrease across hydraulic jumps. As we have already seen, systems with outcropping are of interest, so the numerical scheme should be able to cope with 'drying' of one of the layers.

We sketch in the following the main ideas of constructing well-balanced entropy-satisfying finite-volume schemes for RSW equations. Such schemes are used in all RSW simulations presented in the book. The purpose of this section is to explain the guidelines leading to reliable numerical schemes allowing us to treat topography, rotation, and drying correctly. The technical details can be found in specialised literature.

One-dimensional shallow water with topography

The fundamental building block for what follows is a numerical scheme for one-layer shallow water with topography in one dimension. The equations of such model in conservative form are

$$\begin{cases} h_t + (hu)_x = 0, \\ (hu)_t + (hu^2 + gh^2/2)_x + hB_x = 0, \end{cases} \tag{7.46}$$

where $B(x)/g$ represents the topography. Due to its presence, the numerical difficulties mentioned earlier arise already in this simplest model. Ensuring $h \geq 0$, maintaining steady states $u = 0$, $gh + B = \text{const}$ ('well-balanced' property), treating of drying $h \to 0$, satisfying a discrete entropy inequality (7.38), where the 'entropy' is the energy in the presence of topography:

$$e = hu^2/2 + gh^2/2 + ghB,$$

with entropy flux $\left(e + gh^2/2\right) u$, are the requirements for the scheme.

A numerical scheme is necessarily discrete in time and space. Discretisation of (7.46) in space is achieved by introducing a regular one-dimensional grid $x_{i+1/2}$, $i = 1, 2, ..., N$ with cells (finite volumes) $(x_{i-1/2}, x_{i+1/2})$ of lengths $\Delta x_i = x_{i+1/2} - x_{i-1/2}$, centred at $x_i = (x_{i-1/2} + x_{i+1/2})/2$. The fields $U = (h, hu)$ and B are discretised over this grid, and at each step n in discrete time the discretised data is (U_i^n, Z_i). The first-order in time scheme approximates the evolution of U as follows (B_i do not evolve):

$$U_i^{n+1} - U_i^n + \frac{\Delta t}{\Delta x_i}(F_{i+1/2-} - F_{i-1/2+}) = 0. \tag{7.47}$$

Here,

$$F_{i+1/2-} = \mathcal{F}_l(U_i, U_{i+1}, \Delta B_{i+1/2}), \quad F_{i+1/2+} = \mathcal{F}_r(U_i, U_{i+1}, \Delta B_{i+1/2}), \tag{7.48}$$

and $\Delta B_{i+1/2} = B_{i+1} - B_i$.

$\mathcal{F}_l$ and $\mathcal{F}_r$ are left and right *numerical fluxes*, which must satisfy two consistency properties:

- consistency with the conservative term:

$$\mathcal{F}_l(U, U, 0) = \mathcal{F}_r(U, U, 0) = F(U) \equiv (hu, hu^2 + gh^2/2), \tag{7.49}$$

- consistency with the source:

$$\mathcal{F}_r(U_l, U_r, \Delta B) - \mathcal{F}_l(U_l, U_r, \Delta B) = (0, -h\Delta B) + o(\Delta B), \tag{7.50}$$

as $U_l, U_r \to U$ and $\Delta B \to 0$.

Required mass conservation and well-balanced properties are expressed, respectively, as

$$\mathcal{F}_l^h(U_l, U_r, \Delta B) = \mathcal{F}_r^h(U_l, U_r, \Delta B) \equiv F^h(U_l, U_r, \Delta B). \tag{7.51}$$

$$F_{i+1/2-} = F(U_i), \ F_{i+1/2+} = F(U_{i+1}), \tag{7.52}$$

whenever $u_i = u_{i+1} = 0$ and $gh_{i+1} - gh_i + \Delta B_{i+1/2} = 0$. A so-called *hydrostatic reconstruction*:

$$\begin{cases} F_l(U_l, U_r, \Delta B) = \mathcal{F}(U_l^*, U_r^*) + \begin{pmatrix} 0 \\ \frac{g}{2}h_l^2 - \frac{g}{2}h_{l*}^2 \end{pmatrix}, \\ \\ F_r(U_l, U_r, \Delta B) = \mathcal{F}(U_l^*, U_r^*) + \begin{pmatrix} 0 \\ \frac{g}{2}h_r^2 - \frac{g}{2}h_{r*}^2 \end{pmatrix}, \end{cases} \tag{7.53}$$

where $U_l^* = (h_{l*}, h_{l*}u_l)$, $U_r^* = (h_{r*}, h_{r*}u_r)$, and

$$h_{l*} = \max(0, h_l - \max(0, \Delta B/g)),$$

$$h_{r*} = \max(0, h_r - \max(0, -\Delta B/g)).$$

satisfies all formulated requirements with any entropy satisfying consistent numerical flux $\mathcal{F}$ of the problem with $B = \text{const}$. Multiple choices for $\mathcal{F}$ are available in the literature (approximate Riemann solvers).

Including rotation

Let us now consider 1.5D shallow water with topography and Coriolis force:

$$\begin{cases} h_t + (hu)_x = 0, \\ (hu)_t + (hu^2 + gh^2/2)_x + hB_x - fhv = 0, \\ (hv)_t + (huv)_x + fhu = 0, \end{cases} \tag{7.54}$$

where $B = B(x)$ and, in general, $f = f(x)$. Stationary solutions are given by $u = 0$, $fv = (gh + B)_x$. A trick, which allows us to circumvent the problem with the stiff Coriolis term, consists in treating it as an *apparent topography*, that is to identify the two first equations in (7.54) as the system (7.46) with a new topography $Z + R$, where $R_x = -fv$. This is always possible in one dimension, in spite of the fact that in two dimensions the Coriolis force is not potential. As v depends on time while R should be time-independent, we take $R_x^n = -fv^n$, and solve (7.46) on the time interval (t_n, t_{n+1}) with topography $B + R^n$. Discretised 1.5D RSW equations are thus obtained:

$$\begin{cases} h_i^{n+1} - h_i^n + \dfrac{\Delta t}{\Delta x_i}(F_{i+1/2}^h - F_{i-1/2}^h) = 0, \\[2ex] h_i^{n+1}u_i^{n+1} - h_i^n u_i^n + \dfrac{\Delta t}{\Delta x_i}(F_{i+1/2-}^{hu} - F_{i-1/2+}^{hu}) = 0, \\[2ex] h_i^{n+1}v_i^{n+1} - h_i^n v_i^n + \dfrac{\Delta t}{\Delta x_i}(F_{i+1/2-}^{hv} - F_{i-1/2+}^{hv}) = 0, \end{cases} \qquad (7.55)$$

with

$$(F_{i+1/2}^h, F_{i+1/2-}^{hu}) = \mathcal{F}_l^{1d}(h_i, u_i, h_{i+1}, u_{i+1}, \Delta b_{i+1/2} + \Delta r_{i+1/2}^n),$$

$$(F_{i+1/2}^h, F_{i+1/2+}^{hu}) = \mathcal{F}_r^{1d}(h_i, u_i, h_{i+1}, u_{i+1}, \Delta b_{i+1/2} + \Delta r_{i+1/2}^n), \qquad (7.56)$$

$$\Delta r_{i+1/2}^n = -f_{i+1/2}\frac{v_i^n + v_{i+1}^n}{2}\Delta x_{i+1/2}/g. \qquad (7.57)$$

and $\mathcal{F}_l^{1d}$ and $\mathcal{F}_r^{1d}$ are some numerical hydrostatic reconstruction fluxes for one-dimensional shallow water. A natural discretisation associated to the conservation of the geostrophic momentum $(h(v + \Omega))_t + (hu(v + \Omega))_x = 0$, where $\Omega_x = f$, which in turn is directly related to the PV conservation, is given by

$$F_{i+1/2-}^{hv} = \begin{cases} F_{i+1/2}^h v_i & \text{if } F_{i+1/2}^h \geq 0, \\[1ex] F_{i+1/2}^h(v_{i+1} + \Delta\Omega_{i+1/2}) & \text{if } F_{i+1/2}^h \leq 0, \end{cases} \qquad (7.58)$$

$$F_{i+1/2+}^{hv} = \begin{cases} F_{i+1/2}^h(v_i - \Delta\Omega_{i+1/2}) & \text{if } F_{i+1/2}^h \geq 0, \\[1ex] F_{i+1/2}^h v_{i+1} & \text{if } F_{i+1/2}^h \leq 0, \end{cases} \qquad (7.59)$$

with

$$\Delta\Omega_{i+1/2} = f_{i+1/2}\Delta x_{i+1/2}. \qquad (7.60)$$

Extension to second order in time and space, and to two spatial dimensions

The first-order in space and time scheme sketched earlier is too (numerically) dissipative and in practice, in order to have good accuracy, second-order schemes are used. We will not dwell on the details, and mention only that in the numerical simulations presented below, the Heun method is used for second-order in time simulations. Second-order accuracy in space needs changing (7.47) to

$$U_i^{n+1} - U_i^n + \frac{\Delta t}{\Delta x_i}(F_{i+1/2-} - F_{i-1/2+} - F_i) = 0, \qquad (7.61)$$

where F_i represents some centred flux. Again, without dwelling on the details, which can be found in the references at the end of this chapter, the choice of the centred flux can be found consistently with the requirements of second-order accuracy, well-balanced property, and hydrostatic reconstruction. Going from one to two spatial directions, we have to deal with the systems of the form

$$\partial_t U + \partial_x (F(U,B)) + \partial_y (G(U,B)) + R_1(U,B)\partial_x B + R_2(U,B)\partial_y Z = 0 \qquad (7.62)$$

schematically represented in Figure 7.2 For any two-dimensional quasilinear system,

$$\partial_t U + A_1(U)\partial_x U + A_2(U)\partial_y U = 0, \qquad (7.63)$$

we can consider planar solutions of the form $U(t,x,y) = U(t,\zeta)$, with $\zeta = xn_1 + yn_2$ where $\mathbf{n} = (n_1, n_2)$ is a unit vector, which leads to

$$\partial_t U + A_n(U)\partial_\zeta U = 0, \qquad (7.64)$$

with

$$A_n(U) = n_1 A_1(U) + n_2 A_2(U). \qquad (7.65)$$

Hence, the notions introduced for one-dimensional systems can be applied to (7.64), and we define hyperbolicity, entropies, etc., in (7.63) by defining them for all directions n. A centred regular two-dimensional mesh should be used in order to apply the finite-volume method. It is made of rectangles with sides: $\Delta x_i = x_{i+1/2} - x_{i-1/2}$, $\Delta y_j = y_{j+1/2} - y_{j-1/2}$, centred at $x_{ij} = (x_i, y_j)$. A solution $U(t,x,y)$ to (7.62) is represented by discrete values U_{ij}^n that are approximations of the mean value of U over the cell ij at time $t_n = n\Delta t$,

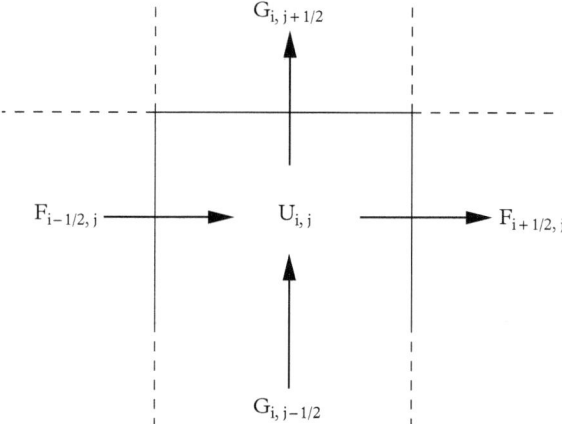

Figure 7.2 *Schematics of the first-order 2D finite-volume scheme.*

$$U_{ij}^n \simeq \frac{1}{\Delta x_i \Delta y_j} \int_{x_{i-1/2}}^{x_{i+1/2}} \int_{y_{j-1/2}}^{y_{j+1/2}} U(t_n, x, y) \, dx dy. \tag{7.66}$$

Finite-volume method for solving (7.62) then is given by

$$U_{ij}^{n+1} - U_{ij}^n + \frac{\Delta t}{\Delta x_i}(F_{i+1/2-,j} - F_{i-1/2+,j}) + \frac{\Delta t}{\Delta y_j}(F_{i,j+1/2-} - F_{i,j-1/2+}) = 0. \tag{7.67}$$

Here, the exchange terms

$$\begin{aligned}
F_{i+1/2\mp,j} &= \mathcal{F}_{l/r}^1(U_{ij}, U_{i+1,j}, B_{ij}, B_{i+1,j}), \\
F_{i,j+1/2\mp} &= \mathcal{F}_{l/r}^2(U_{ij}, U_{i,j+1}, B_{ij}, Z_{i,j+1}),
\end{aligned} \tag{7.68}$$

are defined by numerical fluxes $\mathcal{F}_l^1$, $\mathcal{F}_r^1$, $\mathcal{F}_l^2$, $\mathcal{F}_r^2$, that may be chosen from a variety available in the literature. As in the 1.5D case, this scheme can be rendered second-order in time and space along the same lines.

7.1.7   Illustration: breaking of equatorial waves

Relaxational Riemann solver

In this section, in addition to examples already given, we present another illustration of the capability of the numerical scheme to treat front formation in geophysically relevant configurations. It is obtained by applying the general method explained earlier, where a relaxation scheme is used to calculate numerical fluxes. As we have seen, to calculate these latter it is sufficient to calculate them for the one-dimensional irrotational problem.

$$\begin{cases}
h_t + (hu)_x = 0, \\
(hu)_t + (hu^2 + gh^2/2)_x = 0, \\
(hv)_t + (huv)_x = 0.
\end{cases} \tag{7.69}$$

As was already stated, available Riemann solvers are used for this purpose. The approximate Riemann solver we apply is obtained from the exact solution of the Riemann problem for the relaxation system

$$\begin{cases}
h_t + (hu)_x = 0, \\
(hu)_t + (hu^2 + \pi)_x = 0, \\
(h\pi/\sigma^2)_t + (h\pi u/\sigma^2)_x + u_x = 0, \\
(h\sigma)_t + (hu\sigma)_x = 0,
\end{cases} \tag{7.70}$$

where π is initialised with $\pi_{l(r)} = gh_{l(r)}^2/2$, for the left and right states, respectively, and an auxiliary variable σ is introduced to handle the situations of drying $h \to 0$. The solutions of this Riemann problem are straightforward to compute. It can be shown that such solver satisfies a discrete entropy inequality.

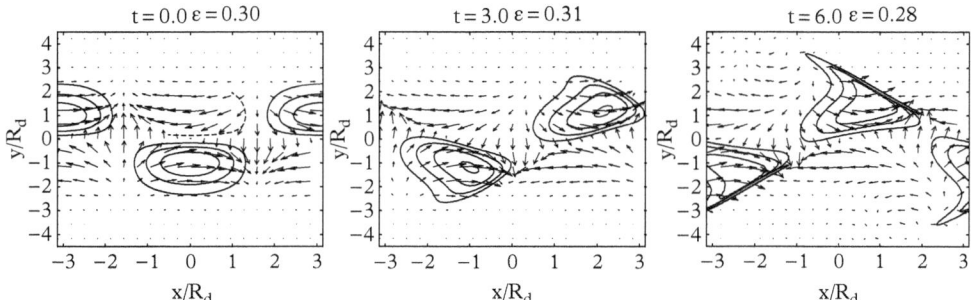

Figure 7.3 *Breaking of a periodic equatorial Yanai wave with non-dimensional amplitude $\epsilon = 0.31$. Zone of enhanced numerical dissipation is indicated in grey. Snapshots at $t = 0, 3, 6\,(\beta R_e)^{-1}$.*

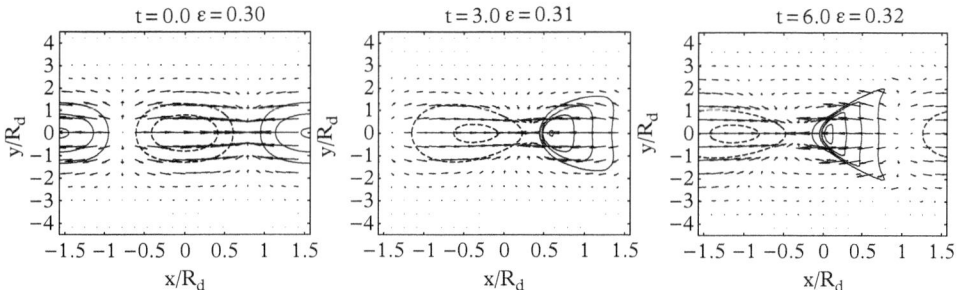

Figure 7.4 *Same as in Fig. 7.3, but for a westward-propagating equatorial inertia–gravity wave with non-dimensional amplitude $\epsilon = 0.3$.*

Wave breaking

As an example we present formation of fronts by equatorial waves, to complement the equatorial Kelvin wave breaking already displayed in Figure 5.8. Figure 7.3 presents Yanai wave breaking and formation of characteristic chevron-shaped fronts. Westward-propagating equatorial inertia–gravity waves also break, forming another characteristic patterns, as shown in Figure 7.4, while eastward-propagating inertia–gravity waves form a symmetric pattern with symmetric, with respect to the equator, chevrons of the type shown in Figure 7.3.

7.2 Two-layer model

7.2.1 1.5 dimensional two-layer RSW

We consider two-layer RSW on the f plane with a constraint of no variation in one of the spatial directions. As we will see, the baroclinic effects due to the difference in motions of the layers lead to considerable complications. As we will be using the same system in Part III in order to introduce the effects of humidity, we will consider it in the atmospheric context, where the role of density is played by potential temperature θ, the layers are inverted, as compared to the oceanic context, in the sense that static stability corresponds to θ increasing with height.

The equations of the atmospheric two-layer 1.5D RSW model are

$$
\begin{cases}
\partial_t u_1 + u_1 \partial_x u_1 - f v_1 + g \partial_x (h_1 + h_2) = 0, \\
\partial_t v_1 + u_1 (f + \partial_x v_1) = 0, \\
\partial_t u_2 + u_2 \partial_x u_2 - f v_2 + g \partial_x (h_1 + \alpha h_2) = 0, \\
\partial_t v_2 + u_2 (f + \partial_x v_2) = 0, \\
\partial_t h_1 + \partial_x (h_1 u_1) = 0, \\
\partial_t h_2 + \partial_x (h_2 u_2) = 0,
\end{cases}
\tag{7.71}
$$

where $u_{1,2}$, $v_{1,2}$ are two components of velocity in the, respectively, lower (1) and upper (2) layers, $h_{1,2}$ are thicknesses of the layers, and $\alpha = \frac{\theta_2}{\theta_1} > 1$ is the stratification parameter. PV in each layer is conserved. The geostrophic equilibria

$$
\begin{cases}
u_1 = 0, & -f v_1 + g \partial_x (h_1 + h_2) = 0, \\
u_2 = 0, & -f v_2 + g \partial_x (h_1 + \alpha h_2) = 0
\end{cases}
\tag{7.72}
$$

are exact stationary solutions of (7.71). The spectrum of small perturbations consists of baroclinic and barotropic inertia–gravity waves propagating in the x direction with standard dispersion relations.

7.2.2   Characteristic equation and loss of hyperbolicity

The system (7.71) can be rewritten in the standard quasilinear form:

$$
\partial_t f + \mathbf{A}(f) \partial_x f = b(f),
\tag{7.73}
$$

where f is a vector with components $(u_1, u_2, v_1, v_2, h_1, h_2)$, $b = (f v_1, -f u_1, f v_2, -f u_2, 0, 0)$, and the entries of the 6×6 matrix $\mathbf{A}$ may be easily recovered from (7.71). For this system to be hyperbolic, the left eigenvalues of the matrix $\mathbf{A}$ should be real and different. In this case they correspond to propagation velocities $c(x, t)$ along the characteristics. The characteristic equation $\det(\mathbf{A} - c\mathbf{I}) = 0$, where $\mathbf{I}$ is the unit matrix is

$$
\left[(u_1 - c)^2 - g h_1 \right] \left[(u_2 - c)^2 - \alpha g h_2 \right] - g^2 h_1 h_2 = 0,
\tag{7.74}
$$

if 'advective' characteristics $c = u_1$ and $c = u_2$ are omitted. This is a fourth-order algebraic equation which may have complex solutions, in which case the system loses hyperbolicity, which was impossible in the one-layer model.

To understand the meaning of the characteristics we linearise (7.74) around a state of rest. We thus obtain an approximation of the characteristics for small perturbations. The solutions are

$$c_{\pm}^2 = g(H_1 + \alpha H_2)\frac{1 \pm \sqrt{\Delta}}{2}, \tag{7.75}$$

where

$$\Delta = 1 - \frac{4H_1 H_2(\alpha - 1)}{(H_1 + \alpha H_2)^2} = \frac{(H_1 - \alpha H_2)^2 + 4H_1 H_2}{(H_1 + \alpha H_2)^2}. \tag{7.76}$$

These solutions are related to the speeds of linear gravity waves in the system. For statically stable stratifications ($\alpha > 1$), it is easy to show that $0 < \Delta < 1$ and $c_{\pm}^2 > 0$, which means that all linear characteristic velocities c_k ($k = 1, \ldots, 4$) are real and different, and thus the linearised system (7.73) is hyperbolic. The slower solutions c_-^2 correspond to baroclinic, and the faster solutions c_+^2 to barotropic waves.

The hyperbolicity of the full nonlinear system is guaranteed if all four solutions c_k ($k = 1, \ldots, 4$) of the characteristic equation (7.74) are real, which is true close to the state of rest, as just shown. To understand what happens in the general case we rewrite the characteristic equation (7.74) in the form

$$(p^2 - 1)(r^2 - 1) = \frac{1}{\alpha}, \tag{7.77}$$

where we defined

$$p = \frac{u_1 - c}{\sqrt{gh_1}} \quad \text{and} \quad r = \frac{u_2 - c}{\sqrt{\alpha gh_2}}. \tag{7.78}$$

Solutions of equation (7.77) can be found as intersections of two curves in the (p, r) plane: one defined by the equation (7.77) and the second being the straight line:

$$r = \sqrt{\frac{h_1}{\alpha h_2}}p - \frac{u^{bc}}{\sqrt{\alpha gh_2}}, \tag{7.79}$$

as follows from (7.78). We define here the baroclinic velocity as $u^{bc} = u_2 - u_1$.

For $p^2 > 1$ and $r^2 > 1$, the solutions of (7.77) are simply

$$r = \pm\sqrt{1 + \frac{1}{\alpha(p^2 - 1)}}, \tag{7.80}$$

and possess straight asymptotes at $p = \pm 1$ and $r = \pm 1$. For $p^2 < 1$ and $r^2 < 1$, the equation (7.77) describes a curve close to the circle $p^2 + r^2 = 1$. The graphical solution

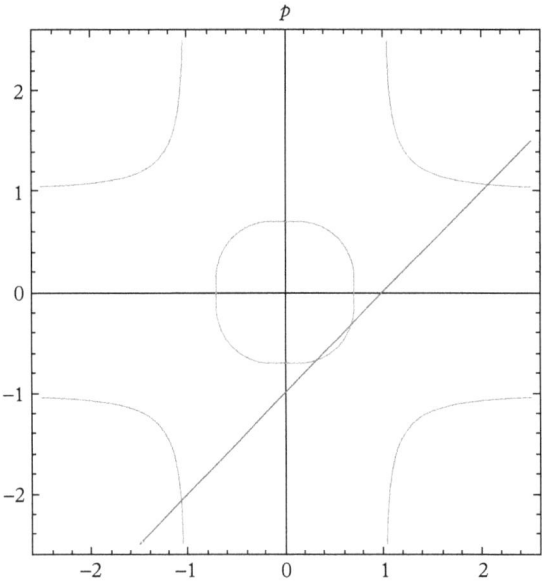

Figure 7.5 *Solutions of the characteristic equation (7.74) obtained as intersections between the curves (7.77) and the straight line (7.79) in the (p, r) plane.* $d = \frac{h_1}{h_2} = 2, \alpha = 2, \text{ and } \frac{u^{bc}}{\sqrt{\alpha g h_1}} = 1.$

of (7.74) is shown in Figure 7.5. A sufficient condition for existence of four intersection points is obtained by a simple geometric argument: the straight line must cross the circle of radius $\sqrt{\frac{\alpha-1}{\alpha}}$. This imposes the following restriction on the baroclinic velocity, that is on the vertical shear:

$$\frac{|u^{bc}|}{\sqrt{g h_1}} \leq \begin{cases} \sqrt{\frac{\alpha-1}{d}} & \text{if } \quad d \leq \alpha, \\ \sqrt{\frac{\alpha-1}{\alpha}} & \text{if } \quad d \geq \alpha, \end{cases} \tag{7.81}$$

where $d = \frac{h_1}{h_2}$. Thus, if the velocity shear between the layers is strong enough, the two-layer RSW loses hyperbolicity. This is an essential difference with its one-layer counterpart, which always remains hyperbolic. The conditions (7.81), in fact coincide with the criterion of the shear instability of Kelvin–Helmholtz (KH) type, which will be analysed in Part II (cf. Problems 7.2, 7.3). Developing KH instability is known to produce roll-up of unstable waves, and formation of KH billows, which are characterised by enhanced mixing and dissipation. Of course, such phenomena can not be captured by essentially two-dimensional RSW models. However, the link with the KH instability provides a useful insight into the physical meaning of the hyperbolicity loss, which is accompanied in simulations by strong numerical dissipation.

7.2.3   Rankine–Hugoniot conditions

The exchange terms appearing in the momentum equations prevent rewriting equations (7.71) in conservative form $\partial_t f_j + \nabla \cdot F_j(f) = d_j(f), j = 1,2$. Only $h_{1,2}$ and the total momentum $u_1 h_1 + u_2 h_2$ are conserved, as already discussed in Chapter 3. The available RH conditions for the 1.5D system (7.71) can be derived by standard rules:

$$
\begin{cases}
-\mathcal{U}[u_1 h_1 + u_2 h_2] + \left[u_1^2 h_1 + u_2^2 h_2 + g\frac{h_1^2}{2} + g\alpha\frac{h_2^2}{2} + gh_1 h_2 \right] = 0, \\[2mm]
-\mathcal{U}[v_1 h_1] + [u_1 v_1 h_1] = 0, \\[2mm]
-\mathcal{U}[v_2 h_2] + [u_2 v_2 h_2] = 0, \\[2mm]
-\mathcal{U}[h_1] + [h_1 u_1] = 0, \\[2mm]
-\mathcal{U}[h_2] + [h_2 u_2] = 0,
\end{cases}
\tag{7.82}
$$

where, as before, $\mathcal{U}$ is the velocity of propagation of the discontinuity and $[\dots]$ is a jump across the discontinuity. As the conditions (7.82) are not complete, with one more condition missing, an additional constraint must be chosen to close the system. This point has been largely discussed in the literature. Since shocks dissipate energy, the additional constraint is usually chosen to represent the energy loss in one or another layer, but no universal rule exists and the closure problem of the RH conditions in a two-layer shallow-water model remains open.

7.2.4 A finite-volume numerical method for two-layer RSW

The possible loss of hyperbolicity and non-closedness of Rankine–Hugoniot conditions represent a serious obstacle for constructing consistent numerical schemes for two- (or more) layer RSW equations. We sketch below a numerical scheme which is used for the numerical illustrations in this book and which correctly treats the hyperbolicity loss, in a sense that the corresponding zones (if any) of the computational domain remain localised, do not grow in size in time, and correspond to enhanced mixing and numerical dissipation, as is expected for the Kelvin–Helmholtz instability. As in the one-layer case we start with the multi-layer one-dimensional shallow-water system with free surface and topography in the conservative form:

$$
\partial_t h_j + \partial_x \left(h_j u_j \right) = 0,
\tag{7.83}
$$

$$
\partial_t \left(h_j u_j \right) + \partial_x \left(h_j u_j^2 + gh_j^2/2 \right) + gh_j \left(b + \sum_{k>j} h_k + \sum_{k<j} \frac{\rho_k}{\rho_j} h_k \right) = 0,
\tag{7.84}
$$

where $h_j \geq 0$, $j = 1, 3, \dots N$ are the layer depths, u_j the layer velocities, $b(x)$ topography, and

$$
0 < \rho_1 \leq \dots \leq \rho_N
$$

are the layer densities. We should stress that imposing a rigid lid leads to the constraint of non-divergent barotropic velocity, that is effective incompressibility of the barotropic component, and hence infinite characteristic velocities, which prevent the application of finite-volume methods. This situation will be not considered.

The convex entropy of the system is, again, the energy. A crucial observation is that the system (7.83), (7.84) may be thought of as a stack of N equivalent one-layer shallow-water systems for $U^j = (h_j, h_j u_j)$, each with its *effective topography*:

$$z^j = z + \sum_{k>j} h_k + \sum_{k<j} \frac{\rho_k}{\rho_j} h_k. \tag{7.85}$$

A finite-volume scheme with numerical fluxes $\mathcal{F}_{l/r}$

$$U_i^{j,n+1} - U_i^j + \frac{\Delta t}{\Delta x_i} \left(\mathcal{F}_l(U_i^j, U_{i+1}^j, z_i^j, z_{i+1}^j) - \mathcal{F}_r(U_{i-1}^j, U_i^j, z_{i-1}^j, z_i^j) \right) = 0. \tag{7.86}$$

can be then applied to each shallow-water layer j with its effective topography, and well-balancing, hydrostatic reconstruction, and apparent topography method go through along the lines of Section 7.1.6. The N layers are coupled, but the equations (7.86) can be solved simultaneously using a so-called sum method. The numerical scheme is consistent with Rankine–Hugoniot conditions layer-wise:

$$\left[h_j u_j^2 + \frac{g h_j^2}{2} \right] + g \langle h_j \rangle \left[z^j \right] - U \left[h_j u_j \right] = 0, \tag{7.87}$$

where $\langle \ldots \rangle$ denote the half-sum of the values of the corresponding quantity on the left and on the right of the discontinuity.

As compared to one-layer schemes, this scheme can not be rigorously justified because of possible hyperbolicity loss. It should be therefore considered as heuristic. Yet, it was checked to give results coinciding with those of the one-layer scheme, which *is* rigorously justified, in the large-depth and/or large density of the lower layer in the two-layer model. The scheme gives physically correct results, as seen in the examples following.

7.3 Summary, comments, and bibliographic remarks

The main purpose of this chapter was to address wave breaking and shock formation inherent in shallow-water models. As usual, restricting the number of spatial dimensions leads to considerable simplifications and allows us to understand these processes by relatively simple technical means. However, in the presence of rotation a pure one-dimensional reduction is not possible, as the Coriolis force couples two horizontal components of velocity, whence the use of 1.5-dimensional model where the velocity component in the second spatial direction is present, although all dependence on the corresponding coordinate is dropped off. While it is well-known that non-rotating shallow water is equivalent to two-dimensional barotropic gas dynamics and, therefore, develops shocks, the question of the influence of rotation upon this process was long-standing. As we showed earlier, rotation does not prevent breaking. The passage to Lagrangian description was crucial in obtaining this result. Hydraulic theory therefore can be applied to RSW similarly to non-rotating shallow water. As in gas dynamics, the standard Rankine–Hugoniot conditions for jumps of dynamical variables allow us to completely quantify shocks (hydraulic jumps) in the one-layer RSW. The gas dynamics condition of entropy increase across the shock is replaced by energy decrease condition. It is important that production by shocks of uttermost important in GFD quantity, the PV, can be calculated from the hydraulic theory. As a particular case of gas dynamics, one-layer RSW is a quasilinear hyperbolic system of partial differential equations. Mathematical theory of such systems, which are the systems of conservation laws, is well developed. Shocks are their weak solutions, and are defined, precisely, by jump conditions. Efficient finite-volume numerical methods exist for such systems, and are therefore transposable to one-layer RSW, provided a satisfactory way to treat the Coriolis term, which is a stiff source, is found. This is achieved by treating this term as apparent topography and applying efficient, well-balanced, consistent numerical schemes developed in hydraulics for inclusion of topographic effects. It should be stressed that in the domain of the present book, that is for applications to large-scale atmospheric and oceanic motions, shocks should not be interpreted literally, but as sharp fronts where dissipation and mixing are mostly concentrated. They appear as such, and not as genuine discontinuities, in the numerical simulations.

The two-layer RSW, which is unavoidable if baroclinic effects are to be included, has two inherent difficulties in this context. The first is that it can change type and become elliptic for certain values of dynamical variables, which is independent of rotation. The second is that even in the domain of hyperbolicity, the standard Rankine–Hugoniot conditions are insufficient in number to fix completely the jumps of all variables across a shock, so weak solutions are not uniquely defined. As we explained, the hyperbolicity loss is directly related to shear (Kelvin–Helmoltz type) instability in the model, which arises when a difference between upper- and lower-layer velocities becomes large. Hence, the hyperbolicity loss reflects a physical reality, and is not a mathematical artefact. Nevertheless, it represents a serious difficulty in numerical modelling. While applying hydraulic theory to the two-layer model, the missing Rankine–Hugoniot condition is to be chosen according to physical nature of the involved processes in each of the layers, which are not resolved in the model, for example, preferential dissipation in one of the layers. In spite of these difficulties, a consistent finite-volume numerical scheme could be developed for two-layer RSW. By construction, it does not have any built-in asymmetry in (numerical) dissipation.

The presentation of 1.5-dimensional RSW model from Eulerian and Lagrangian viewpoints followed Zeitlin, Medvedev, and Plougonven (2003). Material on quasilinear systems and interpretation of non-rotating shallow water as such system is standard (e.g. Whitham 1974). Comprehensive (and not too formal) presentation of the theory of quasilinear systems can be found in Rozhdenstvenskii and Janenko (1983). Analysis of shock formation in terms of derivatives of Riemann invariants is due to Lax (1973). The proof of wave breaking in the presence of rotation was given in Zeitlin, Medvedev, and

Plougonven (2003). Hydraulic theory, as applied to shallow water, is exposed in numerous articles and books. We follow Lighthill (1978), Bouchut, LeSommer, and Zeitlin (2004), and LeSommer, Reznik, and Zeitlin (2004) in presenting it above. The description of finite-volume numerical schemes closely follows (Bouchut 2007). The idea to treat rotation and apparent topography is due to F. Bouchut. The relaxational solver was proposed in Bouchut (2004), and was applied to equatorial wave breaking in Bouchut, LeSommer, and Zeitlin (2005). Analysis of hyperbolicity loss in the two-layer RSW follows the approach of Ovsyannikov (1979) as presented in Lambaerts *et al.* (2011). A demonstration of hyperbolicity loss in two-layer model with a rigid lid, and its link to KH instability was given in LeSommer *et al.* (2003). Discussion of different choices of Rankine–Hugoniot conditions in two-layer shallow water can be found in, for example, Klemp, Rotunno, and Skamarock (1997). The multi-layer generalisation of the numerical scheme for RSW follows Bouchut and Zeitlin (2011).

Figures 7.1, 7.3, and 7.4 are reproduced from Bouchut, LeSommer, and Zeitlin (2005), and Figure 7.5 from Lambaerts *et al.* (2011), with the permission of AIP publishing.

7.4 Problems

7.1 Demonstrate that shocks in one-layer RSW are always compressive, i.e. $[h] < 0$ across a shock.

7.2 Demonstrate equation (7.33).

7.3 Obtain evolution equation for PV in 1.5D RSW in the presence of Newtonian viscosity.

7.4 Consider one- dimensional two-layer RSW model with a rigid lid in the absence of rotation, with layers of densities $\rho_{1,2}$ and thicknesses $h_{1,2}$. Demonstrate that if the velocity shear verifies the relation

$$(u_2 - u_1)^2 > g((\rho_2 - \rho_1) \left(\frac{h_1}{\rho_1} + \frac{h_2}{\rho_2} \right), \tag{7.88}$$

the flow is linearly unstable (KH instability). How this result changes in the presence of rotation in 1.5D model?

7.5 In the equations of 1.5D two-layer model of the previous problem, eliminate pressures in the layers and reduce the system to four equations for variables u_2, h_2, v_2, v_1. Show that this system is quasilinear, find conditions of hyperbolicity, and show that they are equivalent to the KH instability criterion (7.88).

Part II

Understanding Fundamental Dynamical Phenomena with Rotating Shallow-Water Models

8

Geostrophic Adjustment and Wave–Vortex (Non)Interaction

In this chapter we will consider probably the most typical and most important process in geophysical fluid dynamics: the process of geostrophic adjustment. It consists in adaptation of geophysical fields to the state of geostrophic equilibrium by emission of inertia–gravity waves. On the one hand, analysing this process will allow us to justify the quasi-geostrophic models for slow motions which we obtained in Chapter 5 by direct filtering of inertia–gravity waves. On the other hand, it will allow us to understand an important source of inertia–gravity wave emissions in the atmosphere and oceans. We will also get an idea of how to solve the *initialisation* problem, that is to project the initial data in the models onto slow motions, the most important for predicting the long-time behaviour of the system, in order to avoid 'pollution' by fast, less important but consuming much of computational resources, modes. To give an idea of the technique, the main steps of the calculations will be presented in a rather detailed way in the simplest one-layer case in the infinite f plane. In other cases they will be briefly sketched and only the results will be presented, the details relegated to the cited literature.

8.1 Geostrophic adjustment in the barotropic (one-layer) model

8.1.1 Quasi-geostrophic regime

Scaling and formulation of the problem

We start the analysis of the geostrophic adjustment in the simplest case of (quasi-) barotropic one-layer RSW model. We will be working with an infinite domain in the f plane, under the following assumptions characterising the QG regime of Chapter 5:

- Unique horizontal spatial and velocity scales L, V for all motions,
- Small Rossby number ϵ, and small free-surface deviations: $\lambda \sim \epsilon$,
- Simultaneous presence of fast $t \sim f_0^{-1}$ and slow $t_1 \sim (\epsilon f_0)^{-1}$ time scales.

Geophysical Fluid Dynamics. Vladimir Zeitlin,
Oxford University Press (2018). © Vladimir Zeitlin. DOI: 10.1093/oso/9780198804338.001.0001

Under these assumptions the non-dimensional equations of one-layer RSW become

$$(\partial_t + \epsilon \partial_{t_1})\mathbf{v} + \epsilon(\mathbf{v} \cdot \nabla \mathbf{v}) + \hat{\mathbf{z}} \wedge \mathbf{v} + \nabla \eta = 0, \tag{8.1}$$

$$(\partial_t + \epsilon \partial_{t_1})\eta + (1 + \epsilon \eta)\nabla \cdot \mathbf{v} + \epsilon \mathbf{v} \cdot \nabla \eta = 0, \tag{8.2}$$

where η is the non-dimensional deviation of the free surface, and $1 + \epsilon \eta$ is the non-dimensional thickness of the fluid layer. The non-dimensional PV anomaly is $Q = \epsilon \frac{\zeta - \eta}{1 + \epsilon \eta}$, and its conservation is expressed as

$$\partial_t Q + \epsilon \mathbf{v} \cdot \nabla Q = 0. \tag{8.3}$$

Mathematically speaking, the problem of the geostrophic adjustment is a Cauchy problem with *localised* initial conditions:

$$u|_{t=0} = u_I \,, \; v|_{t=0} = v_I \,, \; \eta|_{t=0} = \eta_I, \tag{8.4}$$

where the typical scale (a single one, by hypothesis) of the initial perturbation is L. To solve this problem we will be using a two-time asymptotic analysis and will expand all variables in asymptotic series in the small parameter, the Rossby number:

$$\mathbf{v} = \mathbf{v}_0(x, y; t, t_1, ...) + \epsilon \mathbf{v}_1(x, y; t, t_1, ...) + ...$$

$$\eta = \eta_0(x, y; t, t_1, ...) + \epsilon \eta_1(x, y; t, t_1, ...) + ...$$

An important observation is that any variable can be decomposed in a unique way in to slow and fast parts order by order in ϵ:

$$\eta_i = \bar{\eta}_i(x, y; t_1, ...) + \tilde{\eta}_i(x, y; t, t_1, ...), \quad i = 0, 1, 2, ..., \tag{8.5}$$

and similarly for $\mathbf{v}$. Here,

$$\bar{\eta}_i(x, y; t_1, ...) = \lim_{T \to \infty} \frac{1}{T} \int_0^T \eta_i(x, y, t, t_1, ...)\, dt, \tag{8.6}$$

$$\tilde{\eta}_i(x, y; t, t_1, ...) = \eta_i(x, y, t, t_1, ...) - \bar{\eta}_i(x, y; t_1, ...). \tag{8.7}$$

In the following we will be systematically applying fast–slow decomposition at each order of the asymptotic expansion. As will become clear, it is convenient to use momentum equations and the PV equation (8.3), instead of Eq. (8.2) for η. By definition, fast-time averages of fast variables, as well as averages of the products of fast and slow ones, are identically zero, while fast-time averages of slow variables give these variables themselves.

Lowest-order approximation

In this approximation we have

$$\partial_t \mathbf{v}_0 + \hat{\mathbf{z}} \wedge \mathbf{v}_0 = -\nabla \eta_0 , \tag{8.8}$$

$$\partial_t (\zeta_0 - \eta_0) = 0 , \tag{8.9}$$

where $\zeta_0 = \hat{\mathbf{z}} \cdot \nabla \wedge \mathbf{v}_0$ is zero-order relative vorticity. From (8.4) we get

$$u_0|_{t=0} = u_I , \; v_0|_{t=0} = v_I , \; \eta_0|_{t=0} = \eta_I. \tag{8.10}$$

Rewriting (8.8) in terms of relative vorticity ζ and divergence $D_0 = \nabla \cdot \mathbf{v}_0$ gives

$$\partial_t \zeta_0 + D_0 = 0 , \tag{8.11}$$

$$\partial_t D_0 - \zeta_0 = -\nabla^2 \eta_0 . \tag{8.12}$$

Eq. (8.9) can be immediately integrated in fast time t:

$$\zeta_0 - \eta_0 = \Pi_0 , \tag{8.13}$$

where Π_0 is some yet unknown function of x, y, t_1 (an integration 'constant'). Elimination of ζ_0 and D_0 in favour of η_0 leads to the linear inhomogeneous equation

$$-\frac{\partial^2 \eta_0}{\partial t^2} - \eta_0 + \nabla^2 \eta_0 = \Pi_0(x, y; t_1, t_2, ...) . \tag{8.14}$$

Its slow and fast components obey the following equations, obtained by fast-slow decomposition and fast-time averaging of (8.14):

$$-\frac{\partial^2 \bar{\eta}_0}{\partial t^2} - \bar{\eta}_0 + \nabla^2 \bar{\eta}_0 = 0, \tag{8.15}$$

$$- \bar{\eta}_0 + \nabla^2 \bar{\eta}_0 = \Pi_0 . \tag{8.16}$$

They are, respectively, the Klein–Gordon and Helmholtz equations. It should be emphasised that Π_0 is thus the *geostrophic PV* constructed from the slow component $\bar{\eta}_0$. In order to integrate equations (8.15) and (8.16) we need to solve the *initialisation problem*, in other words, to answer the question how to divide the initial conditions into slow and fast parts. The response, which is unique at $\epsilon \to 0$, is given by the following procedure. By definition,

$$\Pi_0(x, y; 0) = \partial_x v_I - \partial_y u_I - \eta_I \equiv \Pi_I(x, y). \tag{8.17}$$

This allows for determination of the initial value $\bar{\eta}_{0I}$ of $\bar{\eta}_0$ by *inversion*:

$$- \bar{\eta}_{0I} + \nabla^2 \bar{\eta}_{0I} = \Pi_I, \; \Rightarrow \bar{\eta}_{0I} = (\nabla^2 - 1)^{-1} \Pi_I, \tag{8.18}$$

where the inverse operator on the right-hand side is given by the Green's function already considered in Chapters 4 and 5. As a consequence, the initial value $\tilde{\eta}_{0I}$ of $\tilde{\eta}_0$, by definition, is

$$\tilde{\eta}_{0I} = \eta_I - \bar{\eta}_{0I}. \tag{8.19}$$

The second initial condition for $\tilde{\eta}_0$, which is necessary to solve the second-order-in-time equation (8.15), follows from a combination of PV and ζ - D equations:

$$\partial_t \tilde{\eta}_0|_{t=0} = -D_I \equiv \partial_x u_I + \partial_y v_I . \tag{8.20}$$

The Klein–Gordon equation for the fast component is just the wave equation for linear inertia–gravity waves, as is easy to see. The initialisation problem for η_0 is, thus, solved. We then use analogous decomposition for $\mathbf{v}$:

$$\mathbf{v}_0 = \tilde{\mathbf{v}}_0(x,y;t,\dots) + \bar{\mathbf{v}}_0(x,y;t_1,\dots). \tag{8.21}$$

The slow components verify the geostrophic balance relation

$$\bar{\mathbf{v}}_0 = \hat{\mathbf{z}} \wedge \nabla \bar{\eta}_0, \tag{8.22}$$

and the fast ones obey the equations

$$\partial_t \tilde{\mathbf{v}}_0 + \hat{\mathbf{z}} \wedge \tilde{\mathbf{v}}_0 = -\nabla \tilde{\eta}_0 \tag{8.23}$$

with initial conditions

$$\tilde{u}_{0I} = u_I - \bar{u}_{0I} ; \quad \tilde{v}_{0I} = v_I - \bar{v}_{0I}, \tag{8.24}$$

where $\bar{u}_{0I}, \bar{v}_{0I}, \bar{\eta}_{0I}$ verify (8.22). It is to be emphasised that the linearised PV $\tilde{\zeta}_0 - \tilde{\eta}_0$ of the fast component is identically zero, which confirms that the fast component of motion corresponds to inertia–gravity waves, cf. Problem 3.8. The solution of the wave equation, thus, represents inertia–gravity waves propagating out of the initial perturbation; they are created by its *unbalanced* part $\tilde{u}_I^{(0)}, \tilde{v}_I^{(0)}, \tilde{\eta}_{0I}$:

$$\tilde{\eta}_0(\mathbf{x};t) = \sum_{\pm} \int d\mathbf{k}\, H_0^{(\pm)}(\mathbf{k}) e^{i(\mathbf{k}\cdot\mathbf{x}\pm\Omega_k t)}. \tag{8.25}$$

Here, $\Omega_k = \sqrt{k^2 + 1}$ is the non-dimensional frequency of inertia–gravity waves, and

$$H_0^{(\pm)}(\mathbf{k}) = \frac{1}{2}\left(\hat{\tilde{\eta}}_{0I}(\mathbf{k}) \pm i\frac{\hat{D}_I(\mathbf{k})}{\Omega_k}\right), \tag{8.26}$$

and the hat notation is used for the Fourier transformations of the corresponding quantities. Complex velocity $\mathcal{U} = u + iv$ is convenient for representing the solution for $\tilde{v}$. The wave equation for $\mathcal{U}$

$$-\frac{\partial^2 \tilde{\mathcal{U}}_0}{\partial t^2} - \tilde{\mathcal{U}}_0 + \nabla^2 \tilde{\mathcal{U}}_0 = 0 \tag{8.27}$$

is to be solved with initial conditions

$$\tilde{\mathcal{U}}_0\big|_{t=0} = \tilde{u}_I^{(0)} + i\,\tilde{v}_I^{(0)} \equiv \tilde{\mathcal{U}}_{0I}, \tag{8.28}$$

$$\partial_t \tilde{\mathcal{U}}_0\big|_{t=0} \equiv \mathcal{W}_I = -i\tilde{\mathcal{U}}_{0I} - \left(\partial_x \tilde{\eta}_{0I} + i\,\partial_y \tilde{\eta}_{0I}\right), \tag{8.29}$$

and the solution of this initial-value problem is given by

$$\tilde{\mathcal{U}}_0(\mathbf{x}; t) = \sum_{\pm} \int d\mathbf{k}\, U_0^{(\pm)}(\mathbf{k})\, e^{i(\mathbf{k}\cdot\mathbf{x}\pm\Omega_{\mathbf{k}}t)}, \tag{8.30}$$

with

$$U_0^{(\pm)}(\mathbf{k}) = \frac{1}{2}\left(\hat{\tilde{\mathcal{U}}}_{0I}(\mathbf{k}) \pm i\,\frac{\hat{\mathcal{W}}_I(\mathbf{k})}{\Omega_{\mathbf{k}}}\right). \tag{8.31}$$

It is important to emphasise that due to the fact that initial conditions are localised, inertia–gravity waves produced by the adjustment process are radiated to infinity, if the domain of the flow is unbounded. Therefore, by standard stationary-phase argument, at a given point $\mathbf{x} = (x, y)$ the inertia–gravity wave signal decays in time:

$$\tilde{\eta}_0(\mathbf{x}; t) = \mathcal{O}\left(\frac{1}{t}\right), \quad \tilde{\mathcal{U}}_0(\mathbf{x}; t) = \mathcal{O}\left(\frac{1}{t}\right), \quad t \to \infty. \tag{8.32}$$

Thus, the problem of geostrophic adjustment is solved to the lowest order, with the following main results:

- Slow and fast components are unambiguously defined,
- Fast and slow motions are separated dynamically and non-interacting,
- Fast part is completely resolved, and represents an inertia–gravity wave packet propagating out of the initial perturbation, and
- Evolution of the slow part, which bears all of the initial PV anomaly, is still to be determined.

First-order approximation

PV anomaly equation to the first order is

$$\partial_t \left(\zeta_1 - \eta_1 \right) - \Pi_0 \, \partial_t \tilde{\eta}_0 + \tilde{u}_0 \partial_x \Pi_0 + \tilde{v}_0 \partial_y \Pi_0 = - \, \partial_{t_1} \Pi_0 - \mathcal{J}(\bar{\eta}_0, \Pi_0). \qquad (8.33)$$

The first term with the fast-time derivative on the left-hand side is dangerous and, in principle, gives rise to a secular growth, linearly in time. The perturbation theory should be self-consistent, that is first-order corrections should be bounded in time. A consistency (integrability) condition is obtained by time-averaging (8.33) over t. Upon averaging, all terms containing fast contributions vanish, by boundedness condition, or by the definition of averages of products of fast and slow variables. By definition, the first term vanishes, too. Remarkably, there are no fast–fast terms in the equation. We are thus left with the equation of purely slow motion:

$$\partial_{t_1} \Pi_0 + \mathcal{J}(\bar{\eta}_0, \Pi_0) \equiv \partial_{t_1} (\nabla^2 \bar{\eta}_0 - \bar{\eta}_0) + \mathcal{J}(\bar{\eta}_0, \nabla^2 \bar{\eta}_0) = 0. \qquad (8.34)$$

This is nothing else than the QG equation of Chapter 5 written in the f-plane approximation. It arises here from elimination of resonances in the equation for the fast component to order 1. No fast motion drag appears to this order in the evolution of slow motion, and in this way we justify the direct filtering of inertia–gravity waves which has lead to the QG equation previously.

Momentum equations to this order give

$$\partial_t \mathbf{v}_1 + \hat{\mathbf{z}} \wedge \mathbf{v}_1 = -\nabla \eta_1 - \left(\partial_{t_1} + \mathbf{v}_0 \cdot \nabla \right) \mathbf{v}_0. \qquad (8.35)$$

Rewritten in vorticity–divergence variables, these equations result in an inhomogeneous wave equation for thickness and/or velocity with source terms. These forcing terms are defined by both slow and fast zero-order variables, and are due to (1) slow evolution of the zero-order wave field, (2) nonlinear interactions of zero-order fast waves, and (3) interaction of zero-order fast and slow motions. The solution of this equation gives a correction to the zero-order wave field.

Hence, to this order slow motions are governed by the standard QG equation, and are decoupled from the fast ones. Corrections to the zero-order wave field are produced by slow–slow and slow–fast interactions. Thus, the fast field in this approximation does 'feel' the slow one, but does not influence the evolution of the latter.

Brief resumé of second-order corrections

The next-order approximation goes along the same lines, and results in the QG equation for slow motions 'improved' by next-order corrections. These corrections are entirely defined by the previous-order slow fields and are not influenced by the fast motions. The technical reason for this fact is that with localised initial conditions the fast fields at any given point are decaying, cf. (8.32), and the same is true for their bilinear combinations, which leads to vanishing of their contributions after time-averaging. The 'improved'

QG equation is formulated in terms of the synthetic thickness field which comprises the first-order correction $\bar{\eta} = \bar{\eta}_0 + \epsilon\bar{\eta}_1$:

$$\frac{D}{Dt_1}\left[\nabla^2\bar{\eta} - \bar{\eta} - \epsilon\bar{\eta}\left(\nabla^2\bar{\eta} - \bar{\eta}\right) - \epsilon\nabla\bar{\eta}\cdot\nabla\left(\nabla^2\bar{\eta} - \bar{\eta}\right) - 2\epsilon\mathcal{J}\left(\partial_x\bar{\eta}, \partial_y\bar{\eta}\right)\right] = 0. \qquad (8.36)$$

Here

$$\frac{D}{Dt_1}(...) := \partial_{t_1}(...) + \mathcal{J}\left(\bar{\eta} - \epsilon\frac{(\nabla\bar{\eta})^2}{2}, ...\right). \qquad (8.37)$$

This is still an equation of Lagrangian conservation of PV with first-order corrections both in the advection velocity and in PV. An important conclusion can be drawn from this result. As we have seen, the solution of the geostrophic adjustment problem proves that the standard QG equation is fully consistent. At the same time, it establishes the limits of validity of the standard QG model, which is valid for times of the order $(\epsilon f_0)^{-1}$, while the improved QG equations are valid for much longer times $\mathcal{O}\left((\epsilon f_0)^{-2}\right)$.

The fast field in this approximation acquires a correction obeying an inhomogeneous wave equation with the right-hand side defined by fast and slow fields determined in previous orders.

8.1.2 Frontal geostrophic regime

Equations of motion in non-dimensional variables

In the FG regime the scaling, with a unique spatial and a unique velocity scale, remains the same as in the QG regime above, apart from the nonlinearity parameter λ which is now of the order 1. It thus does not make sense to separate the thickness field into unperturbed part and a perturbation. Hence, we will use the full non-dimensional thickness h in the following. The non-dimensional equations of motion take the following form:

$$\begin{cases} (\partial_t + \epsilon\partial_{t_1})\mathbf{v} + \epsilon(\mathbf{v}\cdot\nabla\mathbf{v}) + \hat{\mathbf{z}}\wedge\mathbf{v} + \nabla h = 0, \\ (\partial_t + \epsilon\partial_{t_1})h + \epsilon h\nabla\cdot\mathbf{v} + \epsilon\mathbf{v}\cdot\nabla h = 0, \end{cases} \qquad (8.38)$$

and we proceed with solution of the Cauchy problem with localised initial conditions (8.4) order by order in ϵ.

Zero-order approximation

As stated, in the FG regime the relative surface elevation is of order 1. So we take $\lambda = 1$, without loss of generality, and consider h to be the total depth of the layer. To the lowest order we thus have

$$\begin{cases} \partial_t\mathbf{v}_0 + \hat{\mathbf{z}}\wedge\mathbf{v}_0 = -\nabla h_0, \\ \partial_t h_0 = 0, \end{cases} \qquad (8.39)$$

with the same initial conditions (8.4) as before. Solutions to (8.39) can be readily decomposed into fast and slow components, with the slow one verifying the geostrophic balance (8.22), and the fast one verifying the homogeneous equation (8.39). The entire thickness variable is slow. The initial conditions for both components are

$$\bar{h}_0\big|_{t=0} = h_I \,, \quad \bar{\mathbf{v}}\big|_{t=0} = \mathbf{v}_I^{(g)} \,, \quad \tilde{\mathbf{v}}\big|_{t=0} = \mathbf{v}_I^{(a)} \,, \tag{8.40}$$

where we introduced geostrophic $\mathbf{v}_I^{(g)}$ and ageostrophic $\mathbf{v}_I^{(a)}$ components of the initial velocity with respect to the initial free-surface elevation:

$$\mathbf{v}_I^{(g)} = \hat{\mathbf{z}} \wedge \nabla h_I \,, \quad \mathbf{v}_I^{(a)} = \mathbf{v}_I - \mathbf{v}_I^{(g)} \,. \tag{8.41}$$

Solutions of the homogeneous version of (8.39)

$$\partial_t \mathbf{v}_0 + \hat{\mathbf{z}} \wedge \mathbf{v}_0 = 0 \tag{8.42}$$

are non-propagative inertial oscillations arising in the limit of vanishing wave numbers in the dispersion relation for inertia–gravity waves. In terms of complex velocity $\mathcal{U}$ the solution for the fast velocity field is written as

$$\tilde{\mathcal{U}}_0(\mathbf{x}; t) = \mathcal{A}_0(\mathbf{x}; t_1, \ldots)e^{-it} \,, \quad \mathcal{A}_0(\mathbf{x}; t_1, \ldots)\big|_{t=0} = \mathcal{U}_I^{(a)} \,, \tag{8.43}$$

where $\mathcal{A}_0$ is the complex amplitude of the inertial oscillations, which is, generally, modulated in space and slow time. It turns out that introducing complex notation not only for dependent but also for independent variables is technically advantageous. We thus introduce, as in Chapter 6, the variables $\xi = x + iy$, $\xi^* = x - iy$, with

$$\partial_\xi = \frac{1}{2}\left(\partial_x - i\partial_y\right), \ \partial_{\xi^*} = \frac{1}{2}\left(\partial_x + i\partial_y\right), \ \nabla^2 = 4\partial^2_{\xi\xi^*}, \ \mathcal{J}(\xi, \xi^*) = -2i \,, \tag{8.44}$$

and

$$\mathbf{v}_i \cdot \nabla = \mathcal{U}_i \partial_\xi + \mathcal{U}_i^* \partial_{\xi^*} \,, \quad \nabla \cdot \mathbf{v}_i = \partial_\xi \mathcal{U}_i + \partial_{\xi^*} \mathcal{U}_i^* \,. \tag{8.45}$$

Thus, to the leading order of the perturbation theory in the FD regime, the fast and the slow components of motion are non-interacting. Fast motions are inertial oscillations resulting from the unbalanced part of the initial conditions. Slow motions result from the geostrophically balanced part of initial conditions, and their evolution remains undetermined at this stage.

First-order approximation

For the velocity field we get the same equations (8.35) as in the QG case. The evolution of the thickness field is given by

$$\partial_t h_1 = -\partial_{t_1} \bar{h}_0 - \nabla \cdot \left(\bar{h}_0 \mathbf{v}_0\right) \,. \tag{8.46}$$

Only the unbalanced part of velocity gives rise to the right-hand side, as the balanced part is orthogonal to the gradient of $\bar{h}_0$ and has zero divergence. The first term on the right-hand side is t-independent and leads to a secular linear growth in time of h_1 and hence should vanish, which means that h_0 evolves in t_2, and not in t_1. This scaling being established, equation (8.46) is readily integrated and gives, in complex notation,

$$h_1 = \tilde{h}_1 + \bar{h}_1 = -i\,\partial_\xi\left(\bar{h}_0\,\tilde{\mathcal{U}}_0\right) + i\,\partial_{\xi*}\left(\bar{h}_0\,\tilde{\mathcal{U}}_0^*\right) + \bar{h}_1. \tag{8.47}$$

Injecting this result into the first equation in (8.38), we get the following equation for complex velocity $\mathcal{U}_1$:

$$\partial_t\mathcal{U}_1 + i\mathcal{U}_1 = -2i\partial_{\xi*}\left[-\partial_\xi\left(\bar{h}_0\,\tilde{\mathcal{U}}_0\right) + \partial_{\xi*}\left(\bar{h}_0\,\tilde{\mathcal{U}}_0^*\right) - i\,\bar{h}_1\right]$$
$$- \left(\partial_{t_1}\mathcal{U}_0 + \mathcal{U}_0\partial_z\mathcal{U}_0 + \mathcal{U}_0^*\partial_{\xi*}\mathcal{U}_0\right). \tag{8.48}$$

This is an inhomogeneous equation for inertial oscillations which are non-propagative and non-dispersive, unlike the inertia–gravity waves appearing at the same stage in the QG regime above. Hence, a forcing with the inertial frequency on the right-hand side would lead to a resonance. The resonant forcing terms $\sim e^{-it}$ are those containing $\mathcal{U}_0$ only once on the right-hand side of (8.48). They should vanish in order to avoid secular growth of $\mathcal{U}_1$, and the following modulation equation for the amplitude of inertial oscillations results:

$$\partial_{t_1}\mathcal{A}_0 - 2i\partial_{\xi\xi*}^2\left(\bar{h}_0\mathcal{A}_0\right) + 2i\mathcal{A}_0\partial_{\xi\xi*}^2\bar{h}_0 + 2i\left(\partial_{\xi*}\bar{h}_0\partial_z\mathcal{A}_0 - \partial_z\bar{h}_0\partial_{\xi*}\mathcal{A}_0\right) = 0. \tag{8.49}$$

In terms of real spatial variables it reads

$$\partial_{t_1}\mathcal{A}_0 + \mathcal{J}(\bar{h}_0, \mathcal{A}_0) - \frac{i}{2}\left[\nabla^2(\bar{h}_0\mathcal{A}_0) - \mathcal{A}_0\nabla^2\bar{h}_0\right] = 0. \tag{8.50}$$

After having eliminated the secular growth, the regular solution of (8.48) satisfying zero initial conditions and giving the first-order correction to the inertial oscillations field can be easily found:

$$\tilde{\mathcal{U}}_1 = -\frac{i}{2}\mathcal{C}_1 e^{it} + i\mathcal{C}_2 e^{-2it} + \mathcal{A}_1(t_1, \ldots)e^{-it}. \tag{8.51}$$

Here,

$$\mathcal{C}_1 = \partial_{xy}^2\left(\bar{h}_0\mathcal{A}_0^*\right) + \mathcal{A}_0^*\partial_{xy}^2\bar{h}_0 + \frac{i}{2}\left[(\partial_y^2 - \partial_x^2)(\bar{h}_0\mathcal{A}_0^*) + \mathcal{A}_0^*(\partial_y^2 - \partial_x^2)(\bar{h}_0)\right].$$
$$\mathcal{C}_2 = -\frac{1}{2}\mathcal{A}_0\left(\partial_x\mathcal{A}_0 - i\partial_y\mathcal{A}_0\right). \tag{8.52}$$

It is worth mentioning that a correction to the velocity of slow motion also appears, as compared to QG regime, and contains a contribution arising from the self-interaction of the fast component:

$$\bar{\mathcal{U}}_1 = i\left(2\partial_{\xi*}\bar{h}_1 + \mathcal{J}(\bar{h}_0,\bar{\mathcal{U}}_0) + \mathcal{A}_0^*\,\partial_{\xi*}\mathcal{A}_0\right). \tag{8.53}$$

Thus, to the second order in Rossby number the slow component of motion remains undetermined, being, in fact, even slower than it was initially supposed. The fast component, the inertial oscillations, is slowly modulated, with modulation determined by a linear Schrödinger-type equation (16.78) with coefficients depending on the slow motion.

Second-order approximation

The equations of motion in this approximation are

$$\partial_t \mathbf{v}_2 + \left(\partial_{t_2} + \mathbf{v}_1\cdot\nabla\right)\mathbf{v}_0 + \left(\partial_{t_1} + \mathbf{v}_0\cdot\nabla\right)\mathbf{v}_1 + \hat{\mathbf{z}}\wedge\mathbf{v}_2 = -\nabla h_2, \tag{8.54}$$

$$\partial_t h_2 + \partial_{t_1} h_1 + \partial_{t_2}\bar{h}_0 + \nabla\left(\bar{h}_0\,\mathbf{v}_1 + h_1\,\mathbf{v}_0\right) = 0. \tag{8.55}$$

Time-averaging of (8.55), under hypotheses of boundedness of h_2 in time, gives

$$\partial_{t_1}\bar{h}_1 + \partial_{t_2}\bar{h}_0 + \nabla\cdot\left(\bar{h}_0\,\bar{\mathbf{v}}_1 + \bar{h}_1\,\bar{\mathbf{v}}_0\right) + \overline{\nabla\left(\tilde{h}_1\tilde{\mathbf{v}}_0\right)} = 0. \tag{8.56}$$

Calculation of the first two bilinear combinations of height and velocity in this equation is straightforward:

$$\nabla\cdot(\bar{h}_0\,\bar{\mathbf{v}}_1) = \mathcal{J}(\bar{h}_1,\bar{h}_0) - \mathcal{J}\left(\bar{h}_0,\bar{h}_0\nabla^2\bar{h}_0 + \tfrac{(\nabla\bar{h}_0)^2}{2}\right) - \tfrac{1}{4}\Big[\mathcal{J}(\bar{h}_0,|\tilde{\mathcal{A}}_0|^2$$
$$- i\left[\nabla\bar{h}_0\cdot(\mathcal{A}_0^*\nabla\mathcal{A}_0 - \mathcal{A}_0\nabla\mathcal{A}_0^*) + \bar{h}_0\left(\mathcal{A}_0^*\nabla^2\mathcal{A}_0 - \mathcal{A}_0\nabla^2\mathcal{A}_0^*\right)\right]\Big], \tag{8.57}$$

$$\nabla\cdot(\bar{h}_1\,\bar{\mathbf{v}}_0) = \mathcal{J}(\bar{h}_0,\bar{h}_1)$$

where we used (8.53). For the last term we get

$$\overline{\nabla\left(\tilde{h}_1\tilde{\mathbf{v}}_0\right)} = \frac{1}{4}\Big[\mathcal{J}(\bar{h}_0,|\mathcal{A}_0|^2 - \tag{8.58}$$
$$i\left[\nabla\bar{h}_0\cdot(\mathcal{A}_0^*\nabla\mathcal{A}_0 - \mathcal{A}_0\nabla\mathcal{A}_0^*) + \bar{h}_0\left(\mathcal{A}_0^*\nabla^2\mathcal{A}_0 - \mathcal{A}_0\nabla^2\mathcal{A}_0^*\right)\right]\Big],$$

where we used (8.47) and (8.51). The contributions containing fast–fast terms therefore remarkably cancel, and a purely slow evolution equation for h_0 results:

$$\partial_{t_2}\bar{h}_0 - \mathcal{J}\left(\bar{h}_0,\bar{h}_0\nabla^2\bar{h}_0 + \frac{(\nabla\bar{h}_0)^2}{2}\right) = 0. \tag{8.59}$$

Thus, as in the QG case, there is no fast-motion drag upon the slow component of motion, in spite of the fact that slow velocity does contain such terms, cf. (8.53). At the same time, the fast component is guided by the slow one, as follows from the modulation equation (8.50). One could expect a cubic $\mathcal{A}_0$ correction to (8.50) at this level of approximation, which would result in a celebrated non-linear Schrödinger equation.

This, however, does not happen. Indeed, such a correction could only appear from the elimination of resonances in the equation for the next-order correction to velocity $\mathcal{U}_2$. From (8.54) written in complex notation we see that the resonant cubic terms $\sim e^{-it}$ do appear in the expressions

$$\tilde{\mathcal{U}}_0^* \partial_{\xi^*} \tilde{\mathcal{U}}_1 \quad \text{and} \quad \tilde{\mathcal{U}}_0 \partial_z \tilde{\mathcal{U}}_1 + \tilde{\mathcal{U}}_1^* \partial_z \tilde{\mathcal{U}}_0 + \tilde{\mathcal{U}}_1 \partial_{\xi^*} \tilde{\mathcal{U}}_0 , \qquad (8.60)$$

but their combination vanishes identically:

$$-i\tilde{\mathcal{U}}_0^* \partial_{\xi^*} \left(\tilde{\mathcal{U}}_0 \partial_\xi \tilde{\mathcal{U}}_0 \right) + i\tilde{\mathcal{U}}_0 \partial_\xi \left(\tilde{\mathcal{U}}_0^* \partial_{\xi^*} \tilde{\mathcal{U}}_0 \right) + i\tilde{\mathcal{U}}_0^* \partial_{\xi^*} \tilde{\mathcal{U}}_0 \partial_\xi \tilde{\mathcal{U}}_0 - i\tilde{\mathcal{U}}_0 \partial_\xi \tilde{\mathcal{U}}_0^* \partial_{z^*} \tilde{\mathcal{U}}_0 \equiv 0 . \qquad (8.61)$$

Therefore, there are no nonlinear corrections to (16.78) up to order 4 in ϵ. We can make the resemblance between (8.50) and the classical Schrödinger equation even closer by changing variables: $\mathcal{A}_0 \rightarrow \mathcal{B} = \bar{h}_0 \mathcal{A}_0$, which yields

$$i \left[\partial_{t_1} \mathcal{B} + \mathcal{J}(\bar{h}_0, \mathcal{B}) \right] + \frac{\bar{h}_0}{2} \nabla^2 \mathcal{B} - \frac{\nabla^2 \bar{h}_0}{2} \mathcal{B} = 0 . \qquad (8.62)$$

This is the standard Schrödinger equation with a variable 'mass' $\bar{h}_0$, a 'potential' $\sim \nabla^2 \bar{h}_0$, and additional advection by the geostrophic velocity field produced by $\bar{h}_0$.

We will not continue the calculations of higher-order corrections, which are straightforward and give an evolution equation for the next correction $\mathcal{A}_1$ to the envelope of inertial oscillations.

Comparison of QG and FG adjustments

The frontal geostrophic (FG) results are similar to the QG ones in a sense that dynamical separation of fast and slow components of motion also takes place, and the slow component evolves according to the balanced equation without being influenced by the fast one. However, in the FG regime the fast component is directly affected by the slow component, which is guiding it according to the Schrödinger-type equation (8.50). Still, the energy of inertial oscillations and the energy of slow motions

$$E_{II} = \int dx dy \frac{1}{2} |\mathcal{A}_0|^2 \quad E_{slow} = \int dx dy \frac{1}{2} \bar{h}_0 \left(\nabla \bar{h}_0 \right)^2 \qquad (8.63)$$

are separately conserved. These two conservation laws express dynamical separation of the two components of motion.

The following qualitative picture thus arises in the FG regime: any initial thickness perturbation evolves very slowly according to (8.59). A non-geostrophic part of the velocity perturbation is oscillating with the inertial period and is slowly dispersed according to (8.50), where the coefficients may be considered as being constant in time at the dispersion time scale. It is simultaneously advected by the geostrophic part of velocity. The details of the dispersion process depend on the initial field ('front') $h_I(\mathbf{x})$. As in the standard Schrödinger equation case, it is possible, for particular profiles, that some part of the initial wave packet is trapped by the front ('bound states') and evolves together with it.

8.2 Geostrophic adjustment in the baroclinic (two-layer) model

8.2.1 Quasi-geostrophic regime

Scaling and formulation of the problem

We start from the two-layer RSW model with a rigid lid and will be supposing in this section that the ratio of the layers' depths $d = \mathcal{O}(1)$. We will be also working in the limit of weak stratification $\rho_2 - \rho_1 \to 0$, already used in Chapter 5, and apply the QG scaling of that chapter: unique horizontal velocity scale V, the horizontal spatial scale $L \sim R_d = \sqrt{g'H_e}/f_0$, where R_d is the baroclinic Rossby deformation radius, the pressure scale $P = \bar{\rho}f_0VL$, where $\bar{\rho} = \frac{\rho_1+\rho_2}{2}$, and the scale of the interface variations $\eta^* = \epsilon H_e$. Here $H_e = \frac{H_1 H_2}{H_1+H_2}$, and is called equivalent height. (Note the change of vertical scaling with respect to section 5.3.1.) Time scales, as in the one-layer case, are f_0^{-1} and $t_1 \sim (\epsilon f_0)^{-1}$. Introducing as in section 5.3.1 the (order 1) parameters $\bar{h}_i = \frac{H_i}{H_1+H_2}$, $i = 1,2$ we rewrite the equations of the two-layer RSW model in non-dimensional form:

$$\begin{cases} (\partial_t + \epsilon\partial_{t_1})\mathbf{v}_i + \epsilon\mathbf{v}_i \cdot \nabla\mathbf{v}_i + \hat{\mathbf{z}} \wedge \mathbf{v}_i + \nabla\pi_i = 0, \\ (\partial_t + \epsilon\partial_{t_1})(1 - (-1)^{i+1}\epsilon\,\bar{h}_{i+1}\,\eta) + \nabla \cdot \left((1 - (-1)^{i+1}\epsilon\,\bar{h}_{i+1}\,\eta)\mathbf{v}_i\right) = 0. \end{cases} \tag{8.64}$$

Non-dimensional potential vorticities of the layers are

$$\Pi_i = \frac{\epsilon\,\zeta_i + 1}{1 - (-1)^{i+1}\bar{h}_{i+1}\epsilon\,\eta}, \quad i = 1,2, \tag{8.65}$$

and non-dimensional relation between the interface displacement and pressures in the layers is $\pi_2 - \pi_1 = \eta$, as in Chapter 5. The solution of the equations of motion in the infinite f plane is represented in the form of asymptotic expansions:

$$\begin{cases} \mathbf{v}_i = \mathbf{v}_i^{(0)}(x,y;t,t_1,t_2,...) + \epsilon\mathbf{v}_i^{(1)}(x,y;t,t_1,t_2,...) + ... \\ \eta = \eta^{(0)}(x,y;t,t_1,t_2,...) + \epsilon\eta^{(1)}(x,y;t,t_1,t_2,...) + ..., \end{cases} \tag{8.66}$$

where each dynamical variable at each order may be uniquely split into the slow (denoted by over-bar) and fast (denoted by tilde) parts, as in the one-layer case above. We are looking for a perturbative solution of the Cauchy problem for the system (8.64) with initial conditions

$$\mathbf{v}_i|_{t=0} = \mathbf{v}_{I_i}, \quad \eta|_{t=0} = \eta_I. \tag{8.67}$$

We will not present the details of order-by-order calculations which are similar to those presented above in the one-layer case, although they are, obviously, more involved, as they are to be performed in each layer, and coupled via the dynamical boundary condition at the interface. We thus pass directly to the formulation of the results.

The main results

Up to the third order on Rossby number, the fast component of motion does not influence the slow one. The latter is described by 'improved' quasi-geostrophic PV equations

$$\frac{D_i}{Dt_1}\left[\nabla^2\bar{\pi}_i + (-1)^{i+1}\bar{h}_{i+1}\bar{\eta} + \epsilon(-1)^{i+1}\bar{h}_{i+1}\bar{\eta}\left(\nabla^2\bar{\pi}_i + (-1)^{i+1}\bar{h}_{i+1}\bar{\eta}\right)\right. \tag{8.68}$$

$$\left. - \epsilon\nabla\bar{\pi}_i \cdot \nabla\left(\nabla^2\bar{\pi}_i + (-1)^{i+1}\bar{h}_{i+1}\bar{\eta}\right) - 2\epsilon\mathcal{J}\left(\partial_x\bar{\pi}_i, \partial_y\bar{\pi}_i\right)\right] = 0 \, ,$$

with

$$\frac{D_i}{Dt_1}\,(...) := \partial_{t_1}\,(...) + \mathcal{J}\left(\bar{\pi}_i - \epsilon\frac{(\nabla\bar{\pi}_i)^2}{2}, ...\right), \quad i = 1, 2 \, . \tag{8.69}$$

These equations are written in terms of 'synthetic' pressures in the layers, combining zero- and first-order fields: $\bar{\pi}_i = \bar{\pi}_i^{(0)} + \epsilon\bar{\pi}_i^{(1)}$, $\bar{\eta} = \eta_i^{(0)} + \epsilon\eta_i^{(1)}$ with $\bar{\pi}_2 - \bar{\pi}_1 = \bar{\eta}$. The fast component is the internal (interface) inertia–gravity wave packet determined by the initial conditions and described by the wave equation for the 'synthetic' wave field $\tilde{\eta} = \tilde{\eta}_0 + \epsilon\tilde{\eta}_1$

$$-\frac{\partial^2\tilde{\eta}}{\partial t^2} - \tilde{\eta} + \nabla^2\tilde{\eta} = \epsilon R(x, y; t, t_1, ...) \, . \tag{8.70}$$

The right-hand side of (8.70) is generated, as in the one-layer case, by nonlinear interactions of the lowest-order fast field $\tilde{\eta}_0$ with itself and with the slow component. Again, the key property of this source term is that it is not resonant, having zero fast-time mean, and hence the field $\tilde{\eta}$ (and other fast fields) corresponds to internal inertia–gravity waves propagating out of the localised initial perturbation and decaying in time at any given space location, consistently with radiation boundary conditions. Apart from the secondary inertia–gravity wave emission provided by the slow–fast interaction term on the right-hand side of (8.70), the fast and the slow components of motion feel each other only on the level of initial conditions for their respective dynamical equations. The initialisation problem is solved along the same lines as in the one-layer case. If the terms $\mathcal{O}(\epsilon)$ are dropped off in (8.68) and (8.69), the standard two-layer QG equations of Chapter 5 are recovered. Therefore, as in the one-layer case, the solution of the geostrophic adjustment problem proves dynamical consistency of the QG equation and, on the other hand, shows which corrections are necessary to include, if the evolution of the system on times $\mathcal{O}\left((\epsilon f_0)^{-2}\right)$ is to be considered.

8.2.2 Frontal geostrophic regime

Basic hypotheses and scaling

The FG scaling for the two-layer FG model was described in Chapter 5, where the corresponding balanced dynamics was reviewed. The same scaling is used in the following with an addition of the fast-time scale f_0^{-1}. We rewrite the non-dimensional equations in the FG regime in complex notation:

$$
\begin{cases}
\partial_t \mathcal{U}_i + i\mathcal{U}_i + \epsilon \left(\mathcal{U}_i \partial_\xi \mathcal{U}_i + \mathcal{U}_i^* \partial_{\xi*} \mathcal{U}_i \right) = -2\,\partial_{\xi*}\pi_i, \ i = 1, 2 \\[6pt]
\partial_\xi \left[(1 - \eta)\,\mathcal{U}_1 + (d^{-1} + \eta)\mathcal{U}_2 \right] + c.c = 0, \\[6pt]
\partial_t \eta = \epsilon \partial_\xi \left[(1 - \eta)\mathcal{U}_1 \right] + c.c., \quad \pi_2 = \pi_1 + \eta,
\end{cases}
\tag{8.71}
$$

where $\mathcal{U}_{1,2} = u_{1,2} + iv_{1,2}$ are the complex velocities in respective layers. The same procedure as in Section 8.1.2 is then applied and we pass directly to the results. Two FG subregimes arise, depending on the ratio of the layers' depths: a regime with comparable depths and a regime with shallow upper layer. The second, with the depth ratio $d = \mathcal{O}(\epsilon)$, is most relevant, because in this case the FG assumption that perturbations of the thickness are comparable with the thickness itself is fully justified.

The main results in the case of shallow upper layer

As in the QG regime all fields are split into slow and fast parts. Both the slow and the fast components evolve from uniquely defined respective initial conditions. The fast component consists of inertial oscillations with slowly changing amplitude. The barotropic and the baroclinic complex velocities are expressed as follows:

$$
\mathcal{U}_{bt} = 2i\partial_{\xi*}P, \quad \mathcal{U}_{bc} = -2i\partial_{\xi*}\eta + \mathcal{A}e^{-it},
\tag{8.72}
$$

where the slow functions η and P express the leading-order interface displacement and the barotropic pressure, respectively, and $\mathcal{A} = \mathcal{A}(\xi, \xi^*, t_1, ...)$ is the slowly evolving envelope of the inertial oscillations. The leading-order evolution is determined by two coupled equations for slow P and η, which we already obtained in Chapter 5 by direct filtering of fast motions, and a separate equation for $\mathcal{A}$, of the type of (8.50):

$$
\begin{cases}
\partial_{t_2}\eta + J(P, \eta) - J\left(\eta, (1 - \eta)\nabla^2\eta - \frac{(\nabla\eta)^2}{2} \right) = 0, \\[6pt]
\partial_{t_2}\nabla^2 P + J(P, \nabla^2 P) + J\left(\eta, (1 - \eta)\nabla^2\eta - \frac{(\nabla\eta)^2}{2} \right) = 0, \\[6pt]
\partial_{t_1}\mathcal{A} - J(\eta, \mathcal{A}) - \frac{i}{2}\nabla^2\mathcal{A} + \frac{i}{2}\left[\nabla^2(\eta\mathcal{A}) - \mathcal{A}\nabla^2\eta \right] = 0.
\end{cases}
\tag{8.73}
$$

The most important and nontrivial feature of this system is that, although the inertial oscillations do not run away as inertia–gravity waves in Section 8.2, they still do not make any contribution to the evolution of the slow component. Hence, baroclinicity does not affect this crucial property of the FG dynamics observed previously in the barotropic one-layer case, and the fast oscillations exert no drag on the slow motion. At the same time, the slow modulation of the inertial oscillations is guided by the baroclinic vortex motion, since the coefficients in the modulation equation for $\mathcal{A}$ depend on η.

8.3 Geostrophic adjustment in one dimension and the first idea of frontogenesis

As should be clear from Chapter 7, the 1.5D versions of one- and two-layer RSW models represent a useful laboratory for understanding fundamental properties of these models. In the geostrophic adjustment context they allow us to go beyond the limitations of the asymptotic expansions approach of Sections 8.1 and 8.2. In addition, high-resolution numerical simulations with finite-volume schemes are especially cheap and fast in this case. It is worth emphasising that the very concept of geostrophic adjustment was introduced by Rossby in the framework of 1.5D RSW, although not called this name then. As mentioned in Sections 7.1.1 and 7.2.1, geostrophically balanced states are exact stationary solutions of the 1.5D RSW models. The slow motions in 1.5D models are, thus, 'infinitely slow', and the geostrophic adjustment problem becomes a problem of relaxation of the system to a stationary state which is a typical problem in mechanics and quantum mechanics. In this section we obtain some general theoretical predictions for the geostrophic adjustment in 1.5D models, and then confront them with numerical simulations. Our results on the existence of the adjusted state, which can be rigorously established in the 1.5D case, will allow us to get a first idea of the mechanism of frontogenesis.

8.3.1 Theoretical considerations

To establish general theoretical results on the 1.5D adjustment it is advatageous to use the Lagrangian formulation already presented in Section 7.1.2. By multiplying the left-hand side of equation (7.11) by the initial distribution of thickness h_I we get

$$h_I \left(\ddot{X} + f^2 X - fM \right) + \left(\frac{gh_I^2}{2X'^2} \right)' = 0. \tag{8.74}$$

In accordance with Section 3.4.2 this equation can be obtained from the variational principle with the action

$$S = \int dt \int dx \, \mathcal{L}, \tag{8.75}$$

where the Lagrangian density is given by

$$\mathcal{L} = h_I \left(\frac{\dot{X}^2}{2} - f^2 \frac{X^2}{2} + fMX \right) - \frac{gh_I^2}{2} \frac{1}{X'}. \tag{8.76}$$

Introducing canonical momentum $\mathcal{P} = h_I \dot{X}$ corresponding to the canonical coordinate X we get the Hamiltonian

$$\mathcal{H} = \int dx \left(\frac{\mathcal{P}^2}{2h_I} + h_I \left(f^2 \frac{X^2}{2} - fMX \right) + \frac{gh_I^2}{2} \frac{1}{X'} \right). \tag{8.77}$$

The steady states are extrema of the Hamiltonian, and from the condition $\delta \mathcal{H} = 0$ we obtain

$$h_I \left(f^2 X - fM \right) + \left(\frac{gh_I^2}{2X'^2} \right)' = 0; \ \mathcal{P} = 0. \tag{8.78}$$

For steady states $X = x$, and (8.78) is satisfied by the geostrophic wind expression for v_I in terms of gradient of h_I. A straightforward calculation of the second variation $\delta^2 \mathcal{H}$ shows that it is always positive-definite and, thus, the geostrophic equilibria are formally stable in the 1.5D system (see Chapter 10 for stability analysis in Hamiltonian systems). It should be emphasised that the equation for the geostrophic equilibrium may be obtained by simply omitting the term with time derivatives in (8.74).

Equation (8.74) has an advantage of physical transparency, and explicitly incorporates arbitrary initial conditions. It contains, however, variable coefficients, which complicates the analysis. A further simplification, which was already used in Section 7.1.2, can be achieved by an additional change of variables $x = x(a)$, which 'straightens' the initial elevation profile $h_I(x)$. This means a transformation to a new, more technically convenient, set of Lagrangian labels. We recall that it leads to the following relations: $\mathcal{J} = \partial X/\partial a = H/h(X,t)$, where H is the uniform mean height, and $g\,\partial_X h = \partial_a P$. Lagrangian equations in terms of coordinate a (7.29) are equivalent to a single equation, which may be as well obtained by differentiation and change of variables from (7.11):

$$\ddot{\mathcal{J}} + f^2 \mathcal{J} + \partial_{aa}^2 P = fHq. \tag{8.79}$$

Here, $q(a)$ is PV expressed in terms of the independent variable a:

$$q(a) = \frac{1}{H} \left(\partial_a v(a,t) + f\,\mathcal{J}(a,t) \right) = \frac{1}{H} \left(\partial_a v_I(a) + f\,\mathcal{J}_I(a) \right), \quad \dot{q} = 0. \tag{8.80}$$

The advantage of the present formulation with respect to equation (8.74) is that it highlights the role of PV: for a given set of (localised) initial conditions the adjusted state is a stationary state with the same potential vorticity as the initial one. Stationary solutions $\mathcal{J}_s$ satisfy the equation

$$\frac{gH}{f} \frac{d^2}{da^2} \left(\frac{1}{2\mathcal{J}_s^2(a)} \right) + f\,\mathcal{J}_s(a) = Hq(a). \tag{8.81}$$

This equation can be rewritten in non-dimensional terms (we scale P with gH, a with the deformation radius $R_d^2 = gH/f^2$, and q with f/h) as a non-autonomous ordinary differential equation describing the motion of a material point in a given potential under the action of a time-dependent force, where a is 'time':

$$\frac{d^2 P}{da^2} + \frac{1}{\sqrt{2P}} = q. \tag{8.82}$$

Far from a localised front-like initial perturbation, at $a = \pm\infty$, the 'external force' on the right-hand side of (8.81) or (8.82) exactly equilibrates the 'potential force' (the second term on the left-hand side) Therefore, a nontrivial solution, if it exists, is a separatrix relating two states of unstable equilibrium in the phase space of this simple one-dimensional mechanical system, and thus resembles the soliton or instanton solutions in a number of physical models.

A general question then can be asked: does a physically acceptable, that is smooth with everywhere positive thickness, geostrophically adjusted state exist for any initial PV, and is it unique? To answer this question it is more convenient to come back to the original variables X: $dX = \mathcal{J}da$. The equation for the adjusted state then reads

$$-\frac{g}{f}\frac{d^2 h(X)}{dX^2} + h(X)\, q(X) = f. \qquad (8.83)$$

Here PV is considered as a function of X which is given by the inverse mapping $x = x(X, t)$ (or $a = a(X)$): $q(X) = \dfrac{1}{h_I(x(X))}\left(f + \dfrac{\partial v_I(x(X))}{\partial x}\right)$. The following theorem giving sufficient conditions of existence and uniqueness of the adjusted state can be proved with the help of standard tools of the theory of ordinary differential equations:

Theorem. *For positive $q(X)$ with compact support derivatives and arbitrary constant asymptotics (front) equation (8.83) has unique bounded and everywhere positive solution $h(X)$ at $-\infty \leq X \leq +\infty$.*

We should recall that classical frontogenesis in the full primitive equations model can be understood as 'catastrophic' geostrophic adjustment, when existence of a smooth adjusted state is not guaranteed. A necessary condition for frontogenesis is nonpositiveness of the initial PV. We therefore have a direct analogy of this situation in the RSW model, which will be illustrated by numerical simulations in Section 8.4.

Let us now suppose that an adjusted state exists, and consider small perturbation about such a state. We recall the equation for displacements of Lagrangian parcels from their initial positions (7.12):

$$\ddot{\phi} + f^2\phi + gh_I'\left(\frac{1}{(1+\phi')^2}\right) + \frac{gh_I}{2}\left(\frac{1}{(1+\phi')^{-2}}\right)' = fv_I.$$

If the initial conditions v_I and h_I are in geostrophic balance $gh_I' = fv_I$, then $\phi = 0$ is an exact solution of this equation. Therefore, for small deviations from the exact geostrophy, the equation can be linearised, and solved by the method of small perturbations, as in preceding sections. For this it is necessary that the initial imbalance $A_I = fv_I - gh_I'$ would be small. As is easy to see, this condition is equivalent to the smallness of the Rossby number, $Ro = \frac{U}{fL} \ll 1$, where U and L are characteristic velocity and scale in the x direction. An advantage of the Lagrangian approach is that contrary to the Eulerian one, where it is necessary to attribute the initial imbalance either to v_I or to h_I, the imbalance A_I enters here as a whole. It should be emphasised that there exists yet another source of imbalance, a non-zero u_I.

The linearised equation (7.12) reads

$$\ddot{\phi} + f^2\phi - 2gh'_I\phi' - gh_I\phi'' = A_I. \tag{8.84}$$

Solutions are sought as combinations of slow and fast components $\phi = \bar{\phi} + \tilde{\phi}$, with the same notation as in Section 8.1. For the time-averaged part of the solution, corresponding to the shift of Lagrangian particles which is necessary to arrive at their equilibrium positions, we get

$$f^2\bar{\phi} - 2gh'_I\bar{\phi}' - gh_I\bar{\phi}'' = A_I. \tag{8.85}$$

This is a linear second-order inhomogeneous ordinary differential equation (ODE). By introducing a new variable $\bar{\Phi} = gh_I\bar{\phi}$ and dividing by gh_I it can be cast in canonical form:

$$-\bar{\Phi}'' + \left(\frac{f^2 + gh''_I}{gh_I}\right)\bar{\Phi} = A_I, \tag{8.86}$$

where the geostrophic PV $Q^{(g)} = \frac{f^2 + gh''_I}{h_I}$ constructed from the initial height perturbation appears in the second term. The solution of the linear inhomogeneous ODE (8.86) can be obtained by the standard method of variation of constants, once solutions of the homogeneous equation are known. As earlier, we are interested in localised front-like initial conditions, where A_I is a compact support function and h_I is monotonous and has constant asymptotics at infinity. This means that

$$h_I|_{\pm\infty} = h_{\pm}, \quad q^{(g)}\big|_{\pm\infty} = \frac{f}{h_{\pm}} > 0. \tag{8.87}$$

Solutions of the homogeneous equation are then either exponentially growing or exponentially decaying at both spatial infinities. It is not obvious that a decaying at both spatial infinities solution of the inhomogeneous equation can be found for arbitrary $q_I^{(g)}(x)$. By the same method which was used in the full non-perturbative proof earlier it can be shown that a unique solution of (8.85) exists for non-negative initial geostrophic PV $q_I^{(g)} \geq 0$. Hence, at least for non-negative initial PV one can always find an adjusted state and thus solve the problem for the time-independent 'slow' variable.

For the variable in time fast part of the solution, a homogeneous equation,

$$\ddot{\tilde{\phi}} + f^2\tilde{\phi} - 2gh'_I\tilde{\phi}' - gh_I\tilde{\phi}'' = 0, \tag{8.88}$$

results. By introducing, as above, a new variable, $\tilde{\Phi} = gh_I\tilde{\phi}$, this equation may be rewritten as

$$\ddot{\tilde{\Phi}} + (f^2 + gh''_I)\tilde{\Phi} - gh_I\tilde{\Phi}'' = 0, \tag{8.89}$$

which should be solved with initial conditions $\tilde{\phi}_I$, $u_I = \dot{\tilde{\phi}}_I$, where $\tilde{\phi}_I$ is yet to be determined (the initialisation problem). Introducing the Fourier transform $\tilde{\Psi}(x,\omega) = \int_{-\infty}^{\infty} e^{+i\omega t} \tilde{\Phi}(x,t)\, dt$, we get

$$-\tilde{\Psi}'' + \left[\frac{f}{g} \left(\frac{f + \frac{g h_I''}{f}}{h_I} \right) - \frac{\omega^2}{g h_I} \right] \tilde{\Psi} = 0. \tag{8.90}$$

This can be interpreted as a Schrödinger equation, where $q_I^{(g)}$ plays a rôle of potential. As $\phi_I = \tilde{\phi}_I + \bar{\phi}_I = 0$ the initial condition for $\tilde{\phi}$ and, hence, $\tilde{\Phi}$ follows once $\bar{\phi}$ is found: $\tilde{\phi}_I = -\bar{\phi}$. As is known from the general theory of the Schrödinger equation, solutions of (8.89) represent waves (one-dimensional inertia–gravity waves) propagating out of the initial localised perturbation, plus, possibly, some bound states (trapped modes). The presence of the trapped modes would mean non-attainability of the adjusted state, as part of the energy of the initial perturbation which is projected onto these modes can not be evacuated by radiation. However, trapping by an isolated front is impossible in the one-layer model.

Let us first show that trapped modes should be sub-inertial, that is they have frequencies below f. In the case of a front-like initial configuration the asymptotics of $\tilde{\phi}_I$ and u_I at infinity are zero, and those of h_I and $q^{(g)}$ are constant, cf. (8.87). Hence, at $x \to \pm\infty$ (8.90) becomes

$$-\tilde{\Psi}'' + \frac{f^2 - \omega^2}{g h_\pm} \tilde{\Psi} = 0, \tag{8.91}$$

and in order to have bound states, which are by definition decaying at spatial infinity, we should have $\omega < f$. However, this is impossible. Let us consider the Fourier transformation of $\tilde{\phi}$: $\tilde{\phi} = \int d\omega \left(\psi(\omega,x) e^{-i\omega t} + c.c \right)$. Then for each Fourier component $\psi(\omega,x)$ we get from (8.88)

$$g h_I \psi'' + 2g h_I' \psi' + (\omega^2 - f^2)\psi = 0, \tag{8.92}$$

which is equivalent to (8.90). By multiplying (8.92) by $g h_I \psi^*$, where the asterisk denotes the complex conjugation, we obtain

$$\left(g^2 h_I^2 \psi^* \psi' \right)' - g^2 h_I^2 \psi'^* \psi' + \left(\omega^2 - f^2 \right) g h_I \psi^* \psi = 0, \tag{8.93}$$

and for decaying at $\pm\infty$ states the estimate

$$\omega^2 = f^2 + \frac{\int_{-\infty}^{+\infty} dx\, g^2 h_I^2 \left| \psi' \right|^2}{\int_{-\infty}^{+\infty} dx\, g h_I \left| \psi \right|^2} \geq f^2 \tag{8.94}$$

follows by integration. The fact that frequencies are supra-inertial contradicts the initial hypothesis and, hence, sub-inertial trapped modes do not exist, and the frequency spectrum is continuous. (As we will see in Chapter 10, this also means that inertial instability does not exist in one-layer RSW on the f plane, but this result changes in the two-layer model.) Therefore, all of the initial $\tilde{\phi}$ will be dispersed leaving only its stationary part $\bar{\phi}$ in the vicinity of the initial perturbation. The same considerations as in Section 8.1 show that the outgoing waves do not exert any drag upon the stationary state in lowest orders in Ro and, thus, slow and fast variables are dynamically split, at least for non-negative PV. The rate of the relaxation towards the adjusted state depends on further details of the potential $q^{(g)}$ in the Schrödinger equation (8.90). If quasi-stationary states—those which decay only by a sub-barrier tunneling—are present, the decay rate will be exponential, as is well known from quantum mechanics. Otherwise the decay will be dispersive and follow the $t^{-\frac{1}{2}}$ law in one spatial dimension. Here, by decay we mean the time decrease of the amplitude of a spatially localised perturbation.

In order to improve our understanding of the details of the relaxation process, it is useful to return to the system (7.29) and linearise it about a stationary state v_s, $\mathcal{J}_s$:

$$u = \tilde{u}, \quad v = v_s + \tilde{v}, \quad \mathcal{J} = \mathcal{J}_s + \tilde{\mathcal{J}},$$

which gives

$$
\begin{cases}
\tilde{u}_t - f\tilde{v} - gH(\tilde{\mathcal{J}}/\mathcal{J}_s^3)_a = 0, \\
\tilde{v}_t + fu = 0, \\
\tilde{\mathcal{J}}_t - u_a = 0.
\end{cases}
\tag{8.95}
$$

By using the fact that fast motions are PV-less (i.e. $f\tilde{\mathcal{J}} + \tilde{v}_a = 0$), it is easy to get from (8.95) a single equation for $\tilde{v}$:

$$\tilde{v}_{tt} + f^2\tilde{v} - gH(\tilde{v}_a/\mathcal{J}_s^3)_a = 0. \tag{8.96}$$

Let us consider harmonic in time solutions $\tilde{v} = \hat{v}(a)e^{-i\omega t} + $ c.c. Then

$$(gH_s\hat{v}_a)_a + (\omega^2 - f^2)\hat{v} = 0, \tag{8.97}$$

where $H_s = H/\mathcal{J}_s^3$. Note that supra-inertiality of ω and, hence, the absence of trapped states follow trivially from (8.97). Introducing a new dependent variable ψ: $\hat{v} = \frac{\psi}{gH_s^{1/2}}$ we cast (8.97) to the two-term canonical form

$$\frac{d^2\psi}{da^2} + \left[\frac{\omega^2 - f^2}{gH_s} - \frac{1}{4}\left(\frac{(H_s)_a}{H_s}\right)^2 - \frac{1}{2}\left(\frac{(H_s)_a}{H_s}\right)_a \right]\psi = 0. \tag{8.98}$$

Rewritten as

$$\frac{d^2\psi}{da^2} + k^2\psi = 0, \tag{8.99}$$

this equation can be interpreted as that of an oscillator with variable frequency $k(a)$ (or, as before, as a Schrödinger equation with a potential V and an energy E such that

$k^2 = E - V(a)$). It is clear that k^2 can be negative even if $\omega > f$ for H_s of certain form. This means that for some intervals on the x axis the wave number k can be imaginary, corresponding to tunnelling which was mentioned earlier. This, in turn, means that quasi-stationary states slowly decaying by the tunnelling mechanism can exist.

8.3.2 Numerical simulations: Rossby adjustment

The classical Rossby problem of adjustment of unbalanced jet provides a good illustration of the theoretical considerations of Section 8.2. It will be treated numerically, with the well-balanced shock-resolving numerical scheme of Chapter 7, which is particularly simple to implement in the 1.5D case. The initial conditions are taken in the form

$$h(x, 0) = H = \text{const}, \quad u(x, 0) = 0, \quad v(x, 0) = VN_L(x), \qquad (8.100)$$

where the normalised jet profile $N_L(x) = \dfrac{(1 + tanh(4x/L + 2)) \cdot (1 - tanh(4x/L - 2))}{(1 + tanh(2))^2}$

is shown in the left panel of Figure 8.1. The Rossby and Burger numbers are $Ro = \dfrac{V}{fL}$, $Bu = \dfrac{gH}{f^2L^2}$. A typical evolution of the height field $h(x, t)$ in the Rossby adjustment problem is shown in Figure 8.2 for $Ro = 1$ and $Bu = 0.25$. As we see, two shocks are formed in the zone of anticyclonic shear of the initial jet. They propagate out of the jet, decreasing in intensity, and transporting and dissipating the excess of energy with respect to stationary adjusted state which is visible in the central part of the lower curve of the figure, and consists of a central jet with two side counter-jets. This process, in fact, is an illustration of the 'frontogenesis' in RSW, according to the results of Section 8.2, as at this Rossby number the PV in the zone of anticyclonic shear is negative, as follows from the right panel of Figure 8.1. Note that although the zone of negative PV shrinks as a result of dissipation provoked by the passage of the shock, it does not disapper. There is also a shift of the whole PV distribution due to advection by high-amplitude wave

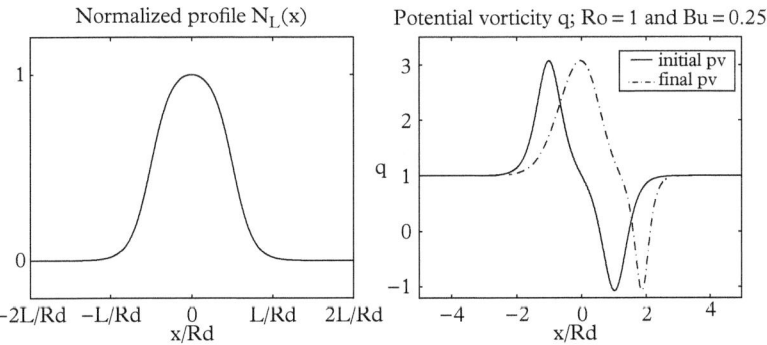

Figure 8.1 *Normalised profile of velocity of the jet (left panel), and initial (solid) and final (dashed) PV distributions in the Rossby adjustment problem (right panel).*

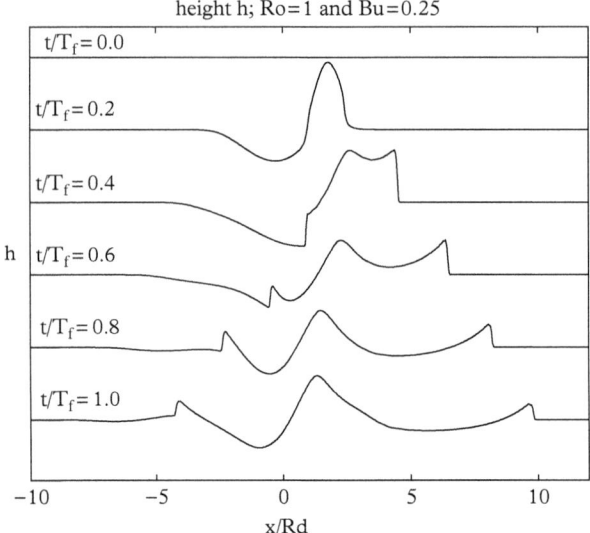

Figure 8.2 *Rossby adjustment of a jet. Two shocks are formed at t = 0.3, and propagate to the left and to the right from the jet, respectively. One of the shocks is formed within the jet core, at the anticyclonic side and another one at its periphery. Time in units of* f^{-1}.

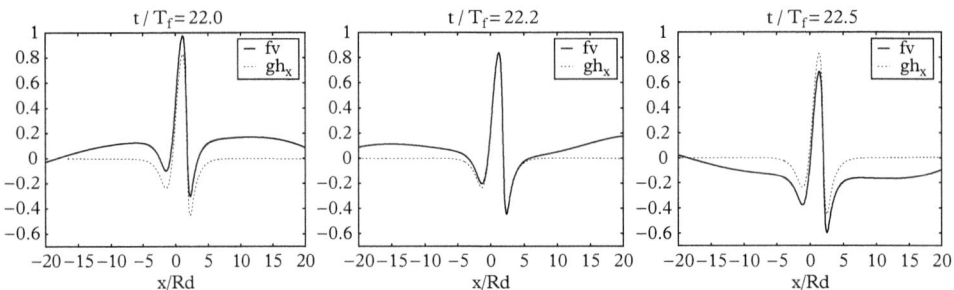

Figure 8.3 *Check of balance between fv and* gh_x *during the Rossby adjustment: although a close to geostrophic balance mean state (middle panel) is rapidly achieved, oscillations persist in the jet core. The amplitude of oscillations is decreasing with time and the period is close to* $2\pi/f$.

produced at the initial stages of adjustment. In spite of the sink of energy provided by the shocks, the adjusted state is rapidly attained only on average, as quasi-inertial oscillations with frequencies close to f remain at the location of the initial jet for a long time, as follows from Figure 8.3. Persistent quasi-inertial oscillations are systematically observed in such numerical simulations, their decay law is close to $t^{\frac{1}{2}}$.

8.4 Geostrophic adjustment in the presence of boundaries, topography, and at the equator

The procedure described in Section 8.1.1 can be extended to configurations with coastal, topographic, and equatorial wave guides, although it becomes more involved due to the presence of wave-guide modes. Analysis of slow motions in these configurations was already done in Section 5.5, and inclusion of fast inertia–gravity waves does not change the resulting slow-motion equations. In all cases considered below, the fast–slow splitting can be proved, if localised initial conditions and radiation boundary conditions are imposed. In the following we will not offer the technicalities of the demonstrations, but only emphasise the main points and give illustrations from numerical simulations.

8.4.1   Geostrophic adjustment with a lateral boundary

In the presence of a lateral boundary the slow motion consists in geostrophically balanced PV-bearing component and the coastal Kelvin waves. The fast inertia–gravity waves split out. Any localised initial perturbation adjusts to a geostrophically balanced state by emitting a packet of inertia–gravity waves and a packet of Kelvin waves running along the coast. Kelvin waves slowly steepen and eventually break. This process is fully confirmed by numerical simulations with the well-balanced code of Chapter 7. A typical result of such simulations is presented in Figure 8.4. The computational domain is a half f plane, with a boundary along the x axis. Initial thickness anomaly is situated close to the boundary (note the difference of scales in x and y direction). It emits a clearly visible packet of inertia–gravity waves, and becomes geostrophically balanced, as follows from alignment of velocity vectors with isobars. A part of the perturbation is moving along the wall with typical characteristics of a Kelvin wave.

8.4.2   Geostrophic adjustment over escarpment

As was shown in Section 5.5, the slow component in the presence of escarpment consists of geostrophically balanced motions and topographic Rossby waves. Any initial perturbation located over escarpment produces inertia–gravity wave emission, and also emission of a packet of topographic waves, as part of the initial signal is projected on the topographic modes. To illustrate this process we have chosen initial perturbation in a form of unbalanced pressure front perpendicular to the escarpment. In the absence of topography, this situation falls in the category of 1.5D flows considered in Section 8.3, and its adjustment would lead to formation of an along-front jet by emission of inertia–gravity waves at both sides of the front. In the presence of escarpment the process is totally different, as a localised packet of topographic modes forms at the intersection of the front with the escarpment, and starts moving along the escarpment. This phenomenon is illustrated in Figures 8.5 and 8.6, where results of the corresponding numerical simulations are displayed. As follows from Figure 8.5, after emission of inertia–gravity waves at the

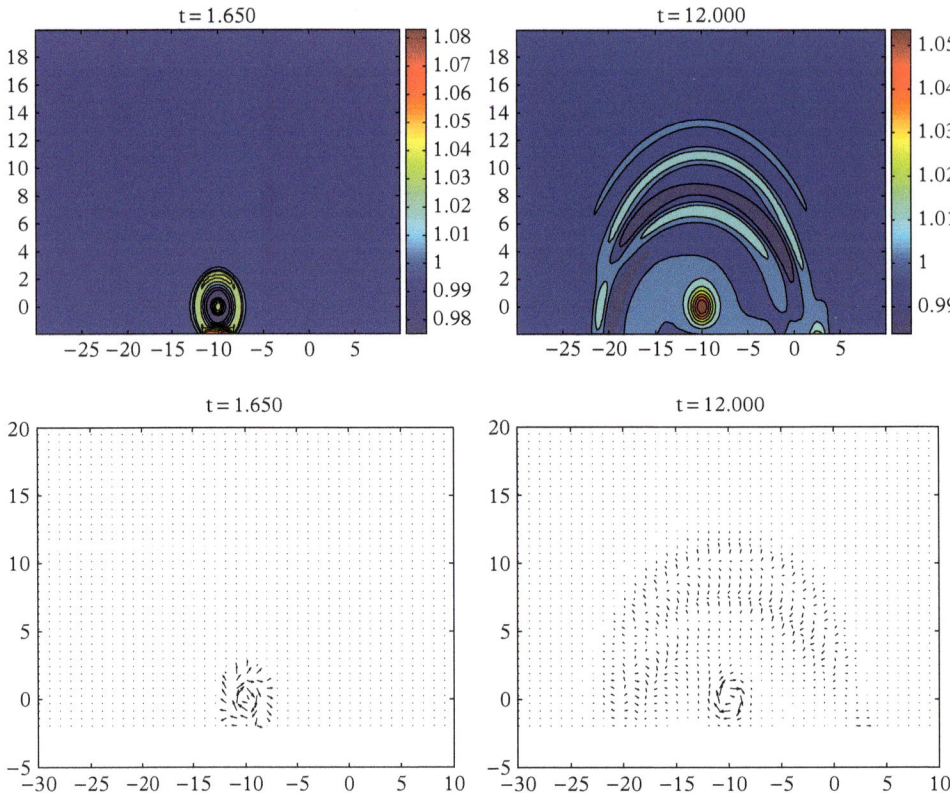

Figure 8.4 *Numerical simulations of the geostrophic adjustment. Left panel: initial stage of adjustment; right panel: advanced stage of adjustment. Upper row: pressure (thickness) field. Lower row: corresponding velocity field. Time is measured in f^{-1} and length in R_d. Initial perturbation consists of a bump in thickness, with no velocity.*

initial stages, the evolution mostly consists in the formation of a localised topographic wave, which is clearly visible in Figure 8.6. What is important for applications is that this localised topographic wave packet has sufficient amplitude to capture and transport fluid and therefore, tracers, along the escarpment, as can be shown be seeding a tracer field in the simulations (not shown).

8.4.3   Geostrophic adjustment in the equatorial beta plane

The slow component in the long-wave sector of shallow-water motions at the equator consists of Rossby and Kelvin waves, as was shown in Section 5.5. So a long-wave initial perturbation, with meridional scale much smaller than zonal scale, adjusts by forming a packet of Rossby waves moving westward, and a packet of Kelvin waves moving eastward. Long inertia–gravity waves are emitted during this process, but due to the fact

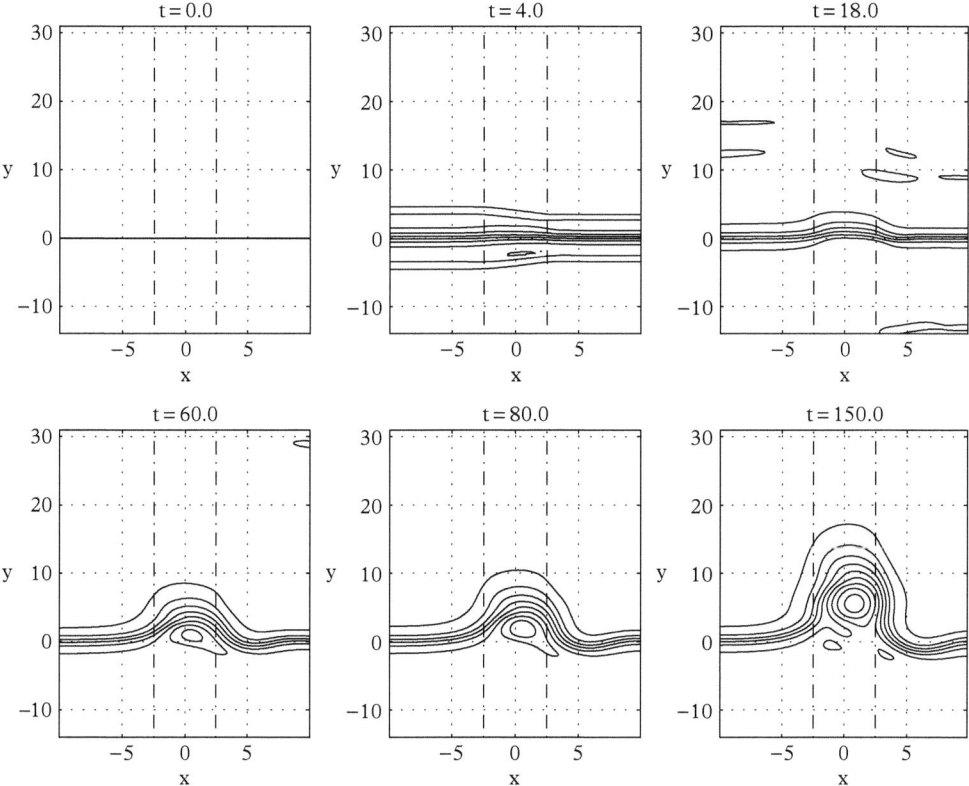

Figure 8.5 *Evolution of the isobars in the x–y plane during adjustment of the pressure front perpendicular to the bottom escarpment. Time is measured in f^{-1} and length in R_d. The limits of the escarpment are indicated by dashed lines.*

that their group velocities are small they are staying for the long time at the location of the initial perturbation, similarly to what we have seen in the numerical simulations of 1.5D adjustment in Section 8.3. This process is illustrated in Figure 8.7 where we present the results of corresponding numerical simulations. It is to be stressed that, unlike the previous two subsections, the geostrophically balanced motion on the equatorial beta plane is represented by propagative Rossby waves, and not by stationary vortices. It should be remembered that numerical simulations displayed in Figure 8.7 are initialised with a symmetric with respect to equator perturbation, which explains the absence of any signature of Yanai waves. The Kelvin wave propagating eastward in the figure has a positive pressure anomaly and steepens at the front. In the opposite case of negative pressure anomaly it would steepen at the rear, in agreement with predictions of hydraulic theory of Chapter 7.

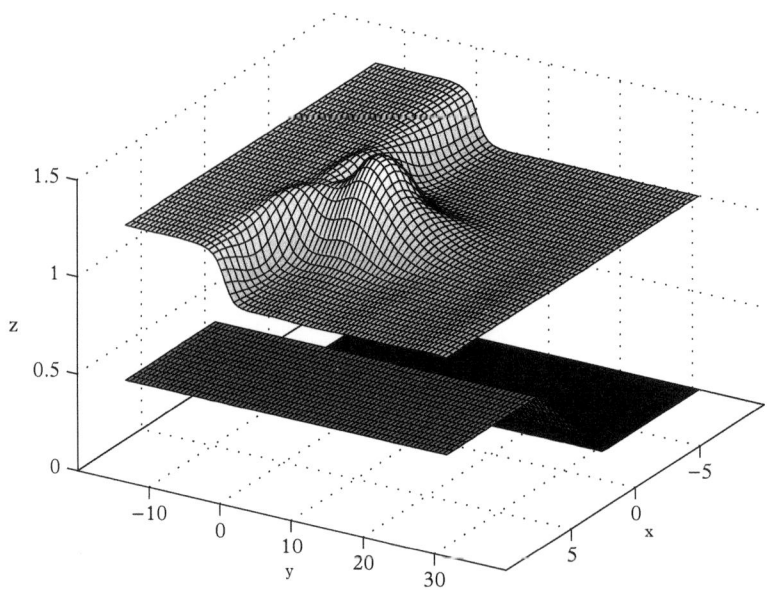

Figure 8.6 *A snapshot at the late stage of geostrophic adjustment of a pressure front perpendicular to the bottom escarpment, as seen in the pressure (thickness) field. A packet of topographic waves starting along the escarpment is visible in a form of a bump.*

8.5   Summary, comments, and bibliographic remarks

Geostrophic adjustment is a fundamental process in geophysical fluid dynamics. The notion of geostrophic adjustment was introduced in Rossby (1938) and developed in classical papers and books (Monin and Obukhov 1958, Gill 1982). As we have seen, the problem of the geostrophic adjustment can be completely solved, at low Rossby numbers, both in one- and two-layer shallow-water models by multi-time scale asymptotic expansions, under the simplifying hypothesis of single spatial-scale motions. Moreover, the solution of the geostrophic adjustment problem provides a solid justification of QG models and wave–vortex splitting in the case of motions with spatial scales of the order of the deformation radius. It also gives a solution of the initialisation problem, allowing us to correctly project any initial conditions onto the slow and fast components of motion. In the case of motions with scales much larger than deformation radius, the solution of the adjustment problem leads to frontal geostrophic equations, different from the standard QG, but still having no fast-motions drag, and to a modulation equation for quasi-inertial oscillations, which are guided by slow motions. The FG equations for slow motions contain multiple spatial derivatives, which can lead to ill-posedness of corresponding initial-value problems for some initial conditions (Oliver and Vasylkevich 2016). In the presence of coastal, topographic, or equatorial wave guides, the adjustment process results in emission of both inertia–gravity and wave-guide modes. Inertia–gravity waves

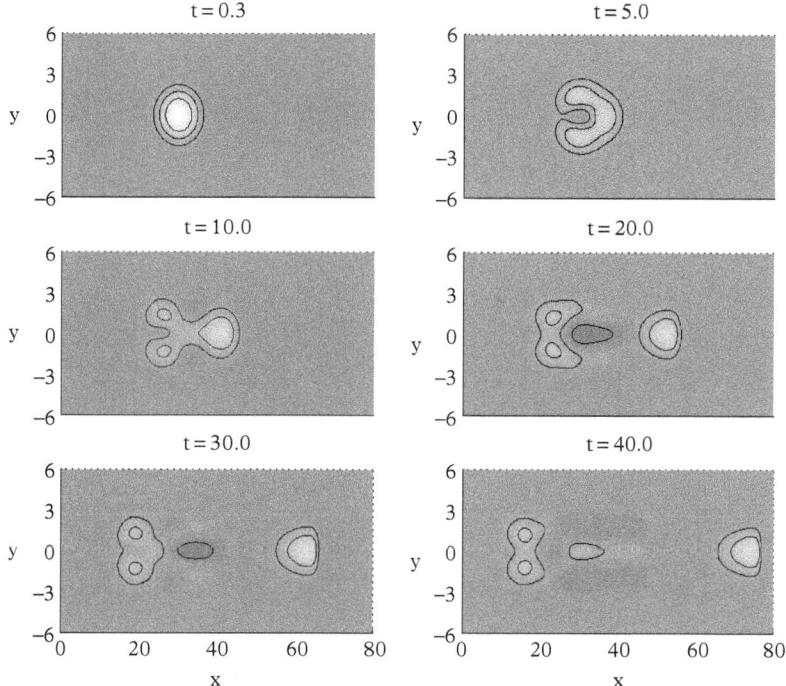

Figure 8.7 *Adjustment of a positive pressure anomaly of large zonal scale at the equator, with formation of Rossby and Kelvin waves, and of a slowly dispersing packet of equatorial inertia–gravity waves. Yanai waves are absent by symmetry. Levels of grey: layer thickness. Time is measured in $(\beta R_e)^{-1}$ and distances in R_e.*

are decoupled from the geostrophically balanced motions, as well as Kelvin waves, while topographic and equatorial Rossby waves are part of the balanced slow motions themselves. It should be stressed that the demonstrations in this chapter were made under hypothesis of *radiation boundary conditions* for fast inertia–gravity waves, and localised initial conditions at the lateral boundary, if appropriate. Relaxing this hypothesis can lead to violations of fast–slow splitting (e.g. Thomas 2016), and to recoupling of Kelvin wave to the geostrophic motions at the coast (Reznik and Grimshaw 2002). It should be also emphasised that the method of straightforward asymptotic expansions used in this chapter does not capture inertia-gravity waves of exponentially small amplitude which can be generated by balanced motions (Vanneste and Yavneh 2004).

The presentation of the theory of the geostrophic adjustment above follows the papers by Reznik, Zeitlin, and BenJelloul (2001) and (Zeitlin, Reznik, and BenJelloul 2003) in 2D case (cf. the last paper for generalisation to continuously stratified flows), and the paper by Zeitlin, Medvedev, and Plougonven (2003) in the 1.5D case. High-resolution numerical simulations of the geostrophic adjustment in 1.5D RSW were first performed in Kuo and Polvani (1999), the presentation above follows

Bouchut, LeSommer, and Zeitlin (2004). The problem of adjustment near the idealised straight coast in RSW was solved in Reznik and Grimshaw (2002), and on the equator in LeSommer, Reznik, and Zeitlin (2004). A theory and laboratory experiments of adjustment process over escarpment can be found in the classical paper by Gill *et al.* (1986). Adjustment of a front over escarpment was studied numerically in the paper by Bouchut, Scherer, and Zeitlin (2008).

Figures 8.1, 8.2, and 8.3 are reproduced from Bouchut, LeSommer, and Zeitlin (2004), Figure 8.7 is reproduced from LeSommer, Reznik, and Zeitlin (2004), all with permissions. Figures 8.5 and 8.6 are reproduced from Bouchut, Scherer, and Zeitlin (2008) with the permission of AIP publishing.

9

RSW Modons and their Surprising Properties: RSW Turbulence

In this chapter we come back to the modon solutions of quasi-geostrophic (QG) models which were introduced in Chapter 6, and show that they have analogues in full rotating shallow-water (RSW) equations, not only at small but also at moderate Rossby numbers. Relaxation of dipolar perturbations towards these solutions of the RSW equations at large Rossby number provides an example of *ageostrophic adjustment*, as compared to geostrophic adjustment at small Rossby numbers of Chapter 8. Solutions themselves display some rather surprising properties, like quasi-elastic collisions, indicating that there probably exist some hidden symmetries and related conservation laws in the model.

9.1 QG vs RSW modons: one-layer model

9.1.1 General properties of steady solutions

A natural question regarding the modon solutions of QG equations considered in Chapter 6 is about their counterparts in the parent one- and two-layer RSW models. One of the main properties of QG modons is that they are moving steadily along rectilinear trajectories. If we are looking for steady-translating solutions with constant zonal velocity U in the RSW model on the f-plane, they obey the equations

$$\begin{cases} (u-U)u_x + vu_y - fv + gh_x = 0, \\ (u-U)v_x + vv_y + fu + gh_y = 0, \\ ((u-U)h)_x + (vh)_y = 0. \end{cases} \tag{9.1}$$

The last equation suggests introduction of the co-moving stream function χ:

$$(u-U)h = -\chi_y, \quad vh = \chi_x. \tag{9.2}$$

Geophysical Fluid Dynamics. Vladimir Zeitlin,
Oxford University Press (2018). © Vladimir Zeitlin. DOI: 10.1093/oso/9780198804338.001.0001

In terms of χ the potential vorticity (PV) is expressed as follows:

$$q = \frac{1}{h^2}\nabla^2\chi - \frac{1}{h^3}\left(h_x\chi_x + h_y\chi_y\right) + \frac{f}{h}. \tag{9.3}$$

Together with PV, another important quantity is Bernoulli function, which is related to the energy density. Calculated in the frame of reference moving with the velocity U, it is

$$B = \frac{1}{2}\left(u - U\right)^2 + \frac{1}{2}v^2 + gh + U\,fy. \tag{9.4}$$

For steadily moving solutions it is expressed in terms of χ as

$$B = \frac{(\chi_x)^2 + (\chi_y)^2}{2h^2} + gh + U\,fy. \tag{9.5}$$

As follows from (9.3) and (9.5), the Bernoulli function and PV are functionally related. Although it is not easier to solve the equations (9.3) and (9.5) for χ and h at given q and B than to solve the original system (9.1), the fact that q and B are functionally dependent is useful for diagnostics of numerical simulations, as a clear functional relation between q and B, which may be detected by analysing so-called *scatter plots* of one as a function of another, allows us to identify a steady solution. In the case of functional dependence between the two, the scatter plot appears as a well-defined curve, while in the opposite case it is, instead, a cloud of points. Using this technique, we can immediately test the QG modon solution in RSW, taking the thickness as geostrophic stream function. Figure 9.1 shows scatter plots of Bernoulli function vs PV obtained from velocity and thickness distributions of QG modons of Section 6.2.2 with small and moderate Rossby numbers. The Rossby number is calculated using the translation velocity U, which is directly related to the peak velocity of the modon. As follows from the figure, while at very small *Ro* the QG modon is close to a solution of the full RSW model, as expected from the nature of the QG approximation, at higher Rossby number this is not at all the case.

9.1.2   'Ageostrophic adjustment' of QG modons

In order to understand whether modon-like solutions of RSW equations exist at non-negligible Rossby numbers, we start with a QG modon as initial configuration, and follow numerically its evolution, using the numerical scheme explained in Chapter 7. We will see in this way that a dipolar vortex moving steadily without changing form does emerge in such numerical experiments in the full RSW model.

Figure 9.2 represents the evolution of PV in a simulation initialised with the QG modon corresponding to the right panel of Figure 9.1. As seen in the figure, the dipole moves along the curved trajectory spilling a part of cyclonic PV, and losing the cyclone–anticyclone symmetry. Nevertheless, at late stages the dipole has a sharp scatter plot, as follows from Figure 9.3, and thus represents a solution of the RSW equations.

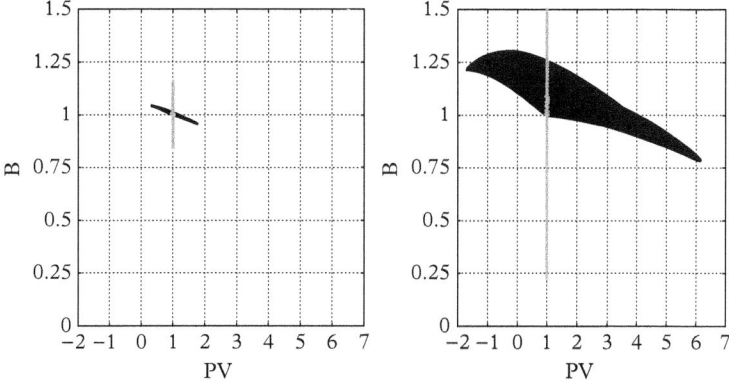

Figure 9.1 *Scatter plots of Bernoulli function vs PV calculated with the distributions of thickness and velocity of a QG modon with Ro = 0.04 (left panel), and Ro = 0.2 (right panel). Black (grey): inner (outer) region inside (outside) the separatrix.*

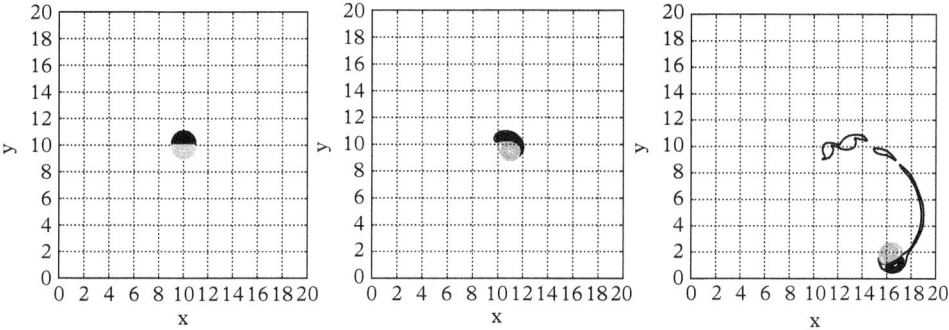

Figure 9.2 *Adjustment of a QG modon with Ro = 0.2: Snapshots of PV at t = 0, 10, 100f^{-1}, from left to right. Black: positive (cyclonic) PV, grey: negative (anticyclonic) PV.*

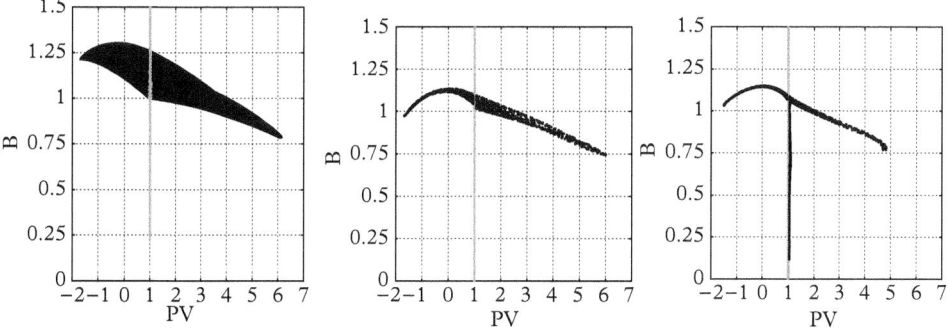

Figure 9.3 *Adjustment of QG modon, as seen in the scatter plot Bernoulli function vs PV at times corresponding to snapshots of Figure 9.2, from left to right. Black: modon core, grey: outer region.*

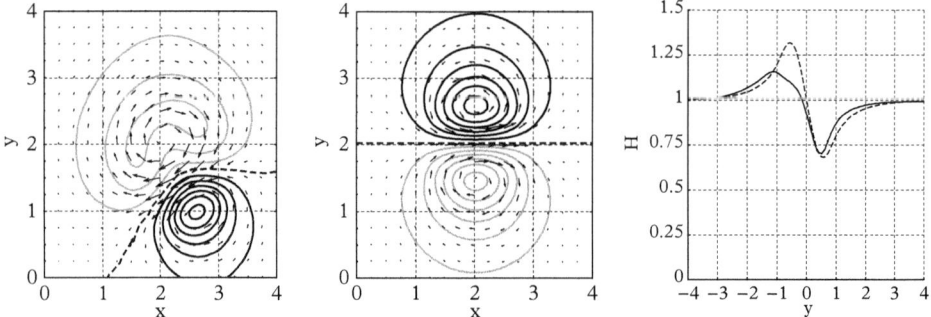

Figure 9.4 *Comparison of thickness and velocity distributions of the RSW modon, left panel, and the QG modon, middle panel in the x − y plane, and comparison of their thickness profiles, right panel: solid (dotted) line, RSW (QG) modon.*

A comparison of phase portraits of the resulting RSW modon and the QG modon is presented in Figure 9.4, together with a comparison of respective thickness profiles, as seen in a section across the dipole axis.

The RSW modon is clearly asymmetric, and this asymmetry leads to bending of its trajectory. The process of adaptation (adjustment) of the initial QG modon configuration to the RSW modon state is accompanied not only by ejection of 'redundant' cyclonic PV but also by emission of inertia–gravity waves, as follows from Figure 9.5, which is a fundamental property of the geostrophic adjustment process studied in Chapter 8. Yet, in the present case it is rather an 'ageostrophic adjustment', as the Rossby number is far from being negligible. As follows from the results of Chapter 8, inertia–gravity waves can be diagnosed with the help of the divergence field. Figure 9.5 clearly

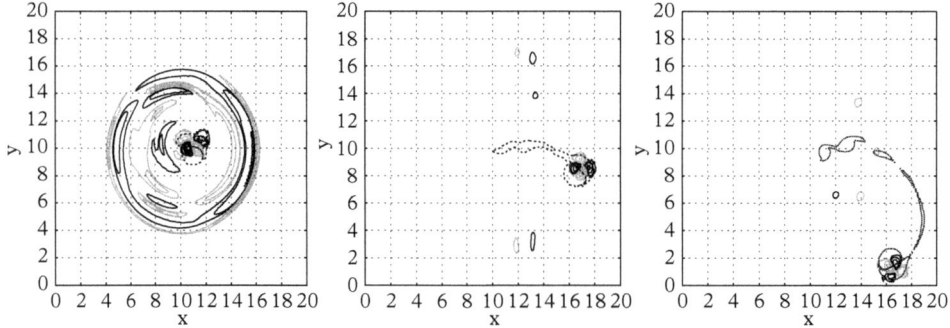

Figure 9.5 *Adjustment of the QG modon, as seen in the divergence field. Snapshots at $t = 5, 45, 100f^{-1}$, from left to right. A train of concentric inertia–gravity waves is clearly seen in the left panel. It rapidly rans out of the calculational domain, leaving a quadrupolar divergence pattern moving with the end-state vortex, which confirms its ageostrophic character. Black: divergence, grey: convergence.*

shows a wave packet running out of the vortex. Thus emitted inertia–gravity waves are then evacuated with the help of numerical sponges, absorbing boundary conditions at the boundaries of the computational domain. The domain is chosen to be sufficiently large compared to the vortex size, not to disrupt the vortex itself by the sponges.

9.1.3 Properties of RSW modons

We illustrate in this section some salient properties of the RSW modons, as follows from numerical simulations. As its QG counterpart, the RSW modon transports fluid (and, hence, tracers) in its core. Figure 9.6 illustrates this property, showing that the modon seeded with a passive tracer transports this latter along its trajectory.

An illustration of robustness of the ageostrophic modon is provided by its interaction with topography presented in Figure 9.7. We take the simplest escarpment topography, which was used already in previous chapters. The escarpment is chosen to be steep and is given by the step-function of x. The initial modon is oriented in a way to approach the escarpment in the normal direction, as shown in the lower panel of the figure. For a low escarpment with relative height $\frac{\Delta h}{H_0} = 0.05$ corresponding to the figure, the modon crosses the topography along an inflected trajectory but maintains its coherence and structure. The form of trajectory can be explained in standard terms from the PV conservation and topographic compression of fluid columns. A critical relative height of the escarpment is about 0.2. Beyond this value the modon disaggregates after encountering the step.

An extreme case of a high step is the lateral boundary (coast). We have already seen in Chapter 6 (see Problem 6.6) that a point vortex moves along

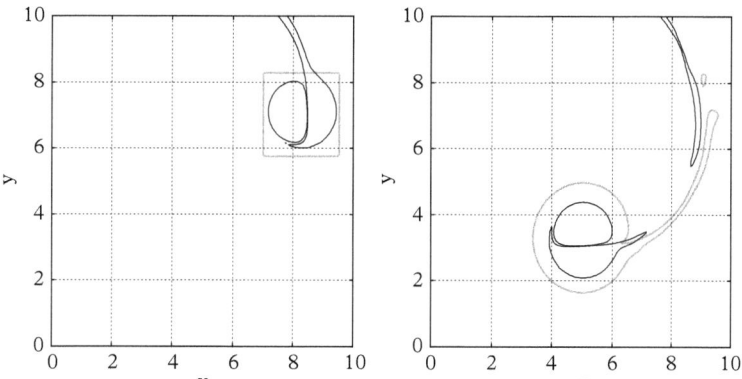

Figure 9.6 *Transport of a passive tracer (inside the grey contour) by RSW modon. Left panel: t = 0, tracer is uniformly seeded in a rectangle containing the vortex cores. Right panel: t = 40f^{-1}, most part of the tracer is captured in the vortex cores and transported along the vortex trajectory, while some part of the tracer is detrained with the vortex filament. The modon is represented by the isolines of PV with |Q| = 0.05 (black).*

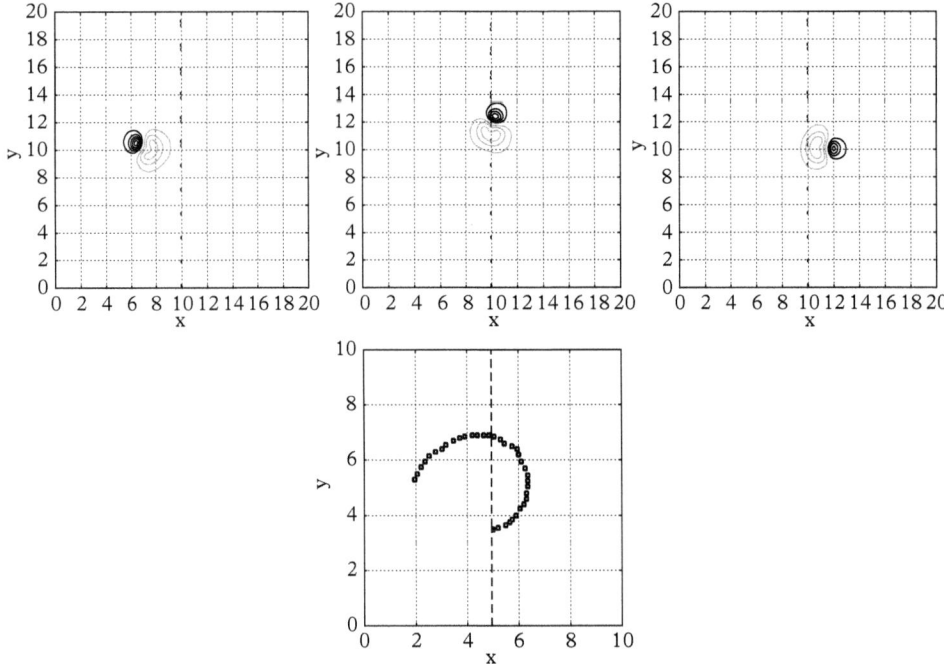

Figure 9.7 *Upper row: snapshots at $t = 10, 40, 60f^{-1}$, from left to right, of thickness anomaly of a modon crossing step-like topography indicated by dashed line from deep (left) to shallow (right) side. The relative amplitude of the step is 0.05, Black: cyclone, grey: anticyclone. Lower panel: trajectory of the modon, points at time interval $2f^{-1}$).*

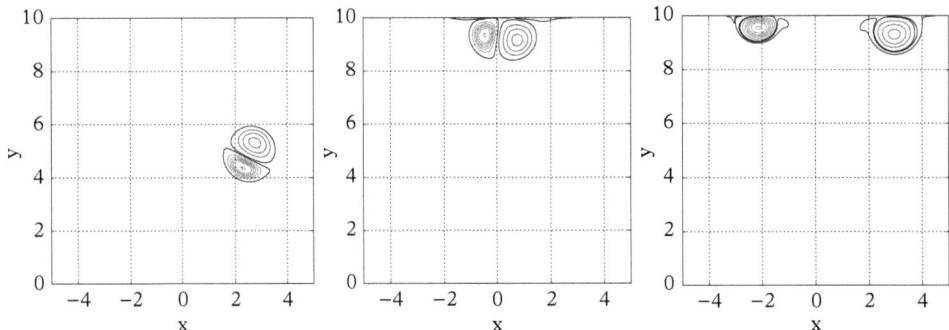

Figure 9.8 *Snapshots at $t = 10, 40, 60f^{-1}$, from left to right, of the interaction of an ageostrophic modon with a wall situated at the top of the domain.*

the boundary with constant velocity. When a RSW modon hits the wall, its anticyclonic and cyclonic components split apart and move along the wall in opposite directions according to the point-vortex prediction, as follows from Figure 9.8.

9.2   QG vs RSW modons: two-layer model

9.2.1   Adjustment of barotropic QG modons

The same programme as in Section 9.1 can be accomplished in the two-layer case. If the quasi-barotropic two-layer modon of Figure 6.3 is used as initial condition in numerical simulations with the two-layer free-surface RSW, at moderate Rossby numbers the scenario of its evolution is very similar to that of the one-layer modon, as follows from Figure 9.9. The evolution starts by 'ageostrophic adjustment' and emission of internal (baroclinic) and barotropic inertia–gravity waves, with a net signature in the divergence fields, cf. Figure 9.10. The typical phase velocities of the waves are, respectively, 0.2 and 1 in units of $\sqrt{gH_0}$, in a good agreement with theoretical values for both types of gravity waves in the two-layer model. A weak filamentation of cyclonic vorticity at the rear of the dipole is observed, as in the one-layer case. It is a typical feature of the vortex dipole evolution observed both in numerical and laboratory experiments. The resulting at late stages modon is still quasi-barotropic, but asymmetric, and has sharp scatter plots in

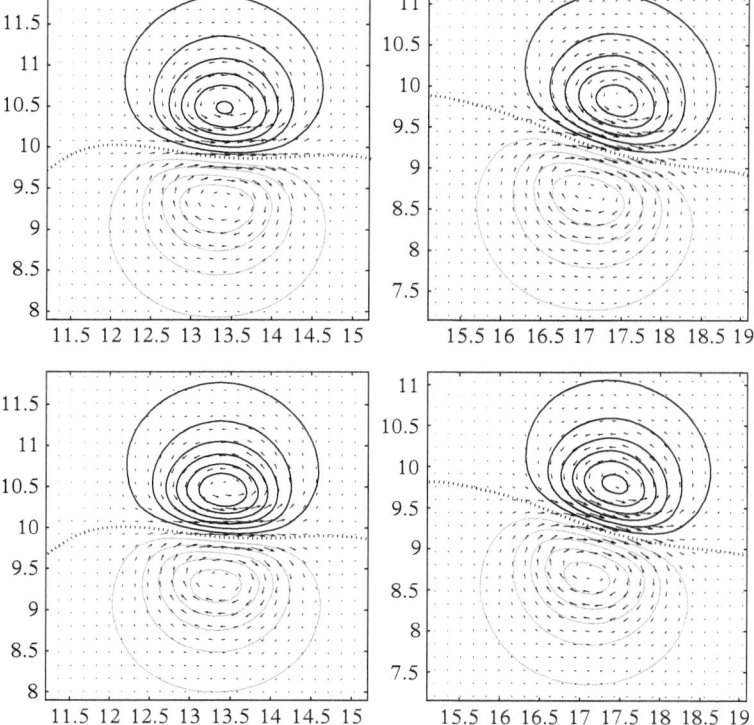

Figure 9.9 *Pressure with superimposed velocity in the first (top) and second (bottom) layers at t = 88 (left column) and t = 200f^{-1} (right column) in the end state of adjustment of a quasi-barotropic modon with initial Ro = 0.2.*

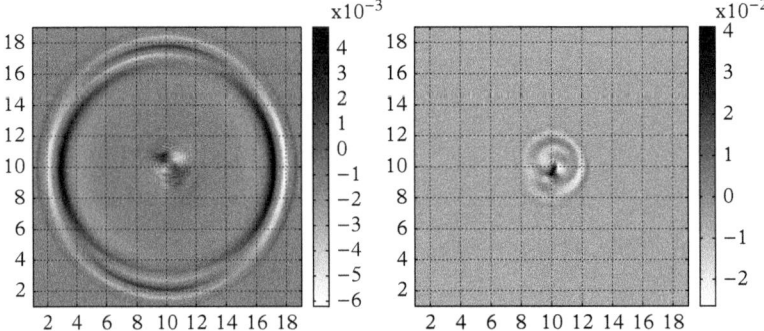

Figure 9.10 *Barotropic (left) and baroclinic (right) divergence fields at t = 8f^{-1} during adjustment of the two-layer quasi-barotropic modon with initial Ro = 0.2.*

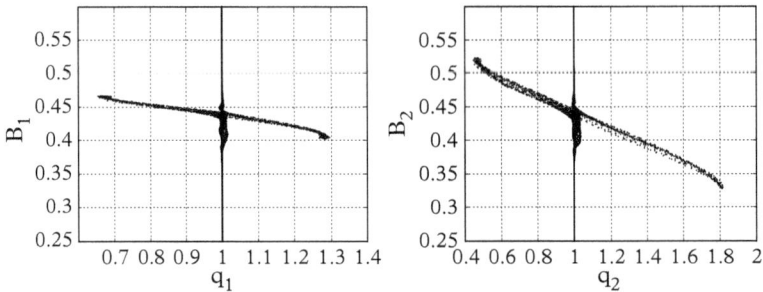

Figure 9.11 *Scatter plots of Bernoulli function vs PV for the adjusted quasi-barotropic modon with initial Ro = 0.2 at t = 180f^{-1}. Left (right) panel: upper (lower) layer.*

both layers, which are shown in Figure 9.11 and confirm that an exact steady solution is being approached.

9.2.2   Adjustment of baroclinic QG modons

The evolution of the baroclinic modon of Figure 6.3 follows a similar scenario, although cyclone–anticyclone asymmetry in the adjusted dipole is stronger, cf. Figure 9.12, and, as a consequence, the trajectory is stronger curved.

9.2.3   Adjustment of essentially ageostrophic modons

At higher Rossby numbers, the barotropic modons still follow the same scenario as in the one-layer case, which is illustrated in Figure 9.13, although a strong numerical dissipation zone appears close to the axis of the dipole. We will see the meaning of this phenomenon in Section 9.3. The evolution of the baroclinic modons in the two-layer

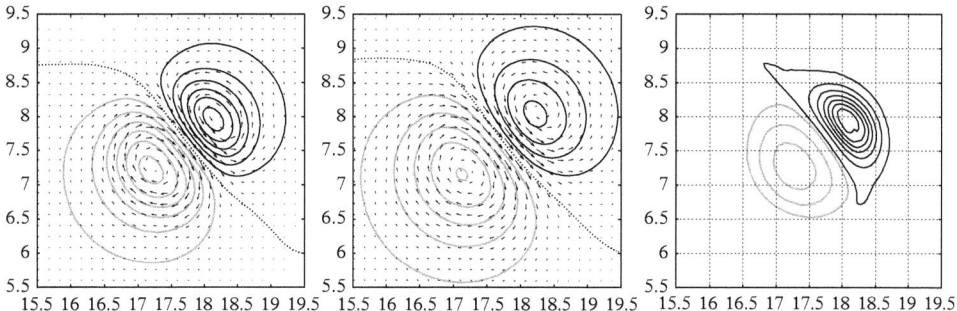

Figure 9.12 *Pressure anomaly with superimposed velocity fields in the upper (left panel) and lower (middle panel) layers, and the PV in the upper layer (right panel) at $t = 252f^{-1}$ of the adjusted baroclinic modon with initial $Ro = 0.2$. Black: cyclone, grey: anticyclone.*

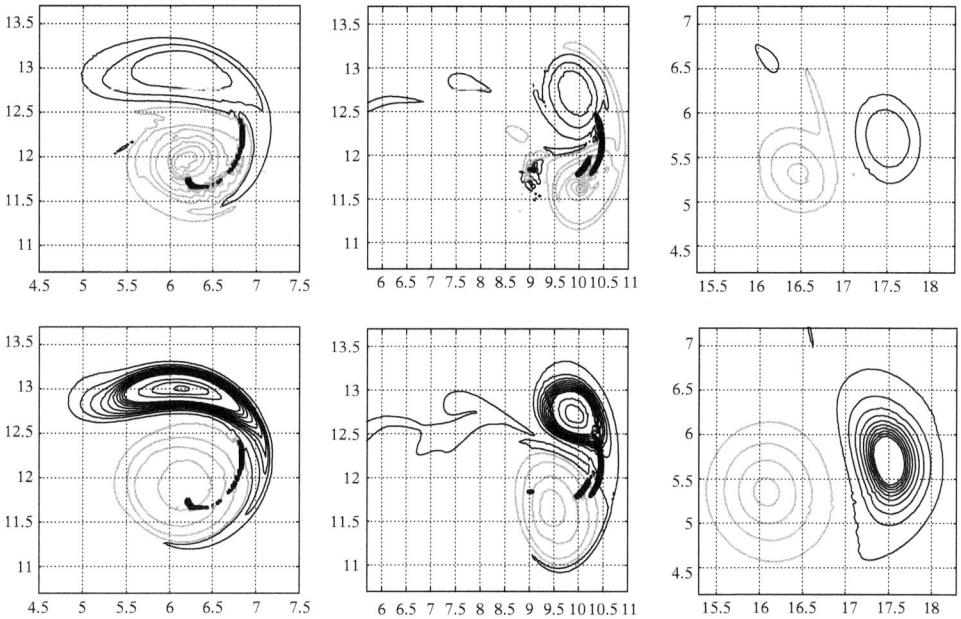

Figure 9.13 *Evolution of PV in a simulation initialised with QG modon with $Ro = 0.4$ and stratification ratio 0.8. From left to right: snapshots at $t = 8, 40, 200f^{-1}$. First row: layer 1; second row: layer 2. Black: cyclone, grey: anticyclone. A zone of high dissipation is indicated by a thick black line.*

model at higher Rossby numbers exhibits a variety of scenarios, depending on the values of stratification and aspect ratio. As an illustration, we show in Figure 9.14 a formation of a so-called rodon configuration from an initial baroclinic QG modon which happens at stratification ratio 0.6. The upper-layer dipole in this case splits, leaving the anticyclonic vortex behind, while the cyclonic vortex is driven by the lower-layer dipole.

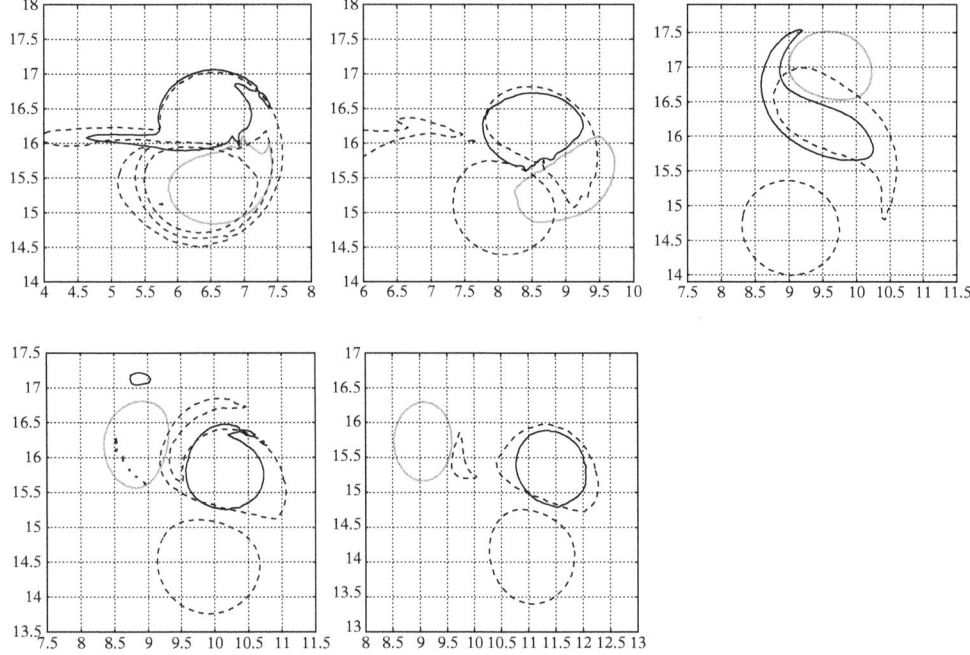

Figure 9.14 *Evolution of PV during adjustment of the baroclinic QG modon with initial Ro = 0.4 and stratification ratio 0.6. Dotted line: lower layer, continuous black (grey) lines: cyclone (anticyclone) in the upper layer. From left to right and up to down: t = 20, 40, 60, 80, 100f^{-1}.*

9.3 Shock modons

As we have seen in Section 9.2, relaxation of a QG modon with sufficiently high Rossby number produces front-like structures at its axis. To understand this phenomenon better, and be sure that it is not a numerical artefact, we repeat the adjustment experiment with essentially ageostrophic modon in one-layer RSW, where the numerical scheme is surely shock-capturing, and its consistency can be rigorously proved, unlike the semi-heuristic two-layer scheme, cf. Chapter 7.

In Figure 9.15 we present the evolution of a QG modon with Rossby number $Ro = 0.4$ in the one-layer configuration. Compared to the case of a smaller Rossby number of Figure 9.2, the evolution in this case leads to stronger ejection of the positive vorticity, forming a secondary cyclone in the wake of the dipole. At the same time, a stronger cyclone – anticyclone asymmetry of the dipole arises and, as a consequence, a stronger curvature of its trajectory. As a result of the combination of these effects, the dipole collides (cf. the upper-right panel of the figure) with the secondary cyclone and exits more compact after the collision (cf. the lower-left panel). A localised PV anomaly appears at the axis of the dipole, which was never the case for the lower Rossby number modons analysed earlier. This anomaly corresponds to a sharp front (shock), as is clear from

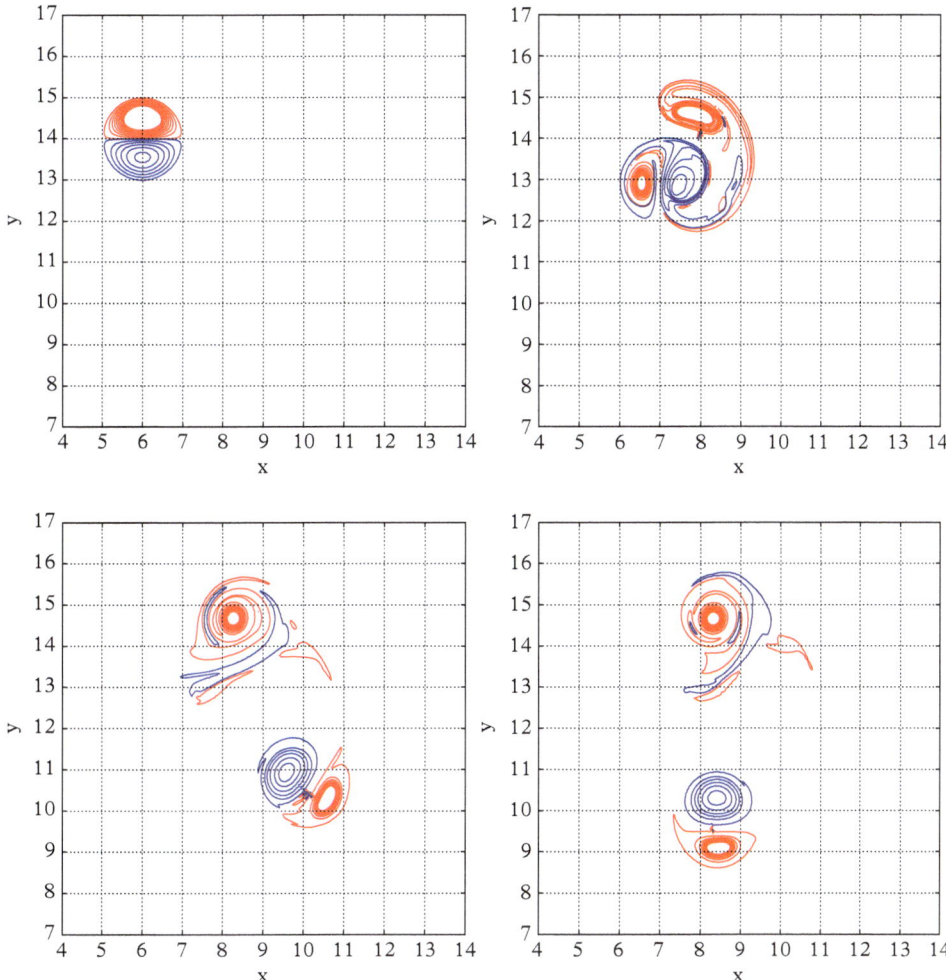

Figure 9.15 *Snapshots of the evolution of PV in the simulation initialised with QG modon with Ro = 0.4 in one-layer RSW at (from left to right, and from top to bottom) t = 0, 12, 36, 46f*$^{-1}$.

the Figure 9.16, and can be also seen in the cross-sections of thickness, divergence, and velocity along the modon's axis.

Thus, we get a modon with associated shock across its axis, forming a coherent structure, as follows from Figure 9.17. As the shock induces enhanced numerical dissipation, it disappears after some time, leaving an 'ordinary' modon, with a sharper scatter plot. It should be emphasised that during the whole above-described process, including collision, the energy of the system changes very little. We will see that such quasi-elastic collisions, in the sense of energy conservation, are proper to modons. Coming back to the evolution of the two-layer barotropic ageostrophic modon of Section 9.2.3, a similar to the scenario which we have just described takes place, with a shock forming in the upper layer at the axis of the modon, and leading to enhanced numerical dissipation.

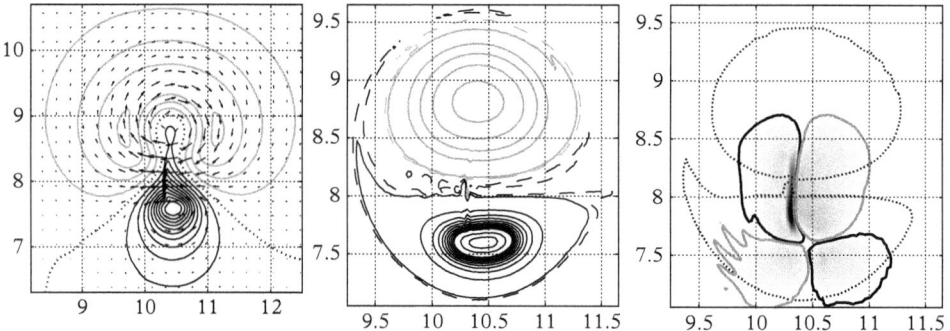

Figure 9.16 *Zoom of the shock modon: isobars with superimposed velocity field (left panel), relative vorticity isolines (middle panel), and divergence (levels of grey, right panel). Thin black lines in the middle and left panels: PV = ±0.1 contours roughly corresponding to the dipole's boundary (the separatrix).*

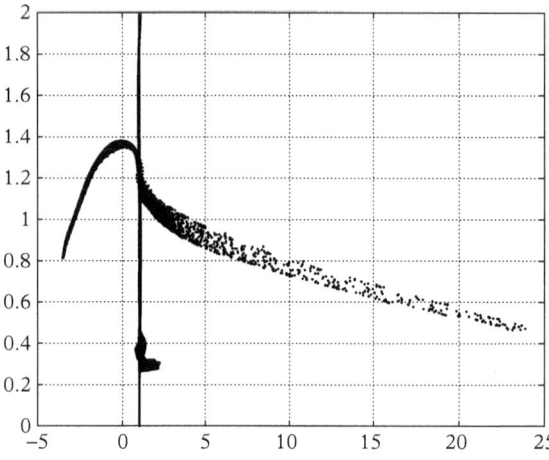

Figure 9.17 *Bernoulli function vs PV for the shock modon of Fig. 9.15 at t = $46f^{-1}$.*

9.4   Interactions of RSW modons

As we have seen in Section 9.3, a whole class of essentially non-linear coherent structures, which are exact solutions (or at least are close to such) exist in one- and two-layer RSW. They are localised, have finite energy, and can move far from their initial location and transport tracers. A natural question arises, what happens during encounter of such structures? The answer is obtained by initialising numerical simulations with pairs of RSW modons, which are oriented in a way to meet each other while moving along their respective trajectories. As we have seen already in Section 9.3, a collision of a modon with a monopolar vortex is quasi-elastic; both vortices survive it and there is no specific energy loss associated with the collision. We will see this rather surprising property of modon collisions repeatedly in the following. A rough classification of outcomes of such collisions is as follows. Four main types of collisions were observed:

- 2 RSW modons → 2 RSW modons: quasi-elastic collision with partner exchange,
- 2 RSW modons → 2 'loose' RSW modons: inelastic collision,
- 2 RSW modons → 1 tripole: inelastic collision with fusion of anticyclones,
- 2 RSW modons → 1 tripole + 1 monopole: inelastic collision with fusion of cyclones and strong filamentation of vorticity.

We briefly describe each of them in the following. The simulations in all cases are initialised by RSW modons obtained by 'ageostrophic' adjustment of QG modons. A portion of the computational domain containing RSW modon after adjustment is cut-off and 'pasted' to initial configuration at different orientations. Typical Rossby numbers of the RSW modons are between 0.2 and 0.3.

9.4.1 2 modons → 2 modons collision

The first type of collision is illustrated in Figure 9.18. It is elastic, with the modons exchanging partners, that is the cyclone of one of them coupling with the anticyclone of another one, the same for the remaining vortices. The 'newborn' modons exit the collision having trajectories at almost right angle with respect to initial ones. The elastic character of the collision is confirmed by the behaviour of energy presented in Figure 9.19.

9.4.2 2 → 2 'loose' modon collision

The second type of collision is illustrated in Figure 9.20. It produces a pair of modons of a new type. As follows from Figure 9.21, where distributions of pressure and PV in this new structure are displayed, the new modon is 'loose' in a sense that its cyclonic and

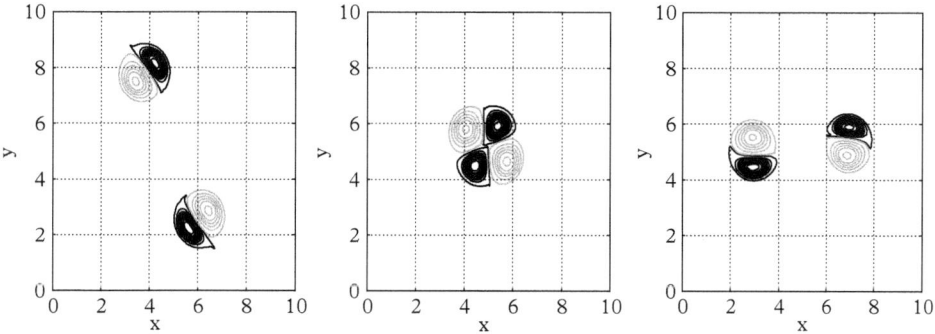

Figure 9.18 *Snapshots of PV at $t = 25, 40, 50f^{-1}$, from left to right, during collision of two RSW modons. Black: cyclone, grey: anticyclone. Contour interval: 0.5, the lowest displayed value of the PV is 0.05.*

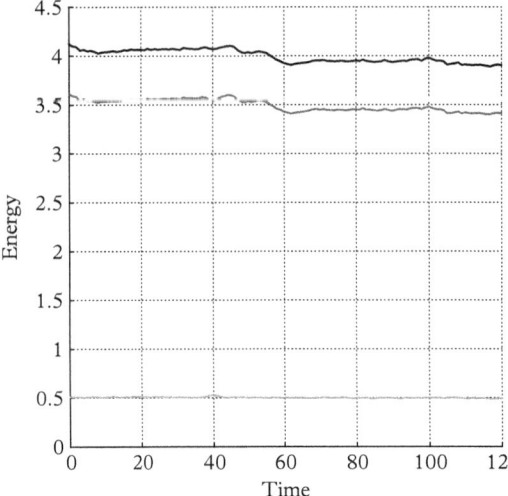

Figure 9.19 *Evolution of the total (upper curve), kinetic (middle curve), and potential (lower curve) energy during the collision displayed in Fig. 9.18.*

anticyclonic PV anomalies are much farther apart than in the 'normal' RSW modon. This new modon may be called nonlinear, or essentially ageostrophic, in a sense that even at small Rossby numbers its PV distribution is very different from that of the QG modon. Yet, as is clear from its scatter plot, which is presented in the right panel of the figure 9.21, it is a coherent structure, although the scatter curve is nonlinear, which again explains the name. Numerical dissipation is enhanced during the collision time, as follows from Figure 9.22, which shows the evolution of energy. This is consistent with substantial reorganisation of the component vortices, necessitating reconnection of vortex lines, which is impossible without dissipation. This type of collision is, therefore, not elastic in the sense of energy conservation.

9.4.3   2 modons → tripole collisions

Another type of inelastic collision of two RSW modons leads to formation of tripoles, as illustrated in Figure 9.23. If the modons hit each other with their anticyclonic components, the tripole forms by merging of these latter. Such process requires reconnections of vorticity isolines and is accompanied by a transient increase of numerical dissipation. The tripole resulting from this process is presented in Figure 9.24 and is coherent, as follows from its scatter plot.

9.4.4   2 modons → tripole + monopole collisions

There is a fourth scenario of collisions of RSW modons which leads to formation of a tripole and monopole vortices. It is realised when colliding modons hit each other with their cyclonic components. Unlike anticyclonic components in the previous case, these latter do not merge but produce a strong filamentation, which leads to formation of a

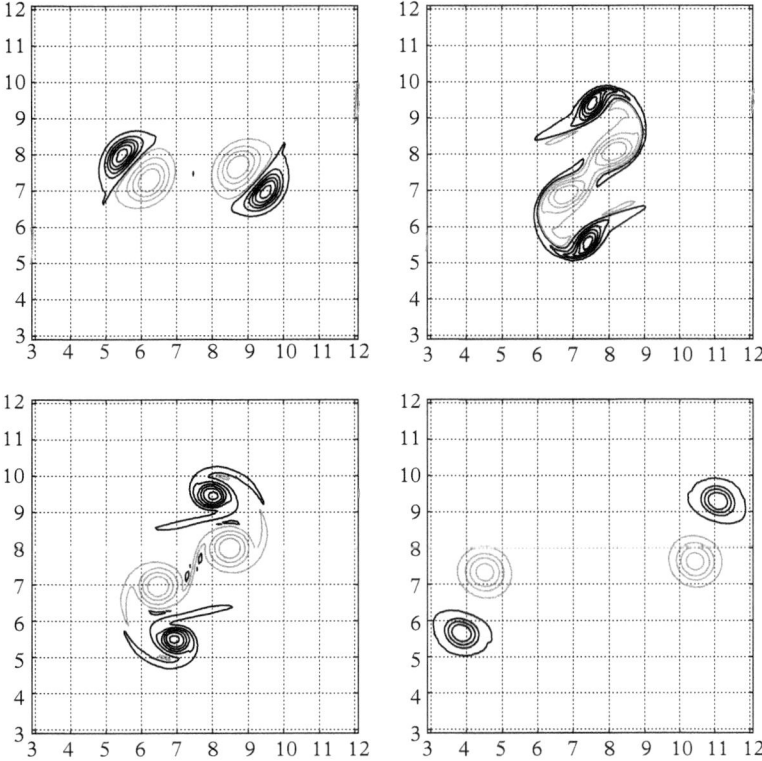

Figure 9.20 *Evolution of PV during collision of two modons producing two 'loose', or nonlinear modons: $t = 18, 34, 40, 100f^{-1}$, from left to right, and from top to bottom.*

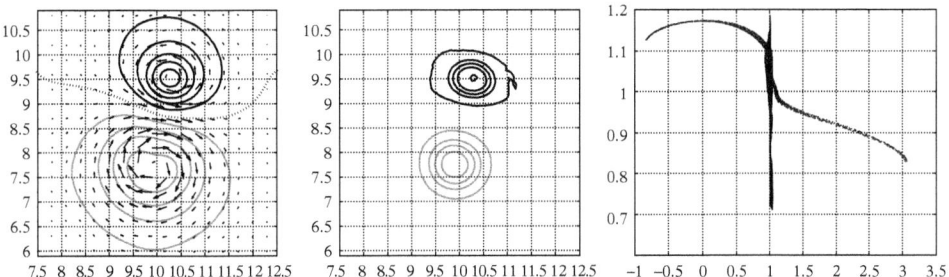

Figure 9.21 *Thickness (left panel) and PV (middle panel) anomalies, and the scatter plot (right panel) of a nonlinear modon.*

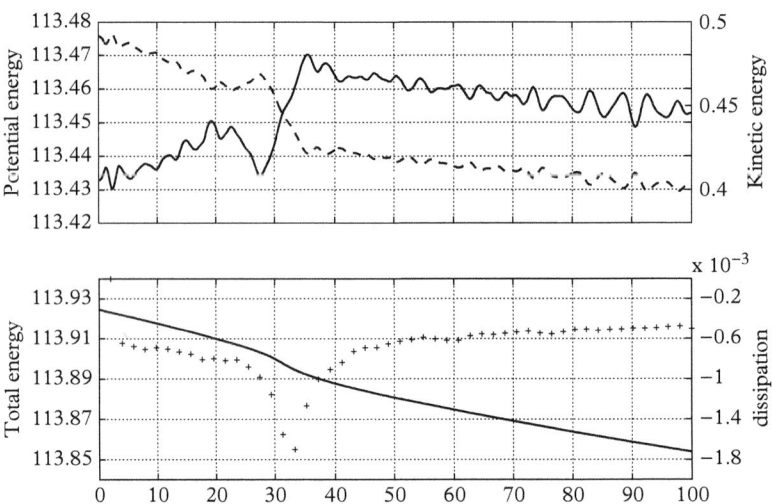

Figure 9.22 *Energy budget during the collision leading to formation of nonlinear ('loose') modons. Upper panel: kinetic (solid) and potential (dashed) energy vs time measured in f^{-1}; lower panel: total energy (solid), and numerical dissipation (crosses).*

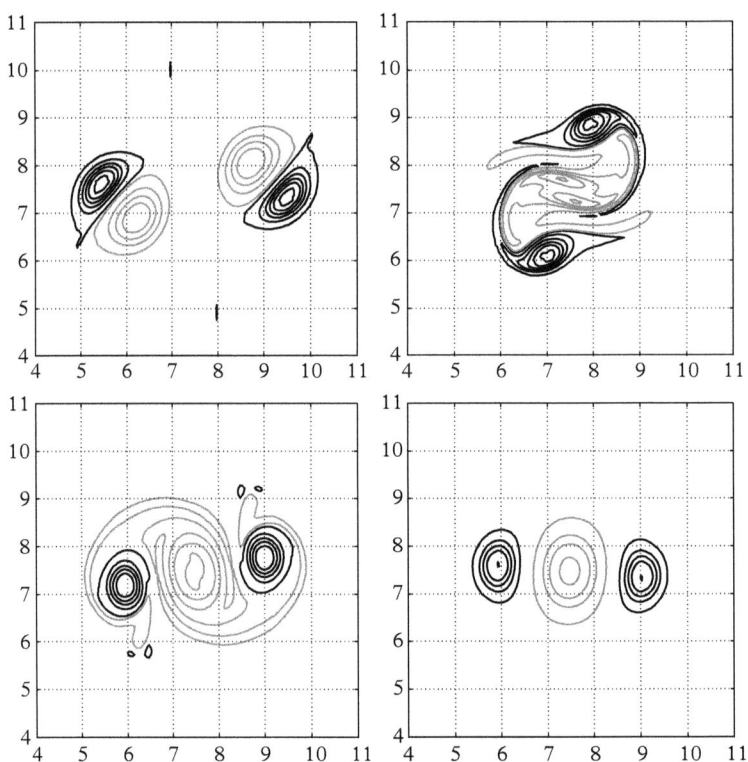

Figure 9.23 *Evolution of PV during collision of two RSW modons producing a tripole: $t = 18, 34, 50, 100 f^{-1}$, from left to right, and from top to bottom.*

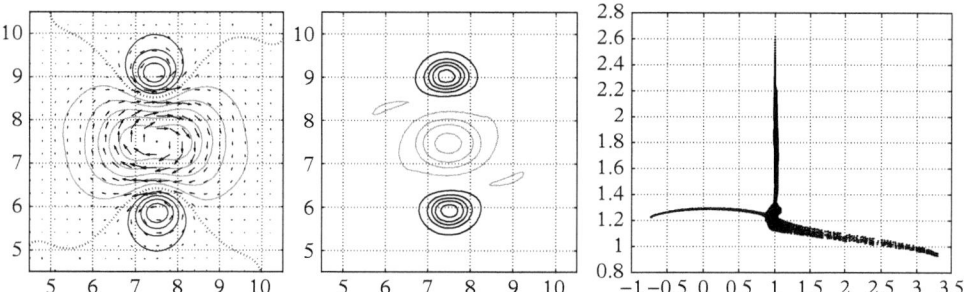

Figure 9.24 *Thickness and velocity (left panel) and PV (middle panel) anomalies, and the scatter plot (right panel) of the tripole of Fig. 9.23 at* $t = 76f^{-1}$.

secondary rather loose dipole. It is a secondary collision of a surviving the first collision modon with this latter which leads to capture of the cyclonic component of the loose dipole and formation of a tripole and an anticyclonic monopole surrounded by a thin band of cyclonic vorticity (not shown). This process is also inelastic, with stronger, as compared to other scenarios, numerical dissipation related to strong filamentation of cyclonic vortices.

9.4.5 Collisions of shock modons

Probably, the most surprising property of shock modons discussed in Section 9.3 is that they also undergo quasi-elastic collisions, in a sense that both modons emerge after collision with the same structure, including the central shocks. Due to these latter, the energy dissipation is constantly present but is not particularly enhanced during collision. This process is illustrated in Figure 9.25.

9.5 2D vs RSW turbulence

As mentioned in Chapter 6, one of the reasons of interest to 2D turbulence is a similarity between 2D Euler and one-layer QG equations on the f-plane and, thus, potential relevance for geophysical applications. It is instructive to compare 2D turbulence with turbulence in RSW model, and see how inclusion of the wave component influences the results. The wave turbulence proper will be studied in Chapter 13. Here we will concentrate on vortex turbulence. The standard way of initialising numerical simulations in 2D turbulence is to use random vorticity fields. The modon solutions of RSW equations, which were studied earlier in this chapter, provide an alternative way of initialising the simulations, with an ensemble of randomly oriented modons. Compared to the standard ones, this is a qualitatively different initialisation which allows us to analyse dependence of the results of simulations of decaying turbulence on initial conditions. We present a comaprison of such simulations of decaying, non-forced, turbulence performed with alternative initialisations.

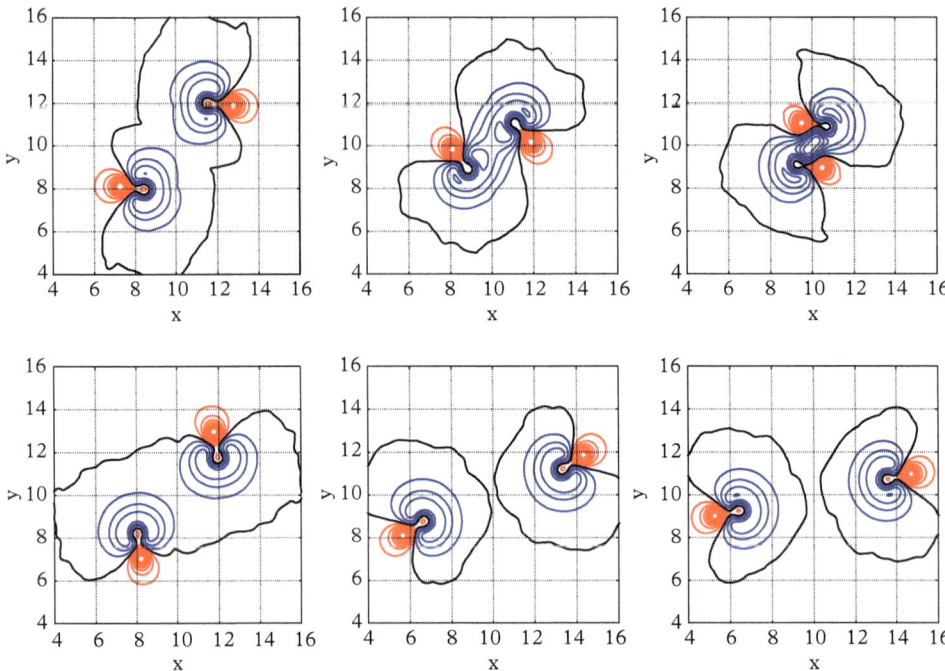

Figure 9.25 *Evolution of thickness anomaly during the quasi-elastic collision of two shock modons: t = 18, 26, 34, 46, 58, 62f^{-1}, from left to right, and from top to bottom. Black contour: zero PV level.*

9.5.1   Set-up and initialisations

Numerical experiments on decaying RSW turbulence are performed with the well-balanced entropy-satisfying finite-volume scheme of Chapter 7, and different types of initialisations: the standard for turbulence simulations ones, with random velocity and pressure fields, and 'non-standard' ones, with randomly oriented coherent dipoles, the modons, where the individual dipole fields were imported from the numerical experiments described in Section 9.1.2 of the present chapter. The obvious difference between two types of initialisation is that in the first the velocity, pressure (height), and vorticity fields are randomly distributed, and decorrelated, while they are much strongly correlated in the second case.

For simulations with coherent-structure initialisations presented below we use a 8 × 8 ensemble of ageostrophic modons. The calculational domain in the simulations with modon initialisation has the size $(32\ R_d)^2$, while the grid size is $dx = dy = 0.05\ R_d$. As already stated, this method of initialisation is essentially different from what is used in most of the studies of two-dimensional turbulence. The initial state is not in geostrophic but in ageostrophic balance. The divergence field associated with each dipole is non-negligible, and has a characteristic quadrupolar form, cf. Section 9.1.2.

In simulations with random initialisations, the thickness anomaly field is taken as a Fourier series with random phases and fixed energy spectrum with a 'spectral bump' of the form $\dfrac{k^{m/2}}{(k + k_0)^m}$, where k_0 is a characteristic wave number corresponding to the maximum of the energy density and m is a 'spread' constant. The velocity field was initialised using the geostrophic wind relation and above-defined thickness field. Thus, the initial state is geostrophically balanced in such initialisations, which makes them a proxy of standard initialisations in 2D turbulence. It is to be emphasised that, although the divergence field is absent at $t = 0$, it appears in the form of random IGW through the initial geostrophic adjustment process due to discretisation errors, and because the geostrophic flow is not a stationary solution of the full RSW equations. The numerical domain for the simulations with this kind of initial conditions is $(8\ R_d)^2$ with $dx = dy \sim 0.0156\ R_d$. It is to be emphasised that the calculational domains are smaller in the random initialisations because of the presence of smaller scales in the initial fields. The typical length scale of the initial configuration is chosen to be smaller than the Rossby deformation radius in order to check the influence of the latter on the energy cascade.

Several parameters characterise a given initialisation. The typical initial length scale enters via the Burger number $Bu = (R_d/L)^2$. The local initial Rossby number is the ratio of the maximal absolute value of relative vorticity and the planetary vorticity: ζ_0/f. The global Rossby number is defined either as the maximum value of the modulus of the velocity divided by f and multiplied by the energy centroid, or as the mean absolute value of the vorticity divided by f. We recall that the centroid of energy is defined as $\langle k \rangle = \int kE(k)\,dk / \int E(k)\,dk$. The mean energy anomaly and the enstrophy $\Omega = \int \int hQ^2 dxdy$, where Q denotes PV anomaly, are also relevant. We will describe the results from the runs called C (coherent-structure initialisation), R1 (random one with a Rossby number close to C) and R2 (random one with initial mean energy close to C). The corresponding parameters are presented in Table 9.1, and the vorticity fields illustrating the two types of initialisation are displayed in Figure 9.26. Here and in the following, the length scale is non-dimensionalised by the Rossby deformation radius, while the time and the vorticity are given in units of f^{-1} and f, respectively.

9.5.2   General features of the evolution of the vortex system

The following common features of freely decaying vortex turbulence were observed in numerical simulations. The evolution of the vorticity field, visualised in Figure 9.27, shows formation of larger coherent vortex monopoles through mergers and filamentation of vorticity, which are typical features the 2D turbulence as was explained in Section 6.3.2. Yet, it should be stressed that they are obtained here with a compressible fluid, according to the gas-dynamics analogy for shallow-water model. The evolution of the

Table 9.1 *Parameters of initialisations of different runs.*

Initialisation	Label	Ro	$\max(|\zeta|)/f$	Bu	Energy anomaly	Ω
Coherent structure	C	0.3	3.5	0.16	$2.4 \cdot 10^{-2}$	0.43
Spectral bump	R1	0.45	5	2.5	$4 \cdot 10^{-3}$	1.37
($m = 25$)	R2	1	9	2.5	$1.6 \cdot 10^{-2}$	5.73

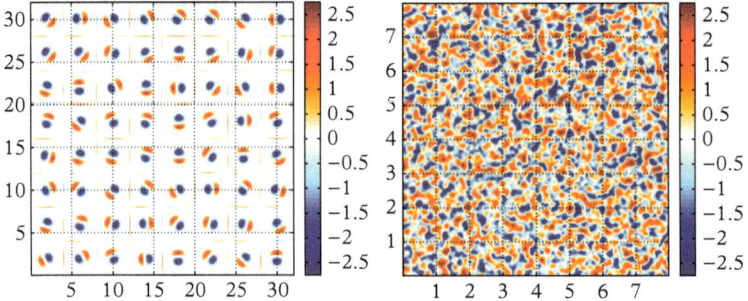

Figure 9.26 *Initial vorticity fields for runs C (left panel) and R1 (right panel), cf. Table 9.1.*

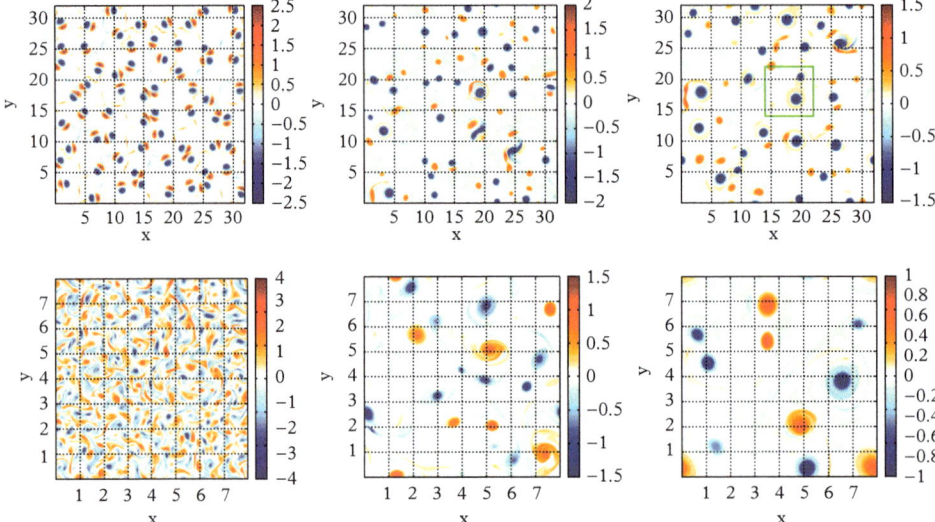

Figure 9.27 *Evolution of the vorticity field in runs C (top row) and R1 (bottom row) at t = 10 (left), 250 (middle) and 500/600f⁻¹ (top/bottom). The square in the middle of the upper right panel gives the size of the domain for runs R.*

energy of the system exhibits a transfer from kinetic to potential energy. This process is associated with formation of coherent structures with stronger and wider surface deviations. It is to be emphasised that the relative energy dissipation rate $\frac{1}{E} \cdot \frac{\partial E}{\partial t}$ is weak, of the order 10^{-4} throughout the simulations. Different initialisations produce qualitatively similar states at the late stages, although the mechanisms involved are clearly different: in one case these are complex interactions between the modons, with collisions and fusions of vortices. In another case, coherent structures emerge from a chaotic flow through organisation of the vorticity field and inverse energy cascade. The vorticity field at late stages consists of an ensemble of near-axisymmetric well-separated vortex monopoles (cf. Figure 9.27, right panels). It should be stressed that in simulations with coherent-structure initialisations, like the run C presented here, the late stages exhibit clear cyclone–anticyclone asymmetry in favour of anticyclones, in the sense that they are greater in number, and more symmetric than the cyclones. This asymmetry is specific to the shallow-water dynamics at sufficiently high Rossby numbers ($Ro > 0.1$), while pure two-dimensional incompressible Euler equations are cyclone–anticyclone symmetric. Evidence of this asymmetry is manifest in the evolution of the skewness, the normalised third-order moment of vorticity, as seen in Figure 9.28. The evolution of the centroid of energy is also presented in this figure (left panel). It is clear that the emergence and strengthening of coherent structures is associated with the decrease of the energy centroid, which may be considered as an evidence of the inverse energy cascade. We see also a more or less distinct saturation of skewness and centroid, which is a manifestation of the inhibition of the cascade processes due to the coherent structures, and of the impact of the finite deformation radius (here we use the notions of direct/inverse energy cascades in a broad sense, meaning the energy transfer towards smaller/larger scales, respectively). Indeed, the typical size of the vortices at this stage is of the order of R_d, and looking at the evolution of the vortex ensemble we see that neighbouring vortices of the same sign practically do not merge, nor produce much filamentation of vorticity. The explanation is, most probably, the screening effect of the finite deformation radius. Nevertheless, the run with coherent initial conditions produces larger structures, with typical sizes of the order of $2\,R_d$.

The energy spectra exhibit a maximum at the location in the k space corresponding to the centroid of energy, and a steep slope beyond it with a power law estimated at k^{-6}, see Figure 9.29. As compared to the well-known enstrophy cascade power law k^{-3} described in Section 6.3, this one is far steeper. It should be noted, however, that the evolution of the system with random initial conditions, cf. the right panel of Figure 9.29, exhibits a transient 'inertial regime', corresponding to a period of active cascade processes, during

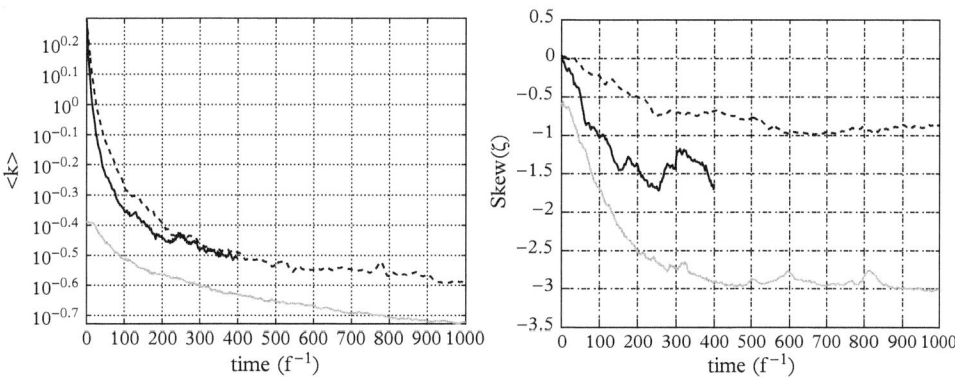

Figure 9.28 *Evolution of the centroid of energy (left panel), and the skewness (right panel) of the vorticity in runs R1 (black dashed), R2 (black continuous), and C (grey).*

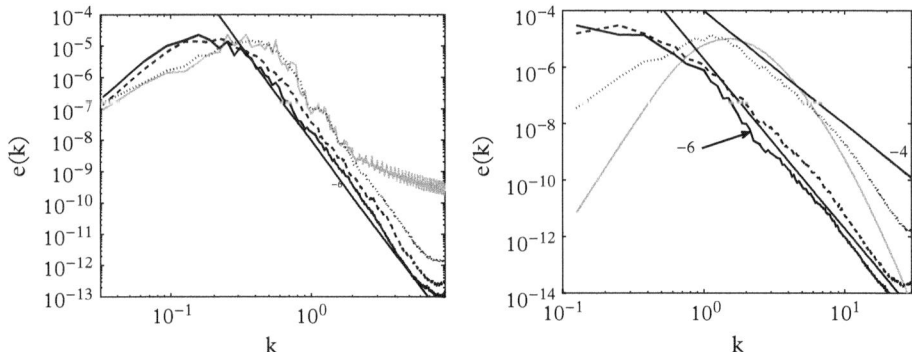

Figure 9.29 *Energy spectrum at t = 0 (grey), 10 (black dotted), 250 (black dashed), and 500f^{-1} (black continuous) for runs C (left panel) and R1 (right panel).*

which the spectrum has a gentler slope close to k^{-4}. Finally, we should emphasise that the divergence field at the end of the simulations has no correlation with the vorticity field, nor with the modulus of its gradient (not shown), being a random wave field of rather weak magnitude (of the order 10^{-2}). As already mentioned, the initial divergence field for the ensemble of modons is driven by vorticity, each vortex dipole bearing a quadrupole of divergence. The correlation of the divergence and vorticity disappears in course of evolution, as the modons are destroyed, and we thus have a manifestation of the wave–vortex decoupling. This might seem not surprising in view of relatively small global Rossby numbers at the late stages. Yet, the local Rossby numbers remain high (greater than 1) at the late stages of the simulation (cf. Figure 9.27, right panels).

9.5.3   Non-universality of RSW turbulence

As already seen, the different initialisations described above permit us to check the influence of initial conditions upon the evolution of RSW turbulence and, thus, a degree of universality of the latter. In order to quantify the vortex system we use a vortex census method. It is based on the Okubo–Weiss criterion which uses the difference between rotation and strain rates: $\zeta^2 - \sigma^2$, with $\sigma = \sqrt{\sigma_n^2 + \sigma_s^2}$ where $\sigma_n = \partial_x u - \partial_y v$ and $\sigma_s = \partial_x v + \partial_y u$ are, respectively, the normal and shear strains. A simply connected domain where this quantity is positive, and the mean vorticity is greater than some fixed threshold, is considered as a coherent vortex. The proper value of the dimensionless threshold is of the order of the initial global Rossby number of the flow for coherent-vortex initialisations, and of the global Rossby number at the beginning of the inertial regime for simulations with random initialisations. We analyse the vortex density ρ, defined as a number of coherent vortices per unit area, and the normalised mean area of the vortices $\langle A \rangle$ defined as the total area occupied by coherent vortices divided by the number of vortices, and normalised by the area of the domain. The scaling exponents in time for these quantities, and other relevant quantities, as follows from the simulations, are summarised in

Table 9.2 *Power laws of integral quantities as functions of time, in the 'inertial regime'.*

Run	ρ	$\langle A \rangle$	A	$\langle \Omega \rangle$	$\langle |\vec{\nabla} \cdot \vec{v}| \rangle$	accuracy
C	-0.45	$0.38 \; ; 0.6$	$-0.14 \; ; 0.08$	-0.55	-0.3	good
R1	-0.72	0.65	-0.2	-0.6	-0.43	good
R2	-1.0	0.6	-0.75	-0.62	-0.45	good

ρ, $\langle A \rangle$ and A are respectively the vortex density, and the mean and the total area of the vortices, and $\langle \Omega \rangle$ is the mean enstrophy of the flow. The 'accuracy' row is an indication of whether the vortex census method does or does not permit us to identify the coherent vortices clearly. Two numbers in a row mean a change of slope in course of evolution.

Table 9.2. The simulations show that evolution of the ensemble of vortices roughly follows power laws, at least once the coherent vortices have emerged, but that the relations between the exponents of different quantities are different, and vary from one run to another. We observe a change in the slope of different quantities in time. The change in the slopes of the decay laws is intriguing, and especially the fact that it does not affect all quantities.

We thus have a clear indication that RSW turbulence does not decay according to a universal law. This is not surprising, from the general point of view, as infinity of conservation laws related to Lagrangian conservation of PV are present in the RSW system. (This consideration is also valid for 2D and QG turbulence.) In the inviscid limit the whole phase space of the system is foliated into subspaces of constant integrals of motion, and the motion is confined to a subspace defined by initial conditions. It is reasonable to think that at small dissipation, which was the case in the simulations presented above, this foliation still constrains the motion. However, in spite of this consideration, theoretical analyses of 2D turbulence are often based on the universality hypothesis.

9.6 Summary, comments, and bibliographic remarks

We have seen in this chapter that dipolar vortices in RSW exhibit quite surprising properties. First, they arise from quasi-geostrophic modons via the process of ageostrophic adjustment, which thus generalises the classical geostrophic adjustment. Although corresponding analytic solutions are not available, the semi-analytic and numerical ones indicate that RSW modons are indeed some exact solutions of the equations of motion. Moreover, their surprising quasi-elastic collisions, even in the case of high intensity and associated shocks inside, make one think about some hidden symmetries in the system. Tripolar vortices arising in the modons collisions turn to be very robust, too. Dipolar and tripolar vortex solutions exist not only in the barotropic, but also in the baroclinic two-layer RSW, and can have essentially baroclinic structure. One of the applications of these particular solutions is for initialisations of direct simulations of RSW turbulence. Initialisations with random ensembles of modons are radically different from standard initialisations and allow us to study dependence of turbulence decay on initial conditions. In this way a non-universality of the decay is demonstrated.

Analysis of steady-translating solutions of RSW equations goes back to Malanotte-Rizzoli (1982). A semi-analytic construction of a modon in RSW, starting from the QG modon, iteratively solving (9.3) and (9.5) for solutions steady moving along the straight line, and matching them across a circular separatrix was proposed in the pioneering paper by Kizner *et al.* (2008). Its comparison with a modon resulting from the relaxation of the QG solution in numerical simulations was made in Ribstein, Gula, and Zeitlin (2010). Adjustment of QG modons and formation of two-layer RSW modons is presented following Lahaye and Zeitlin (2012). Collisions of modons were studied in Ribstein, Gula, and Zeitlin (2010) and Lahaye and Zeitlin (2011). Shock modons were discovered in Lahaye and Zeitlin (2012b). Turbulence in RSW initialised with random ensembles of modons was investigated in Lahaye and Zeitlin (2012c), with comparisons to standard initialisations. The presentation here followed these papers. Studies of RSW turbulence are rare. Farge and Sadourny (1989) simulated the evolution of the forced RSW turbulence using a pseudo-spectral code, and obtained power spectra following a k^{-4} law. Later, Yuan and Hamilton (1994) observed a k^{-3} power spectrum, fairly similar to what is expected in the enstrophy cascade regime of two-dimensional turbulence. Simulations of decaying RSW turbulence were performed in Polvani *et al.* (1994), where the emergence of a cyclone–anticyclone asymmetry during the evolution of an initial Gaussian random-phase vorticity field was reported. For vortex census and Okubo–Weiss criterion see Weiss (1981) and McWilliams (1984). The theory of universal decay of 2D turbulence was proposed in Carnevale *et al.* (1991).

Figures 9.1 to 9.7 and Figures 9.18 and 9.19 are reproduced from Ribstein, Gula, and Zeitlin (2010) with the permission of AIP publishing. Figures 9.9 to 9.14 are reproduced from Lahaye and Zeitlin (2012), with permissions. Figures 9.16 and 9.17 are reproduced from Lahaye and Zeitlin (2012b), with permissions. Figures 9.20 to 9.24 are reproduced from Lahaye and Zeitlin (2011), and Figures 9.26 to 9.29 are reproduced from Lahaye and Zeitlin (2012c) with the permission of AIP publishing.

10

Instabilities of Jets and Fronts and their Nonlinear Evolution

As we have seen in Chapter 8, the universal process of geostrophic adjustment tends to produce a geostrophically balanced state out of arbitrary initial conditions. In translationally or rotationally symmetric cases such balanced states, respectively, jets and vortices, are stationary solutions of the equations of motion. A question then arises: does the evolution ends after the system arrives to such state? This would be the case if such steady solutions were *stable*. It is, however, well-known that it is often not the case, and that stationary solutions of a dynamical system can be *unstable* with respect to small perturbations. Instabilities of geophysical flows are ubiquitous, and knowing their character and evolution is of utmost importance as this evolution leads to reorganisations of the flow and appearance of new structures, changing, for example, its transport and mixing properties. Knowing characteristic patterns of key instabilities is necessary for interpreting observational data and results of numerical simulations. In this and the next chapter we present a gallery of typical geophysical fluid dynamics instabilities with their characteristic evolution patterns, and we start with instabilities of jets.

10.1 Instabilities: general notions and techniques

10.1.1 Definitions and general concepts

In this section we recall general concepts and methods of the stability analysis.

Linear vs nonlinear (in)stability

For an arbitrary dynamical system

$$\dot{\mathcal{U}} = \mathcal{M}\left[\mathcal{U}\right], \tag{10.1}$$

with dynamical variable(s) $\mathcal{U}$ and the operator $\mathcal{M}$ determined by the structure of the model, solutions are trajectories in the space of $\mathcal{U}$:

$$\mathcal{U}(t_0) \longrightarrow \mathcal{U}(t) \tag{10.2}$$

Geophysical Fluid Dynamics. Vladimir Zeitlin,
Oxford University Press (2018). © Vladimir Zeitlin. DOI: 10.1093/oso/9780198804338.001.0001

In hydrodynamics in general, and geophysical fluid dynamics in particular, $\mathcal{U} = (\mathbf{v}, \rho, p, s, \ldots)$. If $\mathcal{U}_0$ is an exact solution of (10.1), for example the state of rest $\mathcal{M}[\mathcal{U}_0] = 0$, or some other one, the *linearisation* procedure consists in representing solutions as $\mathcal{U} = \mathcal{U}_0 + u$, with $||u|| \ll 1$, where $||..||$ denotes a norm in the functional space to which u belongs, and neglecting the terms $\mathcal{O}(||u||^2)$. The linear equations follow:

$$\dot{u} = \hat{\mathcal{L}}[\mathcal{U}_0] \circ u, \tag{10.3}$$

where $\hat{\mathcal{L}}[\mathcal{U}_0]$ is a *linear operator* depending on the solution $\mathcal{U}_0$ and acting on u. The linearised system, if it has no explicit time dependence, allows for the Fourier transformation in time, $u(t) \rightarrow \hat{u}(\omega)e^{i\omega t}$, and a linear eigenproblem for eigenvalues ω and eigenvectors $\hat{u}$ follows, which gives a *spectrum* of ω or, in other words, a *dispersion relation*. The eigenvalues, in general, are complex: $\omega = \omega_r + i\omega_i$, and *linear stability* or *instability* correspond to $\omega_i > 0$, or $\omega_i < 0$, respectively, with exponential growth of small perturbations of the solution in the second case.

The condition of nonlinear, or Lyapunov, stability which, roughly speaking, corresponds to a situation where all trajectories starting in the vicinity of the solution whose stability is investigated, stay close to it, is formulated as follows: if for any ϵ there exist δ such that at all times any perturbed solution does not deviate farther than ϵ from $\mathcal{U}_0$, if initially it was not farther than δ, that is

$$\forall \epsilon \ \exists \delta : ||u|||_{t=0} < \delta \ \Rightarrow \ ||u|||_{\forall t>0} < \epsilon, \tag{10.4}$$

then $\mathcal{U}_0$ is stable.

Stability in Hamiltonian systems

As was already mentioned in Section 2.2.1, any Hamiltonian system is defined in a phase space of dynamical variables $\mathcal{U}$ supplied with a Poisson bracket $\{\ldots, \ldots\}$, and a Hamiltonian $H(\mathcal{U})$. The equations of motion are $\dot{\mathcal{U}} = \{\mathcal{U}, H\}$. Stationary solutions are equilibria of the dynamical system. Thus, $\mathcal{U}_0$ is an extremum of H, and hence the first variation of the Hamiltonian with respect to $\mathcal{U}$ vanishes at the solution: $\delta_{\mathcal{U}} H|_{\mathcal{U}_0} = 0$. The following *sufficient* stability conditions can be established for any Hamiltonian system:

- If the second variation of H $\delta_{\mathcal{U}}^2 H|_{\mathcal{U}_0}$ is sign-definite, then the solution $\mathcal{U}_0$ is *linearly (or formally) stable*:

- If for all $\delta\mathcal{U}$:

$$0 < \text{const} \leq H(\mathcal{U}_0 + \delta\mathcal{U}) - H(\mathcal{U}_0) - \delta_{\mathcal{U}}H(\mathcal{U}_0) \cdot \delta\mathcal{U}, \tag{10.5}$$

then the solution $\mathcal{U}_0$ is *nonlinearly stable*.

It should be emphasised that in the presence of *integrals of motion* (Casimirs) C_α, $\alpha = 1, 2, \ldots$, the Hamiltonian in the stability criteria should be replaced by

$$H_C = H + \sum_\alpha \lambda_\alpha C_\alpha, \tag{10.6}$$

where λ_α are Lagrange multipliers.

10.1.2   (In)stability criteria for plane-parallel flows

We now pass from general considerations to their applications in fluid dynamics, and start with plane-parallel flows which are stationary solutions with the simplest geometry of the equations of motion. We have already encountered baroclinic instability of this kind of flow in Chapter 5. General criteria of (in)stability of such flows can be established analytically in the framework of quasi-geostrophic (QG) and rotating shallow-water (RSW) models. Although we will be systematically applying a direct approach to linear stability analysis, consisting of numerical calculations of eigenfrequencies and eigensolutions of the equations of the model linearised about a steady state, such criteria are useful to bear in mind for diagnosing unstable flows, and benchmarking and analysing results of direct calculations.

Rayleigh and Rayleigh–Kuo criteria

A classical example is the famous Rayleigh criterion. Let us consider the QG equations on the f plane and take, for simplicity, a limit of infinite deformation radius (finite deformation radius can be easily restituted):

$$\nabla^2 \eta_t + \mathcal{J}(\eta, \nabla^2 \eta) = 0. \tag{10.7}$$

Consider a shear flow $\mathbf{v} = U(y)\hat{\mathbf{x}}$ with geostrophic stream function $\eta_0 = -\int^y dy'\, U(y')$, which, as is easy to see, is an exact solution of these equations. Linearisation about this solution: $\eta = \eta_0 + \phi$ leads to

$$\nabla^2 \phi_t + U(y)\nabla^2 \phi_x - \phi_x U''(y) = 0, \tag{10.8}$$

which, after the Fourier transformation: $\phi(x, y, t) \to \hat{\phi}(y)e^{ik(x-ct)} + c.c.$, becomes

$$\hat{\phi}''(y) - \left[k^2 + \frac{U''(y)}{U(y) - c} \right] \hat{\phi}(y) = 0. \tag{10.9}$$

Here the prime means an ordinary derivative with respect to the corresponding argument. Considering the system either in a channel with free-slip boundary conditions, or in the whole plane with decay boundary conditions at $y \to \pm\infty$, multiplying (10.9) by

$\hat{\phi}^*(y)$, the complex conjugate of $\hat{\phi}(y)$, and integrating in y, leads, after integration by parts, to

$$\int_{y_1}^{y_2} dy \left(\hat{\phi}^{*\prime}(y)\hat{\phi}'(y) + \left[k^2 + \frac{U''(y)}{U(y) - c} \right] \hat{\phi}^*(y)\hat{\phi}(y) \right) = 0, \tag{10.10}$$

where we suppose that there are no critical levels: $\forall y :\ U(y) - c \neq 0$. Expression in the left-hand side of (10.10) has real and imaginary parts, which have to be zero separately. Considering the imaginary part of (10.10), we see that if c has a non-vanishing imaginary part, than the flow has an inflection point: $\exists y_0 :\ U''(y_0) = 0$, otherwise the integrand is sign-definite, and the integral cannot vanish. In non-dissipative systems complex eigenvalues always appear in pairs with their conjugates. Thus we get a *necessary* condition of instability: if the flow is unstable it has an inflexion point, which is the famous Rayleigh criterion. It is easy to see that if we consider the same flow on the beta plane, $U''(y)$ is replaced by $U''(y) - \beta$ which should be zero somewhere in the flow, and the so-called Rayleigh–Kuo criterion arises, correspondingly.

The most important fact following from these criteria is that they are related to a change of sign of the gradient of absolute vorticity at some location in the flow in the case of instability. Indeed, the relative vorticity of the plane-parallel flow is $\zeta = -\partial_y U$, and the absolute vorticity under adopted hypotheses is $\zeta + f_0 + \beta y$. As was already explained in Chapter 5, Rossby waves exist due to gradients of potential vorticity (PV). The sign of the PV gradients determines the sign of propagation of the waves. For the barotropic system (10.7) absolute vorticity plays the role of PV. Therefore, if the flow is unstable, Rossby waves can propagate in it in opposite directions. We will see the significance of this fact shortly.

Ripa's stability criteria for zonal flows in RSW models

While the Rayleigh criterion gives a necessary condition of instability, *sufficient* stability conditions can be established on the basis of the Hamiltonian description of the RSW model in terms of Lagrangian variables, cf. section 3.4.1, and of (10.5) and (10.6) while taking into account that Lagrangian conservation of PV leads to appearance of Casimir invariants in Eulerian terms, cf. Section 3.3.1. These criteria of stability of plane-parallel zonal flows in the N-layer RSW model are as follows. A zonal flow $U_i(y)$, $H_i(y)$ in geostrophic equilibrium, which represents a stationary solution of the N-layer RSW equations,

$$fU_i(y) + gH_i'(y) = 0, \ i = 1, 2, \ldots, N, \tag{10.11}$$

with potential vorticity $Q_i(y) = \frac{f(y) - U_i'(y)}{H_i(y)}$, is stable if for all y there exists $\alpha = \text{const}$, such that

1.

$$(U_i(y) - \alpha)\, Q_i'(y) < 0, \ i = 1, 2, \ldots, N, \tag{10.12}$$

   and

2.

$$1 > \sum_{i=1}^{N} \frac{(U_i(y) - \alpha)^2}{gH_i(y)}. \tag{10.13}$$

The first sub-criterion links stability to sign-definiteness of the PV gradient, in accordance with the necessary condition of instability above, while the second, roughly speaking, means that the flow velocity is 'subsonic', that is its velocity layer-wise is less that the phase speed of gravity waves $c_i = \sqrt{gH_i}$.

10.1.3   Direct approach to linear stability analysis of plane-parallel and circular flows

The above-described stability criteria are useful but not constructive, in a sense that they give no information on the structure of unstable perturbations and their growth rates. The situations, like that in the Phillips model of baroclinic instability of Section 5.3.3, where the linear stability analysis can be accomplished by hand, are rare. This is why in what follows we adopt a direct approach, which will be repeated in all examples presented further in this and subsequent chapters. For plane-parallel flows, including localised jets, we will exploit the translational invariance in the along-flow direction and, after having linearised the equations of motion about a chosen solution of the equations of the model (one-layer for barotropic, or two-layer for baroclinic flows), we will make a partial Fourier transformation in this direction and in time, as in deriving the Rayleigh criterion in Section 10.1.2. The system of linearised differential equations in partial derivatives for the perturbations becomes thus a system of ordinary differential equations, where dependent variables are functions of the across-flow coordinate. This system, at a given stream-wise wave number, is an eigenproblem for the eigenvalues of frequency (or stream-wise phase velocity) and corresponding eigenmodes. Unstable modes correspond to complex eigenvalues. The eigenvalue problem can be reduced to an algebraic system by direct discretisation in the span-wise direction, and solved by standard routines. In this way we obtain the growth rates, the phase velocities, and the structure (phase portraits in terms of velocity and pressure/thickness) of unstable modes. Direct discretisation can lead to numerical problems (Runge phenomenon), and a more sophisticated discretisation with the help of Chebyshev polynomials (pseudo-spectral collocation) will be used. It should be emphasised that, even in its pseudo-spectral implementation, the discretisation has caveats, and the results should be treated with care, especially in the presence of critical levels leading to singularities, in order to filter solutions having no physical meaning. This is achieved by systematic refining of the discretisation grid, application of alternative discretisation schemes, and careful analysis of the eigenmodes. The eigenmodes obtained by such direct stability analysis are then used for initialisation of direct numerical simulations with the finite-volume schemes described in Chapter 7.

The approach is the same for circular vortex flows treated in Chapter 11, the stream-wise direction becoming the azimuthal one, and the eigenproblems becoming a system of ordinary differential equations where all dependent variables are functions of the radial coordinate.

10.2   Geostrophic barotropic and baroclinic instabilities of jets

We will start by applying the approach described in Section 10.1.3 to jets with small Rossby numbers. Although the problem can be treated within the QG model, as in Chapter 5, for the sake of continuity with what follows we will address it in the framework of full one- and two-layer shallow-water models on the f plane. We choose to work with the Bickley jet, which is a standard configuration used in numerous studies of hydrodynamic instabilities.

10.2.1   Barotropic instability of a Bickley jet on the f plane

Set-up and parameters

The Bickley jet in one-layer case, cf. Figure 10.1, is given by the following velocity and thickness profiles (the jet profiles, here and below, are denoted by capitals and/or overbars):

$$U = 0, \quad \bar{v} = V = -\frac{g\Delta\eta}{fL}\operatorname{sech}^2\left(\frac{x}{L}\right), \quad \bar{\eta} = H = H_0 + \bar{\eta} = H_0 - \Delta\eta\tanh\left(\frac{x}{L}\right). \quad (10.14)$$

Here L is the width of the jet, and H_0 is the unperturbed thickness of the fluid layer. The jet is in geostrophic equilibrium (and therefore is, at the same time, a pressure/thickness front), and it is easy to see that it is a solution of the one-layer RSW equations. Orientation of the jet on the f plane does not matter, as there is a symmetry with respect to rotations in the x - y plane in the absence of β- effect. The jet is characterised by its Rossby $Ro = \frac{V_0}{fL}$ and Burger $Bu = \frac{R_d^2}{L^2} = \frac{gH_0}{f^2L^2}$ numbers, where R_d is the deformation radius, and $V_0 = \frac{g\Delta\eta}{fL}$ is the peak velocity of the jet. We choose $Ro = 0.1$, $Bu = 10$, in order to place ourselves in the range of parameters of atmospheric mid-latitude jets. As follows from Figure 10.1, the gradient of PV of the jet clearly changes sign, and thus instability is expected.

Results of the linear stability analysis

Following the algorithm of Section 10.1.3, we linearise the equations of the model about (10.14), use the above-described scaling, and obtain dimensionless equations for the perturbation (u, v, η) of the jet (10.14):

$$\begin{cases} Ro\left(\partial_t u + \bar{v}\partial_y u\right) - v + \partial_x\eta = 0, \\ Ro\left(\partial_t v + u\partial_x\bar{v} + \bar{v}\partial_y v\right) + u + \partial_y\eta = 0, \\ Ro\left(\partial_t\eta + \partial_x(u\bar{\eta}) + \bar{v}\partial_y\eta + \bar{\eta}\partial_y v\right) + Bu\left(\partial_x u + \partial_y v\right) = 0. \end{cases} \quad (10.15)$$

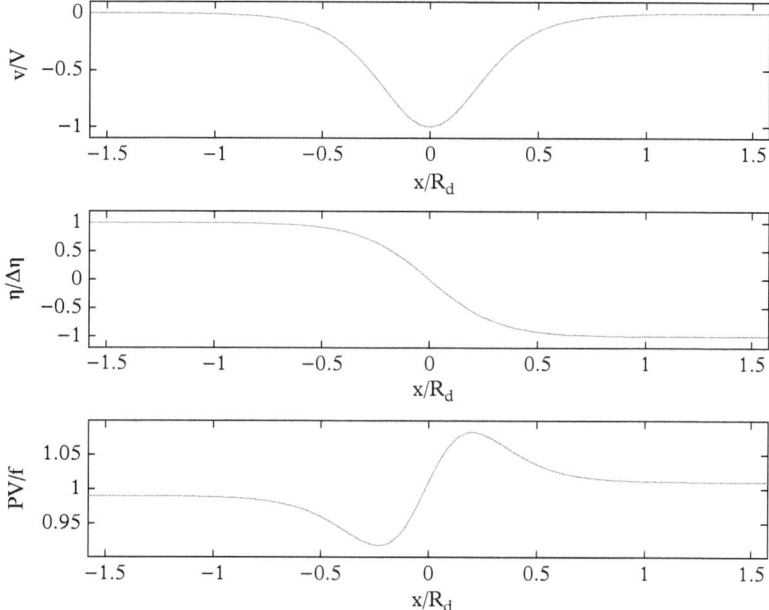

Figure 10.1 *Span-wise profiles of velocity (upper panel), thickness (middle panel), and PV (lower panel) of the Bickley jet in one-layer RSW.* ©*American Meteorological Society. Used with permissions.*

Using decomposition in Fourier modes in stream-wise direction $(u, v, \eta) = (ik\hat{u}, \hat{v}, \hat{\eta}) \exp\{i(ky - \omega t)\}$ + c.c., we arrive at the eigenvalue problem $\mathcal{M}\boldsymbol{a} = c\,\boldsymbol{a}$, with $\boldsymbol{a} = (ik\hat{u}, \hat{v}, \hat{\eta})$ and

$$
\mathcal{M} = \begin{pmatrix}
\bar{v} & \dfrac{1}{Rok^2} & -\dfrac{1}{Rok^2}\partial_x \\[2mm]
\dfrac{1}{Ro} + \partial_x\bar{v} & \bar{v} & 1 \\[2mm]
(\partial_x\bar{\eta} + \bar{\eta}\partial_x) + \dfrac{Bu}{Ro}\partial_x & \bar{\eta} + \dfrac{Bu}{Ro} & \bar{v}
\end{pmatrix}.
\tag{10.16}
$$

The eigenvalues $c = \omega/k$, and corresponding eigenvectors of this problem, are found numerically with the help of the pseudospectral collocation method. The results for the long-wave (small k) part of the spectrum are presented in the left panel of Figure 10.2. We see that there are two unstable modes in the long-wave spectrum. The fact that they are two is due to a change of sign of the gradient of PV, which takes place twice, and hence there are different possibilities for counter-propagating Rossby waves in the system. The most unstable mode is the most important in any system, because if a broad-spectrum perturbation is imposed upon the background jet, it is this one which will be growing with the highest growth rate. As the growth is exponential, the difference

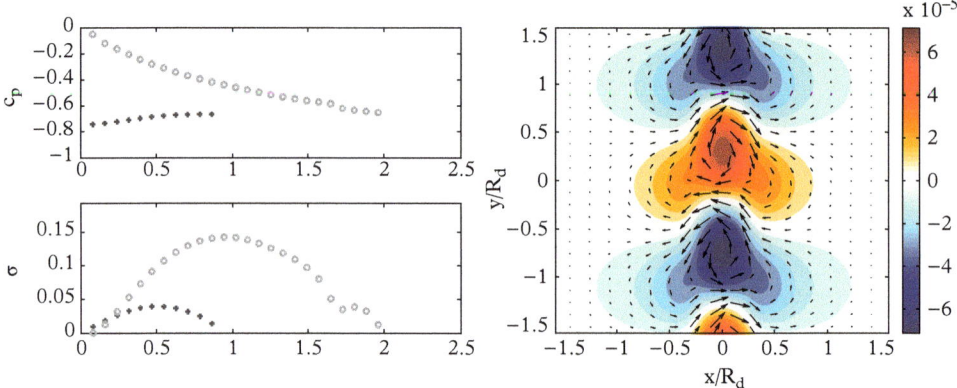

Figure 10.2 *Left panel: phase velocity c_p (top) and growth rate $\sigma = Im(\omega)$ (bottom) of unstable modes of the barotropic Bickley jet on the f plane. Right panel: the mode corresponding to the maximum of the growth rate in the left panel. Colors: pressure (thickness) anomaly; arrows: velocity anomaly with respect to background jet.* © *American Meteorological Society. Used with permissions.*

with the modes with lower growth rates will be increasing also exponentially, and thus the most unstable mode 'wins'. The phase portrait of the most unstable mode is displayed in the right panel of Figure 10.2. As follows from the figure, the velocity field is aligned with the isobars, and hence the unstable mode represents a series of vortices of alternating signs in approximate geostrophic equilibrium.

Nonlinear saturation of the instability

Nonlinear evolution of the instability is studied by imposing the most unstable mode with small amplitude (≈ 1 % of the typical jet values) onto the background jet, and following its evolution, as given by numerical simulation with the well-balanced finite-volume code of Chapter 7. In this, and similar simulations in subsequent sections, the boundary conditions, typically, are 1) periodic in the along-jet direction, with the wavelength (or several wavelengths) of the unstable mode, 2) numerical sponges in the across-jet direction, which are realised by imposing Neumann condition on the velocity in adjacent to the boundary of the computational domain grid cells. The sponges are placed sufficiently far (several jet widths). The results of such simulation in terms of relative vorticity are presented in Figure 10.3 and show formation of secondary vortices by the meandering jet. As is clear from the figure, the evolution of the barotropic instability can be thus regarded as breaking of an unstable Rossby wave.

10.2.2   Baroclinic instability of a Bickley jet in the f plane

Set-up and parameters

As in the previous subsection, we consider a Bickley jet, but this time in the two-layer RSW model in the atmospheric context. In order to mimic the atmospheric situation

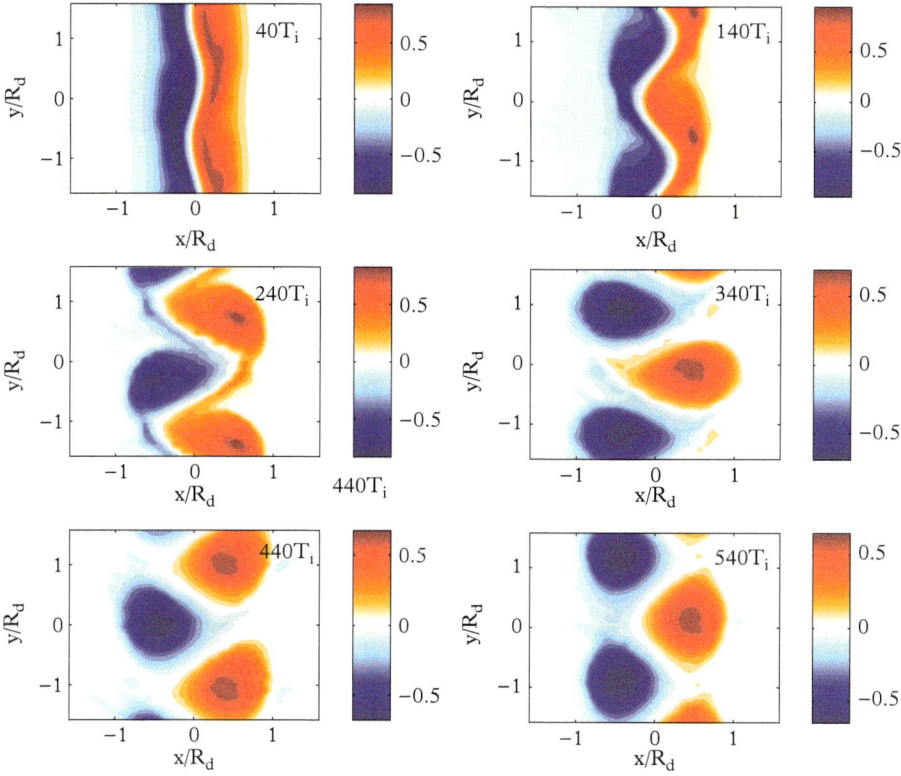

Figure 10.3 *Evolution of relative vorticity and formation of secondary vortices during nonlinear saturation of the barotropc instability of the Bickley jet on the f plane.* ©*American Meteorological Society. Used with permissions.*

with an upper tropospheric zonal jet in mid-latitudes (e.g. Figure 2.1), we will consider a zonal Bickley jet in the upper layer, with the following non-dimensional profiles of velocity and anomaly of geopotential in layers 1 and 2:

$$U_1 = 0, \quad \bar{\eta}_1 = \frac{1}{\alpha - 1} \tanh(y),$$

$$U_2 = \operatorname{sech}^2(y), \quad \bar{\eta}_2 = \frac{-1}{\alpha - 1} \tanh(y). \tag{10.17}$$

There is no anomaly of geopotential on the ground (which plays a role of the free surface): $\bar{\eta}_1 + \bar{\eta}_2 = 0$, in order to annihilate the velocity in the lower layer. The profiles of velocity, pressure, and PV layer-wise are presented in Figure 10.4. The jet is in geostrophic equilibrium and is a solution of the two-layer free-surface RSW equations (3.25). Parameters $Ro = 0.1$, and $Bu = 10$ are the same as in the previous subsection and are typical for atmospheric jets.

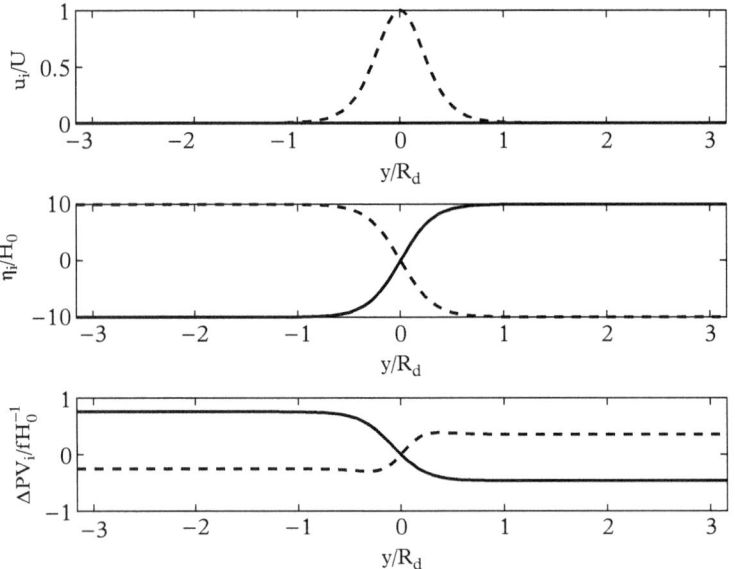

Figure 10.4 *Baroclinic upper-layer Bickley jet. Span-wise profiles of: normalised zonal velocity (upper panel), thickness anomaly (middle panel), and PV anomaly (lower panel) of the baroclinic Bickley jet. Lower (upper) layer: solid (dashed) curves.* © *American Meteorological Society. Used with permissions.*

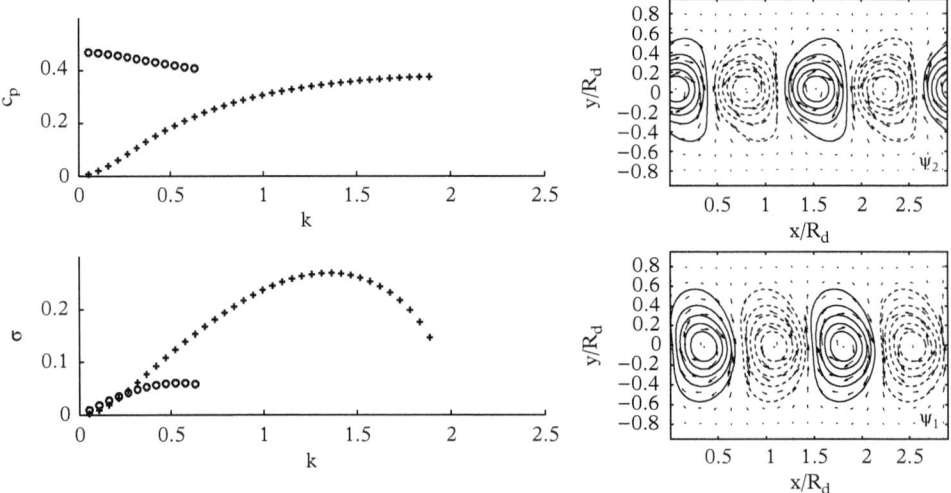

Figure 10.5 *Left panel: Linear stability diagram of the upper-layer Bickley jet: phase velocity (top) and growth rate (bottom) of unstable modes. Right panel: The most unstable mode of the upper-layer Bickley jet. Pressure anomalies (contours) of positive (solid) and negative (dashed) sign, and velocities (arrows) in the upper (top) and lower (bottom) layers.* © *American Meteorological Society. Used with permissions.*

Results of the linear stability analysis

We will not repeat the derivation of the linear eigenproblem, which follows from the linearisation of two-layer RSW equations about the jet (10.17). It is straightforward to obtain, along the same lines as (10.16), and is of similar structure, although more cumbersome, as the number of equations doubles. The numerical solution of this problem by the collocation method gives the dispersion diagram, which is displayed in the long-wave sector in the left panel of Figure 10.5. The quasigeostrophic baroclinic instability revealed by the figure, like its barotropic counterpart of the previous subsection, is a *long-wave* instability, which shuts down at large wave numbers. It has a well-defined most unstable mode, corresponding to the maximum of the growth-rate curve in Figure 10.5. This mode is shown in the right panel of Figure 10.5 and represents a series of quasi-geostrophic vortices of alternating signs in each layer, with a characteristic vertical tilt between the layers.

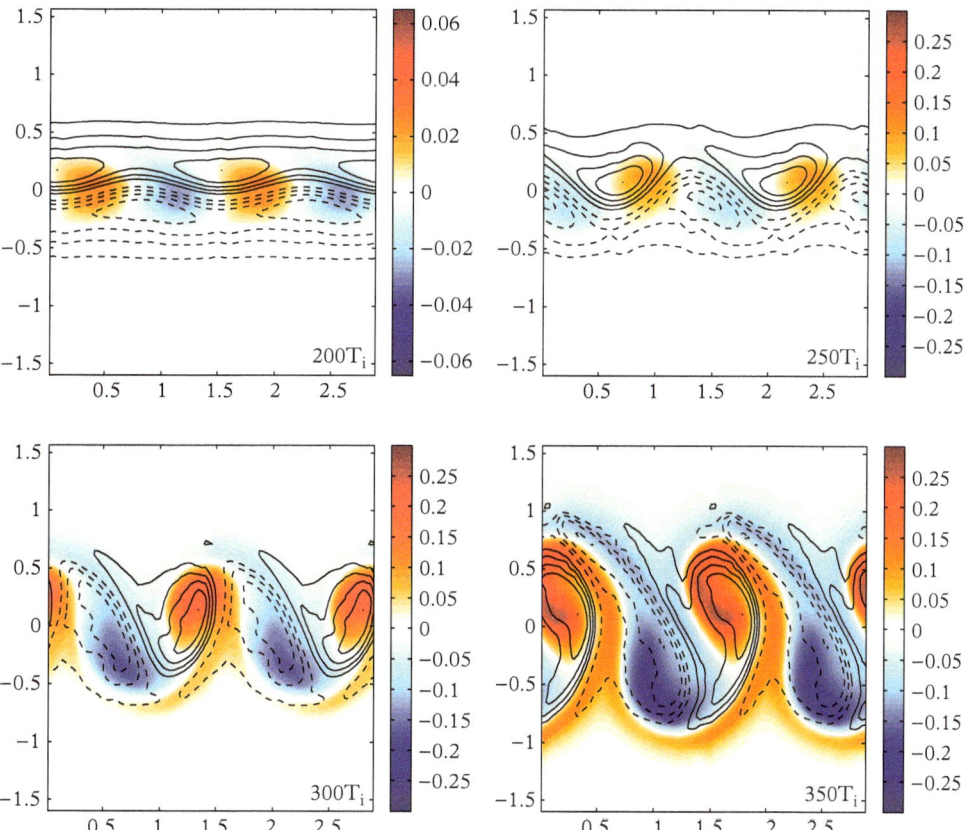

Figure 10.6 *Evolution of the relative vorticity in the lower (colours) and upper (contours) layers during nonlinear saturation of the baroclinic instability of the upper-layer Bickley jet, as follows from direct numerical simulations.* ©*American Meteorological Society. Used with permissions.*

Nonlinear evolution of the instability

Initialising fully nonlinear numerical simulations with this unstable mode, superimposed with small amplitude onto the background jet, allows us to trace the nonlinear evolution of the instability. It consists in formation of typical comma-shaped vortices in the lower layer, as follows from Figure 10.6, where we display the evolution of relative vorticity in both layers.

10.3 Ageostrophic instabilities in the Phillips model: Rossby–Kelvin and shear instabilities

The instabilities considered in Section 2.1 are geostrophic, in a sense that both background flow and unstable modes are in approximate geostrophic equilibrium. They could be analysed directly within the QG approximation, as it was done for the baroclinic instability in Section 5.3.3 where it was treated in the framework of the Phillips model. A question then arises: are these instabilities the only ones in the flows in question, or there are others, appearing with increasing Rossby number? To answer this we will again put ourselves in the framework of the Phillips model but will embed it into the full two-layer RSW equations with a rigid lid. In order to be an exact solution, a zonal flow should be in a geostrophic equilibrium layer-wise:

$$U_j(y) = -\frac{1}{\rho_j f}\partial_y \Pi_j, \, j = 1, 2, \tag{10.18}$$

where Π_j is pressure in the layer j. In the Phillips model, U_j are constant, and $H_j(y)$, the unperturbed thicknesses of the layers, which are related to pressures via dynamic boundary condition at the interface, are linear functions of y. By introducing small deviations of thicknesses and pressures from the equilibrium $h_j = H_j(y) + (-1)^j\eta(x, y, t)$, $\pi_j \to \Pi_j(y) + \pi_j(x, y, t)$, linearising and non-dimensionalising, the following linear system for small perturbations results:

$$\begin{cases} \partial_t u_j + FU_j\partial_x u_j - v_j = \dfrac{1}{\rho_j}\partial_x \pi_j, \\[2mm] \partial_t v_j + FU_j\partial_x v_j + u_j = \dfrac{1}{\rho_j}\partial_y \pi_j, \\[2mm] \partial_t \eta + FU_j\partial_x \eta = (-1)^{j+1}F\left(H_j\partial_x u_j + \partial_y(H_j v_j)\right), \\[2mm] \pi_2 - \pi_1 = \dfrac{2}{F}\eta. \end{cases} \tag{10.19}$$

Here the Froude number $F = \frac{U_0}{\sqrt{gH}} = \frac{U_0}{fR_d}$ is synonymous with the Rossby number, as all distances are scaled with the baroclinic deformation radius R_d, U_0 is velocity scale, $H_0 = H_2(0)$ is vertical scale, and the pressure scales are chosen to be $\rho_j U_0 fR_d$, as in Chapter 3. Under this scaling, $H_j(y) = H_j(0) + (-1)^j F y$. We will be considering the Phillips model in a zonal channel of finite width, which gives another parameter, the

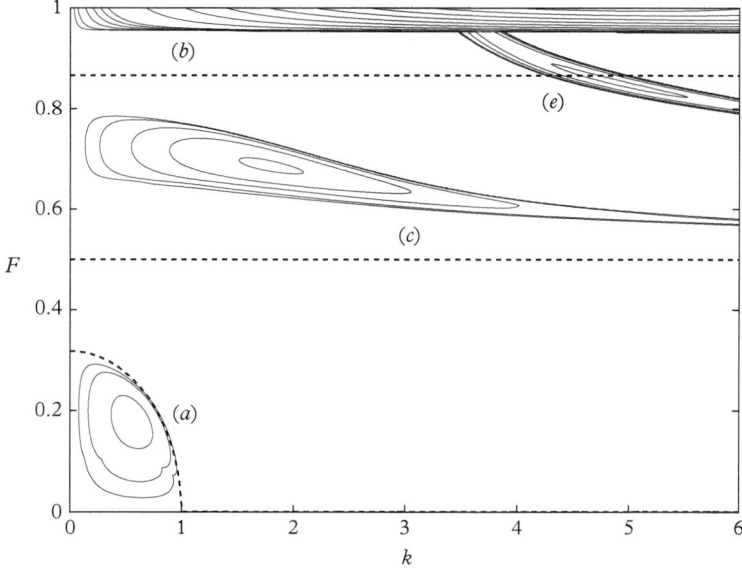

Figure 10.7 *Stability diagram in the Ro − k plane of the Phillips model on the f plane embedded in full 2-layer RSW with equal thicknesses of the layers.*

Burger number *Bu*. To reduce the number of parameters a traditional for such study choice $Bu = F^2$ is made. After Fourier transformation in x and t the system (10.19) becomes a system of ordinary differential equations, and the corresponding eigenproblem is solved by the collocation method. We present the results of the linear stability analysis in the form of stability diagram representing the isopleths of the growth rate in the wave number–Froude (Rossby) number plane. Such diagram in the case of the layers of equal depth $H_1(0) = H_2(0)$ is displayed in Figure 10.7. The zonal phase velocity ω/k and the growth rate as functions of k, but at the same time as functions of F, are presented in the lower panel of Figure 10.8, in order to give an idea of both dependences. The figure corresponds, in fact, to a diagonal section throughout Figure 10.7 and contains a precious information on the nature of the instabilities. First of all, the character of the modes corresponding to each branch of the dispersion diagram can be established from their velocity and thickness distributions (phase portraits). As indicated in the figure, the inertia–gravity waves propagating in both directions with significant positive and negative phase velocities, Kelvin waves propagating along both lateral boundaries in respective directions, and Rossby waves appearing due to the gradients of PV induced by the inclination of the interface in the background flow, are thus identified. The two displayed dispersion curves of inertia–gravity waves propagating in each direction correspond to modes with different span-wise structure. As seen in the figure, there are three wave-number zones containing unstable modes, each of them corresponding to an *intersection of dispersion curves* of above-described 'pristine' modes. Hence, the unstable modes are *hybrid*. The eigenmode corresponding to the long-wave small-Rossby number

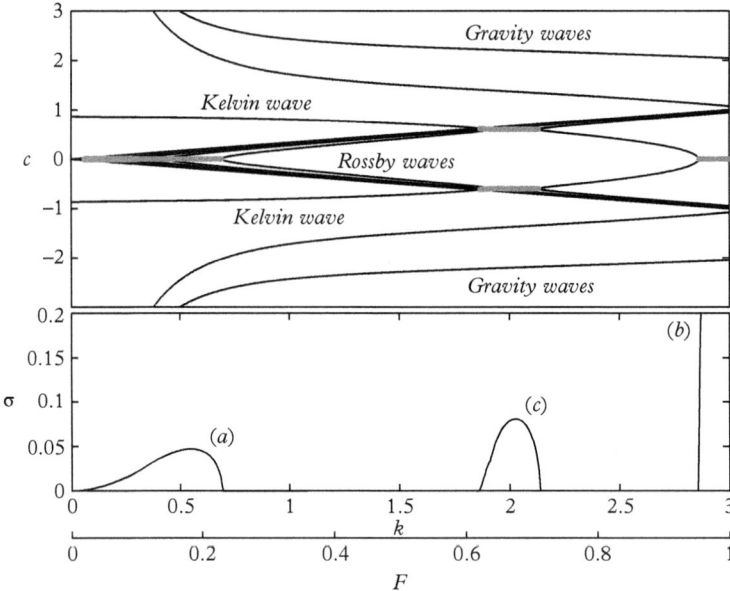

Figure 10.8 *Dispersion and growth rates of the eigenmodes of the linear stability problem in the Phillips model in the f plane embedded into full 2-layer RSW with equal thicknesses of the layers.*

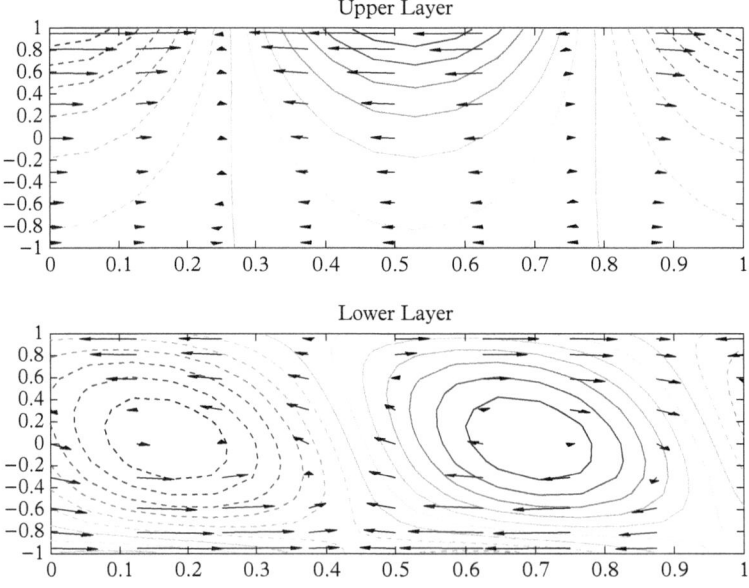

Figure 10.9 *Structure of a typical unstable Rossby–Kelvin mode in ageostrophic Phillips model in the f plane. Contours: pressure (thickness) anomaly, continuous- positive, dashed - negative. Arrows: velocity.*

instability, denoted by (a) in Figure 10.7, is a hybrid of two Rossby waves propagating in the opposite directions, upper-layer and lower-layer Rossby waves, as the gradients of upper- and lower-layer PVs are opposite. This is the classical baroclinic instability of Section 5.3.3. A hybrid mode corresponding to the intersection of the dispersion curves of Rossby and Kelvin modes is given in Figure 10.9. It corresponds to medium-wavelength, medium Rossby-number, Rossby–Kelvin instability denoted by (c) in Figure 10.7. Finally, the vigorous short-wave instability denoted by (b) in Figure 10.7 corresponds to a hybrid of two Kelvin modes running in opposite direction along the opposite lateral boundaries. This is, in fact, a shear, Kelvin–Helmholtz-type instability, and we will find it in other examples following. If the stability diagram is continued further right to larger wave-numbers, intersections of Kelvin and inertia–gravity curves appear, giving other modes of shear instability (zone (e) in Figure 10.7).

10.4   Ageostrophic instabilities of jets and their nonlinear evolution

10.4.1   Linear stability

Set-up and parameters

The results of Section 10.3 show that there are totally new instabilities which are 'lost' in the classical quasi-geostrophic Phillips model, and appear already at moderate Rossby numbers. The set-up of the Phillips model is, however, somewhat artificial, as the system evolves in a box, and lateral walls lead to the appearance of Kelvin waves generating new instabilities. (Nevertheless, we will see in Chapter 11 that such configuration is easily reproducible in laboratory in cylindrical geometry.) Instabilities engendered by Kelvin waves are, indeed, important in practice in the context of coastal currents which will be addressed later in this chapter. Meanwhile, one can ask a question how the quasi-geostrophic instabilities of localised jets studied in Section 10.2 change with increasing Ro. To answer this question for both baroclinic and barotropic instabilities we choose as a background state a barotropic Bickley jet configuration in the two-layer model, that is a jet in geostrophic equilibrium on the f plane with the same profile of mean velocity in both layers, and corresponding distributions of thicknesses of the layers:

$$H_1(x) = H_{10}, \; H_2(x) = H_{20} + \delta \tanh\left(\frac{x}{L}\right), \; U_{1,2}(x) = 0, \; V_{1,2}(x) = \frac{g\delta}{fL}\left(1 - \tanh^2\left(\frac{x}{L}\right)\right).$$
$$(10.20)$$

We work with the two-layer model in the oceanic context configuration, with upper layer 1, and lower layer 2. This configuration is presented in Figure 10.10. The jet is characterised by the unperturbed thicknesses of the layers: H_{10}, $H_{20} = const$, $H_0 = H_{10} + H_{20}$; its width and intensity L and δ, respectively; the aspect and density ratios $d = \frac{H_{20}}{H_{10}}, r = \frac{\rho_1}{\rho_2}$. The peak velocity of the jet is $V_0 = \frac{g\delta}{fL}$, and its Burger and Rossby numbers are $Bu = \frac{gH_0}{f^2L^2}, Ro = \frac{g\delta}{(fL)^2}$.

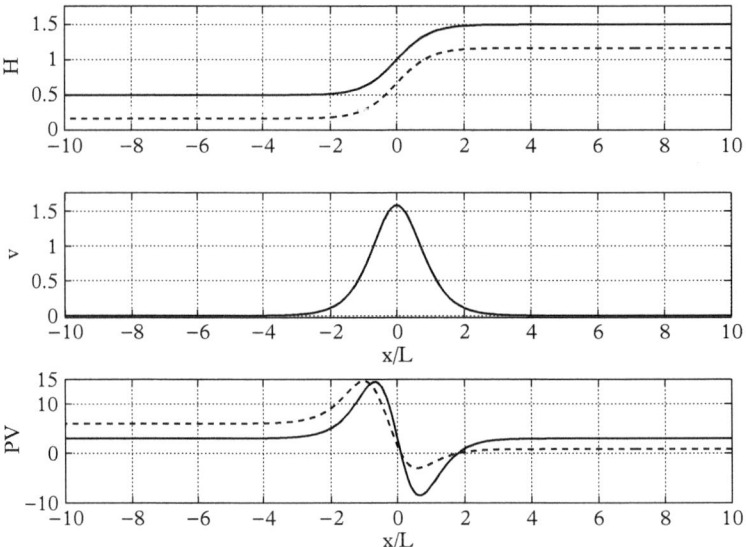

Figure 10.10 *Span-wise structure of the barotropic Bickley jet with $\frac{\delta}{L} = \sqrt{\frac{5}{2}}$ in the 2-layer model. Upper panel: thickness of the layer 2 (dashed) and the total thickness (solid). Middle panel: velocity V in both layers. Lower panel: PV of the layer 2 (dashed) and of the layer 1 (solid). Note a zone of negative PV at the anticyclonic side of the jet.*

Results of the linear stability analysis at small Ro

We will not write down the equations of the two-layer free-surface shallow-water model linearised about the jet (10.20), and pass instead directly to the results of the linear stability analysis obtained by the pseudo-spectral collocation method. They are presented in the form of dispersion diagrams for eigenfrequencies as functions of the stream-wise wave number k, and phase portraits of corresponding eigenmodes. To benchmark our analysis we make a comparison with already studied instabilities of geostrophic jets and start from relatively small Rossby numbers. The corresponding stability diagram for long-wave perturbations of the barotropic Bickley jet is presented in Figure 10.11. Note that we show only the higher growth-rate results; modes with lower and lower growth rates become visible with increasing resolution, but are not shown. In order to understand the nature of unstable modes corresponding to different branches of instability in the right panel of the Figure, we give their phase portraits in Figures 10.12, 10.13, and 10.14. As is clear from the figures, the upper branch in the right panel of Figure 10.11 corresponds to the barotropic instability, with identical perturbations of pressure and velocity in upper and lower layers, while the middle branch corresponds to the baroclinic instability, with a typical tilt of pressure and velocity of the perturbation between the layers which was already found in Section 10.2. The lower branch corresponds to barotropic modes with dipolar across-jet structure, and a node (zero) near the jet axis, as compared to the monopolar

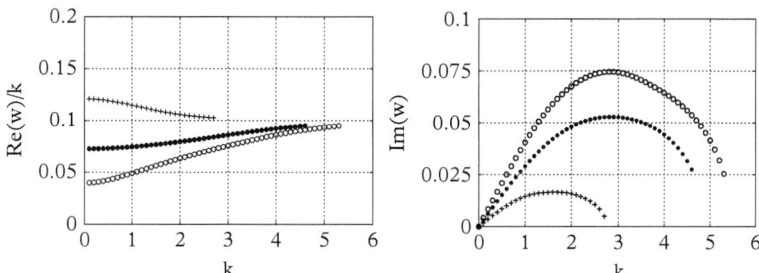

Figure 10.11 *Left panel: phase velocity $Re(\omega)/k$. Right panel: growth rate $Im(\omega)$, as functions of k for unstable modes of a quasi-geostrophic jet with $H_0 = 1, Bu = 10, Ro = 0.5, d = 2, r = 0.5$.*

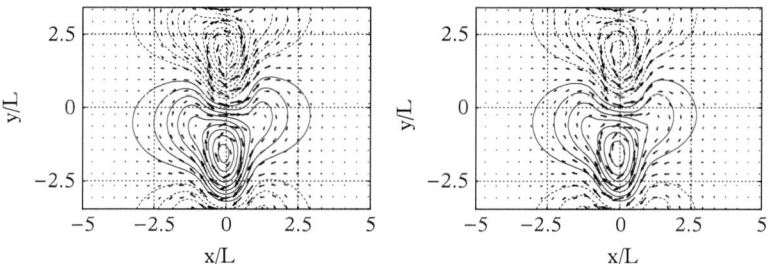

Figure 10.12 *2D structure of the most unstable mode with $k = 2.85$, $Im(\omega) = 0.075, Re(\omega)/k = 0.073$ of the upper branch of the right panel of Fig. 10.11. Left (Right) panel: upper (lower) layer. Pressure (contours, continuous: positive, dashed: negative) and velocity (arrows) anomaly.*

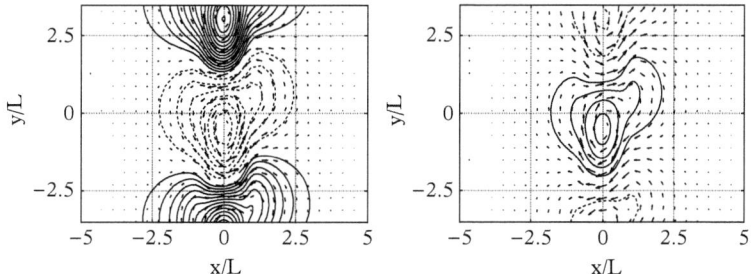

Figure 10.13 *2D structure of the most unstable mode with $k = 2.8$, $Im(\omega) = 0.05, Re(\omega)/k = 0.085$ of the middle branch of the right panel of Fig. 10.11. Left (Right) panel: upper (lower) layer. Pressure (contours, continuous: positive, dashed: negative) and velocity (arrows) anomaly.*

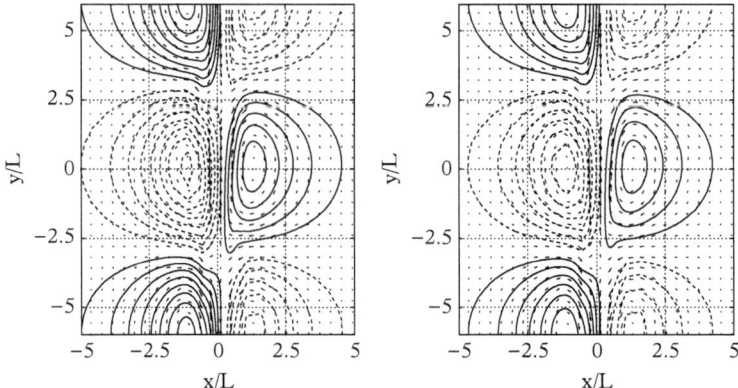

Figure 10.14 *2D structure of the most unstable mode with k = 1.65, Im(ω) = 0.016, Re(ω)/k = 0.073 on the lower branch of the right panel of Fig. 10.11. Left (Right) panel: upper (lower) layer. Pressure (contours, continuous: positive, dashed: negative) and velocity (arrows) anomaly.*

structure of the modes belonging to two higher branches. The even lower branches of instability, not presented in Figure 10.11, correspond to higher span-wise modes of baroclinic and barotropic instabilities, that is they have more nodes in the across-jet direction.

We can thus draw two conclusions, which are, in fact, independent of the particular jet structure: at small Rossby numbers barotropic instability has a higher growth rate than the baroclinic one, if they are both present, and the growth rates of instability modes are decreasing with increasing number of nodes in the across-jet direction.

Results of the linear stability analysis at high Ro

The stability diagram drastically changes at large Rossby numbers, as follows from Figure 10.15. Only the branches corresponding to highest growth rates are shown and, as previously, we concentrate on the long-wave part of the spectrum. The most unstable modes on each branch are indicated by a cross and a dot, respectively, in Figure 10.15, and have very different with respect to the small-*Ro* case structure, as follows from Figures 10.16 and 10.17. The most unstable mode marked by a cross is clearly baroclinic, but it loses approximate symmetry with respect to the jet axis which characterised the unstable modes observed at small *Ro*. The mode is centred at the anticyclonic (negative vorticity) part of the jet. The most unstable mode of the lower branch of the stability diagram is clearly barotropic but is not symmetric either. We also observe two striking features of the stability diagram in the right panel of Fig. 10.15. First, unlike all stability diagrams we have seen up to now, the growth rate on the upper branch has non-zero limit at zero wave number. Second, the two curves of the growth rates cross at some

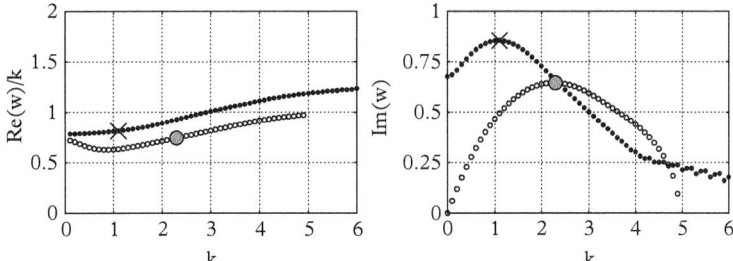

Figure 10.15 *Left: phase velocity $Re(\omega)/k$ and Right: growth rate $Im(\omega)$ as functions of k for the unstable modes of a strongly ageostrophic jet with $H_0 = 1, Bu = 10, Ro = 5, d = 2, r = 0.5$.*

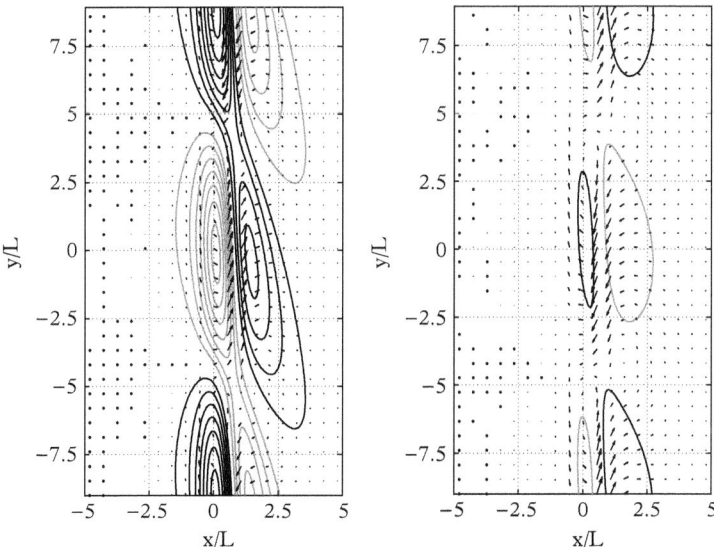

Figure 10.16 *2D structure of the most unstable mode with k = 1.1, $Im(\omega) = 0.86$, $Re(\omega)/k = 0.82$ marked by a cross in Fig. 10.15. Left (Right) panel: Pressure (contours, black: positive, grey: negative) and velocity (arrows) anomaly in upper (lower) layer.*

critical wave number. The crossing means that there is a swap of instabilities, with the most unstable mode changing nature (and structure) at this wave number. Although we do not present these results, the analysis at intermediate Rossby numbers shows that with increasing *Ro* the growth rates of the baroclinic branch of the stability diagram with dipolar across-jet structure get larger and finally overcome the barotropic growth rates,

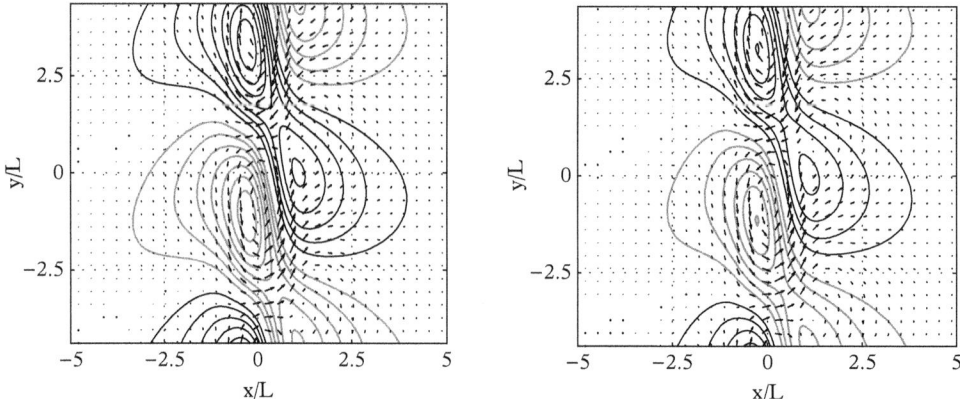

Figure 10.17 *2D structure of the most unstable mode with $k = 2.25$, $Im(\omega) = 0.65$, $Re(\omega)/k = 0.75$ marked by a dot in Fig. 10.15. Left (Right) panel: Pressure (contours, black: positive, grey: negative) and velocity (arrows) anomaly in upper (lower) layer.*

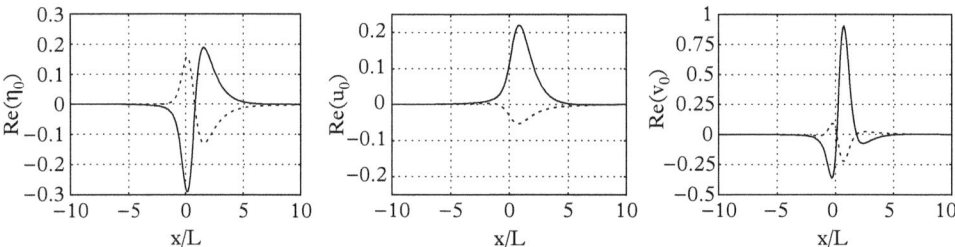

Figure 10.18 *Cross-section of the translationally symmetric unstable mode with $k = 0$ of Figure 10.15. Dashed: layer 2; solid: layer 1. From left to right: amplitudes of the interface deviation, cross-stream and stream wise velocities.*

which leads to the structure shown in Figure 10.17. In order to display unambiguously the non-zero limit of the growth rate at $k \to 0$ we presented in Figure 10.15 the case of exaggeratingly a large value of Rossby number. In fact, the instability at $k = 0$ arises at much smaller $Ro \approx 1.5$.

The symmetric with respect to stream-wise translations instability arising at $k \to 0$ is the classical *inertial instability*, as explained in Section 10.5. Thus, we see that this instability is, in fact, a long-wave limit of essentially ageostrophic baroclinic instability. The across-jet structure of the unstable mode with $k = 0$ is displayed in Figure 10.18. The cross-section at $y = 0$ of the most unstable mode marked by a cross is practically the same (not shown). This means that the upper branch of instability represents *asymmetric inertial instability* which generalises the classical inertial instability to $k \neq 0$.

10.4.2   Nonlinear saturation of essentially ageostrophic instabilities

Nonlinear evolution of the two leading long-wave instabilities at large Ro, asymmetric inertial and ageostrophic barotropic, which were identified above, is different. It is analysed using the same methodology as earlier, by superimposing the small-amplitude most unstable modes corresponding to the cross and the circle in Figure 10.15, respectively, onto the background jet. As we will see, evolution at high Rossby numbers leads to local hyperbolicity loss and considerable dissipation. The numerical model, however, copes well with this difficulty, the zones of hyperbolicity loss remaining limited in space and time. Let us recall that there is no explicit dissipation, only a numerical one, in the well-balanced finite-volume numerical code of Chapter 7. The results of numerical simulations of the evolution of asymmetric inertial instability are presented in Figure 10.19. As follows from the figure, the instability produces strong mixing of PV and leads to appearance of strong velocity shears between the layers, of zones of enhanced dissipation, and zones of hyperbolicity loss. These latter also appear, but are transient during the evolution of the barotropic instability (not shown), while they are persistently regenerated by the inertial instability. The evolution of the barotropic instability produces less shear and dissipation than that of inertial one. It should be emphasised that ageostrophic instabilities produce small-scale localised vortices in the lower layer, as seen from the lower-right panels of Figure 10.19.

An important question in the context of sources of inertia–gravity waves in the atmosphere and the ocean is emission of waves by developing instabilities. This emission is weak by the very nature of quasi-geostrophic instabilities of Section 10.2. It does not make much sense to discuss inertia–gravity wave emission in the channel geometry of Section 10.3. On the contrary, in the configuration of this section the question of emission is natural to ask. As follows from Figure 10.20, inertial instability is an important source of baroclinic inertia–gravity waves. The wave activity is diagnosed with the help of baroclinic divergence field.

It should be stressed that nonlinear evolution of the inertial instability depicted in Figure 10.19 leads to its 'self-healing'. Let us recall that the unstable mode of this instability is localised in the antisymmetric shear of the jet. This zone of the jet is characterised by the negative relative vorticity $\zeta_{jet}(x) = \partial_x V(x)$, where $V(x)$ is given by (10.20), and the corresponding absolute vorticity $f + \zeta_{jet}(x)$ becomes negative for strong shears. In turn, absolute vorticity is equivalent to the x- derivative of the geostrophic (absolute) momentum of the jet $M = fx + V(x)$, which was introduced in Chapter 8. The standard semi-heuristic argument, which can be found in textbooks (to be discussed in Section 10.20) says that at negative (if f is supposed to be positive) gradients of the geostrophic momentum the fluid parcels accelerate out of the jet axis, if displaced, which explains the inertial instability. We present in Figure 10.21 the evolution of the stream-wise averaged geostrophic momentum during the saturation of inertial instability of Figure 10.19. As is clear from the figure, a 'self-healing' of the instability (i.e. straightening of negative gradients of M in span-wise direction) is taking place. We will see this phenomenon again shortly in the translationally symmetric case.

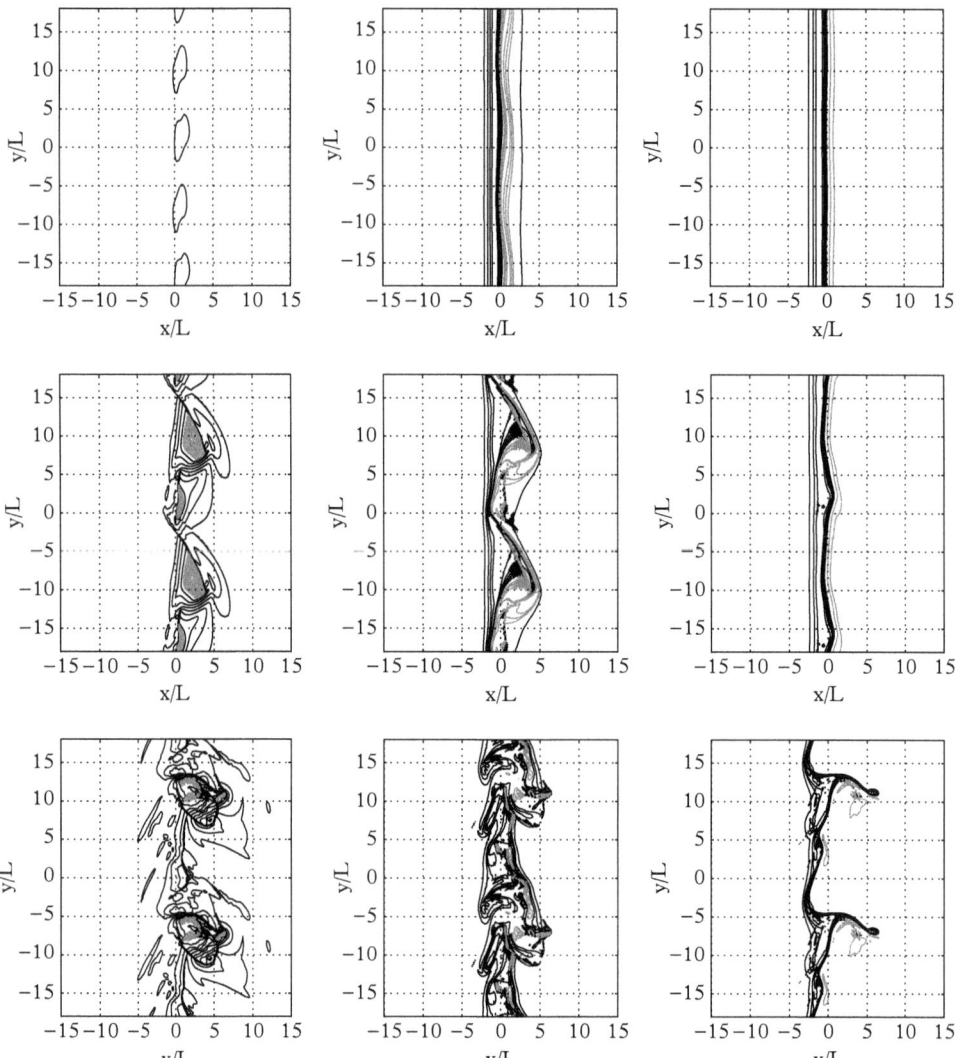

Figure 10.19 *Nonlinear evolution of the most unstable mode of asymmetric inertial instability of the barotropic Bickley jet. Left column: Isolines of the velocity shear $|u_2 - u_1|$ (thin grey lines), enhanced dissipation zones (black) and zones of hyperbolicity loss (grey). PV isolines in layer 1 (middle column) and in layer 2 (right column). Positive/negative PV: black/gray. Snapshots at $t = 4, 8, 14f^{-1}$, from top to bottom.*

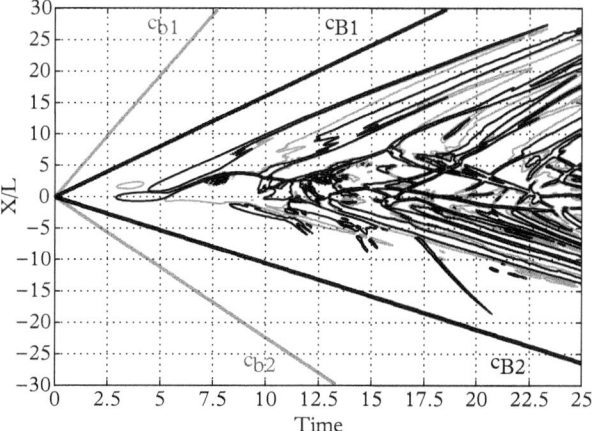

Figure 10.20 *Evolution in space and time (Hovmöller diagram) of the baroclinic divergence field calculated at y = 0. Typical values shown: +0.1 (black), −0.1 (grey). Slopes corresponding to phase velocity of linear barotropic (grey) and baroclinic (black) inertia–gravity waves are shown for comparison.*

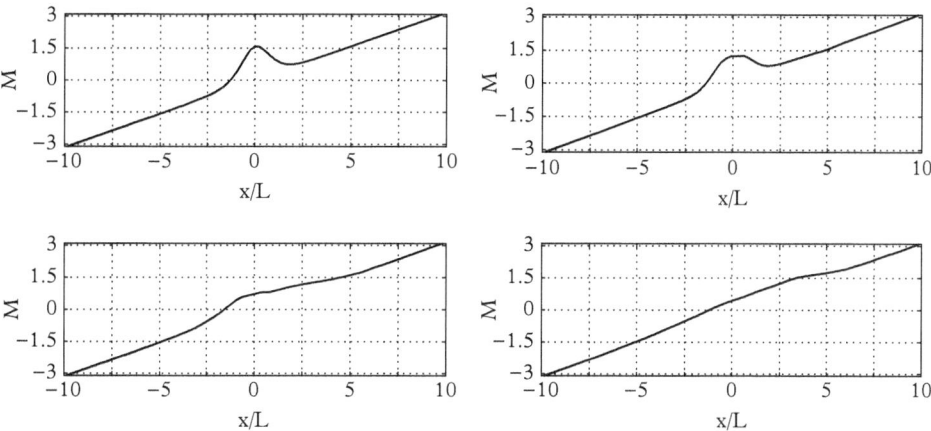

Figure 10.21 *Evolution of stream-wise averaged geostrophic momentum for the developing asymmetric inertial instability. Snapshots at $t = 4$, $t = 8$, $t = 14$, $t = 22 f^{-1}$, from left to right and from top to bottom.*

10.4.3   A brief summary of the results on essentially ageostrophic instabilities of mid-latitude jets

Thus, we have seen that with increasing Rossby number, the classical barotropic and baroclinic instabilities of jets undergo qualitative changes. Not only the growth rates of these instabilities increase, which was expected, but the structure of the most unstable modes changes. The barotropic instability which is dominant at small Rossby numbers, becomes sub-leading, in favour of ageostrophic baroclinic instability. Moreover, the classical bell-shaped curve of growth rates as a function of stream-wise wave number, with growth rates tending to zero at small wave numebrs, drastically changes and gives non-zero growth rates in this limit. A new translationally invariant (symmetric in this sense) instability thus appears. This is the well-known inertial instability which is, therefore, an emanation of strongly ageostrophic baroclinic instability. The corresponding eigenmode is localised in the zone of anticyclonic shear of the jet. However, the asymmetric version of this instability, with small but non-zero wave number has always higher growth rates in the two-layer model. The interpretation of inertial instability in terms of trapped modes will be given in the next Section.

10.5   Understanding the nature of inertial instability

To understand the nature of the symmetric with respect to translations inertial instability revealed in the previous Section, let us consider the 1.5D situation, where all dependence on the along-flow coordinate, say y, is removed. The equations of the two-layer model with a free surface in this configuration were already considered in Section 7.2.1. Here, for simplicity, we will be working with the two-layer RSW with a rigid lid:

$$
\begin{cases}
\partial_t u_1 + u_1 \partial_x u_1 - f v_1 + \rho_1^{-1} \partial_x \pi = 0, \\
\partial_t v_1 + u_1 (f + \partial_x v_1) = 0, \\
\partial_t u_2 + u_2 \partial_x u_2 - f v_2 + \rho_2^{-1} \partial_x \pi + g' \partial_x \eta = 0, \\
\partial_t v_2 + u_2 (f + \partial_x v_2) = 0, \\
\partial_t (H_1 - \eta) + \partial_x ((H_1 - \eta) u_1) = 0, \\
\partial_t (H_2 + \eta) + \partial_x ((H_2 + \eta) u_2) = 0,
\end{cases}
\tag{10.21}
$$

where $(u_1,\ v_1)$, $(u_2,\ v_2)$ are velocity components in the upper and lower layer, respectively; π is barotropic pressure; η is displacement of the interface; H_1 and H_2 are thicknesses of the layers at rest; and $H = H_1 + H_2 = const$, g' is the reduced gravity: $g' = g(\rho_2 - \rho_1)/\rho_2$. Geostrophic equilibria

$$
v_1 = \frac{1}{f \, \rho_1} \, \partial_x \pi, \quad v_2 = \frac{1}{f \rho_2} \, \partial_x \pi + \frac{g'}{f} \partial_x \eta.
\tag{10.22}
$$

are exact solutions of (10.21). Under the standard scaling, the corresponding non-dimensional expressions are

$$V_{1g} = \partial_x \Pi, \quad V_{2g} = r \, \partial_x \Pi + Bu \, \partial_x h_{2g}, \tag{10.23}$$

where $r = \frac{\rho_1}{\rho_2}$ and the Burger number is $Bu = \frac{g' H_2}{f^2 L^2}$, and we introduce $h_{2g} = H_2 + \eta$. In the following we are considering geostrophic equilibria corresponding to localised jets, with velocity rapidly decaying with $|x|$.

Linearisation about a state of geostrophic equilibrium gives

$$\begin{cases} \partial_t u_1 - v_1 + \partial_x \Pi = 0, \\ \partial_t v_1 + u_1 (1 + \partial_x V_{1g}) = 0, \\ \partial_t u_2 - v_2 + r \partial_x \Pi + Bu \partial_x \eta = 0, \\ \partial_t v_2 + u_2 (1 + \partial_x V_{2g}) = 0, \\ \partial_t \eta - \partial_x (h_{1g} u_1) = 0, \\ \partial_t \eta + \partial_x (h_{2g} u_2) = 0. \end{cases} \tag{10.24}$$

The constraint of absence of divergence of the barotropic velocity, which follows by taking the sum of the last two equations in (10.21), after integration in x gives

$$((H_1 - \eta) u_1) + ((H_2 + \eta) u_2) = H U_b(t). \tag{10.25}$$

The right-hand side of this equation represents a total mass flux across the jet due to perturbations and will be set to zero, $U_b = 0$, which means that perturbations are localised in the vicinity of the jet. Using (10.25) and introducing a new variable $U = h_{2g} u_2 = -h_{1g} u_1$, the system (10.24) is reduced to a single equation:

$$Bu \, \partial_{xx}^2 U - \left[\frac{r h_{2g} + h_{1g}}{h_{1g} h_{2g}} (\partial_{tt}^2 + 1) + \frac{r \, \partial_{xx}^2 \Pi}{h_{1g} h_{2g}} + Bu \, \frac{\partial_{xx}^2 h_{2g}}{h_{2g}} \right] U = 0. \tag{10.26}$$

It is easy to see from this equation that if the anticyclonic shear of the jet is strong enough, *unstable trapped modes* are possible. Indeed, by making a Fourier transformation $U(x, t) = \int d\omega \tilde{U}(\omega, x) e^{-i\omega t} + c.c.$, and introducing auxiliary functions

$$F(x) = \frac{r h_{2g} + h_{1g}}{h_{1g} h_{2g}}, \quad G(x) = \frac{r \, \partial_{xx}^2 \Pi}{h_{1g} h_{2g}} + Bu \, \frac{\partial_{xx}^2 h_{2g}}{h_{2g}}, \tag{10.27}$$

we get the following equation for $\tilde{U}(\omega, x)$:

$$Bu \, \partial_{xx}^2 \tilde{U} - \left(F(1 - \omega^2) + G \right) \tilde{U} = 0. \tag{10.28}$$

Multiplication by $\tilde{U}^*$ and integration in x, assuming that solutions are *localised*, give

$$\omega^2 = 1 + \frac{Bu \int |\partial_x \tilde{U}|^2 \, dx + \int G|U|^2 \, dx}{\int F|\tilde{U}|^2 \, dx}. \tag{10.29}$$

The function F is, by definition, positive, but G can be negative, especially in *anticyclonic regions* where $\partial_{xx}^2 \Pi < 0$, cf. (10.27). Hence, for strong enough anticyclonic shears, it is possible that $\omega^2 < 1$, and can even become negative, the latter situation obviously corresponding to an instability. Let us recall that for the classical inertial instability to take place, it is necessary that the product of PV and planetary vorticity f be negative somewhere in the flow (see the comments at the end of this chapter), which means precisely that anticyclonic shear with negative vorticity should be sufficiently strong at some locations (for positive Coriolis parameter f). It should be emphasised that such situation is impossible in the one-layer model, because of the absence of trapping, as was shown in Section 8.3.1.

As an illustration, let us consider a barotropic jet with flat interface $\eta = 0$, cf. (10.22). The equation for U becomes

$$Bu \, \partial_{xx}^2 U - \left[(\partial_{tt}^2 + 1) \, H_e^{-1} + r\partial_{xx}^2 \Pi \, (H_1 H_2)^{-1} \right] U = 0, \tag{10.30}$$

where $H_e = \frac{H_1 H_2}{H_1 + H_2}$ is the equivalent height. After Fourier transformation in time we obtain

$$\partial_{xx}^2 \tilde{U} + \frac{1}{Bu} \left[\omega^2 H_e^{-1} - (H_e^{-1} + (H_1 H_2)^{-1} r\partial_{xx}^2 \Pi) \right] \tilde{U} = 0. \tag{10.31}$$

It is easy to recognise in (10.31) the Shrödinger equation of quantum mechanics:

$$\partial_{xx}^2 \psi + (E - V(x))\psi = 0 \tag{10.32}$$

for a particle with energy $E = \omega^2 (H_e \, Bu)^{-1}$ moving in the potential $V(x) = Bu^{-1} \, (H_e^{-1} + (H_1 H_2)^{-1} r\partial_{xx}^2 \Pi)$. As is well-known, if a potential well is deep enough, the particle is trapped. For even deeper potential wells, nothing prevents the eigenvalues of 'energy' to become negative, thus leading to purely imaginary ω. The corresponding eigenmodes are, thus, non-propagative, which is a particular feature of such symmetric instability. The inertial instability of Section 10.4 is precisely of this nature, although it was identified within a slightly different model, the free-surface two-layer RSW. It is easy to generalise the preceding analysis to this case. Considering small perturbations of the jet (10.20),

$$h_i = H_i(x) + \eta_i'(x,t), u_i = u_i'(x,t), v_i = V(x) + v_i'(x,t), \qquad i = 1, 2, \tag{10.33}$$

and looking for harmonic solutions $(u_i', v_i', \eta_i') = (u_{0i}(x), v_{0i}(x), \eta_{0i}(x)) \, e^{-i\omega t} + \text{c.c.}$, we obtain, along the same lines as earlier, a pair of coupled Schrödinger equations for the across-front velocities of the layers:

$$\left(-\omega^2 + f(f + \partial_x V)\right)\begin{pmatrix} u_{01} \\ u_{02} \end{pmatrix} - g\partial_x^2 \begin{pmatrix} H_1 u_{01} + H_2 u_{02} \\ rH_1 u_{01} + H_2 u_{02} \end{pmatrix} = 0. \qquad (10.34)$$

Rewriting these equations in terms of barotropic and baroclinic velocity components,

$$\begin{pmatrix} u_b \\ u_B \end{pmatrix} = \begin{pmatrix} u_{02} - u_{01} \\ \frac{H_1 u_{01} + H_2 u_{02}}{H_1 + H_2} \end{pmatrix}, \qquad (10.35)$$

we obtain the following integral estimate for the eigenfrequencies ω:

$$\omega^2 = f\frac{\int (f + \partial_x V) H_e |u_b|^2}{\int H_e |u_b|^2} + g(1 - r)\frac{\int |\partial_x (H_e u_b)|^2}{\int H_b |u_b|^2}$$

$$+ g(1 - r)\frac{\int H_e u_b^* \partial_x^2 (H_1 u_B)}{\int H_e |u_b|^2}. \qquad (10.36)$$

As is easy to see (for example, for purely baroclinic modes with $u_B = 0$), if the relative vorticity of the jet $\partial_x V$ is sufficiently negative, ω^2 can be negative, and hence the (symmetric) inertial instability arises. It should be emphasised that it is the product $f(f + \partial_x V) = f\partial_x M$, where M is the geostrophic momentum, which controls the instability. Let us recall that the standard criterion of the inertial instability is, precisely, non-positivity of this expression.

To summarise, inertial instability is due to standing modes trapped in the anticyclonic shear of the jet, with squares of eigenfrequencies diminishing with increasing strength of the shear, and becoming negative at strong enough shears. It is interesting to see how the unstable mode evolves nonlinearly in such symmetric case. As usual, by superimposing the translationally symmetric unstable mode of Figure 10.18 onto the barotropic jet in the 1.5D RSW model, and performing numerical simulations, which are especially easy in this reduced model, we observe a fast breaking of the growing unstable mode, which is shown in Figure 10.22. Such behaviour illustrates the case of flows with non-positive PV where existence of adjusted state is not guaranteed, as was shown above in Chapter 8. Breaking leads to formation of a large-amplitude shock (it should be emphasised that,

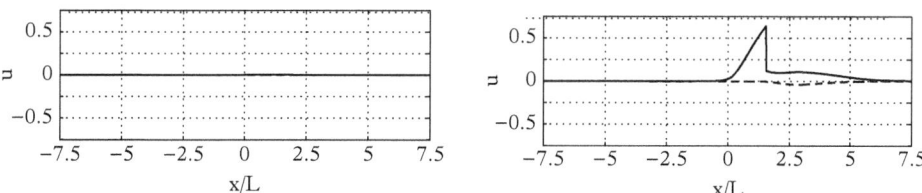

Figure 10.22 *Across-jet velocity u in upper (solid) and lower (dashed) layers at t = 0 (left panel), and t = 7.75f^{-1} (right panel) during development of inertial instability in 1.5D model, as follows from numerical simulation. Parameters of the jet: $H_0 = 1$, $Bu = 10$, $Ro = 5$, $d = 2$, $r = 0.5$.*

at the same scale, the initial perturbation is practically invisible in the left panel of the figure), which triggers strong dissipation, which, in turn, results in a 'corrected' jet configuration, which is inertially stable. We have seen in the preceding sections how such 'self-healing' of the instability is realised in full two-dimensional simulations, and will see below other examples of this feature of inertial instability, and of its close relative, the centrifugal instability of vortices.

10.6   Instabilities of jets at the equator and their nonlinear evolution, with emphasis on inertial instability

The discussion of inertial instability of mid-latitude jets immediately makes one think about the equatorial region. Indeed, as have been shown in Section 10.5, the necessary condition of inertial instability is non-positivity of the product of PV and planetary vorticity some place in the flow. However, there is nothing easier than changing the sign of this product at the equator, where planetary vorticity changes sign itself. It is therefore natural to address the question of inertial instability and, more generally, of ageostrophic instabilities of jets in this region. We present results of the study on the equatorial beta plane along the same lines as that of Section 10.4 on the mid-latitude f plane. However, inertial instability appears already in the one-layer model in this case, so we start with this configuration.

10.6.1   Linear stability and nonlinear saturation of instabilities in one-layer RSW model at the equator

Set-up and parameters. General considerations

We will be not rewriting the one-layer RSW model in the equatorial beta-plane approximation, which is obtained from the standard RSW equations by replacement $f \to \beta y$, where y is the meridional coordinate. As in Section 10.4, we are considering a localised jet, which is centred at the equator, so that symmetry properties with respect to reflexions $y \to -y$ could be used to facilitate the analysis. We analyse easterly jets, which exist in the atmosphere and oceans, such as an atmospheric subtropical easterly jet, and choose Gaussian jet profiles for the zonal velocity and thickness anomalies:

$$H(y) = H_0 - \Delta H e^{-(y/L)^2}, \quad U(y) = -2U_0 e^{-(y/L)^2}, \quad V(y) = 0. \quad (10.37)$$

Here U_0 and ΔH are linked by geostrophic balance on the equatorial beta plane in the meridional direction, and thus provide an exact solution of the RSW equations. The equations of motion can be non-dimensionalised by using the width of the jet L as the horizontal length scale, H_0 as the vertical scale, (half of) the maximal jet velocity U_0 as the velocity scale, and $\frac{L}{U_0}$ as the time scale. The dimensionless parameters are then the equatorial Rossby number $Ro = \frac{U_0}{\beta L^2}$ and the equatorial Burger number $Bu = \frac{gH_0}{(\beta L^2)^2}$. We

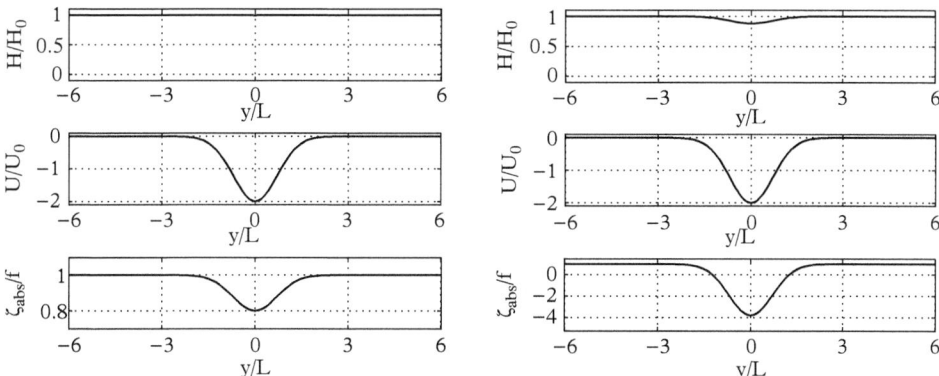

Figure 10.23 *Background easterly equatorial jet with Bu = 10 at Ro = 0.05 (left column) and Ro = 1.2 (right column). Upper row: thickness; λ = 0.005 (0.12) in the left (right) column; Middle row: span-wise zonal velocity profiles; Lower row: span-wise profiles of absolute vorticity βy + ζ normalised by the planetary vorticity βy.*

also introduce the non-dimensional deviation of the free surface $\lambda = \frac{\Delta H}{H_0} = \frac{Ro}{Bu} \le 1$ (this restriction excludes incropping flows), and from now on work with non-dimensional equations, with Ro and Bu as main parameters. Thickness and velocity profiles (10.37), together with absolute vorticity are presented in Figure 10.23, in the cases of geostrophic and ageostrophic jet. It should be emphasised that the normalised absolute vorticity is negative at the central part of the jet in the ageostrophic configuration. We could, thus, expect the appearance of inertial instability at $Ro \ge 1/4$, in accordance with the above-discussed criterion: $f \cdot \zeta \le 0 \Rightarrow Ro \ge 1/4$. Moreover, it can be checked that Ripa's general stability conditions (10.12), (10.13) are satisfied at $Ro \le 1/4$, but not at $Ro \ge 1/4$. However, as we have seen in Section 10.5, inertial instability is due to the modes trapped in the jet. As was shown in Section 8.3, trapped modes are impossible in one-layer RSW on the f plane and, therefore, inertial instability either. Does this mean that the above consideration is misleading? In fact it is not, as shown in the following argument. By applying the Lagrangian approach of Section 8.3 to the 1.5D RSW equations on the the equatorial β plane:

$$\begin{cases} u_t + vu_y - \beta yv = 0, \\ v_t + vvy + \beta yu = -gh_y, \\ h_t + (hv)_y = 0, \end{cases} \tag{10.38}$$

we get their Lagrangian counterpart written in terms of Lagrangian coordinate $Y(y, t)$:

$$\begin{cases} \ddot{Y} + \beta Yu + g\partial_Y h = 0, \\ \dot{u} - \beta Y\dot{Y} = 0, \\ h - h_I(y)\dfrac{1}{\partial_y Y} = 0, \end{cases} \tag{10.39}$$

where h_I is the initial thickness distribution. We linearise around the geostrophically balanced zonal jet with velocity u_I (u_I, h_I replace U, H above), introduce the departures of Lagrangian particles from their initial positions $Y(y, t) = y + \xi(y, t)$, and get the following equation for ξ:

$$\ddot{\xi} + (u_I + \beta y^2)\beta\xi - gh_I\partial_{yy}\xi - 2gh'_I\partial_y\xi = 0, \tag{10.40}$$

whence, introducing the Fourier transform of ξ in time $\xi = \tilde{\xi}e^{i\omega t} + $c.c., we get the integral estimate for the eigenfrequency squared:

$$\omega^2 = \frac{\int_{-\infty}^{+\infty} dy\, \beta(\beta y^2 + u_I)gh_I|\tilde{\xi}|^2 + \int_{-\infty}^{+\infty} dy\, g^2 h_I^2|\tilde{\xi}'|^2}{\int_{-\infty}^{+\infty} dy\, gh_I|\tilde{\xi}|^2}. \tag{10.41}$$

As follows from this formula, unlike its f-plane counterpart (8.94), the Lagrangian estimate (10.41) in the one-layer model does not prevent symmetric instability for strong enough westward flows, when the first integral in the numerator becomes negative. This constitutes an essential difference between the f-plane and the equatorial β-plane RSW models.

Results of the linear stability analysis

As we have seen in Section 10.3, instabilities can be understood in terms of hybrid modes appearing when the dispersion curves of primary waves in the model intersect. This is strictly true for the constant PV flows, which will be treated in subsequent sections, because of the absence of critical levels already discussed in Section 4.3. In the presence of these latter, there may exist some instabilities which result from resonances of regular modes with so-called pseudo-modes, belonging to the singular part of the spectrum of small perturbations. In the dispersion diagram, which represents regular modes only, such a situation is seen as termination of a dispersion curve. Yet, at least a part of the instabilities is always produced by intersections of pairs of dispersion curves of regular modes. As we will see shortly, the most important instabilities of equatorial waves in the RSW model are associated with pairs of regular modes. So, before starting to look for instabilities, we consider first a jet which is stable, according to Ripa's criterion outlined earlier, and identify the primary waves which can propagate on its background. The dispersion diagram of a jet with $Ro = 0.05$ and $Bu = 10$ obtained by such analysis is presented in Figure 10.24 and displays high-speed westward- (left-) and eastward- (right-) propagating inertia–gravity, or Poincaré waves; eastward-propagating non-dispersive Kelvin wave; westward-propagating low-speed dispersive Rossby waves, and dispersive westward-propagating wave intermediate between Rossby and inertia–gravity ones, the Yanai wave. As was shown in Section 4.4.1, a way of deducing the spectrum of equatorial waves over the state of rest is to make a Fourier transformation in x and t, reduce the linearised RSW system to a single equation for $v(y)$, and to solve this latter with the help of decomposition in Gauss–Hermite functions $\phi_n(y)$, where $n = 0, 1, 2, \ldots$, gives the number of nodes in the meridional direction. The resulting cubic dispersion relation gives two singular solutions corresponding to the Kelvin wave

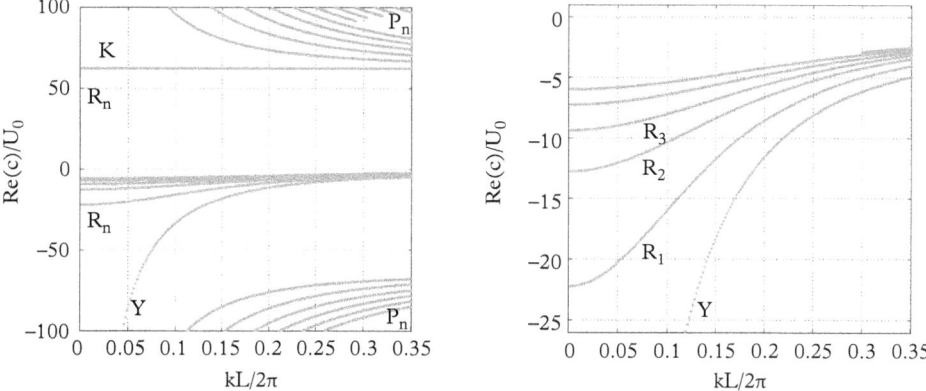

Figure 10.24 *Stability diagram for a background equatorial jet with Ro = 0.05 of the left panel of Fig. 10.23. Left: stable modes. Right: Zoom of the stability diagram at small c. P_n: Poincaré (inertia-gravity) waves with meridional wave number n; K: Kelvin wave; Y: Yanai wave; R_n: Rossby wave with meridional wave number n.*

(K), which can be formally associated with $n = -1$ and has no meridional velocity at all, the Yanai wave (Y) corresponding to $n = 0$, and a triple solution for each $n \geq 1$, consisting of a pair of higher-frequency Poincaré waves (P) and a lower-frequency Rossby wave (R). In the presence of (non-uniform) mean flow this analysis does not hold, yet the number of nodes of the meridional velocity field in the meridional direction can be used to identify the waves in a robust topological way, and we use this diagnostic. In the case of Figure 10.24, the mean current is weak, and the eigenmodes are close in structure to the classical equatorial waves.

With increasing Rossby number of the jet, at fixed Burger number, the instabilities appear beyond $Ro \simeq 0.25$, consistently with Ripa's criterion. Already at $Ro = 0.3$ the stability diagram displayed in Figure 10.25 reveals several different types of instabilities in the long-wave sector we are interesting in (we recall that the classical inertial instability has infinite wavelength). The resonating wave species are identified by the number of nodes in the meridional velocity and by the phase portraits extracted from two branches of the dispersion diagram near the intersection point. This identification, when it is unambiguous, is marked in the figure. Simultaneous presence of multiple waves-species, and corresponding large number of close dispersion curves, as well as hybridisation, which modifies the standard phase portraits, complicate the identification in many cases. However, it should be stressed that, here and below, the number of nodes of the unstable mode is a robust identification, never changing along the instability curve. We recall that the resonance between two Rossby waves is associated with the standard barotropic instability. It is, indeed, present, although the strongest instability is due to the resonance of Yanai waves. This dominant instability shuts down at small albeit finite wave numbers, unlike the classical barotropic one, cf. Figure 10.25. While the unstable modes of the standard barotropic instability branch of the figure are balanced,

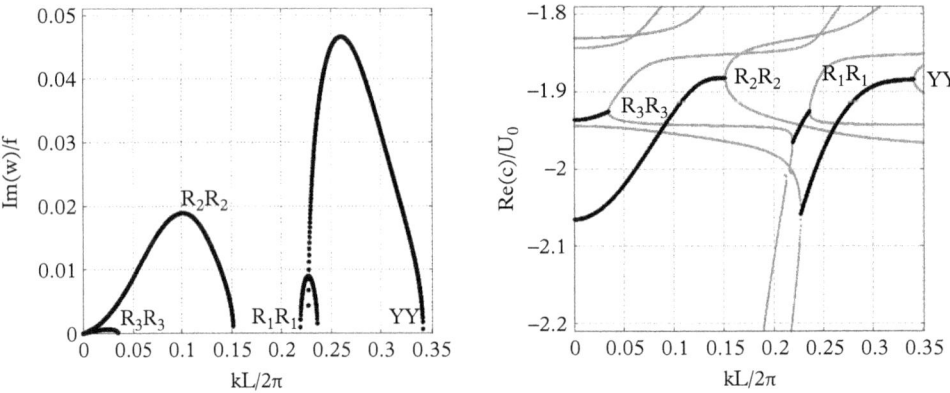

Figure 10.25 *Stability diagram for the background flow with Bu = 10 and Ro = 0.3. Black: unstable modes. Grey: stable modes. R_n: Rossby waves; Y: Yanai waves. Here and below the growth rate is normalised by βL.*

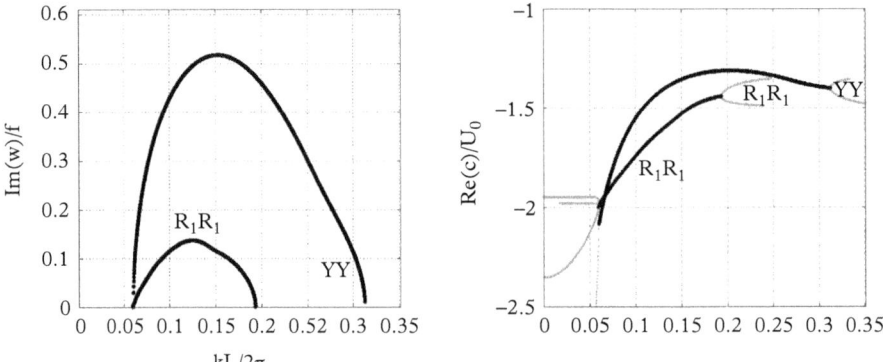

Figure 10.26 *Stability diagram for a background easterly equatorial jet with Ro = 1.2 displayed in the right panel of Fig. 10.23. Same conventions as in Fig. 10.25.*

in the sense that velocity vectors follow the isobars, the unstable modes of the new instability are essentially unbalanced (not shown). The dominant instability is, therefore, ageostrophic, in spite of relative smallness of the Rossby number of the jet.

At essentially ageostrophic Rossby numbers, the stability diagram reveals an increasing number of intersections of the dispersion curves, and related instabilities. Still, the dominant instability with high growth rates is due to the resonance of Yanai waves. We show in Figure 10.26 the stability diagram at $Ro = 1.2$, where the growth rate of this instability is an order of magnitude higher than in $Ro = 0.3$ case of Figure 10.25. The small-amplitude growth rates close to the k axis are removed, as well as non-resonating stable modes (one can get an idea of both from the subsequent figures). The mode having the maximal growth rate is presented in Figure 10.27. As seen in the figure, its

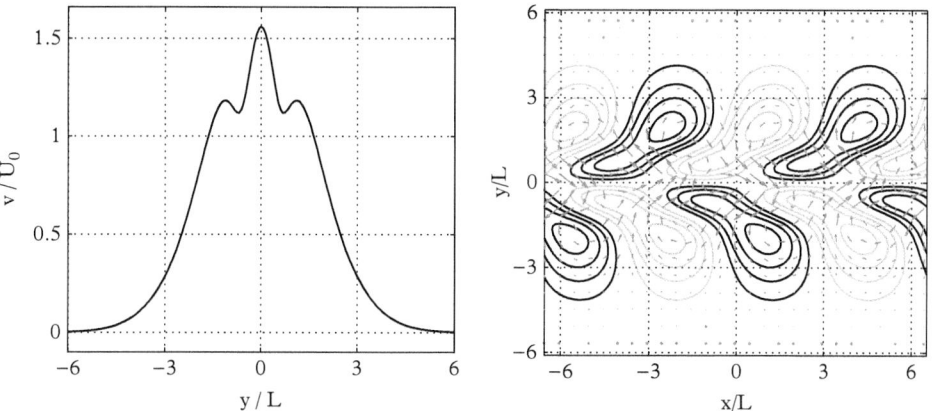

Figure 10.27 *Left panel: profile of $v(y)$. Right panel: two-dimensional structure of the most unstable mode with $k = 0.152(\frac{2\pi}{L})$, $Re(c/U_0) = -1.35$ and $Im(\omega/\beta L) = 0.52$ of Fig. 10.26. Arrows: velocity field, contours: thickness anomaly. Black: positive, Grey: negative.*

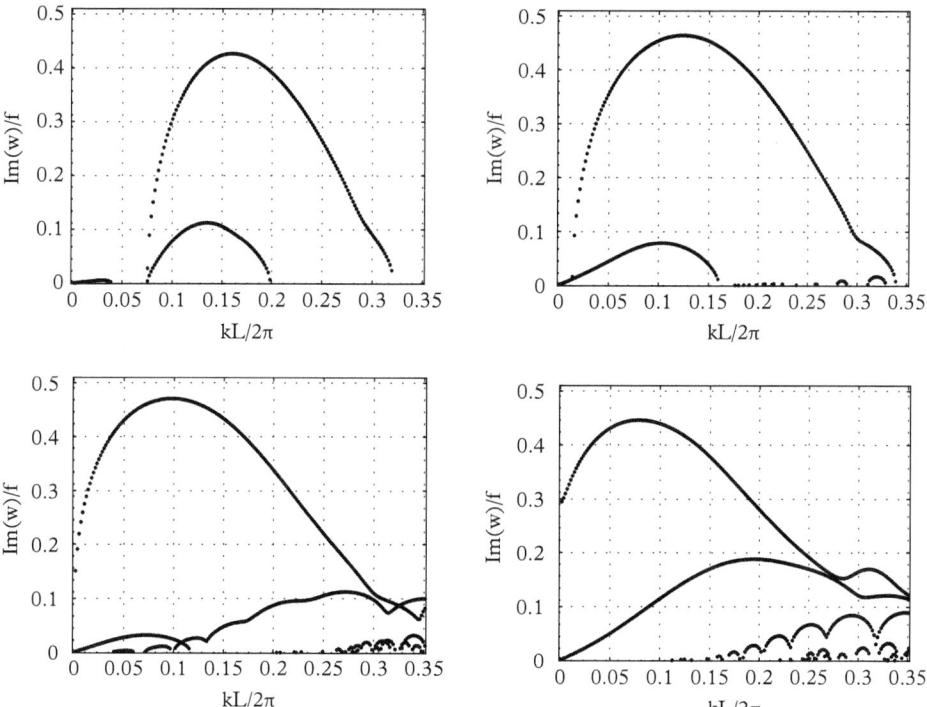

Figure 10.28 *Growth rates of the unstable modes as functions of k for a background easterly equatorial jet with $Ro = 1$, and $Bu = 10$, $Bu = 2$, $Bu = 1.3$, and $Bu = 1.1$, from left to right and from top to bottom. Uppermost curve: YY instability.*

meridional velocity has no nodes, which is typical for Yanai waves. The mode has a structure of pressure perturbation which is dipolar in the meridional direction again like a Yanai wave, cf. Figure 4.12 in Section 4.4.1. The most unstable mode of the second instability branch in Figure 10.26 is tripolar in the meridional direction (not shown). Both structures are consistent with the nature of the corresponding resonating waves indicated in the figure. It should be noted that R_2R_2 instability, next to dominant at lower Rossby numbers, cf. Figure 10.25, gives way to the R_1R_1 one. Here the subscript indicates the index of corresponding Gauss–Hermite function.

Thus, increasing the Rossby number at a fixed large Burger number destabilises the flow, with two main types of long-wave instabilities, RR and YY, the second one being dominant and acting in a limited range of wave numbers bounded from above and below. Interestingly, only westward-propagating waves, which have phase velocity of the same sign as the mean flow, are generating the instabilities, while the eastward-propagating waves, such as the Kelvin waves, play no role.

Although various instabilities are present, we see no sign of inertial instability, which should have non-zero growth rate at $k = 0$. It appears, indeed, but only if the Burger number is low enough, as follows from Figure 10.28, where the evolution of dispersion diagram with Bu at fixed $Ro = 1$ is displayed. As follows from the figure, the lower bound in the range of unstable zonal wave numbers corresponding to the dominant YY instability shifts towards smaller values with decreasing Burger number, and hits zero at $Ro \approx 1.3$, with further decrease in Bu producing nonzero growth rates at $k = 0$. Yet, as in Section 10.4, the asymmetric inertial instability, with similar structure of unstable modes, has higher growth rates.

Nonlinear saturation of the YY instability

Nonlinear saturation of the dominant YY instability, as follows from numerical simulations with well-balanced finite-volume numerical scheme of Chapter 7 initialised with the unstable mode of weak amplitude superimposed onto the background jet, is presented in terms of relative vorticity in Figure 10.29. The boundary conditions are: sponges in the meridional direction imposed at ≈ 10 equatorial deformation radii, and periodicity, with a double period of the unstable mode, in the zonal direction. As follows from the figure, the evolution of the instability exhibits a transformation of the perturbation into a vortex street, a rather standard scenario of developing jet instabilities.

10.6.2   Linear stability and nonlinear saturation of instabilities in two-layer RSW model at the equator

Set-up

In order to investigate the role of baroclinic effects in emergence and nonlinear saturation of above-described instabilities of equatorial jets, and possible appearance of essentially baroclinic instabilities, a similar study to that presented in Section 10.5 can be performed in the framework of the two-layer model. We work with the two-layer RSW with a free surface, in the same configuration as in Section 10.4 but in the equatorial beta-plane. As in that section, we consider a barotropic jet with no velocity shear between the layers,

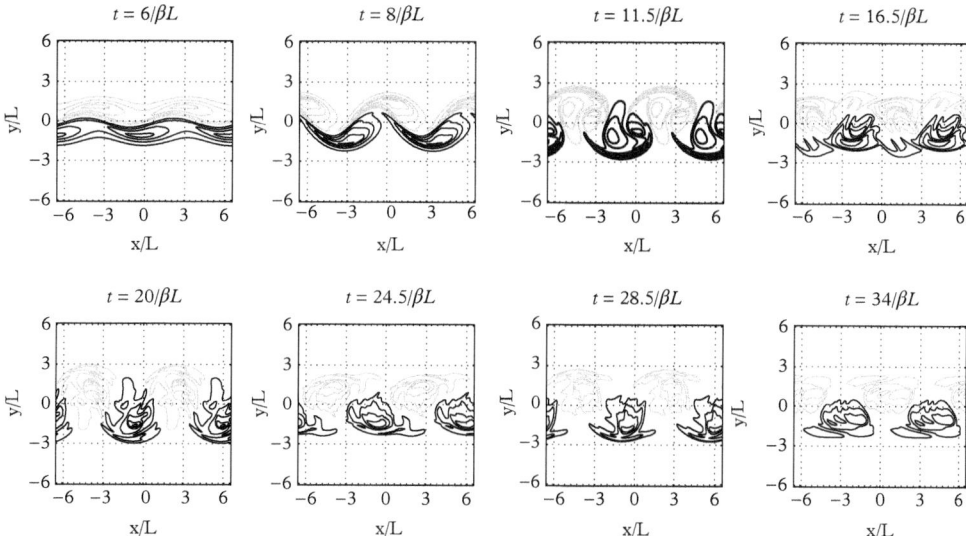

Figure 10.29 *Nonlinear evolution of the most unstable mode of Fig. 10.26, cf. Fig. 10.27. Isolines of relative vorticity. Positive/negative relative vorticity anomaly: black/grey lines.*

which verifies the geostrophic balance relations on the equatorial beta plane layer-wise, is an exact solution of the equations of motion, and has the same Gaussian form as in the one-layer case above:

$$H_1(y) = H_{10}, \quad H_2(y) = H_{20} - \Delta H e^{-(y/L)^2}, \quad U_1(y) = U_2(y) = -2U_0 e^{-(y/L)^2}, \quad (10.42)$$

where $H_{1,2}$ are thicknesses of the layers. This configuration is displayed in Figure 10.30. As in the one-layer case it may be shown that $Ro = 0.25$ is the threshold of instability, flows with lower Ro being stable. We sketch below the results of the linear stability analysis and of simulations of the nonlinear evolution of instabilities of such a flow.

Results of the linear stability analysis

In order to identify the primary waves and, at the same time, benchmark the linear stability analysis procedure, we start with the stable jet, similar to the one-layer case earlier. Dispersion diagram of the flow with $Ro = 0,05$ is presented in Figure 10.31 and displays, as expected, a doubling of wave species observed in Figure 10.24, each wave species having now baroclinic and barotropic variants, as can be checked by looking at their phase portraits. Baroclinic waves are slower, according to the general rule. (Aspect and density ratios are denoted by d and r, respectively, as in Section 10.4.) We can, thus, expect the appearance of both baroclinic and barotropic instabilities due to possible resonances between barotropic and baroclinic waves. We can also expect the appearance of the baroclinic inertial instability, which was discussed in the f plane approximation

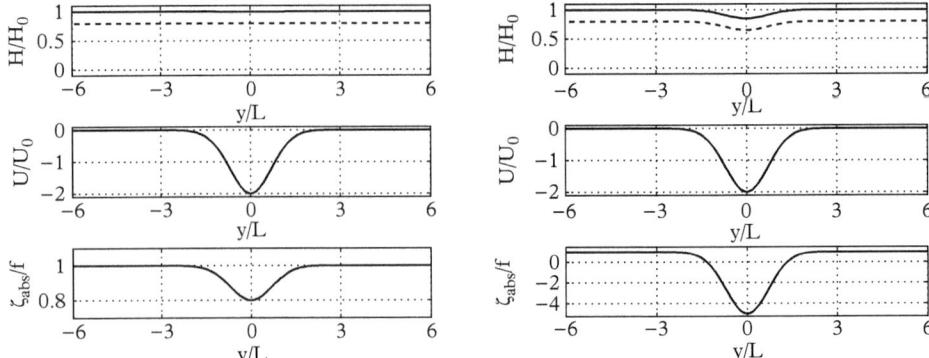

Figure 10.30 *Background easterly equatorial jet in 2-layer configuration with Bu = 10, r = 0.5 and d = 0.25 at Ro = 0.05 (left column) and Ro = 1.5 (right column). Upper panel: thicknesses of the layers; λ = 0.005 (0.15) in left(right) column; Middle panel: zonal velocity profile (same in both layers); Lower panel: profile of the absolute vorticity βy + ζ normalised by the planetary vorticity βy (same in both layers).*

in Section 10.4. The results of the preceding subsection let us expect the barotropic symmetric inertial instability as well. Questions then arise about relative strength of two inertial instabilities, their relation to the asymmetric counterparts, and their zone of residence in the parameter space. We should stress that an intersection of dispersion curves is a necessary but not a sufficient condition for phase-locking and resonance leading to instability. The waves with close frequencies should propagate in opposite directons with respect to the flow, and should be coupled through pressure perturbations in order to produce an instability. The intersections observed in Figure 10.31 are 'harmless' in this sense, and the flow is stable.

In order to detect the key long-wave instabilities we choose a configuration with $Ro = 0.27$, close to the instability threshold. The corresponding dispersion diagram is presented in Figure 10.32, and shows that the leading instability is YY, as in the one-layer case and with similar characteristics, although there are now two unstable modes with very close growth rates in practically the same range of zonal wave numbers. The phase portraits of unstable modes with the same k on each of these two instability curves are presented, respectively, in Figures 10.33 and 10.34, and clearly show that one of them is baroclinic, and another barotropic.

With an increasing Rossby number, and fixed values of other parameters, the lower bounds of unstable wave numbers corresponding to the identified barotropic and baroclinic YY instabilities shift towards zero, but at different rates, the baroclinic instability hitting zero, and thus leading to appearance of symmetric inertial instability, while the barotropic instability being left behind, as follows from Figure 10.35. The most unstable mode of the baroclinic YY instability of the lower-right panel of Figure 10.35 is displayed in Figure 10.36 and confirms its Yanai-wave nature. As in the one-layer case, if Burger number is decreasing at fixed values of the Rossby number and other parameters,

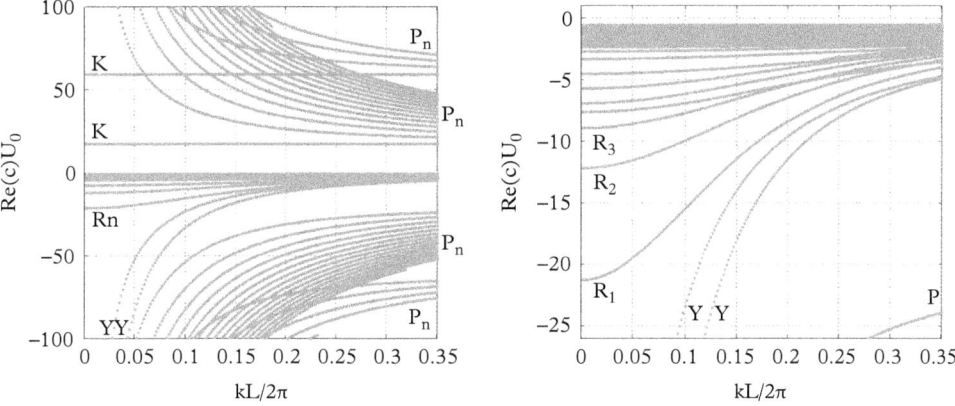

Figure 10.31 *Stability diagram for the easterly equatorial jet in 2-layer configuration with Ro = 0.05 displayed in the left panel of Fig. 10.30. Left: barotropic and baroclinic stable modes; Right: zoom of the stability diagram at small c. Same notation as in Fig. 10.24. Doubling of each type of waves corresponds to baroclinic and barotropic waves of each type. Subscripts indicate meridional structure.*

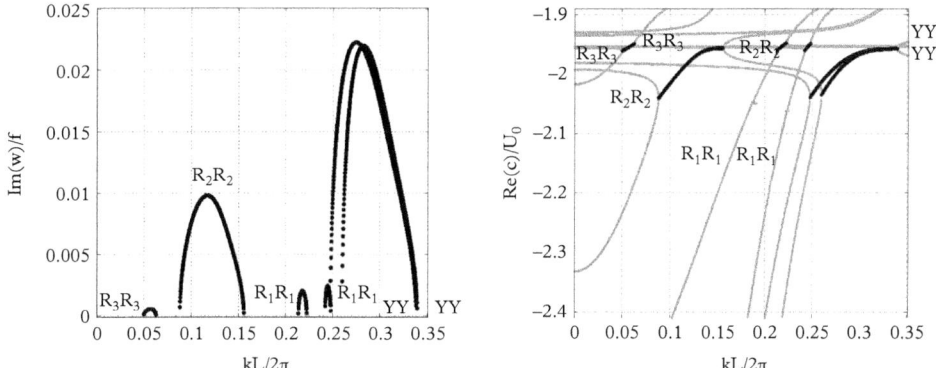

Figure 10.32 *Stability diagram for the easterly equatorial jet in 2-layer configuration with Bu = 10, r = 0.5, d = 0.25 and Ro = 0.27. Black: unstable modes. Grey: stable modes. R_n: Rossby waves; Y: Yanai waves.*

the YY instabilities shift towards lesser zonal wave numbers and hit $k = 0$, the baroclinic instability first.

Thus, we see that baroclinic YY instability is more vigorous than its barotropic counterpart, and it produces symmetric inertial instability even at large Burger numbers, if Rossby numbers are sufficiently high, which is not the case of the barotropic YY instability. At fixed Rossby numbers both baroclinic and barotropic instabilities produce

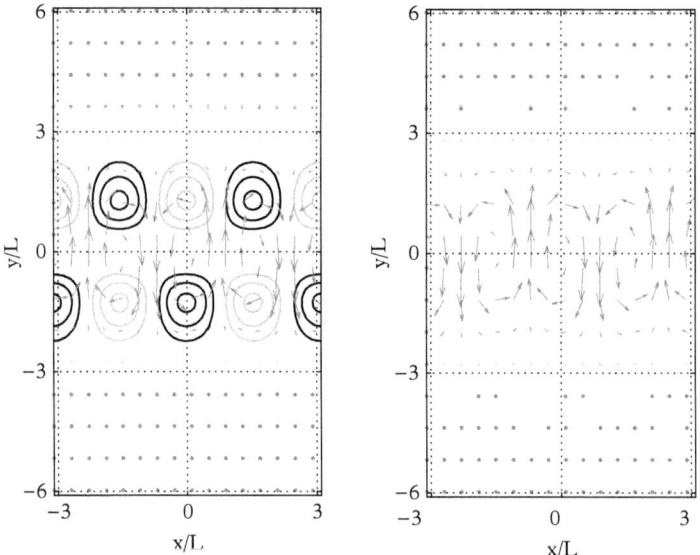

Figure 10.33 *Two-dimensional structure of the unstable mode with $k = 0.32 \cdot \frac{2\pi}{L}$ and $Re(c/U_0) = -1.943$, $Im(\omega/\beta L) = 0.0028$ of Fig. 10.32. Same convention as in Fig. 10.27. Left panel: upper layer. Right panel: lower layer.*

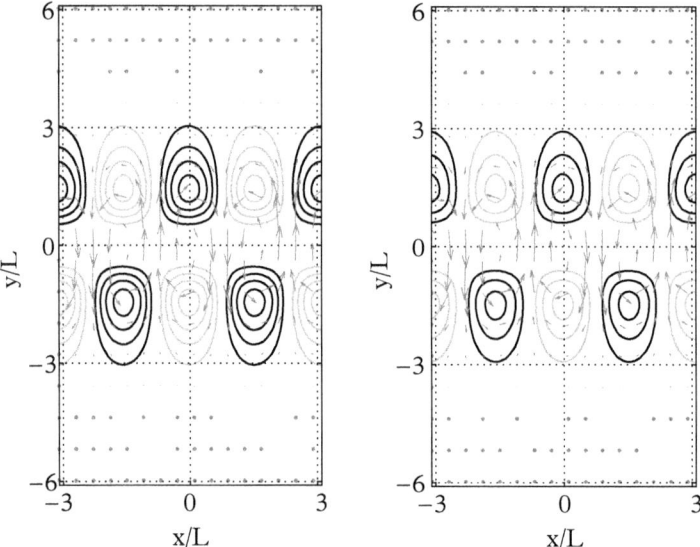

Figure 10.34 *Two-dimensional structure of the unstable mode with $k = 0.32(\frac{2\pi}{L})$ and $Re(c/U_0) = -1.945$, $Im(\omega/\beta L) = 0.0049$ of Fig. 10.32. Same conventions as in Fig. 10.33.*

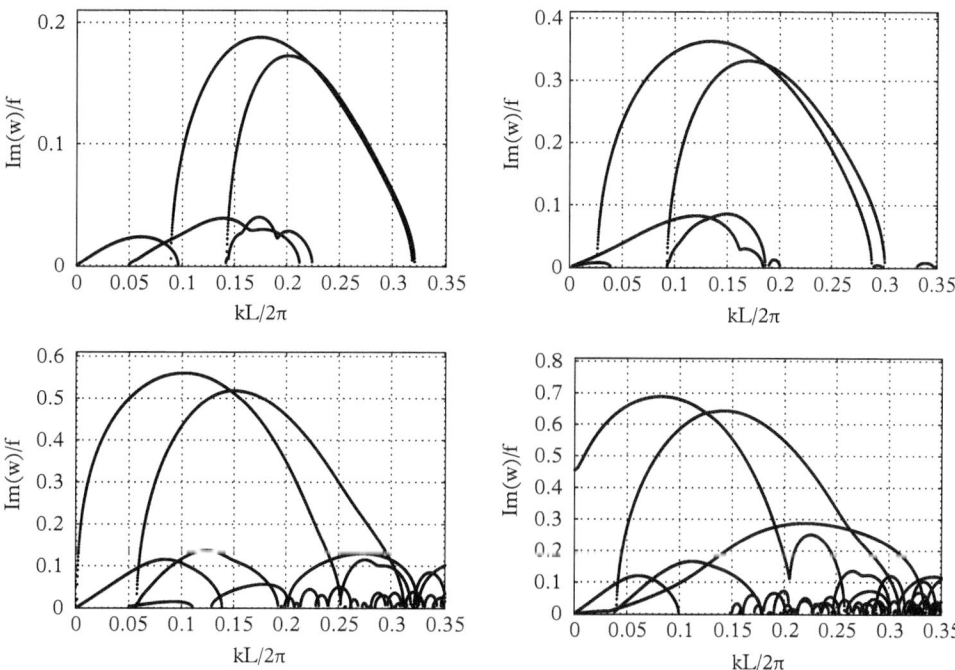

Figure 10.35 *Growth rates of the unstable modes as functions of k for the easterly equatorial jet in 2-layer configuration with Bu = 10, r = 0.5, d = 0.25 and Ro = 0.5 (upper-left panel), Ro = 0.8 (upper-right panel), Ro = 1.2 (lower-left panel), and Ro = 1.5 (lower-right panel).*

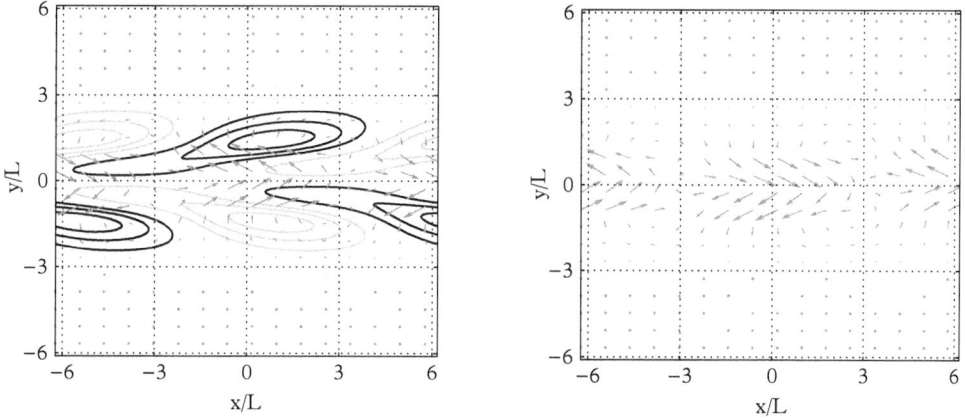

Figure 10.36 *Two-dimensional structure of the unstable mode with $k = 0.08(\frac{2\pi}{L})$ and $Re(c/U_0) = -1.37$, $Im(\omega/\beta L) = 0.69$ of Ro = 1.5 corresponding to the maximal growth rate in the lower-right panel of Fig. 10.35. Same conventions as in Fig. 10.33.*

respective symmetric inertial instabilities, if the Burger number is small enough, the baroclinic one still being more vigorous.

Nonlinear saturation of the leading barotropic and baroclinic instabilities

In what concerns nonlinear saturation, the barotropic inertial instability in the two-layer model saturates in a way similar to the one-layer case displayed in Figure 10.29. No hyperbolicity loss is detected in this case. The saturation of the baroclinic inertial instability, resembles what was observed in the mid-latitude configuration of Section 10.4, cf. Figure 10.19, although the beta effect restrains the lateral spread of secondary vortex structures. Saturation provokes massive mixing, and persistent zones of hyperbolicity loss, which are nevertheless treated by the numerical code in a satisfactory manner.

The strong mixing and dissipation lead to reorganisation of the mean velocity and vorticity of the jet in a way that 'self-healing' of the inertial instability takes place, like in the f-plane configuration, and negative values of the product of planetary and absolute vorticity becomes non-negative, as follows from Figure 10.37.

10.6.3   A brief summary of the results on instabilities of equatorial jets

We have thus seen that the main instability of both barotropic and baroclinic easterly equatorial jets is due to resonance of pairs of Yanai waves, with the structure of the most unstable modes similar to that of Yanai wave itself. Inertial instability, unlike the mid-latitude configuration, appears in the barotropic model, although in the baroclinic model it has higher growth rates and manifests itself in a wider range of parameters. Nonlinear evolution of the inertial instability leads to its 'self-healing', reorganising the initial jet in a way that it becomes inertially stable.

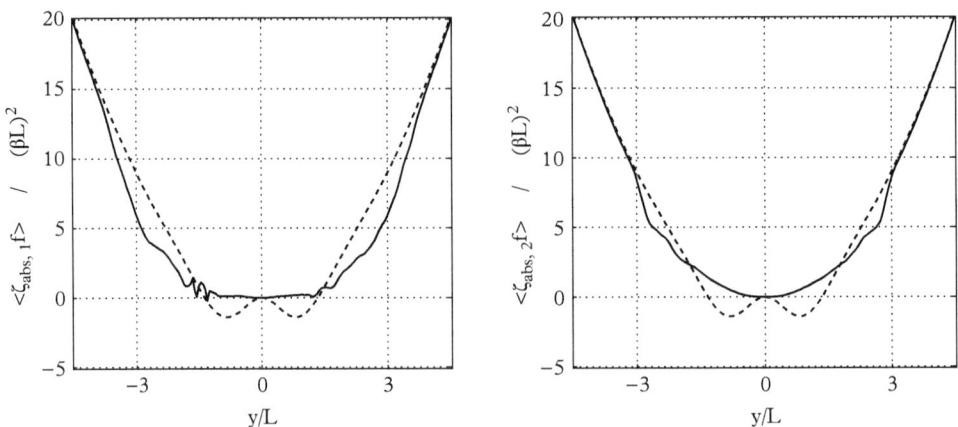

Figure 10.37 *Initial (dashed) vs late ($t = 31.5/\beta L$, solid) meridional profile of $\beta y \cdot \zeta_{abs}$ normalised by $(\beta L)^2$. Left panel: upper layer. Right panel: lower layer.*

10.7   Instabilities of coastal currents and their nonlinear evolution

A special case of jets, especially in the ocean, are those coupled to density and/or temperature fronts evolving in the vicinity of a boundary (coast). Such currents are routinely observed in the ocean. Coastal currents regularly produce meanders, which form vortices, some of them detaching from the current. These phenomena, clearly, result from an instability which we will be studying in the framework of shallow-water models in this section. An idealised set-up for such study, representing a two-layer ocean with a straight coast, is presented in Figure 10.38.

10.7.1   Passive lower layer: results of the linear stability analysis

In the case of a deep and/or high-density lower layer, the latter can be considered, in the first approximation, as passive, and the model becomes the standard one-layer RSW for the active upper layer which was already considered in this configuration in Section 4.3. As was mentioned there, parameter regimes exist where the constant-PV flows are unstable. As in Section 4.3, we find the spectrum of the system (4.52) with the help of pseudo-spectral collocation method. The results for the growth rates are presented in a form of stability diagram in the wave number–Rossby number plane in the left panel of Figure 10.39. As seen in the figure, the flow is stable in the long-wave sector at high enough Rossby numbers. A dispersion diagram for stable waves, arising in such case, was given in Figure 4.9. The highest value of U_0 corresponds to zero current velocity at the coast, and thus instability is possible only in the currents with a return flow at the coast. At the value of the Rossby number indicated by a dashed line in Figure 10.39 the flow is unstable, and three instability zones appear, as follows from the left panel of the figure. This is confirmed by the dispersion diagram presented in the right panel of the figure. As we have already seen, intersections of the dispersion curves indicate instabilities of corresponding hybrid modes. For example, the instability with the highest growth rate in the right panel of Figure 10.39 is the Kelvin–Frontal (KF) one, with the

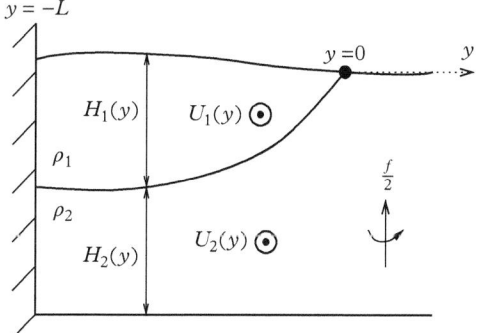

Figure 10.38 *A model of a coastal current in 2-layer RSW. Density front corresponds to outcropping; that is intersection of the interface between the layers and the free surface.*

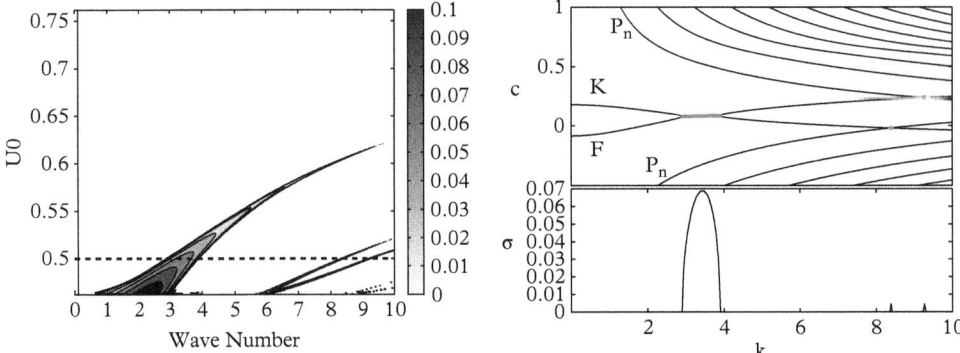

Figure 10.39 *Left panel: Stability diagram of the coastal current with constant PV. Non-dimensional velocity scale U_0 coincides with the Rossby number. Levels of grey: growth rates. Right panel: dispersion diagram of the coastal current with constant PV and Ro = 0.5. Upper panel: phase velocity. Lower panel: growth rate. Wave types: K, Kelvin, F, frontal, P - Poincaré (inertia-gravity), cf Fig. 4.9.*

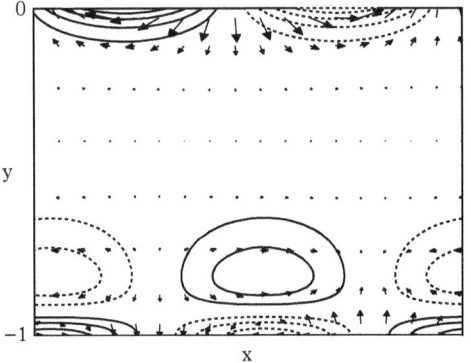

Figure 10.40 *Unstable mode corresponding to the intersection of the dispersion curves of Poincaré and frontal waves. The front (outcropping) is at the top of the frame, and the coast is at the bottom.*

unstable mode which was already displayed in Figure 4.10 of Section 4.3. Other two instabilities, with small growth rates, are Poincaré–Kelvin (PK) and Poincaré–Frontal (PF) frontal waves, respectively. The unstable mode of the latter instability is shown in Figure 10.40. As was discussed in Section 4.3, constant PV flows have an advantage of absence of critical levels, which simplifies identification of instabilities, as all of them are due to intersections of dispersion curves of various primary waves in this case, unlike the situations treated in previous sections.

10.7.2   Passive lower layer: nonlinear evolution of the instability

In spite of simplicity of both the model and the geometry of the flow, nonlinear evolution of its dominant instability reveals a surprising complexity. The developing instability is

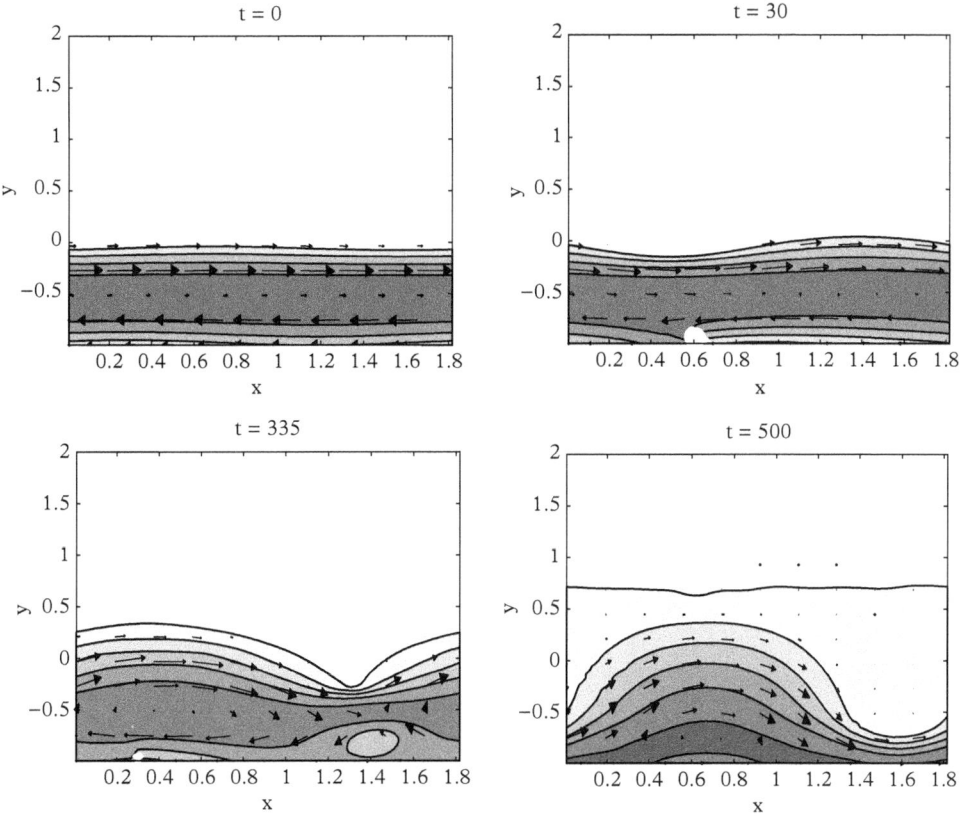

Figure 10.41 *Thickness at the initial (upper row) and late (lower row) stages of evolution of KF instability of the coastal current with a passive lower layer. Snapshots at t = 0, t = 30, t = 335, and t = 500f^{-1} (from left to right, and from top to bottom).*

simulated with the one-layer finite-difference numerical scheme described in Chapter 7, which has no difficulties in treating drying (termination of the fluid layer). Periodic boundary conditions along the coast are imposed, together with no-flux conditions across the boundary. The simulations were initialised with an unstable mode of the dominant KF instability which was superimposed with a weak amplitude (several per cent, as compared to the background values) onto the coastal current. These simulations reveal that nonlinear evolution of the instability has two distinctly different stages which are presented in Figure 10.41. Following upper row of the figure, the unstable KF mode increases in amplitude in time, and at the same time produces strong gradients of thickness and velocity at the boundary, at the location indicated by the white spot. As one could suspect, Kelvin-wave breaking happens in this zone. This conclusion is supported by the phase portrait of the perturbation (with subtracted mean flow) presented in Figure 10.42.

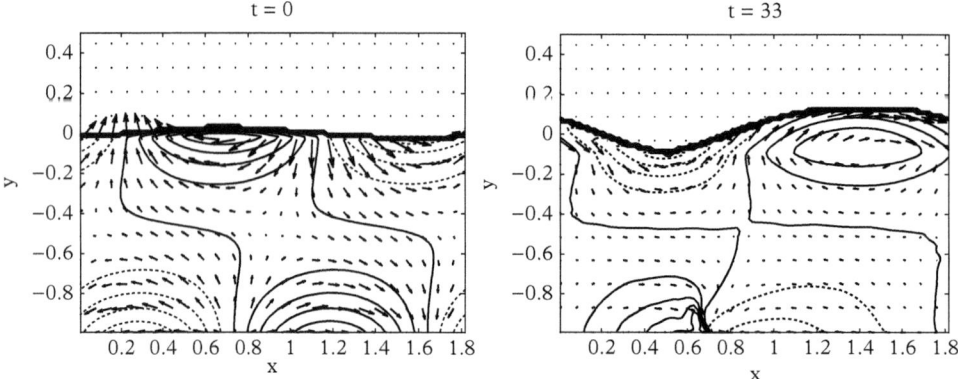

Figure 10.42 *Thickness (contours) and velocity (arrows) of the perturbation of the mean flow corresponding to the upper panels of Fig. 10.41.*

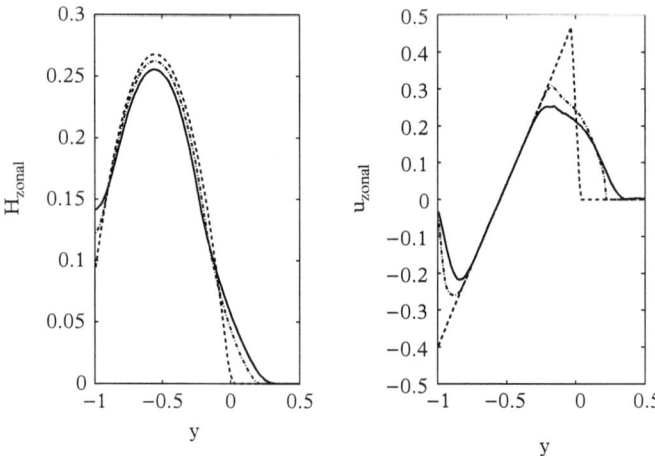

Figure 10.43 *Evolution of the thickness averaged in the stream-wise (along-coast) direction (left), and of the stream-wise velocity (right). Dashed line: t = 0; dashed-dotted line: t = 40, when primary unstable mode is saturated; solid line: t = 300f⁻¹.*

The Kelvin-wave breaking triggers enhanced dissipation and leads to reorganisation of the flow, because breaking waves redistribute PV, as it was shown in Chapter 7. This reorganisation is illustrated in Figure 10.43, and means that development of primary instability leads to appearance of PV gradients in the initially uniform-PV flow. This means, in turn, that a new type of waves, Rossby waves, propagating across PV gradients are now allowed. The phase portrait of the left panel of Figure 10.44, where the perturbations of the mean flow corresponding to Figure 10.41 are displayed, indeed

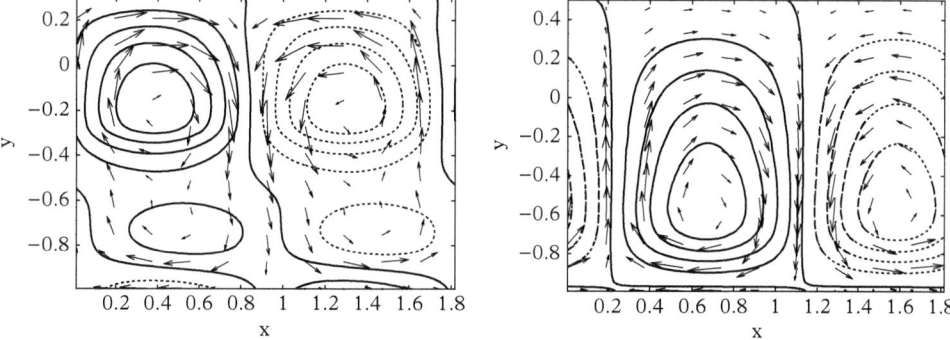

Figure 10.44 *Thickness (contours) and velocity (arrows) corresponding to the left and the right panels, respectively, of the bottom row Fig. 10.41 with subtracted mean flow.*

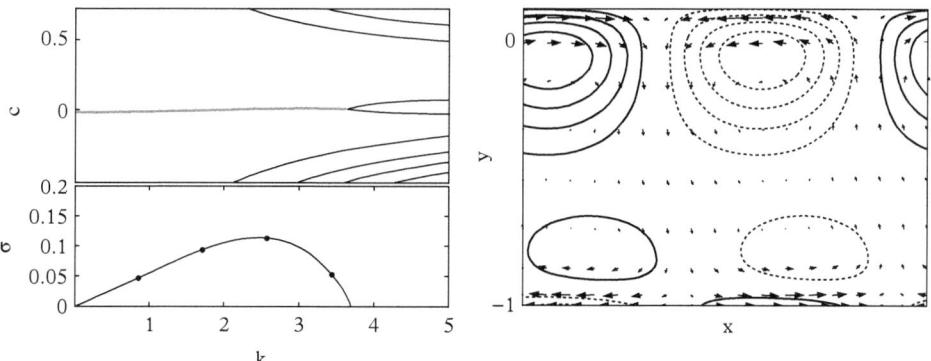

Figure 10.45 *Left panel: dispersion diagram for the mean current resulting from the saturation of primary KF instability. Top: phase velocity. Bottom: growth rate. Right panel: phase portrait of the mode with the highest growth rate in the left panel.*

resembles Rossby waves, with velocity following the isobars in vortex-like structures. The across-flow gradients of the along-coast velocity are opposite at the coast and at the front, cf. Figure 10.43, meaning that the PV gradients are also opposite. Thus, there are two types of Rossby waves propagating in opposite directions, one near the front, and another at the coast. As we have seen in the previous sections of this chapter, waves propagating in opposite directions can phase-lock and grow. We therefore expect an instability. It is easy to repeat the linear stability analysis for the mean flow modified by the primary instability. The results of such analysis for the mean flow at $t = 335f^{-1}$, that is at the onset of the second stage of evolution, are presented in Figure 10.45 and confirm our expectations. There is, therefore a new instability which appears in the flow. It is long-wave, with the most unstable mode displayed in the right panel of the Figure, which is in striking resemblance with the left panel of Figure 10.44, and is, as expected, a hybrid mode combining two Rossby waves: one near the front, and one near the coast. Thus, the two stages of the nonlinear evolution of the instabilities of the coastal current are due to the two successive instabilities developing in the flow: the primary KF one, and the secondary barotropic one (RR, in the hybrid-mode nomenclature), arising due to reorganisation of the flow by saturation of the primary instability.

10.7.3   Active lower layer: results of linear stability analysis

Set-up and boundary conditions at the outcropping

The situation with the lower layer of finite depth is qualitatively different. Indeed, even for the flow in the upper layer with constant PV, as in previous subsections, the PV in the lower layer, even it is motionless, cannot be constant, as the interface between the layers is necessarily curved, and therefore the thickness of the lower layer and, as a consequence, its PV vary across the flow, cf. Figure 10.38. Variable PV allows for Rossby waves in the lower layer which can phase-lock with the waves in the upper layer, and give rise to new, with respect to the passive lower-layer case, instabilities. We will call the passive lower-layer case barotropic and the active lower-layer case baroclinic, for obvious reasons. Therefore, even if the flow is barotropically stable, such as in no-return flows, cf. Figure 4.8, it can be baroclinically unstable, and we show in the following that this is, indeed, the case.

In order to formulate the linear stability problem properly, we have to recall the equations of motion of the two-layer configuration in the f-plane approximation presented in Figure 10.38. The flow consists of an upper layer of density ρ_1 with a free surface terminating at some point (a density front), and a mean steady velocity $U_1(y)$, and a lower layer of density $\rho_2 > \rho_1$. We will be considering a density current with no velocity in the lower layer: $U_2(y) = 0$. The two-layer RSW equations read

$$
\begin{cases}
\left(\partial_t + u_j\partial_x + v_j\partial_y\right) u_j - fv_j = -\dfrac{1}{\rho_j}\partial_x\pi_j, \\[2ex]
\left(\partial_t + u_j\partial_x + v_j\partial_y\right) v_j + fu_j = -\dfrac{1}{\rho_j}\partial_y\pi_j, \\[2ex]
\left(\partial_t + u_j\partial_x + v_j\partial_y\right) h_j + \nabla \cdot (h_j\mathbf{v_j}) = 0,
\end{cases}
\tag{10.43}
$$

where $j = 1, 2$ correspond to upper and lower layers, respectively, (x, y) and $\mathbf{v_j} = (u_j, v_j)$ are zonal and meridional coordinates and velocity components, $h_j(x, y, t)$ are thicknesses of the layers, π_j, ρ_j are pressures and densities of the layers, and f is the constant Coriolis parameter. The pressures and thicknesses are related via the hydrostatic balance relations $\pi_i = \rho_i g \left(r^{j-1}h_1 + h_2\right)$, where $r = \rho_1/\rho_2$ is the stratification parameter, as usual. We linearise these equations about a steady geostrophically balanced state with variable thicknesses $H_j(y)$, and corresponding velocities $U_j(y)$:

$$
\partial_y H_j = (-1)^{j-1}\frac{f}{g'}(U_2 - r^{j-1}U_1),
\tag{10.44}
$$

where $g' = (1 - r)g$ is the reduced gravity. The ratio of unperturbed thicknesses at the coast is $d = H_2(-L)/H_1(-L)$. The linearised equations for small perturbations u'_j, v'_j and h'_j, are

$$\partial_t u'_j + U_j \partial_x u'_j + v'_j \partial_y U_j - f v'_j = -g \, \partial_x (r^{j-1} h'_1 + h'_2),$$
$$\partial_t v'_j + U_j \partial_x v'_j + f u'_j = -g \, \partial_y (r^{j-1} h'_1 + h'_2), \qquad (10.45)$$
$$\partial_t h'_j + U_j \partial_x h'_j + H_j \partial_x u'_j = -\partial_y (H_j v'_j),$$

The boundary condition of no normal flow at the wall is the same as in the one-layer case for both layers: $v_j(-L) = 0$. The boundary condition at the front for the upper layer is also the same as in the one-layer case, because the only constraint to be imposed is the regularity of $(u_1, v_1, h_1 + h_2)$ at $y = 0$. We will exclude from our analysis solutions with free inertia–gravity waves far from the front, and thus focus only on solutions trapped inside the density current. We exclude in this way possible radiative instabilities, with unstable modes comprising inertia–gravity waves propagating out of the front. If the wave structure in x and t is imposed: $(u'_j, v'_j, h'_j)(x, y, t) = (\tilde{u}_j(y), \tilde{v}_j(y), \tilde{h}_j(y)) \, exp [i(kx-\omega t)]$+c.c., then π_2 satisfies the equation $\partial_{yy}\pi_2 = k^2 \pi_2$ in the half plane $y > 0$, where there is no upper layer, nor surface waves by the just formulated hypothesis. Hence, the solution which decays away from the front at $y \to \infty$ has to satisfy the constraint $\partial_y \pi_2 = -k\pi_2$. In order for the solution of π_2 in the region $y < 0$ to match continuously the decaying solution for $y > 0$ we require that the two coincide at $y = 0$. Hence, the boundary conditions at the front in the lower layer give

$$\partial_y (rh_1 + h_2) = -k(rh_1 + h_2) \text{ at } y = 0. \qquad (10.46)$$

We will discuss the physical meaning of such conditions in the next section.

After a proper non-dimensionalisation (we do not present it in detail, although it is worth repeating that the time scale is f^{-1}, and the length scale is the baroclinic deformation radius) we get at a given wave number k a sixth order eigenvalue problem for eigenfrequencies ω and eigenfunctions $(\tilde{u}_j(y), \tilde{v}_j(y), \tilde{h}_j(y))$, which can be solved by applying the pseudo-spectral collocation method once a background coastal current configuration is fixed. We will be considering currents with the upper layer of constant PV, in continuity with the passive lower-layer case above. This means that $1 + H_{1yy} - Q_1 H_1 = 0$, which gives

$$\begin{cases} U_1(y) = U_0 \cosh(\sqrt{Q_1}y) + \dfrac{1}{\sqrt{Q_1}}\sinh(\sqrt{Q_1}y), \\[2mm] H_1(y) = -\dfrac{1}{Q_1}\left[1 - \sqrt{Q_1}\,U_0 \sinh(\sqrt{Q_1}y) - \cosh(\sqrt{Q_1}y)\right], \\[2mm] H_2(y) = (r + d)\,max(H_1) - rH_1(y), \end{cases} \qquad (10.47)$$

where Q_1 is the constant PV in the upper layer and U_0 is the velocity of the upper layer at the front. As already mentioned, in spite of the absence of mean velocity $U_2 = 0$, the bottom layer has variable PV due to the curvature of the interface:

$$Q_2(y) = [(r + d)max(H_1) + \frac{r}{Q_1} \left[1 - \sqrt{Q_1}\, U_0 \sinh(\sqrt{Q_1}y) - \cosh(\sqrt{Q_1}y) \right]]^{-1}. \quad (10.48)$$

It is easy to check that this flow configuration never satisfies Ripa's stability criterion, and therefore, is always unstable, while in the one-layer case stable configurations did exist.

We apply, as usual, the pseudospectral collocation method in order to find eigenfrequencies and eigenmodes of the system (10.45), and present the results in the form of dispersion diagrams giving dependence of real and imaginary parts of the eigenfrequencies ω on stream-wise wave number k. Instead of the real part of ω we will be plotting the phase velocity $c = \omega/k$, as above. The key parameters of the problem are U_0, the non-dimensional velocity of the upper layer at the front location $y = 0$, which is equivalent to the Rossby number, as in the passive lower-layer case, the aspect ratio $d = H_1(-1)/H_2(-1)$, while the non-dimensional width of the current is equal to 1, and the stratification parameter r. The velocity profiles of coastal density currents in the model can be naturally divided into two classes: those which are stable in the limit of the passive lower layer, and those which are not, respectively, barotropically stable or unstable.

Instabilities of barotropically stable currents

We present dispersion diagrams for a barotropically stable flow at given stratification r and different values of the aspect ratio d in Figure 10.46. The flow is chosen to have non-dimensional PV equal to one in the upper layer, and no flow reversal $U(-1) = 0$. In the upper panel of the upper frame of the figure we easily recognise typical dispersion curves of the Poincaré, frontal, and Kelvin waves, which we have already seen in the case of passive lower layer, cf. Figure 10.39, although there are two new curves, indicated by R and I, respectively in the figure. The first corresponds to a Rossby wave propagating in the lower layer due to sign-definite PV gradient introduced by the non-uniform interface. It has a standard structure, with velocity following the isobars, which we have seen many times already. Due to its weak phase velocity, the curvature of the corresponding dispersion curve is not distinguishable with the chosen scale. The second one corresponds to *inertial mode*. This mode is related to inertial oscillations in the lower layer. The absence of its signature in the upper layer explains the fact that we did not see such mode in the spectrum of small perturbations in the one-layer model, with passive lower layer. A peculiar feature of the inertial mode is that it has no signature in pressure. For this reason it can not couple with other waves and form hybrid modes, and even if there are intersection of the dispersion curve of this mode with others, they do not lead to instabilities, as is seen in the lower panel of the upper frame of Figure 10.46. We should recall that, as PV of the upper-layer flow is constant, and lower layer is motionless, there are no critical levels in the system. Therefore, all instabilities are due to intersections of dispersion curves of primary waves, and thus their nature is easily identifiable. As follows from Figure 10.46, when the lower layer is deep the leading instability is due to a hybrid Rossby–Frontal (RF) wave. If we recall that a frontal wave appears due to the PV gradient at the outcropping (i.e. it has the same nature as the Rossby wave, the RF instability can be regarded as a variant of the baroclinic instability. It is long-wave and has dependence of its growth rate on a wave number similar to the standard baroclinic instability. Shorter-scale instabilities in narrow regions of the wave number space which are seen in the figure are due to intersections of Rossby and inertia–gravity (Poincaré) dispersion curves. These RP instabilities have much smaller growth rates.

With diminishing depth of the lower layer the growth rates of both instabilities increase, and related widths of the intervals of unstable wave numbers grow too. When the lower-layer depth becomes smaller than that of the upper layer, the instabilities due to hybrid Poincaré–Frontal (PF), Kelvin–Poincaré (KP), and Poincaré–Poincaré (PP) modes appear, with growth rates of the order 1 in non-dimensional terms (compared to the order 0.01 growth rates of a typical RF instability), and covering a continuous range of wave numbers extending to infinity. These instabilities are all of Kelvin–Helmholtz type. KF instability discovered in the one-layer case also appears, but has much smaller growth rate than others.

Instabilities of barotropically unstable baroclinic currents

In the barotropically unstable configuration, with the same U_0 as in the passive lower-layer case of the previous subsection, the situation is totally different, as follows from Figure 10.47. When the lower layer is deep, the leading instability, as in the passive lower-layer case, is KF (index 1 in the figure indicates hybridisation of the upper-layer modes). However, the baroclinic RF instability, which was seen in the previous example, is also present, having the growth rate of the same order of magnitude. Rossby–Kelvin (RK) instability, which was first discovered in the ageostrophic Phillips model earlier, is also present, but with a much smaller growth rate, as well as PF one. PF and KP instabilities are barely distinguishable. With decreasing depth of the lower layer the strength of RF instability increases and becomes the same as of the KF instability. RK and RP instabilities are also gaining strength. With further decrease of the lower-layer depth, the KF instability disappears, and the same scenario as in the case of barotropically stable flow takes place: instabilities of KH type considerably strengthen and arise at all sufficiently large wave numbers.

In order to confirm the hybrid character on different unstable modes appearing in Figures 10.46 and 10.47 we present their phase portraits in Figure 10.48

10.7.4   Active lower layer: nonlinear saturation of instabilities

In order to follow developing instabilities of coastal currents at the nonlinear stage, we undertake numerical simulations with the well-balanced finite-volume numerical scheme of Chapter 7. They are, as in the examples presented earlier in this chapter, initialised with the unstable modes superimposed with weak amplitude onto background flow. The boundary conditions are: periodicity with a period of the unstable mode in stream-wise, and sponges sufficiently far from the front, in span-wise directions. Nonlinear evolution of the barotropically unstable flow with an active lower layer of significant depth can be anticipated from the results of nonlinear saturation of the KF instability presented above in the case of the passive lower layer. The baroclinic effects in such case are weak, and come into play only after the KF instability fully develops, which is confirmed by

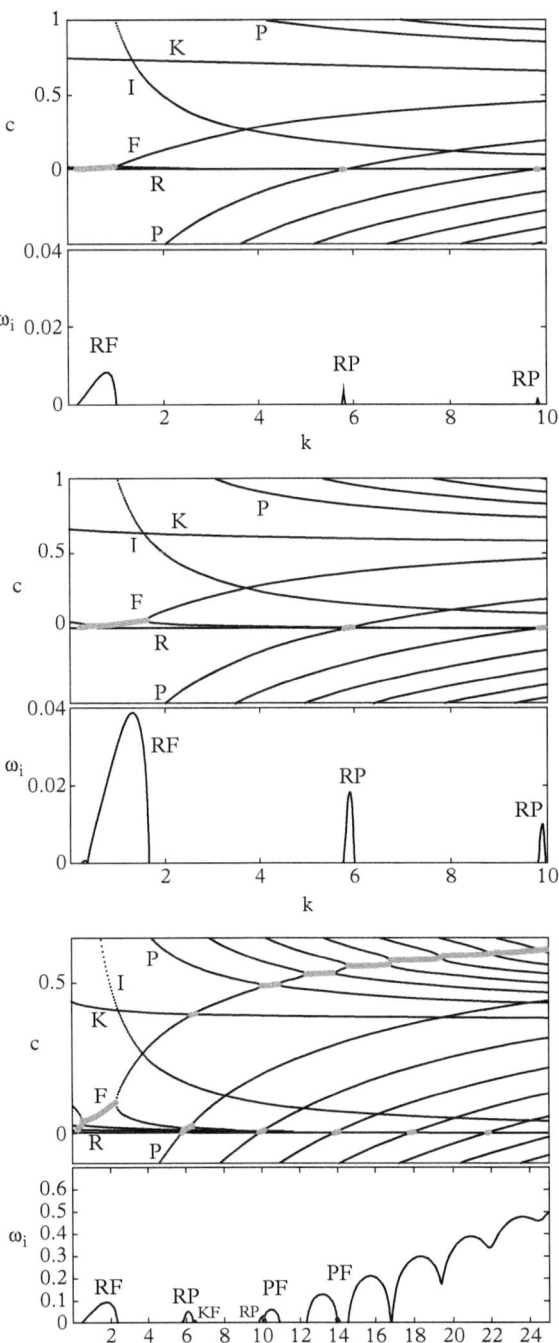

Figure 10.46 *Dispersion diagrams for barotropically stable coastal current at r = 0.5. Upper frame: d = 10. Middle frame: d = 2. Lower frame: d = 0.5. Both horizontal and vertical scales of the bottom frame are shrunk in order to display short-wave Kelvin–Helmholtz instabilities with high growth rates. Upper (lower) panel of each frame: phase velocities (growth rates). Thick grey lines: instability regions*

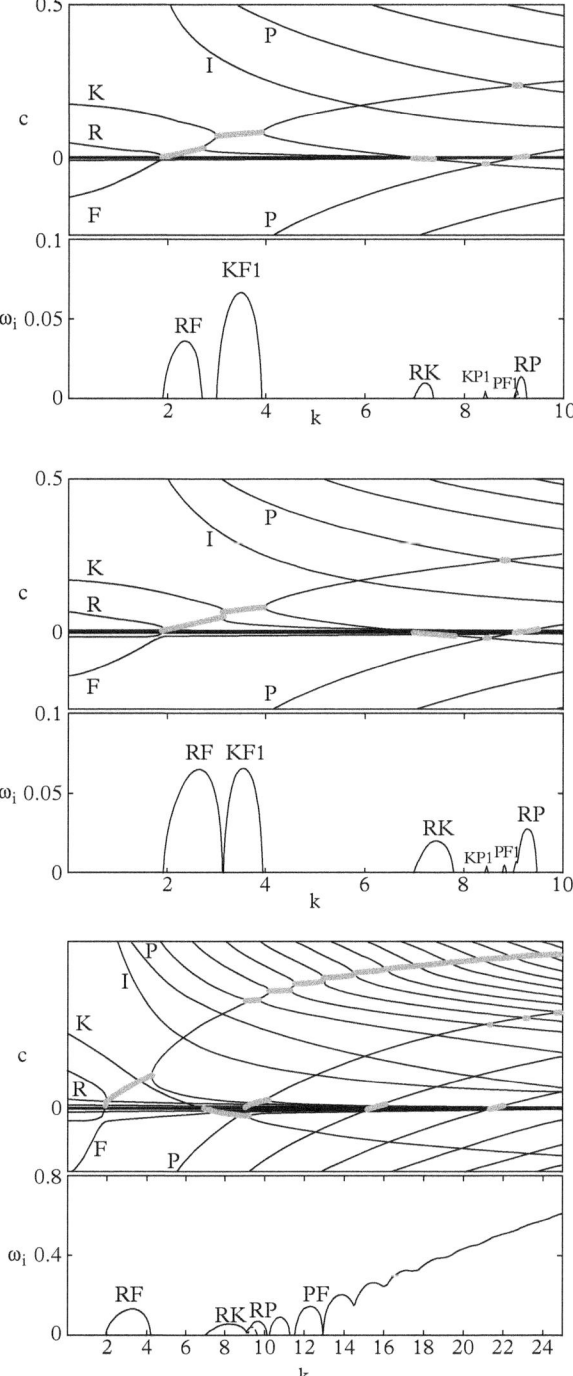

Figure 10.47 *Dispersion diagrams for barotropically unstable coastal current with* $U_0 = 0.5$, $s = 0.5$. *Upper frame:* $r = 10$. *Middle frame:* $r = 5$. *Lower frame:* $r = 2$. *The same conventions as in Fig.* 10.46 *apply.*

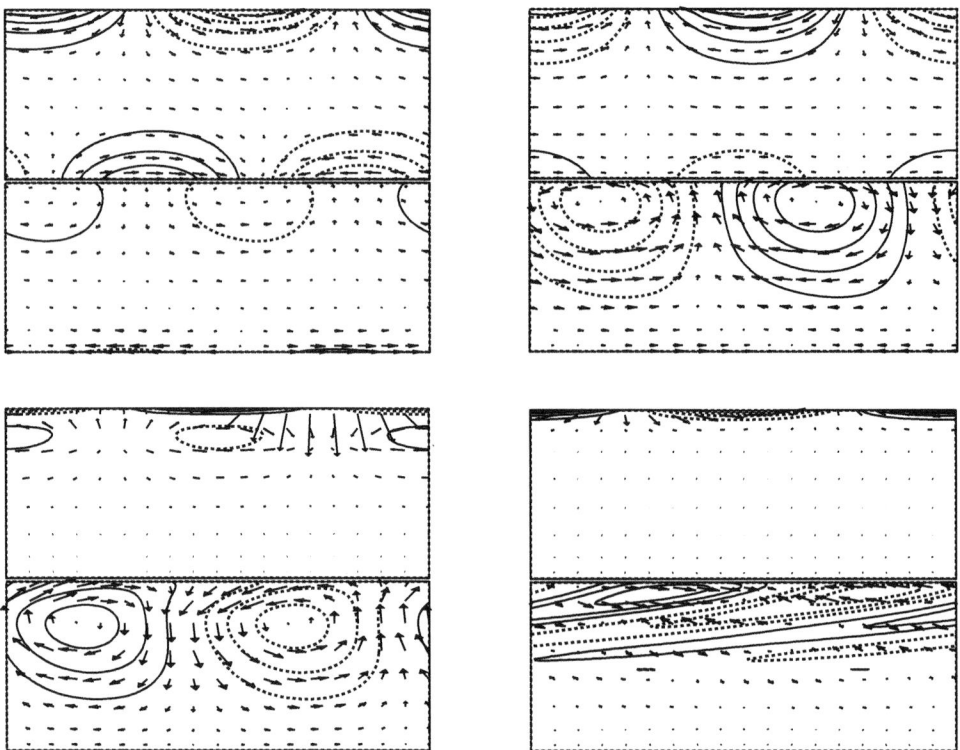

Figure 10.48 *Typical unstable modes of the upper-layer coastal current in two-layer model (left to right and top to bottom): KF1, RF, RP, PF. Top (bottom) part of each panel: upper (lower) layer. Pressure: contours, velocity: arrows. Solid (dashed): positive (negative) values.*

numerical simulations initialised with the KF unstable mode. When the depth of the lower layer decreases, the RF and KF instabilities have comparable strengths and become competing. We will not present this case, which leads to rather complicated behaviour. Instead we concentrate on the barotropically stable configuration with dominant RF instability, which gives rather unexpected results, as Kelvin waves, which are not part of the unstable modes, nevertheless play an important role in the saturation process. Scenario of nonlinear evolution of the RF instability can be summarised as follows. In the *upper layer*, the frontal-wave part of the unstable mode evolves into a series of monopolar vortices at certain spacing. Clipping and reconnection of vortex lines due to formation of Kelvin fronts is responsible for such reorganisation. In the *lower layer* the Rossby-wave part of the unstable mode evolves into a series of vortices of alternating signs, which pair in dipoles. As in the examples we have seen in Chapter 9, these dipoles start moving along their middle axes, which are oriented approximately in the perpendicular to the coast direction. In this way they drive the upper-layer vortices out of the shore and provoke their detachment. Figures 10.49, 10.50, and 10.51 illustrate this scenario, which

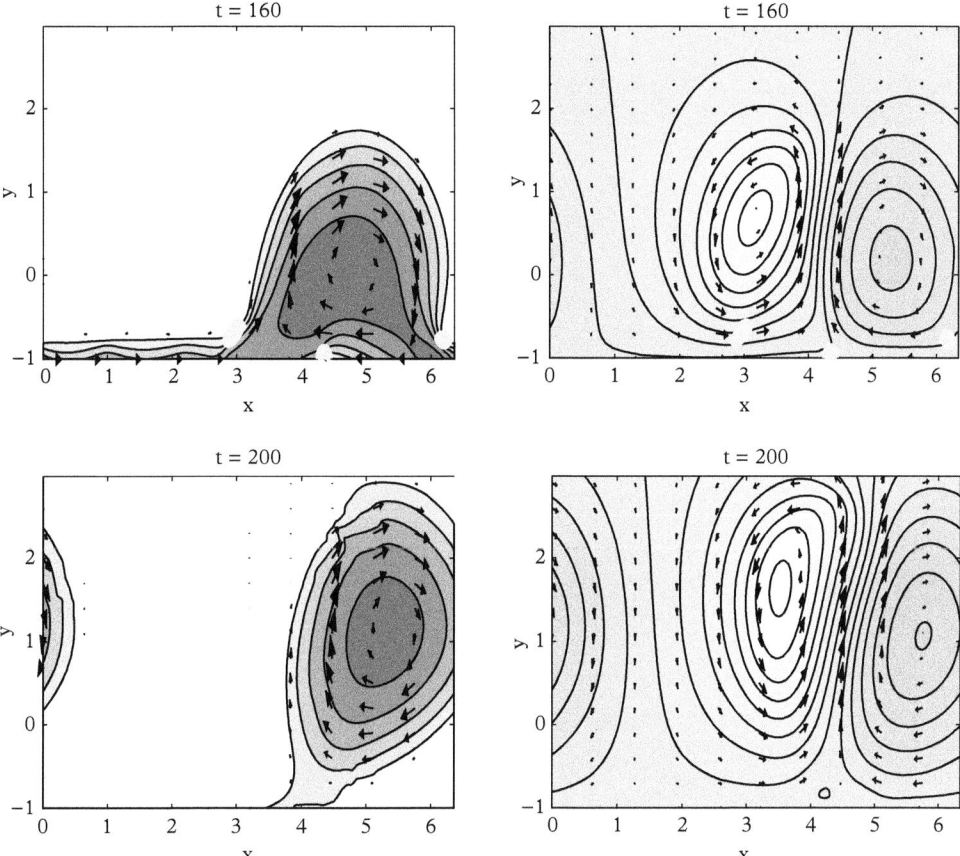

Figure 10.49 *Isolines of the upper-layer thickness (left panel) and isobars in the lower layer (right panel) at $t = 150$ and $t = 200f^{-1}$ during the development of the unstable RF mode superposed onto the upper-layer coastal current with depth ratio $d = 2$.*

gives a possible explanation of the mechanism of vortex detachment from meandering coastal currents in observations.

10.7.5 A brief summary of the results on instabilities of coastal currents

We see that the instabilities of coastal density currents are impressively rich, even in the simplest set-up used here. The hybrid character of unstable modes, first established in Section 10.3 in the rather academic framework of Phillips model, is fully confirmed, and different 'hybridisations' take place depending on the parameters of the flow, and give substantially different scenarios of nonlinear saturation. The breaking Kelvin waves,

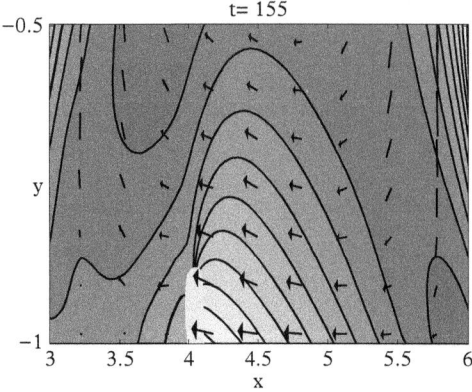

Figure 10.50 *Zoom of the wall region before detachment of the upper-layer vortex during development of the RF instability of the upper-layer coastal current in two-layer model at t = 160f⁻¹cf. Fig. 10.49. Levels of grey: thickness (pressure) in the upper layer. Arrows: upper-layer velocity. Zone of enhanced dissipation: white. A pattern of a breaking Kelvin wave is clearly visible.*

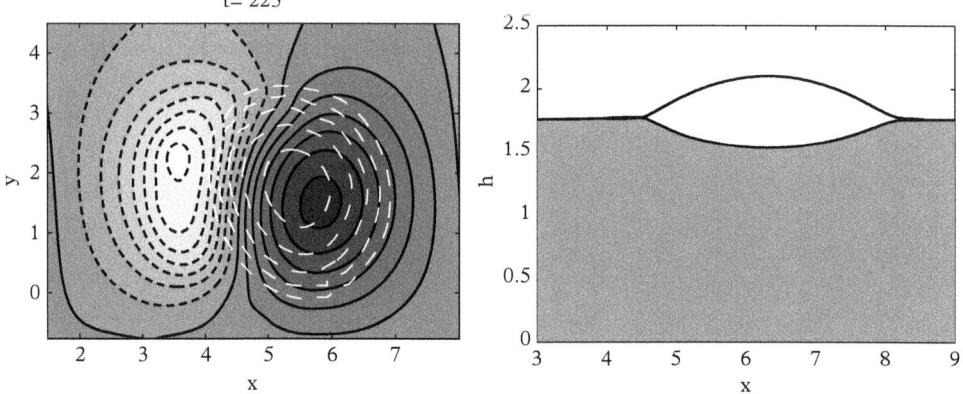

Figure 10.51 *Left panel: Isobars in the upper (white contours) and lower (black contours) layers at t = 250f⁻¹ in the simulation of Figure 10.49. Dark (light) background: anticyclonic (cyclonic) vorticity. Right panel: The cross-sections in the x direction of the detached vortex at t = 300f⁻¹.*

well resolved by the shock-capturing numerical scheme, play crucial role in all of them, creating localised dissipation zones which allow for reconnection of vortex lines and formation of secondary vortices. If we turn to practical implications, the invisible deep layer can have a strong influence through the dipolar vortices which can drive upper-layer structures out of the coast. It should be stressed that the possibility to follow the evolution of the system over a long time period (hundreds of inertial periods), and to detect complicated saturation scenarios is possible thanks to distinguished features of the numerical scheme of Chapter 7, which is quasi-dissipationless, and nevertheless resolves wave breaking. Indeed, for example, for the simulations of Figure 10.49, the energy is practically conserved, with a single relatively important drop at the moment of upper-layer vortex detachments related to Kelvin-wave breaking (not shown).

10.8 Instabilities of double-density fronts and the role of topography

10.8.1 Set-up, scaling, parameters, and boundary conditions

Although already extremely rich, the picture of instabilities of coastal density currents of the previous section is missing an essential ingredient: bottom topography. In order to understand its influence we will consider density currents in a simpler configuration without coast, which is schematically represented in Figure 10.52. Such double-front configuration can be used to model, for example river outflows detached from the coast, or inflows of saltier water at the ocean's bottom (with layers upside-down, with respect to the configuration considered below). The two-layer RSW equations with bathymetry read

$$(\partial_t + u_i \partial_x + v_i \partial_y) u_i - f v_i + \partial_x \Pi_i = 0,$$

$$(\partial_t + u_i \partial_x + v_i \partial_y) v_i + f u_i + \partial_y \Pi_i = 0,$$

$$\partial_t h_i + \partial_x ((h_i - b \delta_{i2}) u_i) + \partial_y ((h_i - b \delta_{i2}) v_i) = 0. \tag{10.49}$$

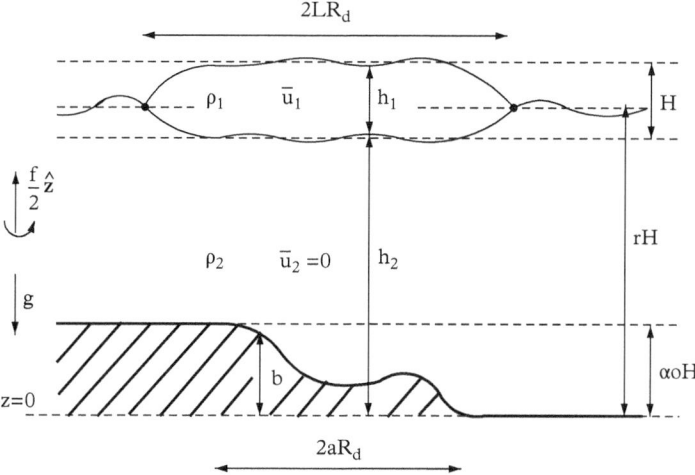

Figure 10.52 *A cross front section of a two layer model of a coupled density front with nontrivial bathymetry. R_d: deformation radius, L and a: non-dimensional half-widths of the balanced current and of the escarpment, respectively. r: depth ratio, α_0: non-dimensional amplitude of the bathymetry variation.*

Here u_i, v_i ($i = 1, 2$) are, as usual, x and y components of the velocity in the layers (layer 1 on top of the layer 2); h_1, $h_2 - b$ are thicknesses of the layers, where b is bathymetry, which will be considered escarpment-like, δ_{ij} denotes Kronecker delta, and we suppose that $f = const$. The geopotentials of the layers Π_i are

$$\Pi_1 = g(h_1 + h_2) \quad , \quad \Pi_2 = g(sh_1 + h_2), \tag{10.50}$$

where $s = \frac{\rho_1}{\rho_2} \leq 1$ will denote the stratification parameter in this section, while r denotes the (inverse) depth ratio, cf. Figure 10.52. Inverse Coriolis parameter f^{-1}, and the baroclinic radius of deformation $R_d = \sqrt{gH(1-s)}/f$ are natural time and length scales of the system. We will be interested in the long-wave part of the spectrum of small perturbations (the short-wave instabilities are always KH-like), and hence introduce the dimensionless wave number $\epsilon = 2\pi R_d/\lambda$, where λ is a typical wavelength of the perturbation. Correspondingly, the following scaling will be used (cf. Figure 10.52): span-wise coordinate y and widths of the current and escarpment are $\sim R_d$; downstream coordinate $x \sim R_d/\epsilon$, time $t \sim 1/\epsilon f$, span-wise velocities $\sim \epsilon\sqrt{gH(1-s)}$, and stream-wise velocities $\sim \sqrt{gH(1-s)}$. Under this scaling $Ro = \frac{1}{2L}$.

The non-dimensional equations of motion thus result:

$$\begin{cases} (\partial_t + u_i\partial_x + v_i\partial_y)u_i - v_i + \partial_x\Pi_i = 0, \\ \epsilon^2(\partial_t + u_i\partial_x + v_i\partial_y)v_i + u_i + \partial_y\Pi_i = 0, \\ \partial_t h_i + \partial_x((h_i - \frac{\alpha_0}{r}b\,\delta_{i2})u_i) + \partial_y((h_i - \frac{\alpha_0}{r}b\,\delta_{i2})v_i) = 0, \end{cases} \tag{10.51}$$

with $\Pi_1 = \dfrac{h_1 + rh_2}{1-s}$, $\Pi_2 = \dfrac{sh_1 + rh_2}{1-s}$. As a background flow we take a stationary geostrophically balanced surface density current, which is parallel to the x axis and is terminating at $\pm L$, with no mean flow in the lower layer:

$$\bar{u}_1 = \bar{u} = -\partial_y\bar{h} \quad , \quad \bar{\Pi}_1 = \bar{h} \quad , \quad \bar{\Pi}_2 = 0 \quad , \quad \bar{u}_2 = \bar{v}_2 = \bar{v}_1 = 0. \tag{10.52}$$

Here, $\bar{h}_1 = \bar{h}$ is the background thickness of the terminating upper layer, $\bar{h}(\pm L) = 0$. There will be no variation of bathymetry beyond the outcroppings $a < L$:

$$b = 1, \text{at } y < -a; \qquad b = 0, \text{at } y > a. \tag{10.53}$$

As in Section 10.6.2, we consider density currents with constant PV $Q = const$, obeying the following relation: $\partial_{yy}\bar{h} - Q\bar{h} + 1 = 0$, which gives, for different Q,

$$Q < 1 : \bar{h} = \frac{1}{Q}\left(1 - \frac{\cosh(y\sqrt{Q})}{\cosh(L\sqrt{Q})}\right), \quad L = \frac{1}{\sqrt{Q}}\ln\left(\frac{1 + \sqrt{Q(2-Q)}}{1-Q}\right),$$

$$Q < 0 : \bar{h} = \frac{1}{Q}\left(1 - \frac{\cos(y\sqrt{|Q|})}{\cos(L\sqrt{|Q|})}\right), \quad L = \frac{1}{\sqrt{|Q|}}\cos^{-1}\left(\frac{1}{1+|Q|}\right), \tag{10.54}$$

$$Q = 0 : \bar{h} = 1 - (y/L)^2, \qquad\qquad L = \sqrt{2}.$$

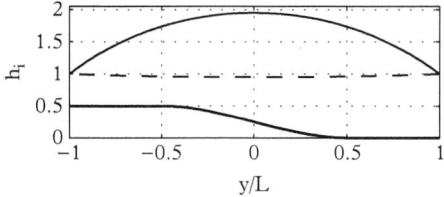 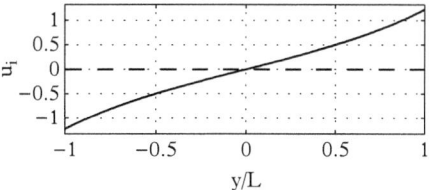

Figure 10.53 *Double-density front with Q = 0.5 in the upper layer. s = 0.5, r = 1, α_0 = 0.5, a = 0.5L. Left panel: interface (dashed), free surface (solid) and bathymetry (bold solid). Right panel: stream-wise velocities of the layers 2 (dashed) and 1 (solid).*

The configuration with $Q = 0.5$, $L \simeq 1.86$ and $Ro \simeq 0.27$, is displayed in Figure 10.53.

As usual, we will linearise (10.51) about the background flow (10.52) and look for eigenmodes of small perturbations (u'_i, v'_i, h'_i) with a given wave number in the stream-wise direction, with eigenfrequencies ω or, equivalently, eigenphase velocities $c = \omega/k$. Proper boundary conditions should be imposed in the presence of two free streamlines limiting the flow span-wise. By definition,

$$\bar{h} + h'_1 = 0 \quad \text{and} \quad \frac{dy}{dt} = v'_1 \quad \text{at} \quad y = \pm L + \lambda_\pm, \tag{10.55}$$

where $\pm L$ are locations of the free streamlines of the balanced flow, and $\lambda_\pm(x, t)$ are perturbations of the free streamlines. In the lower layer, the continuity of the solution is imposed at $\pm L$. Beyond the outcropping we impose an exponential decay of the pressure perturbation on both sides of the double front, which means that we are looking only for instabilities due to the modes trapped in the current. This hypothesis is the same as in Section 10.6.2, and allows us to formulate the eigenproblem in the finite interval $y \in [-L, L]$. It should be stressed that without this hypothesis the technical complexity of the problem would be prohibitive. Physically, by imposing the decay boundary condition we filter out free inertia–gravity waves and concentrate uniquely on the trapped modes, which is consistent with our interest in long-wave instabilities. Only the instabilities resulting from the resonances between the trapped modes will be thus captured, and radiative instabilities due to the resonances with free inertia–gravity waves are excluded. We will come back to such instabilities in the case of high Rossby number vortices treated in Chapter 11. An *a priori* justification of this hypothesis is that condition of efficient emission of inertia–gravity waves by PV anomalies is $Ro \geq 1$, as we have seen in Chapter 6, while we work here with $Ro < 1$. An *a posteriori* justification is that no distinguishable radiative instabilities were observed in fully nonlinear numerical simulations presented in the following.

10.8.2   Linear stability analysis

Wave species and expected resonances

As we have seen many times already, the instabilities of the flow, especially at constant PV and no critical levels, as is again the case here, are produced by hybrid modes resulting from phase-locking of elementary waves propagating in opposite directions. We enumerate the wave 'species', which are present:

- Poincaré (inertia–gravity) modes in both layers, which can propagate in both directions in x;
- Rossby modes in the lower layer (there are no PV gradients in the upper layer), which can also propagate in both directions, as PV gradients due the curvature of the interface are opposite at two outcroppings;
- Frontal modes trapped in the vicinity of the free streamlines in the upper layer and propagating in opposite directions at two outcroppings;
- Topographic waves in the lower layer trapped by the varying bathymetry, which are unidirectional for monotonic bathymetry profiles.

As in the case of coastal currents with active lower layer, there are also inertial oscillations, but they have no signature in pressure and cannot resonate with other modes, as was explained in Section 10.6.2. Therefore, the following hybrid modes (linear resonances) and related instabilities could be expected:

1. the *barotropic* resonances of the upper-layer modes between:
   - two frontal waves (FF);
   - a Poincaré and a frontal wave (P1F);
   - two Poincaré waves (P1P1).
2. the *baroclinic* resonances of the modes in different layers between:
   - an upper frontal wave and a lower Rossby wave (RF);
   - an upper frontal wave and a lower topographic wave (TF);
   - an upper Poincaré wave and a lower Rossby wave (P1R);
   - an upper Poincaré wave and a lower topographic wave (P1T);
   - an upper frontal wave and a lower Poincaré wave (P2F);
   - an upper and a lower Poincaré waves (P1P2).

All of these are confirmed by the direct numerical analysis of the linear stability problem.

Deep lower-layer case

We start with the results of the linear stability analysis for a very deep lower layer ($r = 100$), which we present in the usual form of dispersion diagrams in Figure 10.54. The most unstable mode of this configuration is displayed in Figure 10.55, and clearly shows that the instability is due to phase-locking of a pair of frontal waves propagating along the opposite outcropping lines. This case is, in fact, a benchmark, as such FF instability of coupled density fronts is known theoretically in the case of a passive lower layer (so-called Griffiths–Killworth–Stern instability). The instability is entirely due to

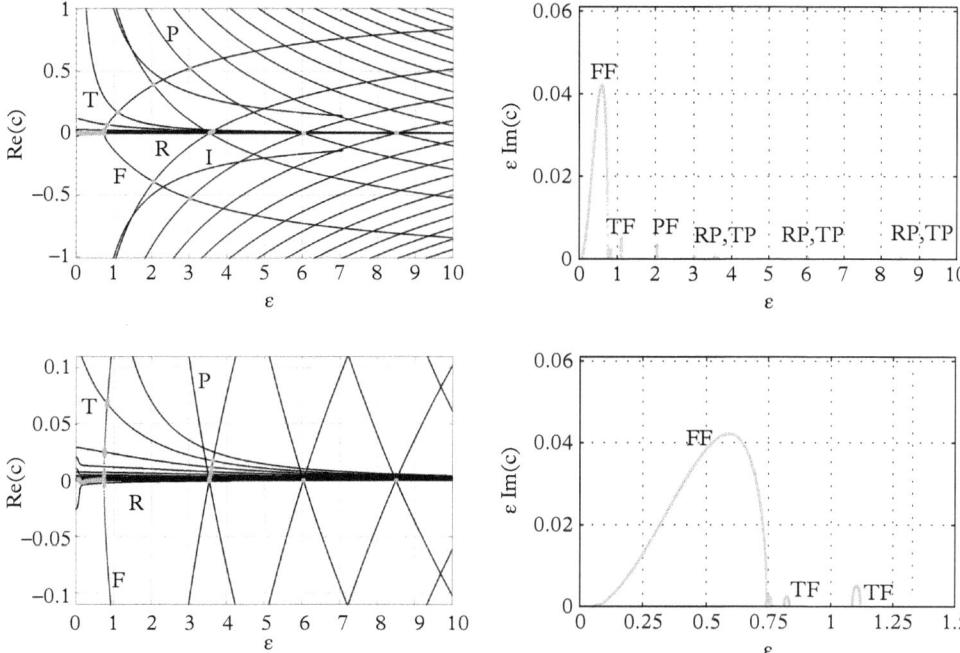

Figure 10.54 *Results of the linear stability analysis of the double-density front with $s = 0.5$, $r = 100$, and $a = 0.5L$, $\alpha_0 = 50$. Left panel: phase velocity. Right panel: growth rate. Grey: unstable. Black: stable. Waves: I, inertial; F, frontal; P, Poincaré; R, Rossby; T, topographic. Bottom: zoom at the long-wave part of the diagram.*

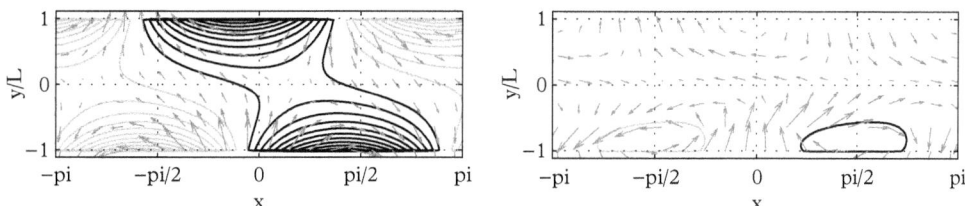

Figure 10.55 *2D structure of the most unstable mode ($\epsilon = 0.59$) of Fig. 10.54: resonance between two frontal waves in the upper layer. Isobars and velocity field of the perturbation in the upper layer, left panel, and lower layer, right panel positive (negative) pressure anomalies: black (grey) contours.*

the waves in the upper layer, the lower layer at such depths is passive. At lesser depths of the lower layer, baroclinic instabilities, owing their existence to waves in the lower layer, appear, as seen in Figure 10.56. The leading instability is still FF, but RF and TF instabilities make their appearance. The corresponding unstable modes are displayed in Figure 10.57 confirming the nature of the instabilities. The second TF instability, with smaller growth rate, in Figure 10.56 is due to the resonance of a frontal wave in the

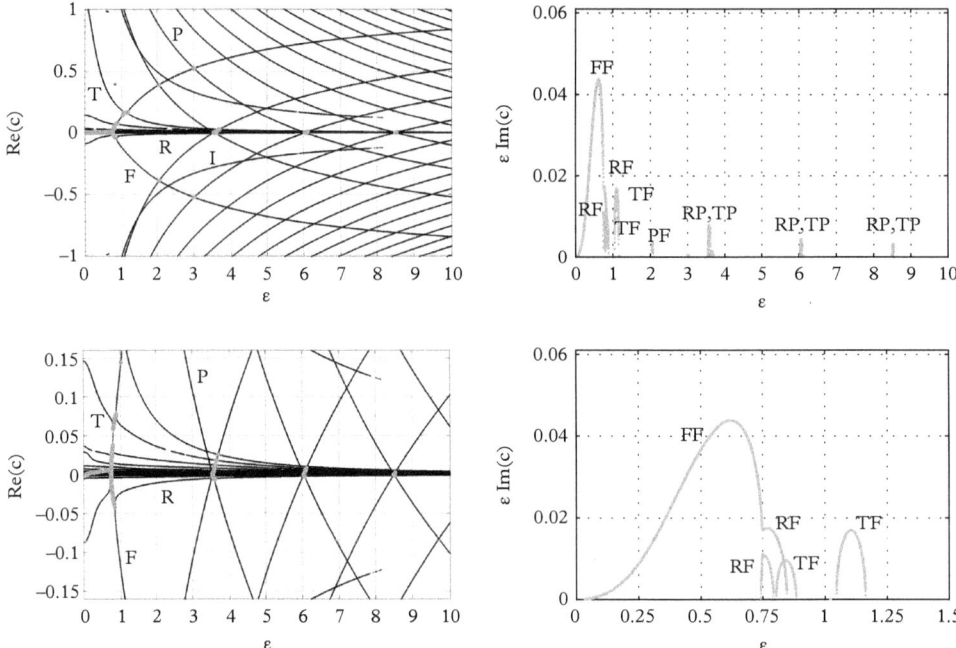

Figure 10.56 *Same as in Fig.* 10.54, *but with depth ratio r =* 10 *and the topography parameter* $\alpha_0 = 5$.

upper layer and the second topographic mode with one node in the span-wise direction. As expected, Poincaré waves can also form unstable hybrid modes, as follows from the figure.

Shallow lower layer

For shallow lower layers, as in the case of coastal currents, the KH-type PF instabilities dominate, and are present at all sufficiently short wavelengths. It is worth emphasising that two leading long-wave instabilities of the system can compete in the case of shallow enough layers. We show this in Figure 10.58, where FF and RF instabilities have the same maximal growth rates.

10.8.3   Nonlinear saturation of the instabilities

As compared to nonlinear saturation of the instabilities of coastal currents in the previous section, we can now address an important question of the role of bathymetry in nonlinear saturation, even if the unstable mode itself is not related to topographic waves, as in the main long-wave FF and RF instabilities. Evolution of instabilities due to topographic waves is also of interest. In nonlinear simulations with well-balanced finite-volume code of Chapter 7, we impose the leading unstable mode of a given instability with a small

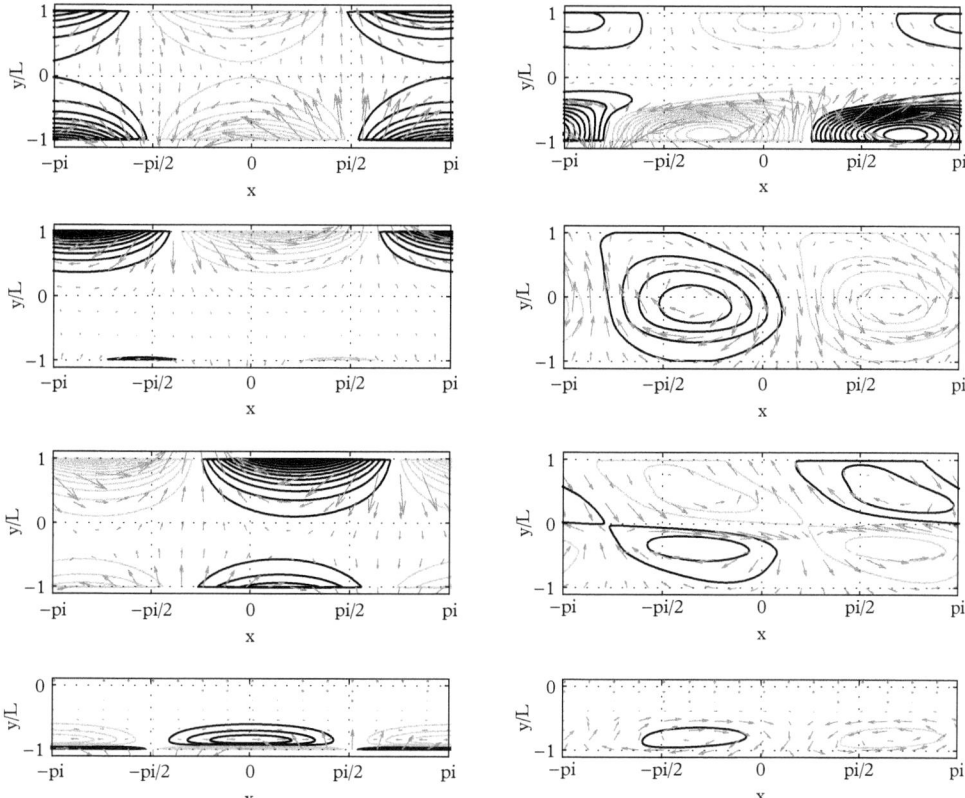

Figure 10.57 *2D structure of unstable modes with 1) $\epsilon = 0.77$, 2) $\epsilon = 1.105$, 3) $\epsilon = 0.842$, 4) $\epsilon = 3.577$, from top to bottom, in Fig. 10.56. Left (right) column: upper (lower) layer. Resonances between 1) a frontal wave in the upper layer and a Rossby wave in the lower layer, 2) a frontal wave in the upper layer and the first topographic mode in the lower layer, 3) a frontal wave in the upper layer and the second topographic mode in the lower layer, 4) a Poincaré mode in the upper layer and a Rossby wave in the lower layer, respectively.*

amplitude of the order of 1% of the unperturbed thickness onto the double-density front, and follow the evolution of the flow. The boundary conditions are periodic stream-wise, and sponges are imposed in span-wise direction on both sides of the jet.

Comparison of the saturation of competing instabilities in the absence of bathymetry

We start with comparison of nonlinear saturation of competing instabilities of Figure 10.58. Saturation of the FF instability is presented in Figure 10.59, and saturation of the RF instability in Figure 10.60. As follows, the saturation patterns of the two instabilities are completely different. While the FF instability ends up by totally disrupting

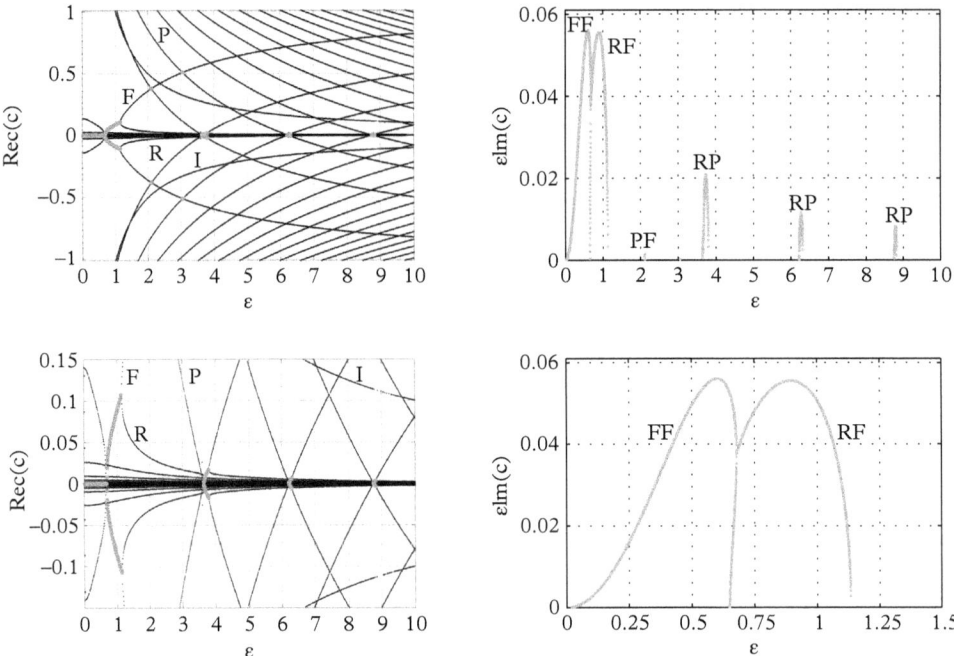

Figure 10.58 *Stability diagram for the background flow with no topography, and Q = 0.6, s = 0.5, r = 2. Same conventions as in Fig. 10.54.*

initial density current and forming a series of co-rotating elliptical vortex lenses, the RF instability follows a scenario of evolution resembling that of the typical baroclinic instability which was considered in Section 10.2 in a similar setting (upper-layer jet with motionless lower layer).

Changes of evolution patterns due to escarpment

The presence of the escarpment beneath the density current drastically changes scenario of saturation of both instabilities. In Figures 10.61 and 10.62 we present the evolution of FF and RF instabilities, respectively, corresponding to Figure 10.56. The series of co-rotating elliptical vortices of the width of the flow transforms in two parallel rows of smaller vortices at both sides of the escarpment during the evolution of FF instability. The RF instability in the presence of escarpment leads, at the late stages, to formation of periodically reconnecting vortices in the upper layer, instead of meandering current in the absence of bathymetry.

Saturation of the instability involving topographic waves

Nonlinear saturation of TF instability, which is presented in Figure 10.63, displays yet another scenario of saturation with a row of vortices of alternating sign over the

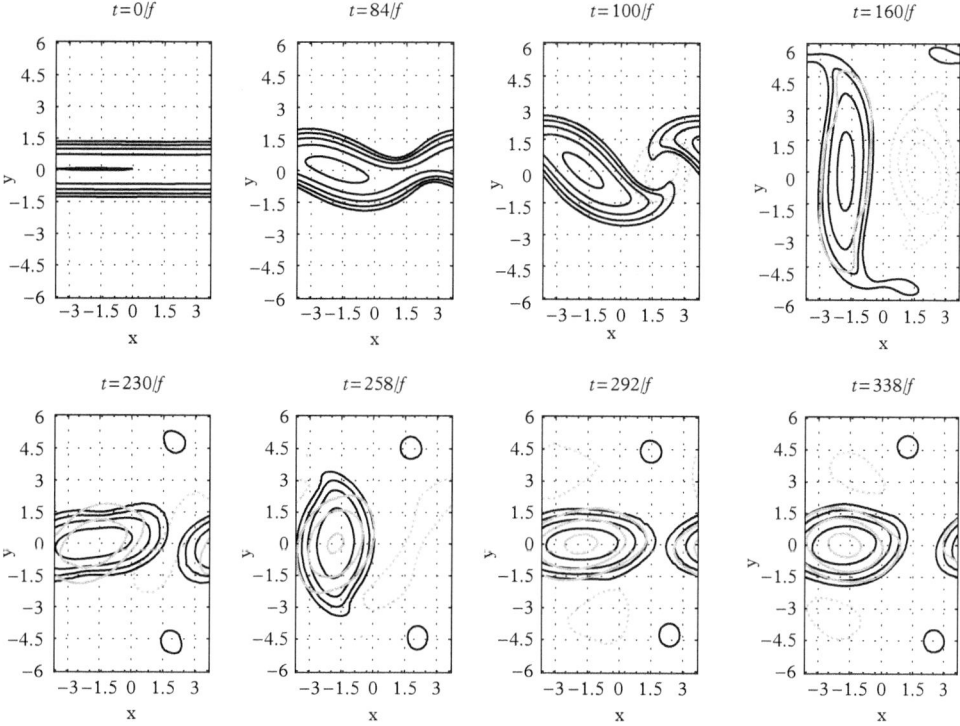

Figure 10.59 *Nonlinear evolution of the FF instability of Fig. 10.58. Thickness h_1 of the upper layer (black). Contour interval 0.2, starting at 0.2. Pressure Π_2 (grey) of the lower layer. Contour interval 0.05, starting at $rH \pm 0.05$ (+/- anomaly: solid/dashed).*

escarpment in the lower layer and asymmetric steady propagating modulation of the density current in the upper layer, which makes one consider the possibility of an exact steady nonlinear wave solution of the system (a nonlinear 'fronto-topographic' wave), although we have no proof of this hypothesis.

10.8.4 A brief summary of the results on instabilities of double fronts over topography

Thus, bottom topography leads to appearance of new instabilities, arising from phase-locking and growth of modes coupling the proper modes of the double front to topographic waves. These instabilities, in general, are weaker than instabilities of the double front proper. Topography also leads to qualitative changes in the scenarios of evolution of the long-wave instabilities, changing topology of the secondary coherent structures resulting from the nonlinear saturation.

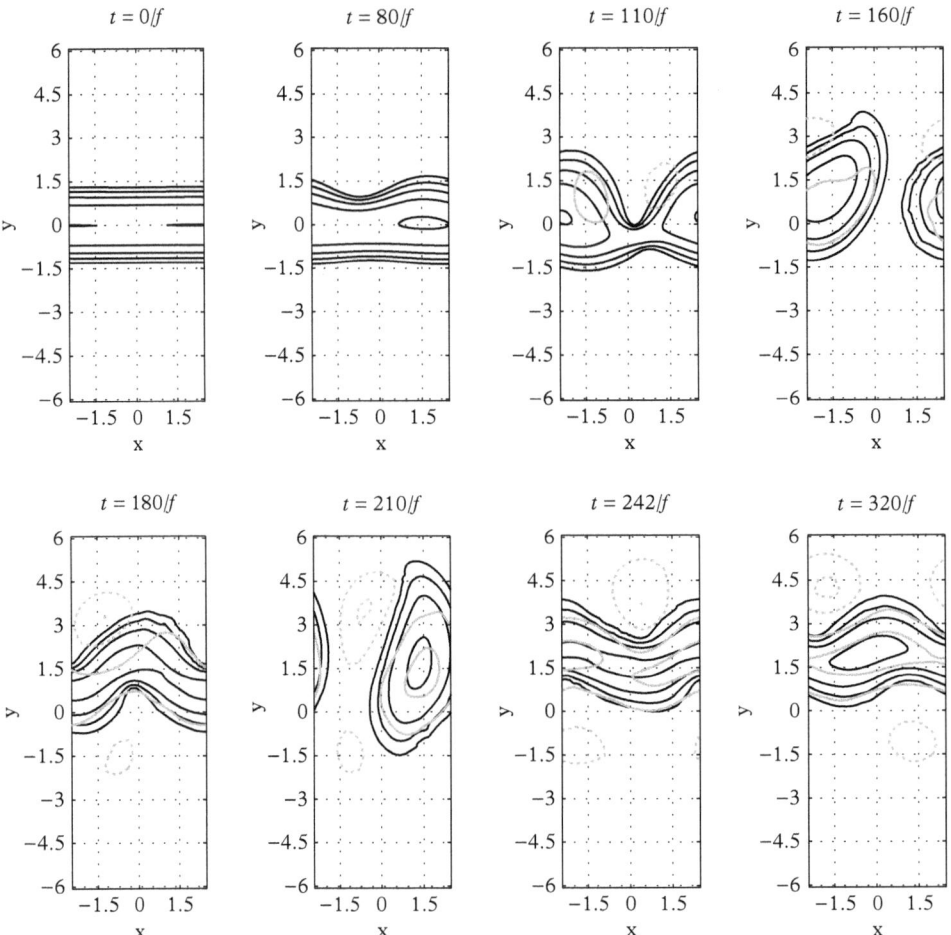

Figure 10.60 *Nonlinear evolution of the RF instability of Fig. 10.58. The same legend as in Fig. 10.59 applies.*

10.9   Summary, comments, and bibliographic remarks

We have seen in this chapter a panorama of jet instabilities in different configurations. We established the nature and origins of these instabilities, and on characteristic patterns of their saturation. Brief summaries on each type of instability were given in corresponding sections, and we will not repeat them. There are, however, several important messages which merit reiteration. First, the classical barotropic and baroclinic instabilities of localised jets are substantially modified with increasing Rossby numbers, and this even for Rossby numbers which are not exaggeratingly high. Second, the classical inertial instability happens to be a small-wave-number limit of ageostrophic baroclinic instability,

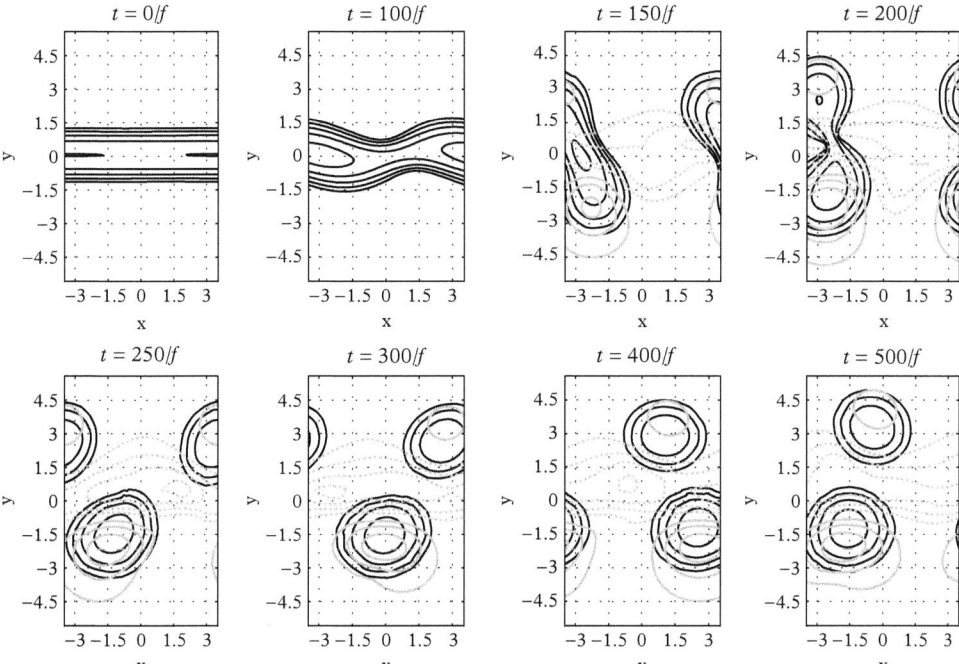

Figure 10.61 *Nonlinear evolution of the FF instability of Fig.* 10.56. *The same legend as in Fig.* 10.59 *applies.*

and always possesses a finite-wave-number counterpart with higher growth rates. Third, the presence of lateral boundaries and/or outcroppings lead to appearance of new instabilities related to coastal and frontal waves. Fourth, bottom topography significantly influences the saturation patterns of the instabilities and introduces new ones. Fifth, inertial instability appears already in one-layer configuration on the equatorial beta plane, which is impossible in mid-latitudes, and is due to resonances between Yanai waves.

In what concerns nonlinear saturation of the instabilities, essentially ageostrophic instabilities produce strong mixing and are efficient sources of inertia–gravity waves. We have also seen that interpretation in terms of linear wave resonances, which is especially easy in flows with constant PV, is helpful in predicting instabilities and their properties. The presented results also demonstrate that rotating shallow water allows for modelling of a plethora of physically relevant situations in a physically transparent way, and by cheap but efficient means.

Hydrodynamic instabilities and, in particular, instabilities in geophysical fluid dynamics represent a vast area of research. Many books and thousands of articles are written on the subject. We limited ourselves by instabilities in shallow-water models, but the literature here is also considerable. Its exhaustive review is impossible within the limits

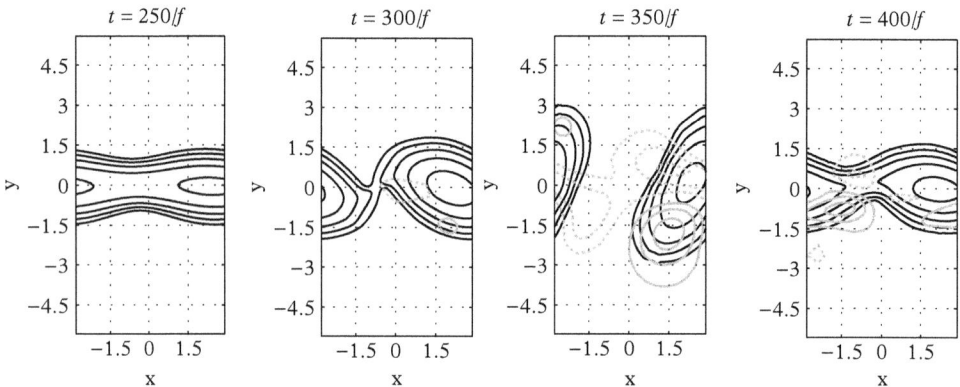

Figure 10.62 *Nonlinear evolution of the RF instability of Fig. 10.56. The same legend as in Fig. 10.59 applies.*

of this book, so the bibliographic comments which follow are necessarily limited. They are concentrated mainly on the key points, and on the source articles of this chapter. Comprehensive list of references can be found in original articles cited below.

The Rayleigh criterion goes back in time for almost 150 years (Rayleigh 1880). Its generalisation to the beta plane was made in (1949). Ripa's stability criterion (Ripa 1991) follows from the application of general stability theory in Hamiltonian systems Holm *et al.* (1985) to shallow water. The pseudo-spectral collocation method is described in Trefethen (2000). Detailed stability analysis of the barotropic instability in one-layer RSW was performed in Poulin and Flierl (2003). The presentation in Section 10.2 follows Lambaerts *et al.* (2011) for geostrophic barotropic and Lambaerts, Lapeyre, and Zeitlin (2012) for geostrophic baroclinic instabilities. The problem of ageostrophic instabilities in the Phillips model was posed in Sakai (1989), where Rossby–Kelvin instability was discovered. The exposition in Section 10.3 follows Gula, Plougonven, and Zeitlin (2009). Section 10.4 is based on Bouchut, Ribstein, and Zeitlin (2011). The standard semi-heuristic explanation of the inertial instability, which is given in textbooks (cf. e.g. Holton 1979), can not explain why this instability does not happen in one-layer model in the f-plane approximation. Interpretation of this instability in terms of trapped standing modes presented in Section 10.5, which allows us to understand this apparent paradox, was first given in LeSommer *et al.* (2003), and further elaborated in Zeitlin (2008). It is to be stressed that this interpretation is not limited by layered models. Exactly the same is valid for continuously stratified primitive equation model (Plougonven and Zeitlin 2009). Existence of asymmetric inertial instability was first evoked in Dunkerton (1982). The process of 'self-healing' of the inertial instability in terms of evolution of the geostrophic momentum was discussed in a number of papers (e.g. Carnevale, Kloosterziel, and Orlandi 2013). Instabilities of equatorial currents were subject of the work by Hayashi and Young (1987) which laid the basis of the analysis of instabilities of jets in terms of wave resonances. The pioneering paper

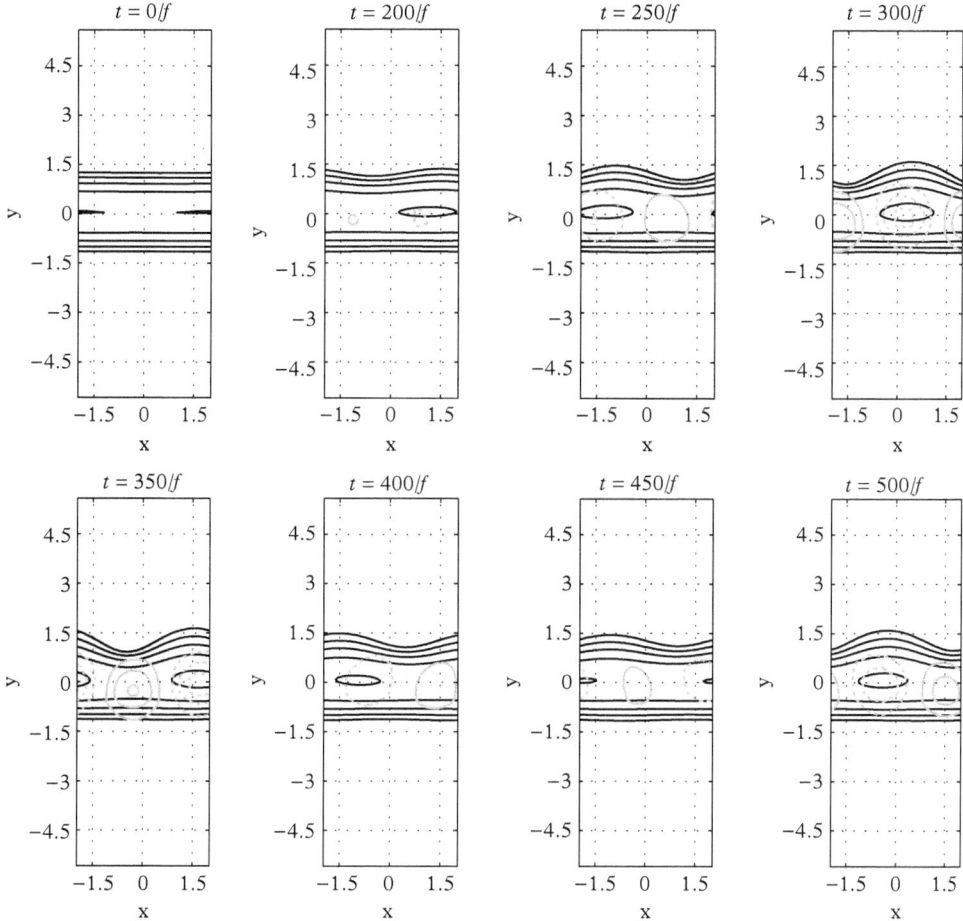

Figure 10.63 *Nonlinear evolution of the TF instability. The same legend as in Fig.* 10.59 *applies.*

by Cairns in 1979 gave interpretation of instabilities in terms of linear wave resonances, and explained the physics of such resonances. Section 10.6 follows Ribstein, Zeitlin, and Tissier (2014). Instabilities resulting from resonances between regular and pseudo-modes were investigated in Iga (2013). Presentation in Section 10.6.2 follows Gula and Zeitlin (2010) and Gula, Zeitlin, and Bouchut (2010). Stability of double-density fronts with passive lower layer was analytically studied in the classical paper by Griffiths, Killworth, and Stern (1982), which was also pioneering in studying ageostrophic in-stability. Instabilities of coupled fronts were studied subsequently by many authors. A corresponding direct stability analysis and nonlinear saturation of the *FF* instability were given in Scherer and Zeitlin (2008). The presentation in Section 10.7 follows Ribstein and Zeitlin (2013). The boundary condition at the outcropping which was used in this chapter was introduced in Paldor and Killworth (1987).

Figures 10.1 to 10.3 are reproduced from Lambaerts, Lapeyre, and Zeitlin (2011), and Figures 10.4 to 10.6 from Lambaerts, Lapeyre, and Zeitlin (2012). Figures 10.7 to 10.9 are reproduced from Gula, Plougonven, and Zeitlin (2009), with permissions. Figures 10.10 to 10.22 are reproduced from Bouchut, Ribstein, and Zeitlin (2011), Figures 10.23 to 10.37 are reproduced from Ribstein, Zeitlin, and Tissier (2014) with the permission of AIP publishing. Figures 10.39 to 10.45 are reproduced from Gula and Zeitlin (2010), Figures 10.38, 10.46 to 10.51 are reproduced from Gula, Zeitlin, and Bouchut (2010), and Figures 10.52 to 10.63 are reproduced from Ribstein and Zeitlin (2013), with permissions.

11

Instabilities in Cylindrical Geometry: Vortices and Laboratory Flows

Axisymmetric vortices, as well as rectilinear jets studied in Chapter 10, represent an important class of exact solutions of the equations of motion in all geophysical fluid dynamics models. However, the geostrophic equilibrium, which guarantees that a jet is an exact solution, should be replaced by *cyclo-geostrophic equilibrium*, which takes into account the centrifugal acceleration, in order for a corresponding vortex to be a solution, as will be shown. Classical quasi-geostrophic barotropic and baroclinic instabilities are well-known for vortices, and an essentially ageostrophic centrifugal instability, which has been known for a century, is a counterpart of inertial instability. As with the jets covered in Chapter 10, we will discuss a range of vortex instabilities, from quasi-geostrophic ones to instabilities of intense hurricane-like vortices. The strategy will be the same: we will first analyse the linear stability of vortices in different parameter regimes with the help of pseudo-spectral collocation method and then obtain characteristic patterns of nonlinear saturation of the instabilities by numerical simulations with well-balanced finite-volume scheme of Chapter 7. In order to capture the details of the background vortex profile, and resolve the eigenmodes of the linear stability problem, it is useful to work with non-uniform grids stretched with respect to the standard Chebyshev grid out of the vortex, which we will do systematically. Treating numerically the limit $r \to 0$ in cylindrical coordinates r, θ can be delicate, due to apparent singularity of curl and divergence operators in this limit. A consistent boundary condition at $r \to 0$ is to require that all components of the perturbation, except for the axisymmetric one, which does not depend on θ, and for which the condition of vanishing pressure (thickness) at zero should be replaced by the condition of vanishing of its radial derivative.

Another relevant configuration in cylindrical geometry is a flow in a rotating annulus frequently used in laboratory experiments. Although experiments are rarely performed with really shallow fluid layers, the shallow-water theory still gives useful information about vertically averaged characteristics of laboratory flows. We will address instabilities in rotating annuli in this chapter as well.

Geophysical Fluid Dynamics. Vladimir Zeitlin,
Oxford University Press (2018). © Vladimir Zeitlin. DOI: 10.1093/oso/9780198804338.001.0001

11.1   Axisymmetric vortex solutions in rotating shallow water

11.1.1   One-layer model

We start with the one-layer RSW model. Anticipating axisymmetric solutions, we rewrite the equations of the model in cylindrical coordinates r, θ in terms of radial and azimuthal components of the velocity $\mathbf{v} = u\hat{\mathbf{r}} + v\hat{\boldsymbol{\theta}}$, where hats denote unit vectors in corresponding directions:

$$
\begin{cases}
\dfrac{du}{dt} - \dfrac{v^2}{r} - fv = -g\,\partial_r h, \\[2mm]
\dfrac{dv}{dt} + \dfrac{uv}{r} + fu = -\dfrac{g}{r}\partial_\theta h, \\[2mm]
\partial_t h + \dfrac{1}{r}\partial_r(hru) + \dfrac{1}{r}\partial_\theta(hv) = 0.
\end{cases}
\tag{11.1}
$$

Here, $\frac{d}{dt} = \partial_t + u\partial_r + (v/r)\partial_\theta$ denotes the Lagrangian derivative in polar coordinates. As is clear from these equations, any axisymmetric flow (vortex) with azimuthal velocity $v = V(r)$ and thickness $h = H(r)$ in cyclo-geostrophic equilibrium (gradient wind balance in the language of meteorology)

$$
\left(\frac{V}{r} + f\right)V = g\,\partial_r H
\tag{11.2}
$$

is an exact solution of (11.1). In what follows we are interested in isolated (shielded) vortices, that is those satisfying (11.2) and having zero circulation at infinity and, hence, finite energy. It is simpler to work with non-dimensional variables so we will be using the scaling $\sqrt{gH_0}$ for velocity, $R_d = \sqrt{gH_0}/f$ for r, and $1/f$ for time, where H_0 is the non-perturbed thickness of the layer. We will be considering so-called α-Gaussian vortices with the following non-dimensional radial profile of azimuthal velocity:

$$
V(r) = \pm\epsilon r^{\alpha/2}\exp\left(-\frac{r^\alpha}{2} + \frac{1}{2}\right), \qquad \alpha \geq 1.
\tag{11.3}
$$

Here, the positive sign corresponds to cyclones and the negative sign to anticyclones. The corresponding profile of $H(r)$, is given by the primitive of the left-hand side of (11.2) which can be found from (11.3) by integration:

$$
H(r) = 1 \mp \epsilon\,\frac{1}{\alpha}\,\sqrt{2e}\,2^{1/\alpha}\Gamma\left(\frac{1}{\alpha} + \frac{1}{2}\right)G\left(\frac{r^\alpha}{2}, \frac{1}{\alpha} + \frac{1}{2}\right),
\tag{11.4}
$$

where $G(x, a) = \frac{1}{\Gamma(a)}\int_x^a e^{-t} t^{a-1}\,dt$, and Γ denotes gamma-function. The α-Gaussian vortices thus have two parameters, α and ϵ, which control the steepness of the azimuthal velocity profile and the amplitude of the velocity, respectively. The parameter ϵ within the chosen scaling is the Rossby number if we consider Burger number to be 1 (see also following). Such profiles provide a simple analytical representation of shielded vortices. The radial distribution of the relative vorticity in the vortex is given by $\zeta = (1/r)d(rV(r))/dr$. The local Rossby number is defined in terms of vorticity as $Ro_{loc} = \zeta/f$. Thus, cyclonic α-Gaussian vortices have a core of positive relative vorticity inside a ring of negative relative vorticity, and vice verse for anticyclonic vortices. Hence, α-Gaussian vortices possess a sign reversal in the vorticity profile, and a barotropic instability is expected, according to the criteria formulated in Section 10.1.

11.1.2   Two-layer model

For studies of baroclinic instabilities the two-layer RSW model with a free surface will be used, and will be mostly considered in the 'oceanic' configuration, cf. Chapter 3. In polar coordinates it is written as

$$
\begin{cases}
\dfrac{\partial \mathbf{v}_i}{\partial t} + \mathbf{v} \cdot \nabla \mathbf{v}_i + \left(f + \dfrac{v_i}{r} \right) \hat{\mathbf{z}} \wedge \mathbf{v}_i + g \nabla (s^{i-1} h_1 + h_2) = 0, \\
\dfrac{\partial h_i}{\partial t} + \nabla \cdot (h_i \mathbf{v}_i) = 0, \quad i = 1, 2.
\end{cases}
\tag{11.5}
$$

Here, $\mathbf{v}_i = (u_i, v_i)$ is velocity in layer i counted from the top, h_i is thickness of the layer i, and in this chapter, in order to avoid confusion with the radial coordinate, we change our notation of the density ratio to $s = \rho_1/\rho_2 < 1$. $d = H_1/H_2$ denotes the thickness (aspect) ratio, and H_i is the thickness of the ith layer at rest. The scaling is the same as in the one-layer case, with $H_0 = H_1 + H_2$.

Stationary solutions are cyclo-geostrophic equilibria layer-wise. In non-dimensional terms they are given by the relations

$$
V_i \left(\frac{V_i}{r} + 1 \right) = \partial_r (s^{i-1} H_1 + H_2), \quad i = 1, 2.
\tag{11.6}
$$

We consider localised vortices with the α-Gaussian velocity profiles layer-wise. In non-dimensional terms they are expressed as

$$
V_i = \pm \frac{Ro}{\sqrt{Bu}} \kappa^{i-1} r_\star^{\alpha/2} e^{-\frac{r_\star^\alpha}{2} + \frac{1}{2}}, \quad i = 1, 2,
\tag{11.7}
$$

where we reconstituted Burger number for completeness. Here, $r_\star = r\sqrt{Bu}$, and the parameters are κ, which controls the intensity of the lower-layer velocity with respect to the upper-layer one, that is the vertical shear, and thus the degree of baroclinicity of the vortex, α, the steepness of the radial profile of velocity/vorticity, which controls the horizontal shear, Ro, the upper-layer Rossby number, $Bu = gH_0/(f^2 L^2)$, the barotropic

Burger number, where L is the distance from the center to the velocity peak, $Bu' = g'H_e/f^2L^2$, the baroclinic Burger number, where g' is reduced gravity, and $H_e = H_1H_2/H_0$ is the equivalent depth.

An example which will be considered in Section 11.2 is an upper-layer vortex, which is a counterpart of the upper-layer jet of Section 10.2. It is obtained by supposing that velocity in the lower layer is zero, and therefore $\kappa \to 0$, and will be considered at $Bu = 1$. Baroclinic vortices with non-zero lower-layer velocity will be treated afterwards.

11.2 Instabilities of isolated quasi-geostrophic vortices and their nonlinear evolution

11.2.1 One-layer configuration, barotropic vortices

Results of the linear stability analysis

The linearisation of the system (11.1) about a vortex solution (11.3) goes along the usual lines. The linearised non-dimensional equations are

$$
\begin{cases}
\left(\partial_t + \dfrac{V}{r}\partial_\theta\right)u - \left(1 + \dfrac{2V}{r}\right)v = -\partial_r\eta, \\[2ex]
\left(\dfrac{\partial V}{\partial r} + 1 + \dfrac{V}{r}\right)u + \left(\partial_t + \dfrac{V}{r}\partial_\theta\right)v = -\dfrac{1}{r}\partial_\theta\eta, \\[2ex]
\left(\partial_t + \dfrac{V}{r}\partial_\theta\right)\eta + \left(H\partial_r + \dfrac{1}{r}\partial_r(rH)\right)u + \dfrac{1}{r}H\partial_\theta v = 0,
\end{cases}
\tag{11.8}
$$

where we denote the perturbations of u and v by the same symbols, and the perturbation of h by η. We look for normal-mode solutions with harmonic dependence on time and polar angle;

$$
[u,\, v,\, \eta](r,\theta,t) = \mathrm{Re}\left\{[i\tilde{u},\, \tilde{v},\, \tilde{\eta}](r)\exp\big(i(l\theta - \omega t)\big)\right\},
\tag{11.9}
$$

where l and ω are the azimuthal wave number and the frequency, respectively. The difference with the case of jets is that the wave number l is discrete, to insure 2π periodicity in θ, as in Section 4.5. The linearised system is discretised over the stretched Chebyshev grid shown in Figure 11.1, where typical profiles of cyclonic and anticyclonic α-Gaussian vortices are displayed. We show in Figure 11.2 a typical output of the linear stability analysis by the collocation method. The unstable mode with the highest growth rate of a cyclonic vortex is presented. Results for the anticyclonic vortex with the same values of α and ϵ are similar. Azimuthal structure of the most unstable mode depends, at a given intensity of the vortex, on the steepness of the profile of azimuthal velocity,

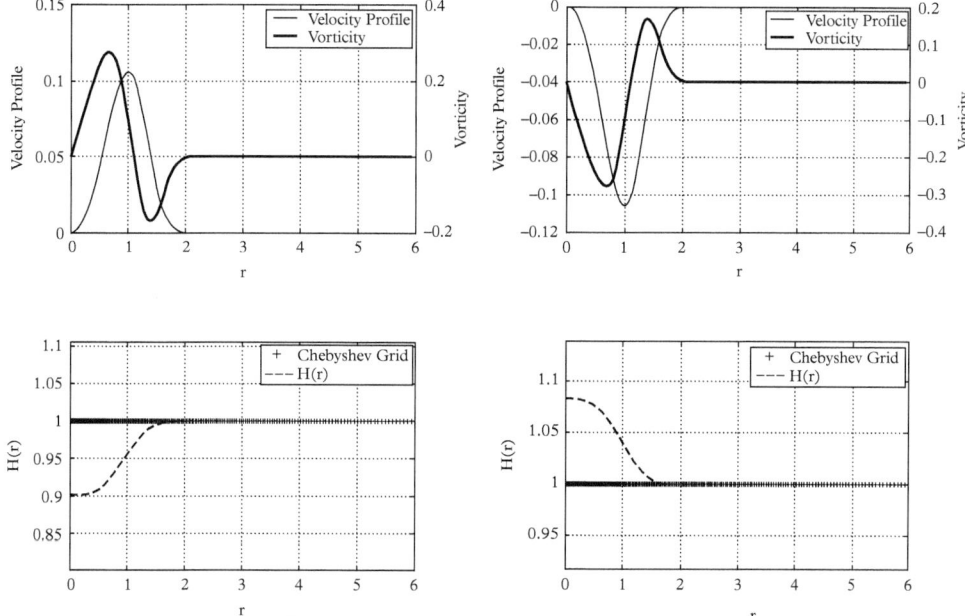

Figure 11.1 *Radial distribution of velocity and relative vorticity (upper row), and thickness (lower row) of cyclonic (left column) and anticyclonic (right column) vortices with azimuthal velocity profile (11.3), α = 4, and ε = 0.1061. Stretched Chebyshev grid (resolution N = 550) is superimposed onto the thickness profile in the lower row.*

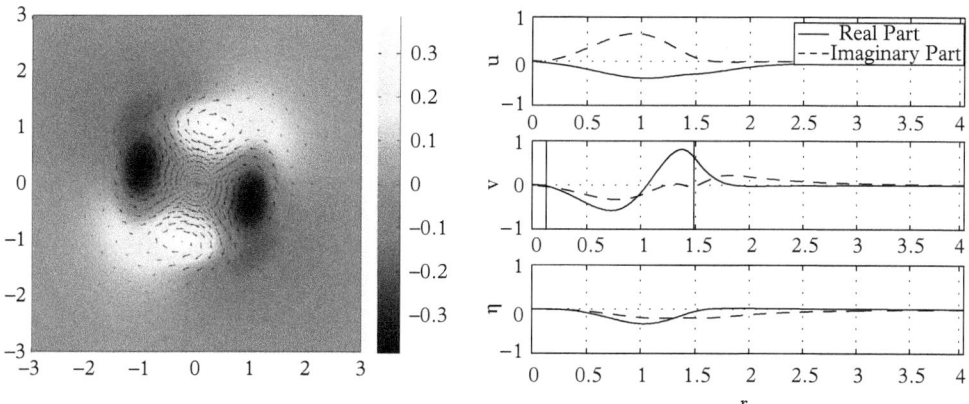

Figure 11.2 *Left panel: thickness (levels of grey) and velocity (arrows) fields of the most unstable mode of the cyclone of Fig. 11.1 in the x − y plane, Right panel: corresponding radial structure of three components (u, v, η)(r) of the unstable mode. Vertical line: position of the critical level r_c at which the phase velocity of the mode is equal to the azimuthal velocity of the vortex.*

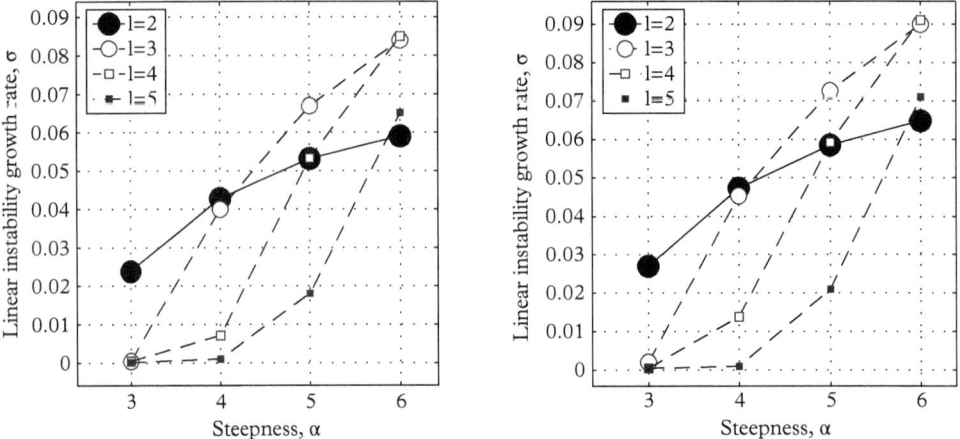

Figure 11.3 *Dependence on steepness parameter α of the linear growth rates of unstable modes with different l for anticyclonic (left panel) and cyclonic (right panel) vortices, both with ε = 0.1414.*

i.e. on the distribution of vorticity of the background vortex, as follows from Figure 11.3. It is worth noting that in the case of $\alpha = 4$ the growth rates of $l = 2$ and $l = 3$ modes are practically the same. Hence, a cross-over of instabilities can take place. As we will see shortly, saturation of the instabilities with different azimuthal wave numbers follows different scenarios, so a small change of parameters can lead to a substantially different evolution.

Nonlinear saturation of the dipolar baroropic one-layer instability

Nonlinear saturation of the leading, at small steepness of the vorticity profile, $l = 2$ instability of both cyclonic and anticyclonic vortices in one-layer model is presented in terms of evolution of PV anomaly in Figure 11.4. The simulations with well-balanced finite-volume code of Chapter 7 were initialised by imposing onto the vortex the velocity and thickness anomaly fields of small amplitude, which correspond to the most unstable mode with azimuthal wave number $l = 2$ found in the linear stability analysis, cf. Figure 11.2. Figure 11.4 clearly shows the so-called dipolar splitting scenario of the evolution of the barotropic instability of the vortex. For both anticyclonic and cyclonic background vortices it consists in several similar typical stages. At the first stage, the growth of the original perturbation leads to formation at the periphery of the background vortex of satellite vortices, with the sign opposite to the main vortex. As time goes on, the core of the main vortex becomes elliptic due to the strain exerted by the satellite vortices which, at the same time, intensify. The satellite vortices continue to strain the core which leads to its splitting. The resulting two vortices originating from the core pair with satellites and produce two vortex dipoles which run in opposite directions from the location of the initial vortex, as they should, cf. Chapter 6.

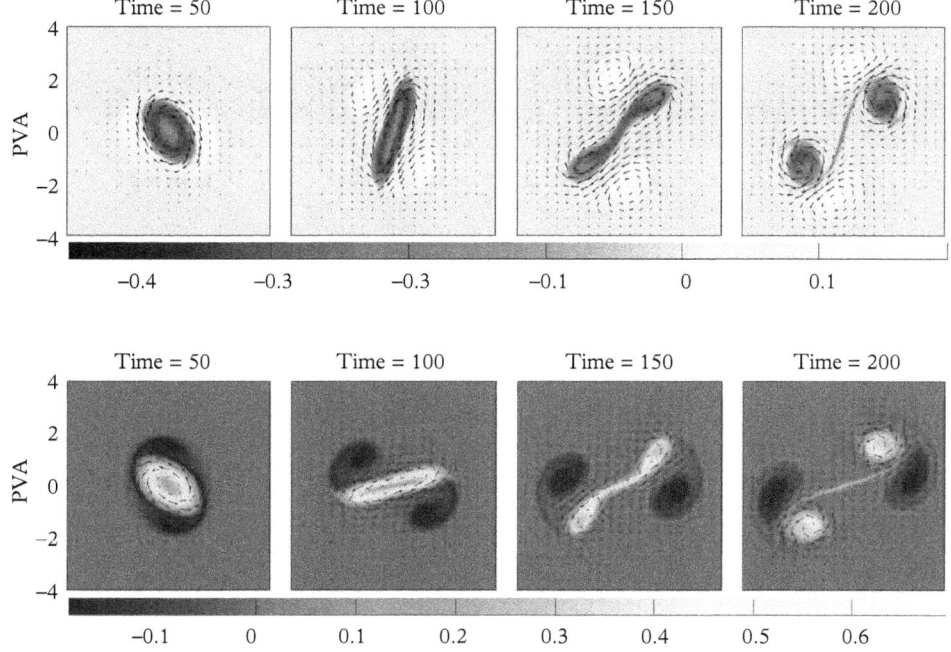

Figure 11.4 *Evolution of PV anomaly of the anticyclonic (upper row) and cyclonic (lower row) vortices with* $\alpha = 4$, $\epsilon = 0.1061$, *during the saturation of the barotropic instability.*

11.2.2 Two-layer configuration, baroclinic upper-layer vortex

Results of the linear stability analysis

The analysis of stability of baroclinic localised vortices in two-layer model goes along the same lines. The non-dimensional linearised system for perturbations of velocity and thickness reads

$$
\begin{cases}
\left(\partial_t + \dfrac{V_i}{r}\partial_\theta\right) u_i - \left(1 + 2\dfrac{V_i}{r}\right) v_i + \partial_r(s^{i-1}\eta_1 + \eta_2) = 0, \\[2ex]
\left(\partial_t + \dfrac{V_i}{r}\partial_\theta\right) v_i + \left(1 + \dfrac{V_i}{r} + \partial_r V_i\right) u_i + \dfrac{1}{r}\partial_\theta(s^{i-1}\eta_1 + \eta_2) = 0, \\[2ex]
\left(\partial_t + \dfrac{V_i}{r}\partial_\theta\right) \eta_i + \left[H_i\partial_r + \dfrac{\partial_r(rH_i)}{r}\right] u_i + \dfrac{H_i}{r}\partial_\theta v_i = 0, \quad i = 1, 2.
\end{cases}
\tag{11.10}
$$

Solutions, as above, are sought in the harmonic form: $(u_i, v_i, \eta_i)(r, \theta, t) = (i\tilde{u}_i, \tilde{v}_i, \tilde{\eta}_i)(r)e^{-i(l\theta - \omega t)}$, where l is the discrete azimuthal wave number. (Corresponding changes should be made in the atmospheric context, cf. Chapter 3.)

The results of the linear stability analysis in the low Rossby number case will be illustrated on the example of an upper-layer vortex mentioned earlier. The radial profiles of

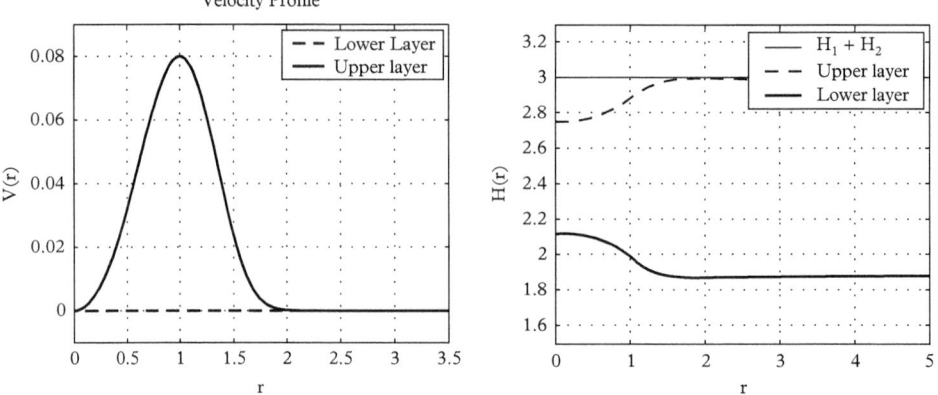

Figure 11.5 *Left panel: radial profiles of azimuthal velocity of the upper-layer atmospheric vortex with $\alpha = 4$ and $s = 1.37$. Right panel: corresponding radial profiles of thickness.*

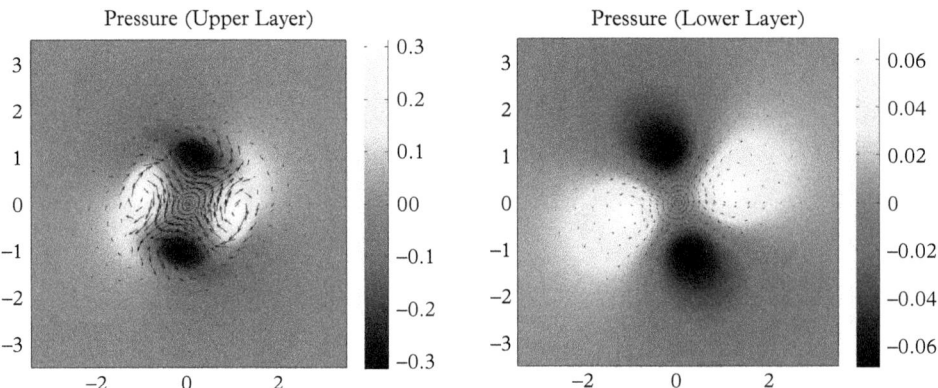

Figure 11.6 *Structure of the most unstable mode of an upper-layer atmospheric cyclone with $\alpha = 4$, $H_2/H_1 = 0.6$, $\epsilon = 0.08$ and $s = 1.37$. Left (Right) panel: pressure and velocity fields in the $x - y$ plane in the upper (lower) layer*

velocity and thickness of the layers of an upper-layer vortex in the atmospheric context are presented in Figure 11.5. Velocity and pressure fields of the most unstable mode of the vortex of Figure 11.5 are displayed in Figure 11.6. As follows from the figure, the most unstable mode of the upper-layer vortex with $\alpha = 4$ has azimuthal wave number $l = 2$, as in the case of barotropic vortex in the previous subsection. As in the case of the barotropic vortex, the azimuthal wave number of the most unstable mode changes with steepness of the azimuthal velocity profile, as shown in the left panel of Figure 11.7. Together with steepness of the velocity profile of the vortex, stratification parameter s

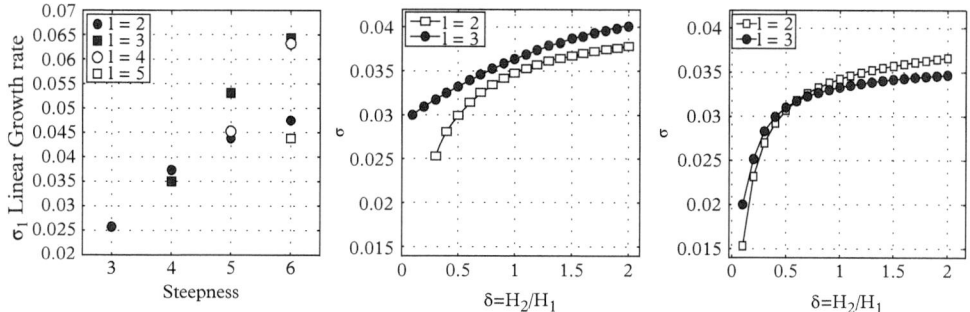

Figure 11.7 *Left panel: dependence on steepness parameter α of the growth rates of unstable modes with different l for upper-layer atmospheric cyclone with $\epsilon = 0.1$, $H_2/H_1 = 0.6$, and $s = 1.37$. Variation of the growth rate of the most unstable mode of the upper-layer cyclone with H_2/H_1 and $\alpha = 4$ at weak ($s = 1.1$, middle panel) and strong ($s = 1.37$, right panel) stratifications.*

and thickness ratio $d = H_2/H_1$ of the model also play an important role. Changing them modifies the structure of the leading instability, as illustrated in the middle and right panels of Figure 11.7. As was already mentioned in the case of barotropic one-layer model, small changes of parameters of the vortex can change a pattern of instability. Here, in the baroclinic two-layer case, small changes of parameters of baroclinicity, stratification, and the aspect ratio, also change the instability pattern.

Nonlinear saturation of the leading instabilities of the upper-layer cyclone

Nonlinear saturation of the $l = 2$ instability of the upper-layer cyclone follows, in the upper layer, the scenario which was observed for the barotropic vortex. At the same time, a signature of the instability develops in the initially motionless lower layer, although with an order of magnitude smaller amplitude. The evolution of this instability, as follows from the simulations with the two-layer code of Chapter 7, is displayed in Figure 11.8 and shows the dipoles running out of the location of the initial vortex in both layers.

The evolution of $l = 3$ instability, which is dominant in a range of stratification parameters with a given velocity profile, cf. Figure 11.7, is totally different. As follows from Figure 11.9, it leads to formation of a vortex tripole with same-sign poles, which further splits in three co-rotating vortices and a central core. Together with dipolar splitting, this is the second known pattern of development of vortex instabilities. It should be emphasised that evolution of the $l = 3$ instability of the barotropic vortices of the previous subsection also leads to the tripole formation, although we did not present it.

A brief summary of the results

We have thus seen that instabilities of localised vortices with low Rossby numbers have most unstable modes with azimuthal wave number $l = 2$ or $l = 3$, and exhibit, correspondingly, dipolar splitting or tripole formation scenarios of saturation. Changing

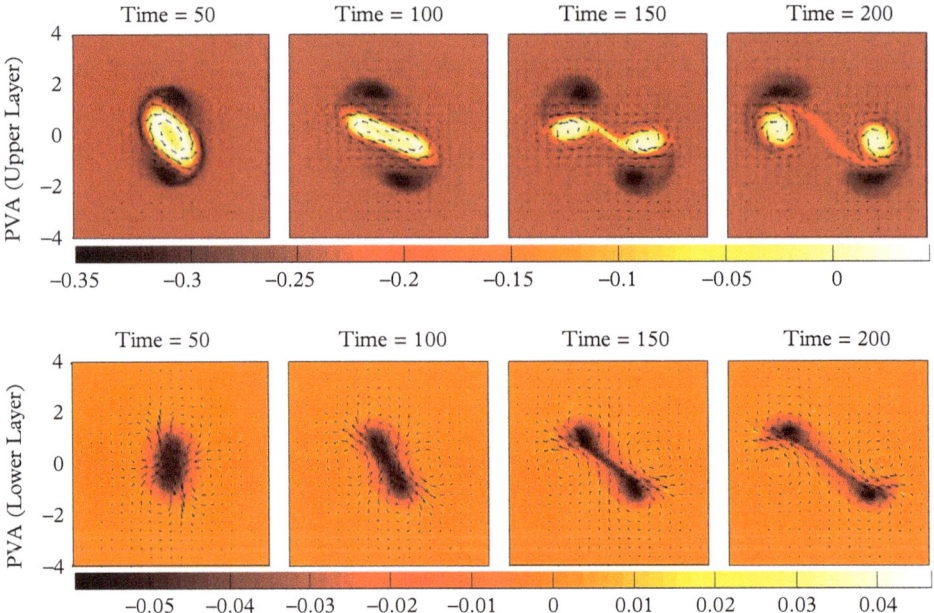

Figure 11.8 *Evolution of PV anomaly in the upper (top row) and lower (bottom row) layers during the evolution of the l = 2 instability of the upper-layer cyclone with $\alpha = 4$, $\epsilon = 0.08$, $s = 1.37$, $H_0 = 3$, $d = 0.6$.*

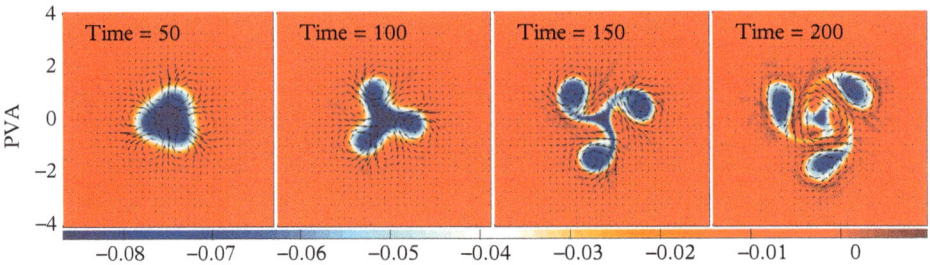

Figure 11.9 *Evolution of PV anomaly in the lower layer in the simulation of the development of the baroclinic instability of upper-layer cyclone with most unstable mode l = 3.*

stratification, and/or aspect ratio, of the background vortex, as well as the steepness of the radial profile of its vorticity, leads to changes of the azimuthal structure of the most unstable mode and, hence, of the saturation scenario. This is useful to bear in mind in the general context of predictability of atmospheric and oceanic flows, as small changes in the parameters of the vortex can lead to qualitative changes in its evolution under the influence of small perturbations.

11.3   Instabilities of ageostrophic vortices and their nonlinear evolution

11.3.1   General considerations

After having studied geostrophic instabilities of vortices in Section 11.2, we can ask the same questions as while studying stability of jets in Section 10.4; namely, what are the ageostrophic counterparts of geostrophic barotropic and baroclinic instabilities, and what is their relation to the centrifugal instability, which is the analogue of inertial instability in cylindrical geometry. We recall that the classical centrifugal instability is azimuthally symmetric and essentially ageostrophic. In order to understand the conditions when such instability could be realised, let us consider an axisymmetric configuration and, for simplicity, a purely barotropic vortex with $\kappa = 1$, and hence $V_1 = V_2 = V$. The linear system of equations (11.10) can be reduced to a single one by straightforward elimination of variables,

$$(-\omega^2 + \overline{\Phi}) \begin{pmatrix} u_1 \\ u_2 \end{pmatrix} = \partial_r \left(\frac{\partial_r}{r} \begin{pmatrix} rH_1 u_1 + rH_2 u_2 \\ rsH_1 u_1 + rH_2 u_2 \end{pmatrix} \right), \tag{11.11}$$

where $\overline{\Phi} = 2\overline{L}_a \overline{\zeta}_a / r^2$, $\overline{L}_a = r^2/2 + rV$ is the non-dimensional absolute angular momentum, and $\overline{\zeta}_a = 1 + \partial_r(rV)/r$ is the non-dimensional absolute vorticity of the background vortex. Following the lines of Section 10.4 we can deduce an estimate for the frequency ω in terms of the baroclinic and barotropic components of the normal modes, which are defined as $u_b = u_1 - u_2$ and $u_B = (H_1 u_1 + H_2 u_2)/H_0$, respectively. By multiplying (11.11) by $H_e u_b^*$, where H_e is equivalent height and integrating over the flow domain, under hypothesis that the velocity field vanishes at infinity, which is consistent with the *trapped* character of the modes we are interested in, we get

$$\omega^2 = \frac{\int \overline{\Phi} \cdot H_e |u_b|^2 dr}{\int H_e |u_b|^2 dr} + (1-s) \left[\frac{\int \left[|\partial_r(H_e u_b)|^2 + \frac{|H_e u_b|^2}{4r^2} \right] dr}{\int H_{eq} |u_b|^2 dr} \right.$$

$$\left. - \frac{\int H_e u_b^* \partial_r \left(\frac{\partial_r(rH_1 u_B)}{r} \right) dr}{\int H_e |u_b|^2 dr} \right]. \tag{11.12}$$

The sign of $\overline{\Phi}$ defines the sign of the first term on the right-hand side of this relation, while the second term is positive-definite. The third term is not sign-definite, but it is the only one containing the barotropic velocity u_B, and thus vanishes for purely baroclinic modes. Hence, we infer that for sufficiently large negative values of $\overline{\Phi}$ there exist trapped baroclinic modes with imaginary eigenfrequencies. We thus recover the classical Rayleigh condition for the centrifugal instability, $\overline{\Phi} \propto \overline{L}_a \overline{\zeta}_a < 0$, where the last product is called Rayleigh discriminant. The whole line of argument parallels what was done in the case of inertial instability in Section 10.4, with $\overline{\Phi}$ playing the role of $\partial_x M$. Useful by

(a)   (b)

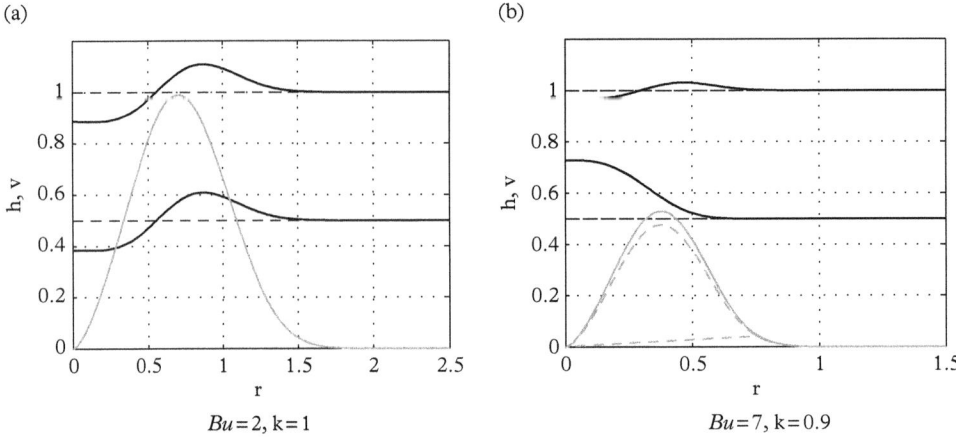

$Bu = 2, k = 1$ $Bu = 7, k = 0.9$

Figure 11.10 *Radial distribution of velocity (grey) and thickness (solid black) for barotropic (left) and baroclinic (right) ageostrophic vortices with* $\alpha = 3$, $d = 1$, $s = 0.9$, $Ro = 1.4$. *Dashed (solid) grey: lower (upper) layer.*

itself, the estimate (11.12) for the eigenfrequencies can be used to benchmark the results of numerical stability analysis in the case $l = 0$.

We will be interested, in the first place, in long-wave, that is small-l unstable modes in the context of centrifugal instability. We will not present the details of linear stability analysis, but rather give a summary of results obtained by investigating a large range of control parameters. Examples of ageostrophic barotropic and baroclinic vortices with velocity profiles obeying (11.7) and subject to such studies are presented in Figure 11.10.

11.3.2   Results of the linear stability analysis

Summary of the results

The main results of the linear stability analysis are as follows. For stratified *barotropic* vortices with different densities of the layers, but without velocity shear between the layers:

- The classical symmetric centrifugal instability with $l = 0$ is dominant at *high Ro* ($\gtrsim 0.8$) and *low Bu* ($\lesssim 10$), and for density and/or depth ratio close to 1,
- At *substantial stratifications*, symmetric centrifugal instability is not the leading one, irrespectively to the values of Ro and Bu; asymmetric centrifugal instability with $l = 1$ dominates at high Ro and low Bu,
- Regardless of stratification, at *high Bu* ($\gtrsim 10$) and/or *low Ro* ($\lesssim 0.7$) the dominant instability is the barotropic one with $l = 2$,
- In between the parameter domains with 1) *high Ro, low Bu*, and 2) *low Ro, large Bu*, asymmetric centrifugal instabilities with $l = 1$ or $l = 2$ are dominant.

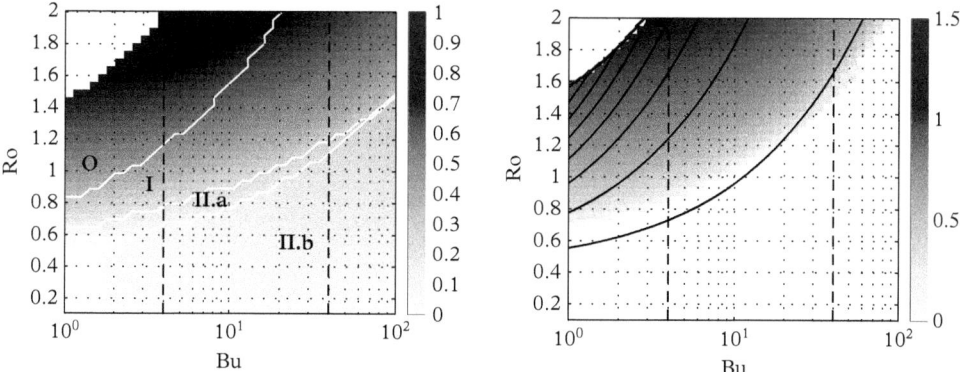

Figure 11.11 *Variation with Rossby and barotropic Burger numbers of the growth rate σ (levels of grey) of the most unstable modes of the barotropic vortex with α = 3, s = 0.99 and d = 1. Baroclinic Burger number Bu′ ≈ Bu/400. Values of Bu′ = 0.01, 0.1: dashed vertical lines. Left: σ of the most unstable modes with l ∈ [0, 4], with separation lines in white between the regimes with different dominant instabilities. O: symmetric centrifugal, l = 0; I: first asymmetric centrifugal, l = 1; II.a: second asymmetric centrifugal, l = 2; II.b: barotropic, l = 2. Right: Maximum value of σ for the unstable modes with l = 0 (levels of grey), and isopleths (black lines) of −Φ̄ for the background vortex with values increasing from 0.1 (lowermost curve) to 3.*

For *baroclinic* vortices, with vertical velocity shear the latter inhibits both axisymmetric and non-axisymmetric centrifugal instabilities, while the barotropic instability remains. It should be emphasised that in this summary the non-axisymmetric instabilities are called centrifugal because of strong similarity, practically coincidental, of their radial structure with that of the standard centrifugal instability, as will be shown in the following.

We summarise the results of the linear stability analysis of weakly stratified barotropic vortices in Figure 11.11 showing the dependence of the growth rates on Rossby and Burger numbers. Calculations were performed at $s = 0.99$, $d = 1$, and the azimuthal wave number l in the range $[0, 4]$. We should stress the qualitative agreement of these results with the the Rayleigh criterion, as the region of dominant symmetric centrifugal instability on the right panel of Figure 11.11 coincides with the zone of strongest negative $\overline{\Phi}$.

Symmetric vs asymmetric centrifugal instability

As is clear from the left panel of Figure 11.11, the maximum of all growth rates is achieved for the axisymmetric centrifugal instability. The radial structure of the corresponding most unstable mode is displayed in the left panel of Figure 11.12. The mode is confined in the zone where the Rayleigh discriminant is negative, and is essentially baroclinic (i.e. with opposite velocities layer-wise and strong deviation of the interface). These properties agree with what can be expected from the qualitative analysis of the expression (11.12). The structure of the second most unstable mode with azimuthal wave number $l = 1$, at the same values of vortex parameters, is also displayed in Figure 11.12.

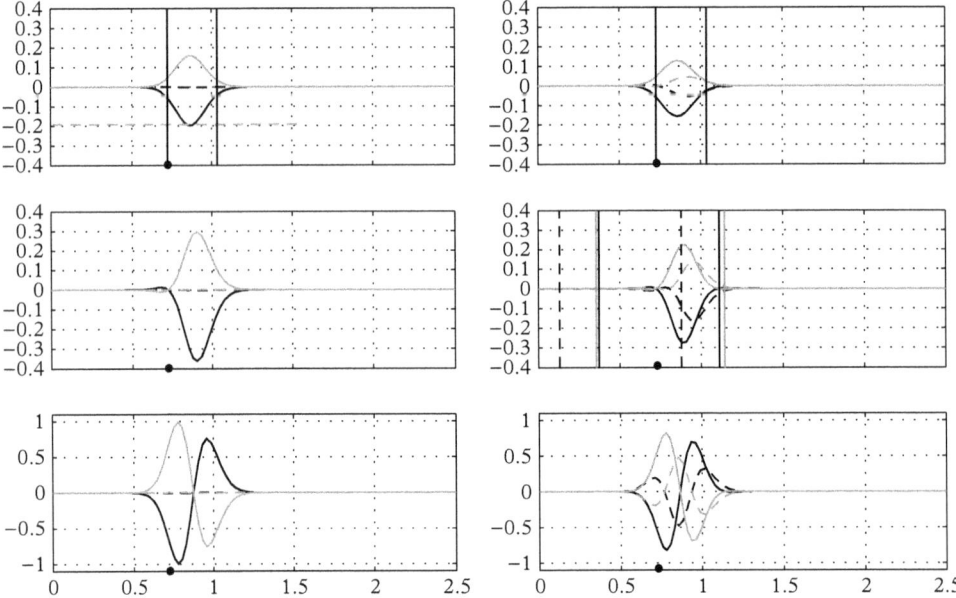

Figure 11.12 *Structure of $l = 0$ and $l = 1$ modes of the centrifugal instability at $Ro = 1.2$, $Bu = 2$, $d = 1$, $\kappa = 1$ and $s = 0.99$, with growth rates 0.76 and 0.73, respectively. First row: radial velocity u, with two vertical lines indicating the boundaries of the area where the Rayleigh discriminant is negative. Second row: azimuthal velocity v, with vertical lines indicating the radius where PV gradient vanishes (black/grey continuous in the upper/lower layer) and the location of the critical radius r_c (black dashed, if applicable). Third row: thickness deviation η. Continuous: real part, dashed: imaginary part, black: upper layer, grey: lower layer. Bullets on the x-axes indicate the location of the velocity maximum in the background flow: $r_{max} = Bu^{-1/2}$.*

Similarities between the two modes are striking: the $l = 1$ mode is trapped in the same zone given by the Rayleigh criterion, and is baroclinic. We can, thus, conclude that the two modes correspond to symmetric and asymmetric versions of the same instability. For this and other reasons, revealed shortly, we call the instability associated with this mode *asymmetric centrifugal instability*. Again, the parallel with symmetric and asymmetric inertial instabilities of Section 10.4 is evident. Unlike its symmetric counterpart, the $l = 1$ centrifugal mode is propagating in the azimuthal direction, and has a critical level which does not coincide with the location where PV gradient of the vortex vanishes, cf. Figure 11.12. We recall that the standard mechanism of the baroclinic instability is the resonance between Rossby waves, which implies that the potential vorticity gradient must vanish at the critical level in the limit of vanishing growth rate (i.e. for the neutral waves undergoing phase locking, according to Lin's theorem, cf. section 4.3). Therefore, the asymmetric centrifugal mode can not be interpreted in terms of the classical baroclinic instability, and represents its ageostrophic generalisation, in parallel to asymmetric inertial instability which is an essentially ageostrophic baroclinic instability of jets.

Asymmetric centrifugal vs barotropic instability

At larger Burger and/or lower Rossby numbers the second azimuthal mode corresponding to another instability becomes dominant, as follows from Figure 11.11. Its structure is presented in the right panel of Figure 11.13. The velocity field of this mode is quasi-barotropic (i.e. the velocites in the upper and the lower layer are close). Growth rates associated with this instability are weaker and increase with Burger number, the opposite case to the centrifugal modes. The mode is almost geostrophically balanced beyond the radius of maximum velocity of the background vortex, and has a critical level located at a radius where the potential vorticity gradient vanishes. This allows us to interpret this unstable mode as a resonance between two non-singular Rossby waves, the condition of existence of a neutral wave mentioned above being verified. Hence, we identify the instability corresponding to this mode as an ageostrophic barotropic. Moreover, when the Rossby number decreases, this mode clearly evolves towards the pattern of the quasi-geostrophic barotropic instability which we have seen in Section 11.2.

At the lower Burger/higher Rossby side of the zone of the dominant barotropic instability in the left panel of Figure 11.11 we see a strip where the most unstable mode has also the azimuthal wave number $l = 2$, but a different structure. As follows from the left panel of Figure 11.13, its radial and vertical structure (trapped in the same zone,

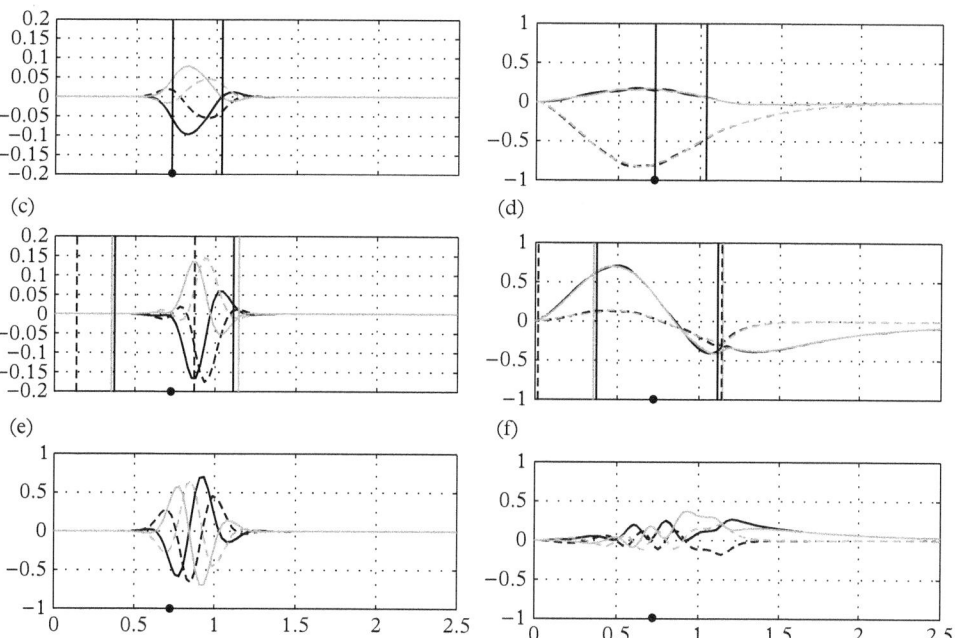

Figure 11.13 *Structure of the $l = 2$ modes corresponding to centrifugal (left) and barotropic (right) instabilities of a vortex with $Ro = 1.2$, $Bu = 2$, $d = \kappa = 1$, $\alpha = 3$, and $s = 0.99$. The growth rates are 0.6 and 0.1, respectively. Same conventions as in Figure 11.12.*

baroclinic) is similar to that of the centrifugal instability modes with $l = 0$ and $l = 1$. This is thus a higher azimuthal mode of asymmetric centrifugal instability. The phase velocities of all centrifugal modes are roughly equal and lie in the range $-1.2 < \omega/l < -0.6$ for the parameters of Figure 11.11, while the phase velocity of the barotropic instability mode lies between -1 and 0.

Dependence on stratification

To understand how the previous results depend on stratification we show in Figure 11.14 the dependence of the growth rates on s, and give a stability diagram in the (Ro, Bu) plane for lesser $s = 0.9$ (stronger stratification), than in Figure 11.11. The vortex is barotropic, with $\alpha = 3$ and $d = 1$. It is to be noted that the number of nodes in the radial profile of the eigenmode decreases as the growth rate increases at fixed value of s in the right panel of the figure. In sharp contrast with the weakly stratified case, the symmetric centrifugal mode is not dominant, although its growth rate is close to the one of the leading first azimuthal mode. As follows from the right panel of Figure 11.14, the growth rates of the centrifugal instability modes decrease with increasing stratification.

Dependence on the aspect ratio and vertical shear

Other parameters which influence the growth rates and the structure of dominant instability are the aspect ratio d and vertical shear, which is controlled by κ. The left panel of Figure 11.15 illustrates the influence of the aspect ratio, showing that at small d symmetric centrifugal instability dominates. The role of vertical shear in selection of the most unstable mode of *baroclinic* vortices is illustrated in the right panel of Figure 11.15 The vortex with the parameters of the figure is stable with respect to symmetric centrifugal instability. The growth rates of all unstable modes decrease with increasing vertical shear (decreasing κ), and those of both centrifugal ones with $l = 1$ and $l = 2$

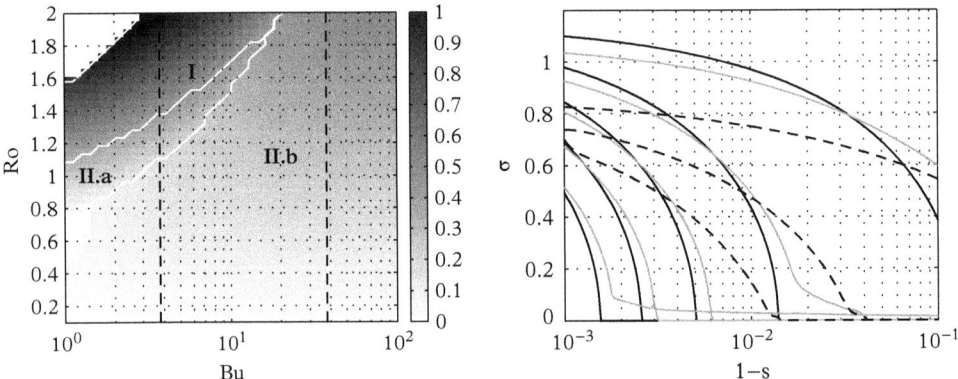

Figure 11.14 *Left panel: same as in Figure 11.11, but for $s = 0.9$. Right panel: dependence of the growth rate of the most unstable modes on $1 - s$. Black continuous: $l = 0$, grey: $l = 1$, black dashed: $l = 2$. $l = 3$ and $l = 2$ modes with smaller growth rates are not shown. Different curves of the same type correspond to different number of nodes in radial direction.*

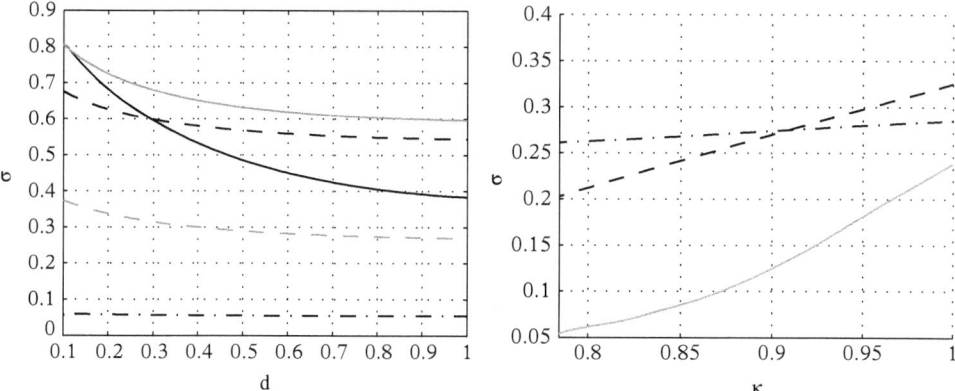

Figure 11.15 *Left panel: dependence of the growth rate of the unstable modes of a barotropic vortex with Ro = 1.4 and Bu = 2 on the aspect ratio d. Right panel: dependence of the growth rate of the unstable modes of a vortex with Ro = 1.4, Bu = 7, d = 1 and s = 0.9 on κ. Black continuous: l = 0 centrifugal mode. Grey continuous: l = 1 centrifugal mode. Black dashed: l = 2 centrifugal mode. Black dash-dotted: l = 2 barotropic mode. Grey dashed: l = 3 centrifugal mode.*

are decreasing faster than that of the barotropic mode with $l = 2$. A swap of instabilities happens around $\kappa = 0.91$, with barotropic instability becoming dominant. This confirms that the centrifugal instability is most relevant for the barotropic vortices, and less relevant for baroclinic ones.

11.3.3 Nonlinear saturation of the instabilities

Evolution of symmetric centrifugal instability and loss of symmetry

The difference in dominant instability modes in different parameter regimes leads to different scenarios of nonlinear saturation which are, as usual, obtained by initialising fully nonlinear numerical simulations by a background vortex with a superimposed small-amplitude unstable mode. Let us start with the axisymmetric centrifugal instability, which is dominant in a wide range of parameters, as we have just seen. At early stages of its evolution the strong shears at the location of the unstable modes produce highly ageostrophic motions with strong divergence, as shown in the upper row of Figure 11.16. Strong shears (i.e. baroclinic velocities) lead to hyperbolicity loss, and related dissipation, according to the explanation in Chapter 7. Yet, the evolution of the unstable mode does not consist just in dissipation through small-scale structures visible at the right panel of the upper row of Figure 11.16. A large-amplitude front forms in this zone, as follows from Figure 11.17. This front then propagates inwards losing its symmetry, as is visible in Figure 11.18. The appearance of growing non-axisymmetric perturbations of the front can be associated with a known folding instability of concentric shocks. One or several azimuthal modes with non-zero l necessarily arise due to this process from the noise, and modulate the front. The particular mode $l = 4$, which can be identified in Figure 11.18 and also in the Fourier spectrum of the signal, is,

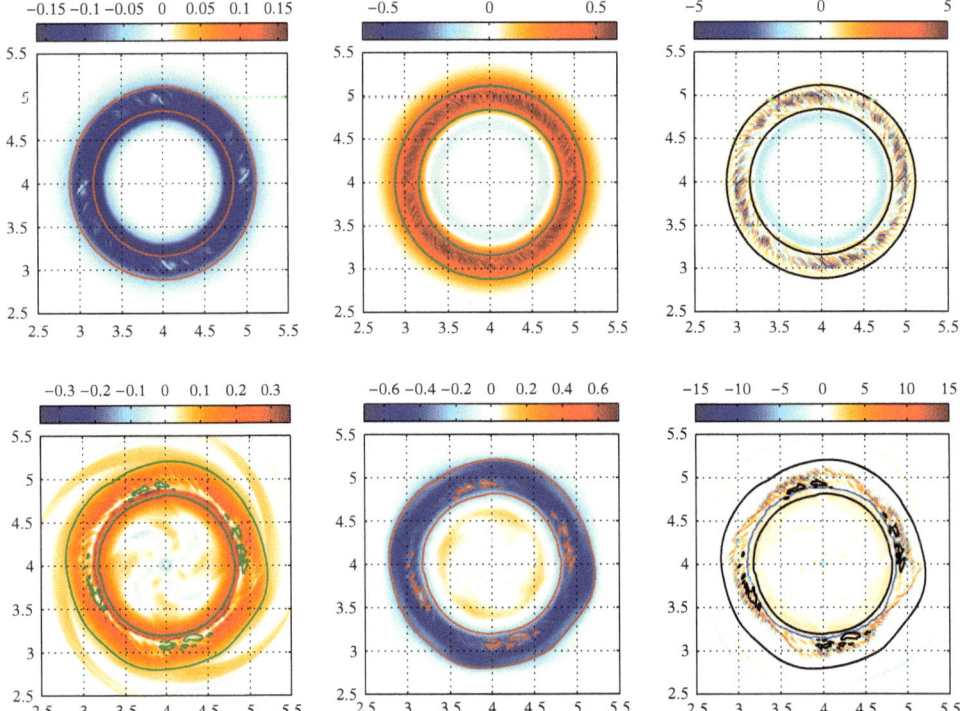

Figure 11.16 *Snapshots at t = 3.5 f^{-1} (upper row) and t = 11.5 f^{-1} (lower row) of radial (left column) and azimuthal (middle column) baroclinic velocities and of baroclinic divergence (right column) during the saturation of symmetric centrifugal instability. A zone of hyperbolicity loss is situated between thin contours.*

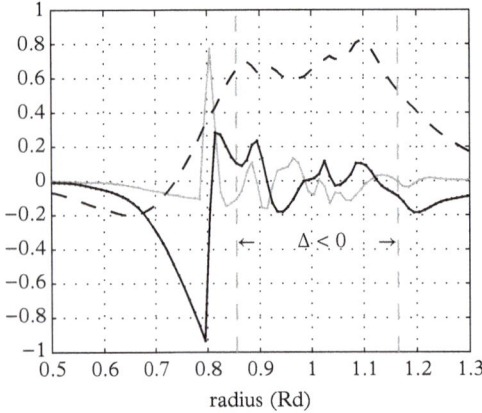

radius (Rd)

Figure 11.17 *Snapshot at t = 4.5 f^{-1}, at a given polar angle, of the radial profiles of radial (black) and azimuthal (grey) components of the baroclinic velocity, and of its divergence (dashed), during the saturation of the axisymmetric centrifugal instability. The zone of negative Rayleigh discriminant of the background vortex is also indicated.*

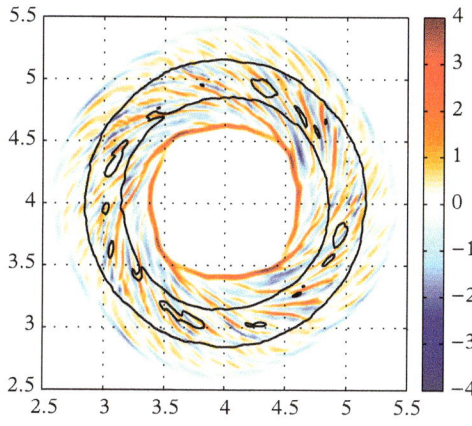

Figure 11.18 *Snapshot at $t = 6f^{-1}$ of divergence of the baroclinic velocity during the saturation of the symmetric centrifugal instability. Inner contour of high divergence corresponds to the inward-propagating front losing its axial symmetry.*

most probably, due to specific errors of discretisation of axisymmetric features on the Cartesian grid. It should however be stressed that while the selection of this particular wave number is apparently due to the chosen discretisation, the process of destabilisation of the front and appearance of an azimuthal structure are universal. Propagation of the front leads to a second episode of strong dissipation in the evolution of symmetric centrifugal instability. The lower row of Figure 11.16 shows the changes in radial distribution of the baroclinic velocity and its divergence generated by the passage of the front. The two above-described episodes of strong dissipation and mixing lead to re-organisation of the mean flow and 'self-healing' of the instability, similar to what we have already seen in Section 10.6 for inertial instability. The evolution of the mean angular momentum and Rayleigh discriminant confirms this conclusion, and are presented in Figure 11.19.

At late stages of the evolution a secondary $l = 4$ instability due to the above-described processes evolves and gives rise to formation of a barotropic tripole which is displayed in Figure 11.20.

Evolution of asymmetric centrifugal and barotropic instabilities

Evolution of the asymmetric centrifugal instability with $l = 1$ resembles that of the symmetric instability at the initial stages. However, the dissipation zones are now clearly non-axisymmetric, and rotating, as shown in Figure 11.21. No clearly distinguishable concentric shock is detected in these simulations. The original vortex deforms increasingly, and finally splits into a well-formed barotropic dipole, as shown in Figure 11.22 (right panel, upper-right corner), and a cyclone with a weaker anticyclonic satellite (right panel, lower-left corner; the anticyclone is even weaker in the lower layer).

Finally, we compare nonlinear evolution of two $l = 2$ instabilities of different nature, each of them may be dominant depending on parameters, as was shown in the previous subsection. In spite of the same azimuthal wave number, the corresponding unstable modes are different, cf. Figure 11.12, and the nonlinear evolution is different as well.

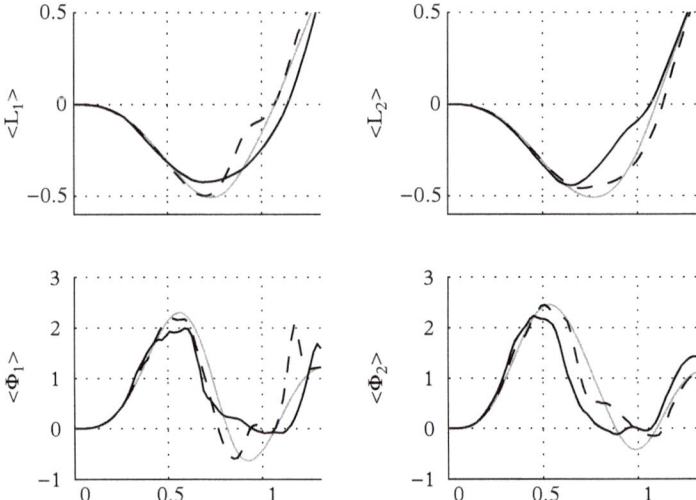

Figure 11.19 *Evolution of azimuthally averaged angular momentum (upper row) and Rayleigh discriminant (bottom row) in the upper (left) and lower (right) layers, $t = 0$ (grey), $t = 8\,f^{-1}$ (dashed), and $t = 18\,f^{-1}$ (solid).*

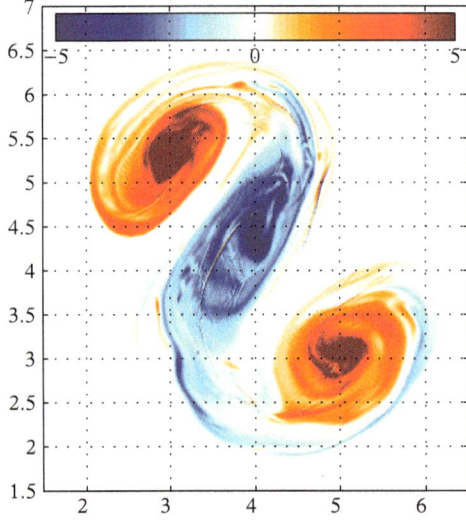

Figure 11.20 *Barotropic tripole, as seen in PV anomaly field in the upper layer: a central anticyclonic and two peripheral cyclonic vortices are formed at $t = 60\,f^{-1}$ during the saturation of symmetric centrifugal instability.*

It is presented in Figures 11.23 and 11.24, respectively, for centrifugal and barotropic instabilities of a barotropic vortex without vertical shear, and with aspect ratio 1. The first produces a vortex tripole, and the second a pair of dipoles, as in the low Rossby number case of Section 11.2.

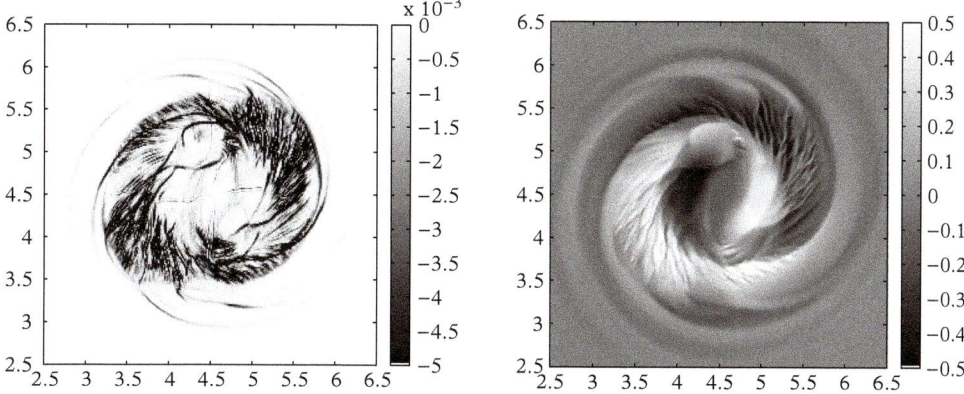

Figure 11.21 *Snapshot at t = 10 of numerical dissipation (left panel) and lower-layer thickness anomaly (right panel) during saturation of asymmetric centrifugal instability with l = 1. High values of the lower-layer thickness anomaly are associated with low values of the upper one, and vice versa.*

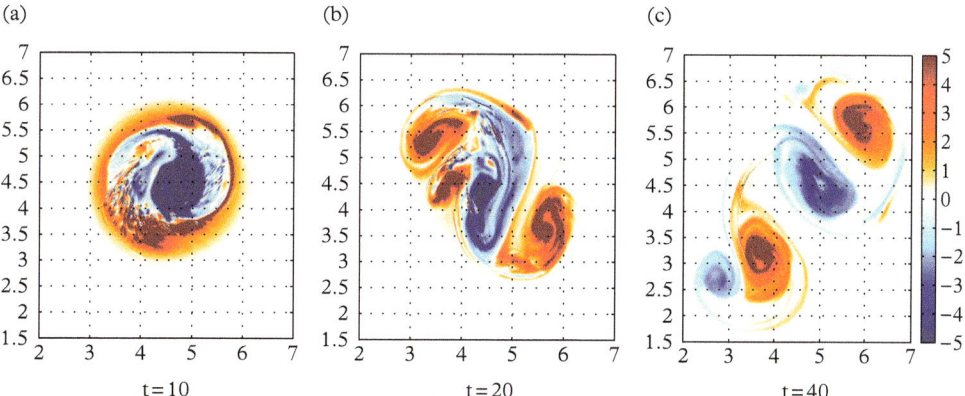

Figure 11.22 *Evolution of PV anomaly in the upper layer, as seen in numerical simulations of the saturation of centrifugal instability with l = 1.*

Inertia–gravity wave emission

As we have seen in preceding chapters, inertia–gravity wave emission accompanies the evolution of vortices. Quantifying such emission associated with the developing instabilities is of importance in the general context of comprehension of sources of inertia–gravity waves. The wave activity can be diagnosed with the help of the divergence field, and is presented, for baroclinic waves at the interface, in Figure 11.25. A circle in Figure 11.24 delimits inner and outer zones used in inertia–gravity wave diagnostics of Figure 11.25. As follows from the figure, centrifugal instability provides an important source of inertia–gravity waves, in accordance to what was observed for inertial instability in Section 10.4, and is much more efficient in this respect than the barotropic one.

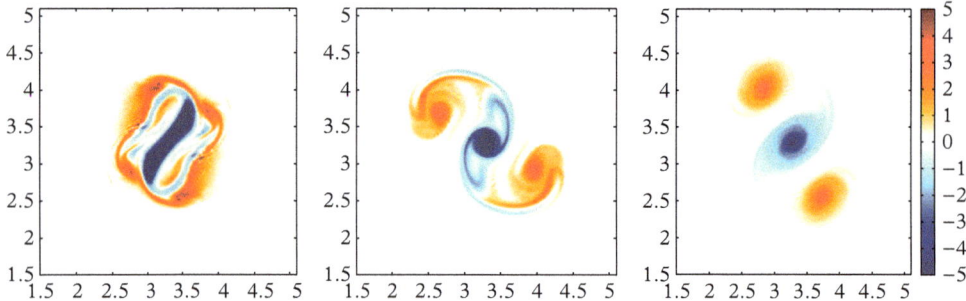

Figure 11.23 *Snapshots at t = 24 (left pane), 48 (middle panel), and 100 f^{-1} (right panel) of the PV anomaly in the upper layer during nonlinear saturation of the l = 2 centrifugal instability of a vortex with Ro = 1.4 and Bu = 7.*

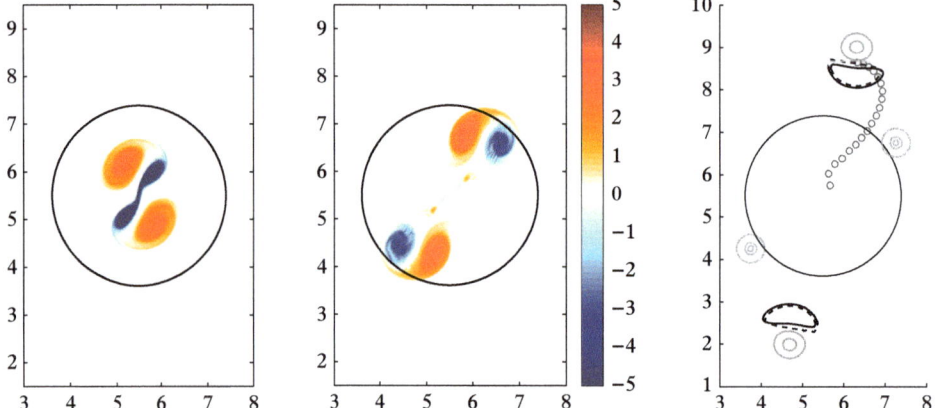

Figure 11.24 *Snapshots at t = 18, 30, 70 f^{-1} (from left to right) of the evolution of l = 2 barotropic instability of a vortex with Ro = 1.4 and Bu = 7, as seen in upper-layer PV anomaly. Up to t ≈ 30 f^{-1} the lower-layer PV is similar. Solid (dotted) lines in the right panel: PVA in the upper (lower) layer, negative values in grey, positive values in black. Trajectory of one of the the upper-layer dipoles is marked by a sequence of circles in the right panel. Large black circle is used for diagnostics of wave activity.*

11.4   Instabilities of intense hurricane-like vortices and their nonlinear evolution

Although we studied ageostrophic instabilities of vortices in the preceding section, we limited ourselves to moderate values of Rossby numbers $Ro \leq 2$, aiming, in the first place, the oceanic vortices. Yet, vortices of synoptic scales (note that we are not talking about small-scale vortices, like tornadoes) with Rossby numbers, which are an order of magnitude greater, exist in the atmosphere. These are tropical cyclones. Obviously, the

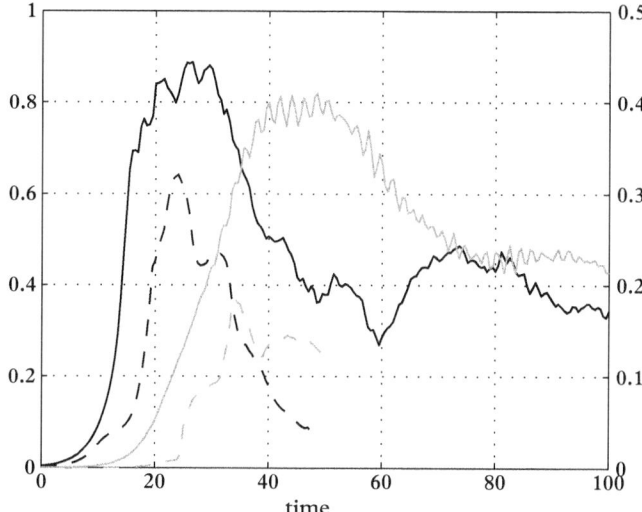

Figure 11.25 *Magnitudes of the divergence of baroclinic velocity inside (black, values on right axis) and outside (grey, values on left axis) a circle of radius $1.9R_d$ around the vortex during the evolution of centrifugal (solid lines) and barotropic (dashed lines) instabilities with $l = 2$.*

RSW models are too poor to grasp the whole complexity of these vortices (we will, however, show in Chapter 15 that it is possible to include the effects of moist convection in the RSW models). Nevertheless, a particular feature of tropical cyclones is that vertical shear of their horizontal velocity is relatively small, which explains why the dynamics of tropical cyclones is often studied with simple barotropic models. It is therefore reasonable to analyse stability of the hurricane-like vortices in the framework of the RSW model which, in addition to capturing vortex dynamics, permits us to describe the related inertia-gravity wave field, while tropical cyclones are known to be efficient sources of these latter. We therefore proceed by fitting velocity and vorticity distributions of typical observed hurricanes and studying the stability of corresponding solutions of the one-layer RSW model and their nonlinear evolution, as in preceding sections of this chapter.

11.4.1 Idealised rotating shallow-water hurricane

General considerations. Scaling

The hydrodynamic characteristics of tropical cyclones are rather special, with the radius of maximum wind of several tens of kilometers, maximum velocity V_{max} of $40 - 60$ m/s, typical value of the Coriolis parameter f (at 20°N) of $5 \cdot 10^{-5}$, relative vorticity ζ up to 100f, and typical Rossby numbers in the interval 10–40. As already mentioned, the vertical wind distribution is approximately barotropic. The radial wind profile is (half-) U-shaped in the core, and decreasing approximately as $1/r$ in the outer region. Such velocity distribution corresponds, approximately, to a constant vorticity core surrounded by a higher vorticity ring, with zero vorticity in the outer region. Already from these qualitative characteristics it follows that the radial gradient of vorticity changes sign, so Rossby waves can propagate in the vortex in opposite azimuthal directions. Thus, the barotropic instability is expected. Saturating barotropic instability will tend to restore a monotonic vorticity profile, so an intensification of the wind in the core can be expected as a result of the developing instability. At high Ro, Lighthill radiation is operational, as we have seen in Chapter 6, so the instability is expected to be *radiative*, associated with a significant gravity-wave field.

To model a hurricane in the framework of RSW equations (11.1) we choose the following scaling: the maximum wind velocity of the hurricane V_{max} is taken as velocity scale, the radius of maximum wind is taken as horizontal scale $L = r|_{V_{max}}$, the time scale is turnover time L/V_{max}, the vertical scale is the unperturbed thickness of the layer H_0, with $h = H_0(1 + \lambda\eta)$, where the parameter λ controls the nonlinearity, as usual. The parameters are standard Rossby and Burger number, and the barotropic Froude number $Fr = \frac{V_{max}}{\sqrt{gH_0}} = Ro/Bu$, which is typically $\mathcal{O}(10^{-1})$ in hurricanes. A non-dimensional axisymmetric vortex solution of RSW equations, with azimuthal velocity $V(r)$ and hydrostatic pressure $gH(r)$, obeys the cyclo-geostrophic equilibrium equation, which takes the form $\left(Ro^{-1} + \frac{V}{r}\right)V = \frac{\lambda}{Fr^2}\frac{dH(r)}{dr}$ with this scaling.

Fitting observed hurricane profiles. Control parameters

In order to represent a typical observed profiles of hurricanes of moderate intensity (e.g. category 3) we use a smoothed piece-wise-constant profile of vorticity with two levels corresponding to the core and the annulus, respectively. Transitions between constant vorticity zones of the vortex are smoothed with the help of a third-order Hermite polynomial. A zone of very weak negative relative vorticity outside the annulus is added, with amplitude and radial extension defined in such a way that the circulation is zero far from the vortex. The resulting vortex profile is shown in Figure 11.26. The most important parameters of the flow are:

- The local Rossby number: maximum value of the relative vorticity in the annulus in units of f;
- The width of the annulus of vorticity, W_r,
- The steepness of the slopes between different zones of constant vorticity, given by their non-dimensional widths d_1, d_2,
- The Froude number Fr.

The Burger number is very high in hurricanes, of the order 10^4. The results we present do not qualitatively change with a variation of Bu. The illustrations below correspond to

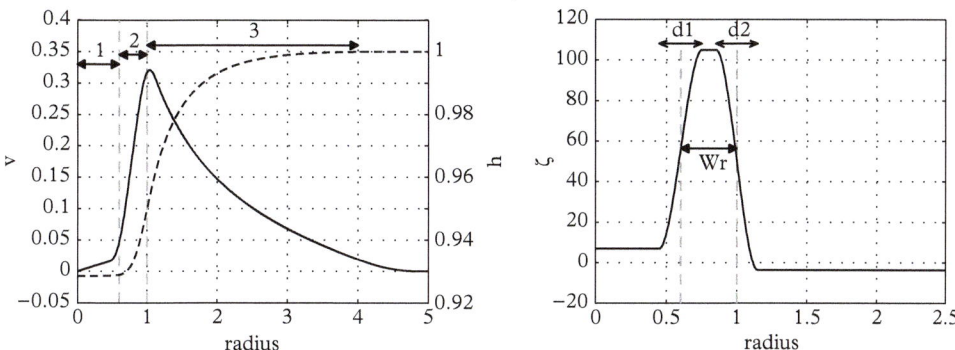

Figure 11.26 *Radial structure of an idealised hurricane. Left panel: non-dimensional velocity* $v/\sqrt{gH_0}$ *(left axis) and thickness anomaly* h/H_0 *(right axis); right panel: relative vorticity* ζ/f. *Zones of different vorticity: core, annulus, and outer region with weak negative vorticity are delimited by vertical dashed lines in the right panel. Widths of transition zones* (d_1, d_2) *and of the annulus* (W_r) *are also indicated.* ©*American Meteorological Society. Used with permissions.*

the choice of local Rossby number $Ro_{\text{loc}} = 105$, of the width of the annulus $W_r = 0.4$ and of the width of transition zones $d_1 = d_2 = 0.15$. The vorticity in the core is $7f^{-1}$. The associated global Rossby number is 32.

11.4.2 Results of the linear stability analysis

The most unstable mode of the idealised tropical cyclone of Figure 11.26 is presented in Figure 11.27. Typically, the range of azimuthal wave numbers of unstable modes is 1 to 6, and there are no unstable modes with distinguishable growth rates out of this range. The growth rate associated with the most unstable mode is $\sigma = 0.24f$, which

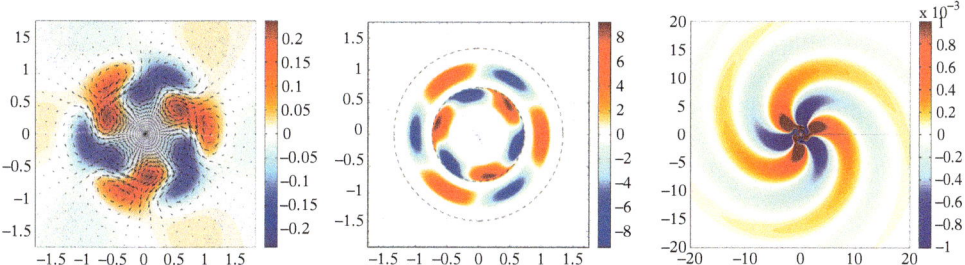

Figure 11.27 *Left: Pressure (colours) and velocity (arrows) associated with the most unstable mode of the idealised hurricane. Middle: Vorticity (colours), with critical radii indicated by dashed lines. Right: Inertia–gravity wave field associated with the instability, as seen in the divergence field (colours). Left and middle panels are zooms at the vortex center.* ©*American Meteorological Society. Used with permissions.*

corresponds to characteristic growth time of about 6 hours, and is in fair agreement with both modelling and observations. The structure of the unstable mode clearly reflects its link to barotropic instability: the velocity of the perturbation follows the isobars, and the patterns of vorticity correspond to two Rossby waves counter-propagating over the two edges of the background vortex annulus. The magnitude of the divergence associated with the unstable mode is two orders smaller than that of vorticity, in agreement with the balanced character of the eigenmode. However, a distinct wave pattern appears away from the balanced part of the unstable mode, and thus the instability is of mixed type and couples a barotropic (Rossby) mode with an outgoing inertia–gravity wave. It should be emphasised that the parameters of the flow—high Rossby number and Froude number below unity—match the conditions of Lighthill radiation in Section 6.4.5.

As already stated, the radiative barotropic instability described here is due to the mutual interaction of Rossby waves counter-propagating at each edge of the annulus of vorticity. The steepness of the vortex profile, and the associated magnitude of the radial gradient of potential vorticity, condition the celerities of the propagating Rossby waves, which in turn determine the efficiency of the phase-locking and growth. As follows from the structure of the linearised problem, the impact of the global Rossby number is weak if its value is large. Results of the linear stability analysis, indeed, confirm that the value of the growth rate of the most unstable mode is insensitive to the global Rossby number (and, hence, the dimensional growth rate scales as the Rossby number, under the scaling we use). In contrast, the growth rates of unstable modes exhibit strong sensitivity to the edge steepness parameters $d_{1,2}$. In particular, a wave number selection mechanism is operational at small values of the edge width (i.e. strong PV gradients) where all wave numbers but one become stable. On the other hand, at weak PV gradients, several wave numbers (from $l = 3$ to $l = 5$, with the vortex parameters used here) are unstable with close growth rates, and may thus compete, if the vortex is subject to a broad-spectrum perturbation. This fact may account for the apparently irregular behaviour of the developing instability in observed hurricanes: sometimes a well-organised inner structure emerges, and sometimes the flow appears to be disorganised. Generally, growth rates decrease with increasing width of the annulus of vorticity. As follows from the structure of the unstable mode in Figure 11.27, this is associated with the diminishing mutual amplification of counter-propagating Rossby waves, as they have weaker spatial overlapping due to a wider annulus in between. This is partially countered by the increasing global Rossby number of the vortex which enters the scaling of the growth rate, so that dimensional growth rate may reach a maximum at some value of the annulus width, before vanishing for large annuli.

11.4.3 Nonlinear saturation of the hurricane's instability

We present here nonlinear saturation of the most unstable mode with wave number $l = 3$, superimposed with small amplitude onto the background vortex, as follows from numerical simulations with well-balanced finite-volume code. The evolution of the relative vorticity is shown in Figure 11.28. The initial perturbation of the vorticity amplifies and gives rise to a pronounced triangular structure with well-defined maxima of vorticity.

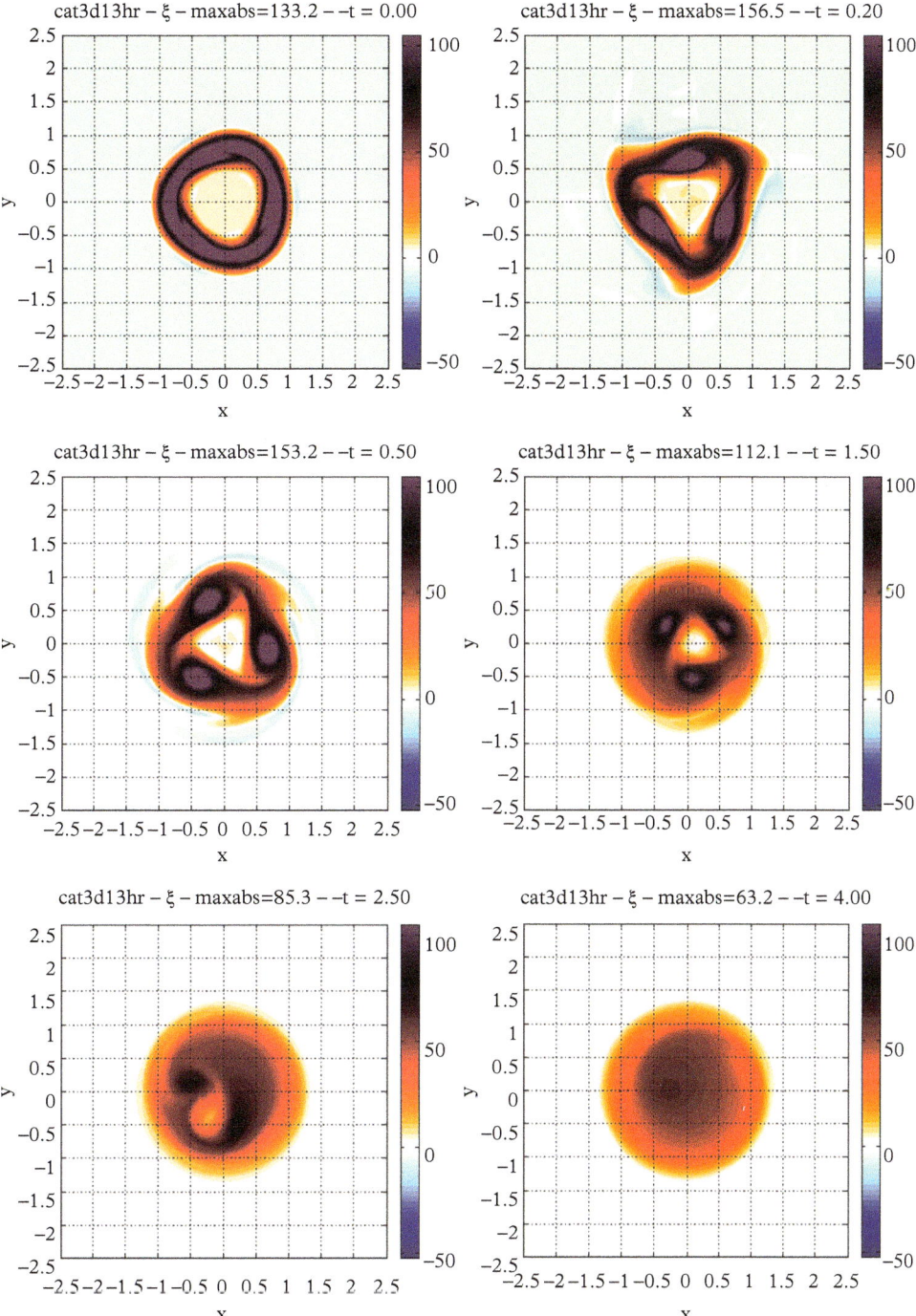

Figure 11.28 *Nonlinear saturation of the unstable mode with azimuthal wave number $l = 3$ of idealised hurricane as seen in evolution of relative vorticity. Snapshots at times $t = 0, 0.2, 0.5, 1.5, 2.5$, and $4f^{-1}$, from left to right, and from top to bottom.* ©*American Meteorological Society. Used with permissions.*

Figure 11.29 *Satellite view of the hurricane Isabel and its meso-vortices. Courtesy NASA.*

Such maxima are called meso-vortices, and are frequently observed in the hurricanes, cf. Figure 11.29. Their number varies in observations, which can be explained by the sensitivity of the azimuthal structure of the most unstable mode to fine details of the hurricane structure discussed earlier. Further evolution of these high-vorticity regions consists of filamentation of vorticity at the outer edge, and merging of vorticity maxima, which in turn leads to the formation of a monopolar, quasi-axisymmetric vortex structure.

Evolution of azimuthally averaged profiles of vorticity, relative vorticity, and pressure are given in Figure 11.30. We see that the vorticity at late stages is monotonically decreasing with radius (same for the PV), so that the new vorticity profile is manifestly marginally stable with respect to the barotropic instability. There are no other types of instability for such a profile (the radiative Rossby-gravity instability is negligible, as follows from the linear stability analysis), and the final monopolar structure is stable. Associated with the new vortex profile, the depression inside the core is amplified in

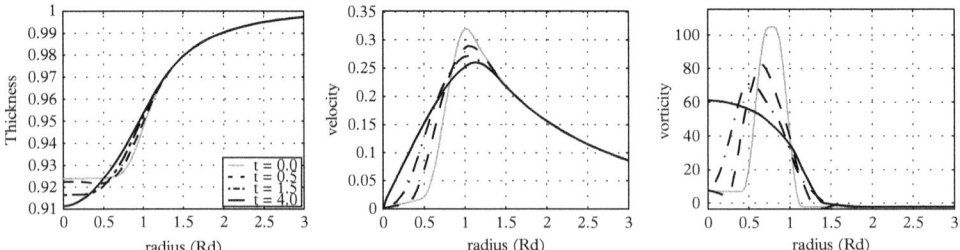

Figure 11.30 *Nonlinear saturation of the hurricane's instability with azimuthal wave number $l = 3$: azimuthally averaged profiles of non-dimensional: thickness h/H_0 (left panel), angular velocity $v/\sqrt{gH_0}$ (middle panel), relative vorticity ζ/f (right panel) at times $t = 0$ (solid grey), 0.5 (black dashed), 1.5 (black dash-dotted), and 4 (solid black).* ©*American Meteorological Society. Used with permissions.*

course of the evolution of the instability, as well as the magnitude of the angular velocity. It should be stressed that appearance of secondary vortices with subsequent symmetrisation of the vortex and intensification of vorticity, with a related fall of pressure in the core, are indeed observed during the life cycle of tropical cyclones. We thus see that the simplest one-layer RSW model permits the understanding of these features from the point of view of developing instabilities of the hurricane-like vortices. We will come back to the evolution of instabilities of such vortices in Chapter 15, after having added simplified thermodynamics of the moist air in the model.

11.4.4   A brief summary of the results on instabilities of idealised hurricanes

We have thus seen that idealised high Rossby number hurricane-type vortices, with a core of constant vorticity surrounded by a high-vorticity annulus, develop specific instabilities of mixed barotropic–radiative type, with a typical structure of the barotropic instability inside the vortex, and inertia–gravity wave field in the outer zone. The azimuthal structure of the dominant unstable mode exhibits high sensitivity to fine details of the vorticity profile. Saturation of the instability leads, first, to appearance of distinct maxima of vorticity in the annulus, in a number determined by the most unstable azimuthal wave number, and then to axisymmetrisation of the vorticity, with its substantial increase in the core, accompanied by a pressure drop.

11.5   Instabilities of laboratory flows in rotating annuli

Annular geometry is the most appropriate and is frequently used in experiments with stratified fluids in rotating tanks. Layer-wise stratification is relatively easy to achieve and maintain, for example with different concentrations of salt, for sufficiently long times. There is a long tradition of experiments of this type motivated, in the first place, by problematics of geophysical fluid dynamics. We will present the results of stability analyses in two typical experimental configurations. Both can be considered, approximately, as being two-layer. The upper layer in the first is under a rigid lid which is super-rotating, in order to produce a velocity shear between the layers. The second configuration is free-surface, with an outcropping interface between the layers. The latter configuration is created, for example, by pouring fresh water over the layer of a salty water, and mimics density currents in nature.

11.5.1   Stability of two-layer flows under the rigid lid

Set-up, scaling, and formulation of the linear stability problem

A typical configuration used in laboratory experiments is presented in Figure 11.31. The annulus has an inner vertical sidewall of radius r_1, an outer vertical sidewall of radius r_2, and a total depth $H_1 + H_2$. The width of the annulus is therefore $r_2 - r_1$,

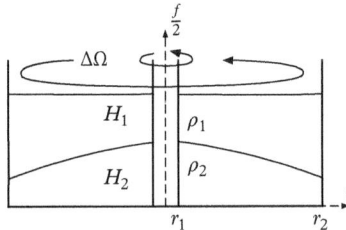

Figure 11.31 *A sketch of a two-layer flow in the annulus with a super-rotating lid.*

and for simplicity we take layers of equal depth H_0 at rest. The base and the lid are both horizontal and flat. The angular velocity about the axis of rotation is $\Omega = f/2$, and the upper lid is super-rotating at $\Omega + \Delta\Omega$. Such differential rotation provides a vertical velocity shear, and a balanced basic state which is close to solid body rotation of each fluid layer, with different angular velocities. In the following, dynamics of the system is approximated with the two-layer RSW model in the f-plane approximation, and this state is represented by velocity and pressure in cyclo-geostrophic equilibrium layer-wise, with linear dependence of the azimuthal velocity on the distance from the axis of rotation. The two-layer RSW equations in polar coordinates adapted to this configuration are

$$\begin{cases} \frac{d_j}{dt}u_j - \left(f + \frac{v_j}{r}\right)v_j - r\Omega^2 = -\partial_r\Pi_j, \\[2mm] \frac{d_j}{dt}v_j + \left(f + \frac{v_j}{r}\right)u_i = -\frac{1}{r}\partial_\theta\Pi_i, \\[2mm] \frac{d_j}{dt}h_j + h_j\nabla\cdot\mathbf{v}_j = 0, \end{cases} \tag{11.13}$$

where $\mathbf{v}_j = (u_j, v_j)$, h_j and Π_j are the velocity (radial and azimuthal), thickness, and pressure normalised by density (geopotential), in the jth layer counted from the top, $j = 1, 2$. D_j denote Lagrangian derivatives in respective layers. Note that, compared to previous examples, we keep the centrifugal acceleration here. The boundary conditions are free-slip: $u = 0$ at $r = r_1, r_2$, their justification will be given below. We choose $1/f$ as time scale, $r_0 = r_2 - r_1$ as horizontal scale, H_0 as vertical scale, and $V_0 = fr_0$ as velocity scale, and use non-dimensional variables without changing notation. Linearisation about a steady state with time-independent azimuthal velocities $V_1 \neq V_2$ gives the following problem, which is cylindrical-geometry version of the ageostrophic Phillips model, considered in Section 10.2:

$$\begin{cases} \partial_t u_j + \frac{V_j}{r}\partial_\theta u_j - v_j - 2\frac{V_j v_j}{r} = -\partial_r\pi_j, \\[2mm] \partial_t v_j + u_j\partial_r V_j + \frac{V_j}{r}\partial_\theta v_j + u_j + \frac{V_j u_j}{r} = -\frac{1}{r}\partial_\theta\pi_j, \\[2mm] \partial_t h_j + \frac{1}{r}\partial_r(rH_j u_j) + \frac{1}{r}H_j\partial_\theta v_j + \frac{V_j}{r}\partial_\theta h_j = 0. \end{cases} \tag{11.14}$$

Here, the pressure perturbations in the layers π_j are related through the interface perturbation η, as usual:

$$\pi_2 - \pi_1 + S(\pi_2 + \pi_1) = Bu\,\eta, \qquad (11.15)$$

where $S = (\rho_2 - \rho_1)/(\rho_2 + \rho_1)$ is the stratification parameter (we do not apply here the simplifying hypothesis $\rho_1 \sim \rho_2$, which was used in Chapter 5), $Bu = (R_d/r_0)^2$ is the Burger number, $Ro = \Delta\Omega/(2\Omega)$ is the Rossby number, $R_d = (g'H_0)^{\frac{1}{2}}/(2\Omega)$ is the Rossby deformation radius, and $g' = 2\Delta\rho g/(\rho_1 + \rho_2) = 2Sg$ the reduced gravity. The depth profiles $H_j(r)$ and respective velocities $V_j(r)$ in (11.14) correspond to steady cyclo-geostrophically balanced state in each layer, which reads, in non-dimensional terms:

$$V_j + \frac{V_j^2}{r} + \frac{r}{4} = \partial_r \Pi_j. \qquad (11.16)$$

The system (11.13) is dissipationless yet in the experiments the mean axisymmetric flow is controlled by friction. Boundary layers are formed in the real fluid at top and bottom boundaries (Ekman layers), and also at lateral boundaries (Stewartson layers). The Stewartson layers are very thin, and it is reasonable to neglect them and to consider configurations where each layer rotates as a solid body with free-slip boundary conditions. Due to the influence of the Ekman friction, rotation rates of the fluid layers lie in the interval between the rotation rate of the base (0 in the rotating frame) and that of the upper lid ($\sim Ro$ in the rotating frame). Therefore, in general,

$$V_2 = \alpha_2\, r, \quad V_1 = \alpha_1\, r, \qquad (11.17)$$

where the coefficients $\alpha_{1,2}$ depend on viscosities of the layers and the Rossby number. Expressions for the heights of the layers for such mean flow are easy to recover from conditions of cyclo-geostrophic equilibrium and (11.15):

$$H_j = H_j(0) + (-1)^j [\alpha_2 + \alpha_2^2 - \alpha_1 - \alpha_1^2] \frac{r^2}{2Bu}$$

$$+ (-1)^j S[\alpha_2 + \alpha_2^2 + \alpha_1 + \alpha_1^2 + 2] \frac{r^2}{2Bu}. \qquad (11.18)$$

Looking, like in studies of vortex instabilities, for solutions in the form

$$(u_j(r,\theta), v_j(r,\theta), \pi_j(r,\theta)) = (\tilde{u}_j(r), \tilde{v}_j(r), \tilde{\pi}_j(r))\, exp\,[ik(\theta - ct)] + c.c.,$$

where k is a discrete azimuthal wave number, and omitting tildes we get from (11.14) an eigenproblem of order 6 for eigenphase velocities c at given k:

$$\begin{cases} k(V_j - rc)iu_j - (r + 2V_j)v_j + r\pi_j' = 0, \\ -(r + V_j + rV_j')iu_j + k(V_j - rc)v_j + k\pi_j = 0, \\ -(rH_j iu)' + kH_j v + k(V_j - rc)(-1)^j\eta = 0, \\ \pi_2 - \pi_1 + S(\pi_2 + \pi_1) = Bu\,\eta. \end{cases} \tag{11.19}$$

It will be solved, as usual, by the pseudspectral collocation method. It should be emphasised that as the radial extent of the calculational domain is finite, no grid stretching, which was applied in the preceding sections, is necessary in this case. As the model is a cylindrical version of the Phillips model, we expect all instabilities found in Section 10.3 to manifest themselves in the present case as resonances between various annular modes found in Section 4.5. This is indeed the case, and we will display the instabilities resulting from the resonance between Rossby waves in upper and lower layers (the baroclinic instability), the resonance between Rossby and Kelvin or Poincaré waves in respective layers (Rossby–Kelvin instability), and the resonances between two Poincaré, or Kelvin and Poincaré, or two Kelvin modes (Kelvin–Helmholtz type instability). Although each instability occupies its proper domain in the parameter space, we will see that there exist cross-over regions, where two different instabilities coexist and may compete.

Results of the linear stability analysis

The overall stability diagram in the space of parameters of the model is obtained by calculating the eigenmodes and the eigenvalues of the problem (11.17) and (11.19) for about 50000 points in the space of parameters (there are typically 200 to 300 points along each axis in the following figures) and then interpolating. Only discrete azimuthal wave numbers correspond to realisable modes in annular geometry. Yet, for better visualisation, we present the results as if the spectrum was continuous. The stability diagrams are presented in left and right panels of Figure 11.32 displaying, respectively, the growth rates and the wave numbers of most unstable modes as functions of Rossby and Burger numbers. We also show in Figure 11.33 how the dispersion diagrams evolve while changing parameters and approaching the instability band spreading from lower-left to upper-right in Figure 11.32. The figure clearly shows how the initially stable flow without imaginary eigenvalues of c develops instabilities of various kinds as parameters change. As shown in the left column of Figure 11.33, a decrease in Burger number leads to distortion of the dispersion curves of Rossby modes and their reconnection leading to Rossby–Rossby (RR)-resonance, that is the baroclinic instability. Distortion of another type takes place if Ro increases at constant Bu, leading to reconnection of dispersion curves of Rossby modes a) with a Kelvin-mode curve resulting in Rossby–Kelvin (RK) resonance and corresponding instability, and b) with a Poincaré-mode curve resulting in Rossby–Poincaré (RP) resonance and corresponding instability. Further increase in Ro leads to reconnection of Kelvin-mode curves and Kelvin–Kelvin (KK) resonance and related shear instability.

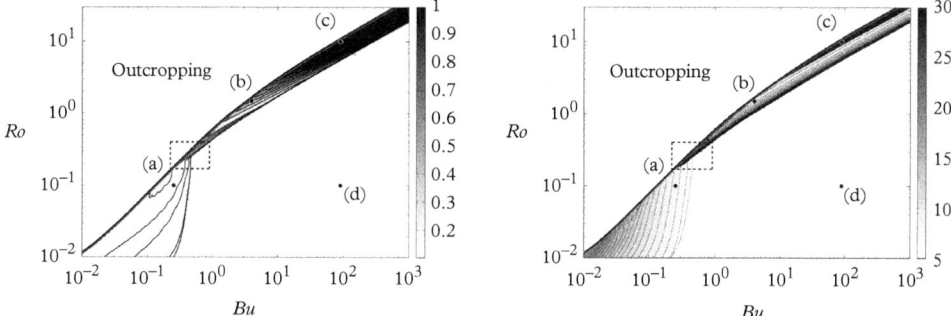

Figure 11.32 *Left panel: growth rates of the most unstable modes as functions of Ro and Bu. Darker zones correspond to higher growth rates. Contours displayed are 0.001, 0.01, 0.02 etc. The thick upper frontier line marks the incropping/outcropping limit when the interface between the two layers intersects the bottom or the top. Right panel: wave numbers of the most unstable modes as functions of Ro and Bu. Darker zones correspond to higher wave numbers. The interval between subsequent contours is 1. Dashed box indicates a cross-over region where baroclinic and RK instabilities coexist.*

Three essential parameters control the wave resonances in the problem: $V = \Delta\Omega r_0$, the velocity shear of the basic flow, $C_R = \Omega\Delta H/H_0 r_0$, the phase velocity of the Rossby waves, and $C_G = \sqrt{g'H_0}$, the phase velocity of the gravity waves. The interpretation of the previous results may be as well done using an alternative set of non-dimensional parameters: $F^* = \frac{V}{C_G} = \frac{\Delta\Omega r_0}{\sqrt{g'H_0}}$, a new Froude number, and $R^* = \frac{V}{C_R} = \frac{g'H}{2\Omega^2 r_0^2}$, a new Rossby number, as is often done in the literature. With these definitions one finds the baroclinic instability at small R^* and Kelvin–Helmholtz instabilities at large F^*, which matches the traditional view of these instabilities.

Thus, the same instabilities as in the ageostrophic Phillips model in a straight channel, cf Section 10.3, are present in the constant-shear flows in annular channels, namely, (a) the baroclinic instability at small values of *Bu* and *Ro* (RR resonance), (b) the Rossby–Kelvin instability (RK or RP resonance) at intermediate values of *Bu* and *Ro*, and (c) the Kelvin–Helmholtz instability (KK or KP resonance), at high values of *Bu* and *Ro*. As usual, the KH instability has highest growth rates and shortest wavelengths, the baroclinic instability is long-wave and low growth rate, and RK instability is intermediate, although spanning a wide range of wave numbers.

Figure 11.34 presents stability diagrams corresponding to different values of *Ro* and *Bu* referring to typical cases (a), (b), (c), respectively, in Figure 11.32. The structure of unstable modes in both layers is displayed in Figure 11.35, together with topographic maps of related interface deviation, which is usually measured in experiments. The upper panel of Figure 11.34 shows a dispersion diagram in the zone of baroclinic instability. Two Rossby waves, each propagating in its layer, are in resonance having the same Doppler shifted phase velocity and giving rise to a baroclinic instability. The middle panel of Figure 11.34 shows a dispersion diagram in a pure Rossby–Kelvin instability zone

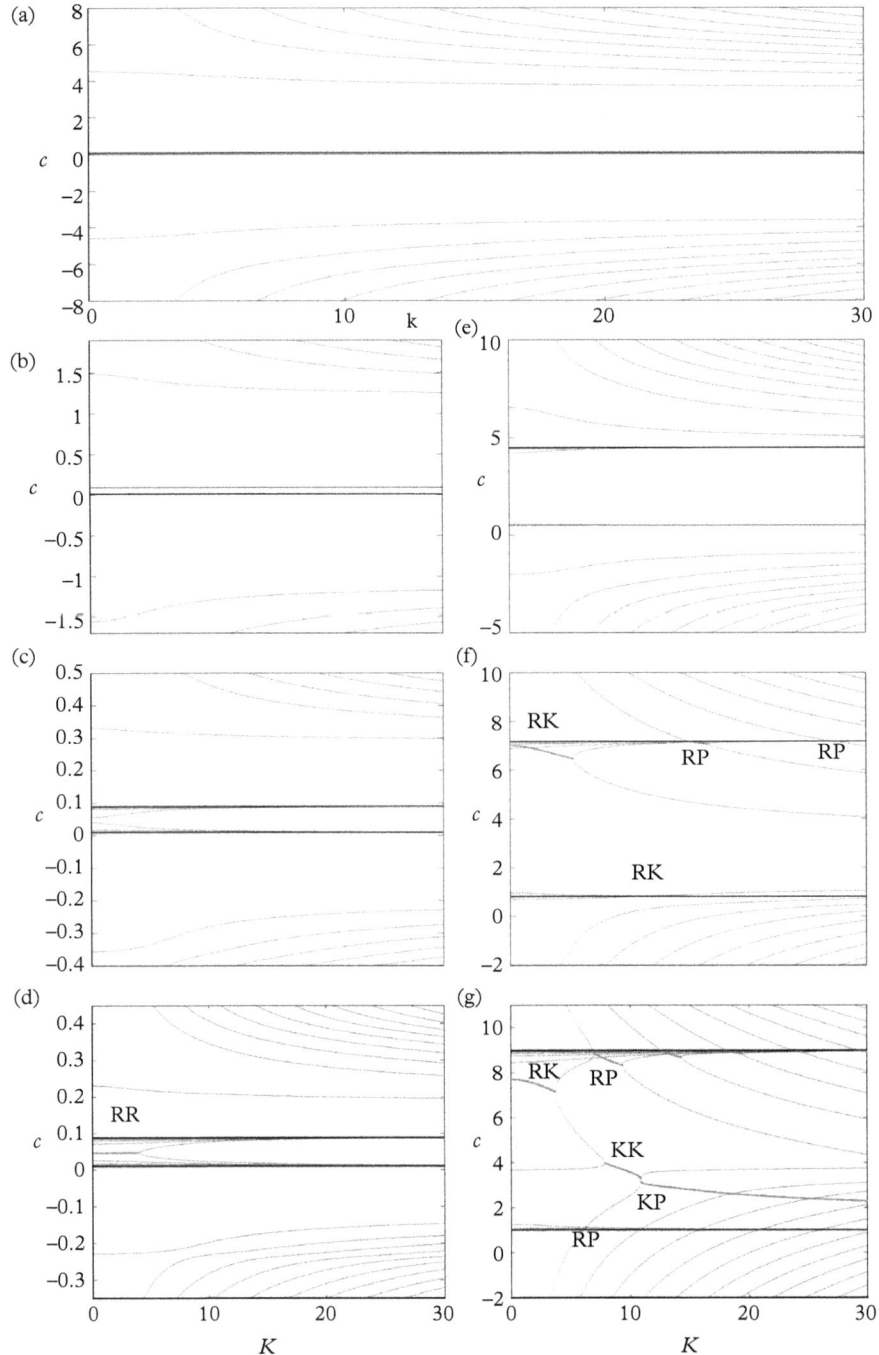

Figure 11.33 *Dispersion diagram in the space c − k, and its evolution with the change of parameters. Upper panel: stable configuration corresponding to point (d) in Fig.* 11.32 *Left column: Ro = 0.1 and Bu decreasing from top to bottom: (a) Bu = 90, (b) Bu = 10, (c) Bu = 0.5 and (d) Bu = 0.25. Right column: Bu = 90 and Ro increasing from top to bottom : (a) Ro = 0.1, (e) Ro = 5, (f) Ro = 8, and (g) Ro = 10. Thick grey lines indicate unstable regions with non-zero Im(c).*

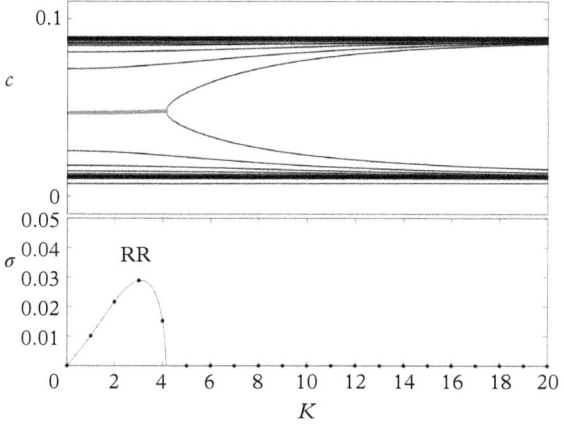

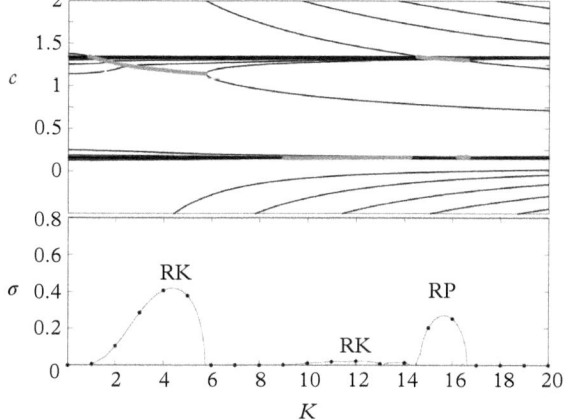

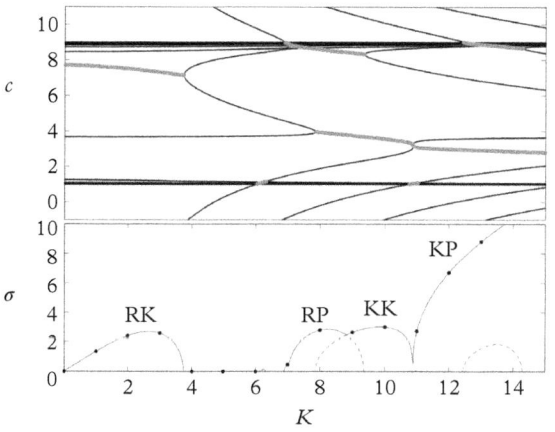

Figure 11.34 *Phase velocity (upper panels) and growth rates (lower panels) of the eigenmodes of the super-rotating lid configuration with Ro = 0.1 and Bu = 0.25, upper frame, Ro = 1.5 and Bu = 4, middle frame, Ro = 10 and Bu = 90, lower frame (cf. Fig. 11.32), respectively, from top to bottom. Thick grey lines in the upper panels of the frames indicate resonances between different modes and respective unstable modes.*

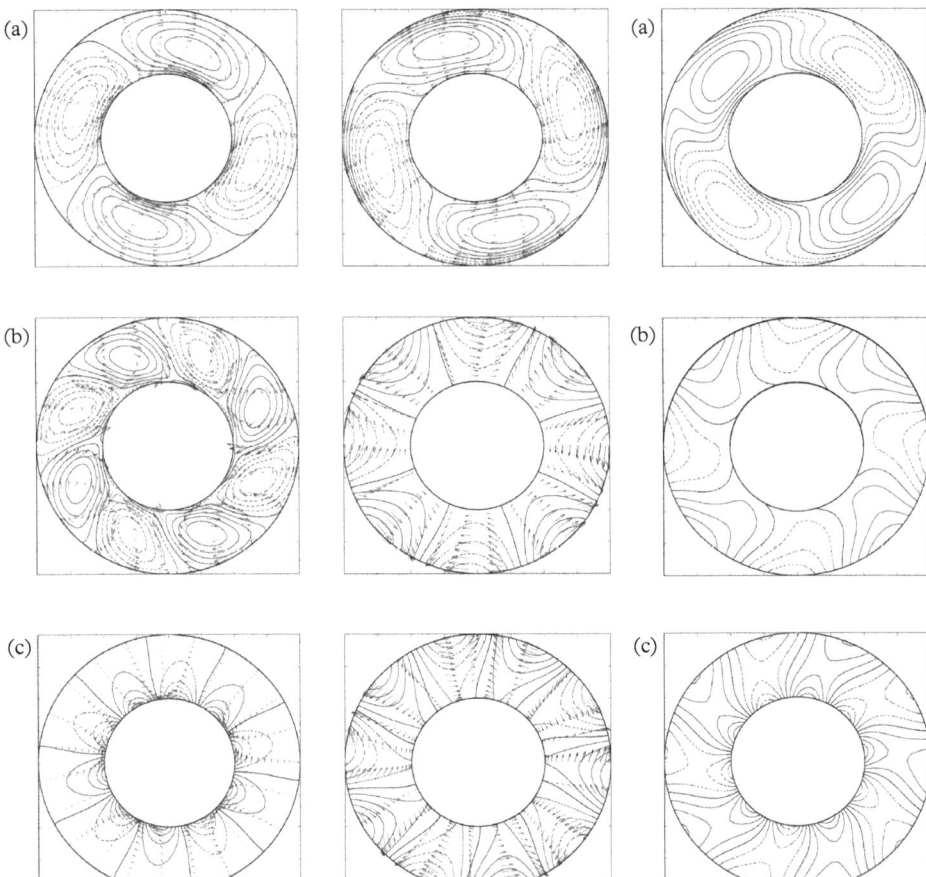

Figure 11.35 *Pressure and velocity fields in the upper (left column) and lower (middle column) layers, and interface height (right column) of (a) baroclinically unstable mode with k = 2, (b) Rossby–Kelvin mode with k = 4, and (c) Kelvin–Helmholtz mode with k = 10 (cf. Fig. 11.34). Solid (dashed) lines: positive (negative) values.*

of parameters. A Rossby wave propagating in the upper layer resonates with a Kelvin wave propagating in the lower layer, which results in RK instability. Rossby wave also resonates with Poincaré and upper-layer Kelvin waves.

The lower panel of Figure 11.34 shows a dispersion diagram in the zone of parameters of KH instability. A Kelvin wave propagating in the upper layer resonates with another Kelvin wave propagating in the lower layer, or with a Poincaré wave, which gives KH instability. For these values of parameters RP and RK instabilities are also present, albeit with lower growth rates. Thus, RK and KH instabilities coexist for large *Bu* and *Ro* having comparable growth rates, although different characteristic wave numbers. As follows from the last figure, and from Figure 11.32, close values of the growth rates may correspond to essentially different wavelengths of the most unstable modes. This means that different instabilities may coexist and compete.

11.5.2   Stability of flows in rotating annulus with outcropping and topography

Set-up, scaling, and boundary conditions

We now consider the outcropping situation, where the interface between the layers joins the free surface forming a surface front, as shown in Figure 11.36, where a nontrivial bottom topography, of a type easily realisable in experiments, is also added. The figure represents an idealised buoyancy-driven coastal current in an annular basin. In experiments a lighter fluid flows above a denser fluid, and is confined between the surface front and the internal, or external cylinder. We consider an upper layer of lighter fluid of density ρ_1 with a free surface terminating at a point $r = r_0 = r_1 + L$, and mean velocity $U_1(r)$, and a lower layer of density $\rho_2 > \rho_1$ with a mean velocity $U_2(r)$. We adapt the shallow-water equations with outcropping of Section 10.7, and perform stability analysis along the same lines. A difference with the previous subsection is that we now consider a free surface instead of a rigid lid.

We use the time scale $1/f$, the horizontal scale L, the vertical scale $H_0 = H_1(r_1)$, and the velocity scale fL, and pass to non-dimensional variables without changing notation, as before. As in Section 10.7, under this scaling the characteristic value of the mean velocity gives the Rossby number. By linearising about a steady state in cyclo-geostrophic equilibrium, we obtain non-dimensional equations identical to equations (11.14), where the pressure perturbations in the layers π_j are now expressed in terms of the layer thicknesses h_j via the hydrostatic relations:

$$\nabla \pi_j = \frac{Bu}{2S} \nabla (s^{j-1} h_1 + h_2), \tag{11.20}$$

where $s = \rho_1/\rho_2$ is the density ratio, $S = (\rho_2 - \rho_1)/(\rho_2 + \rho_1)$, as earlier, and $Bu = (R_d/r_0)^2$ the Burger number. The depths $H_j(r)$, and the corresponding velocities $V_j(r)$ in (11.14) are in cyclo-geostrophic balance layer-wise:

$$V_j + \frac{V_j^2}{r} + \frac{r}{4} = \frac{Bu}{2s} \partial_r (s^{j-1} H_1 + H_2). \tag{11.21}$$

We again look for solutions harmonic in the azimuthal direction and time. The boundary condition of no normal flow at the coast is the same as in the preceding subsection for

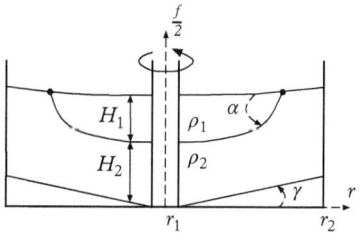

Figure 11.36 *Sketch of a two-layer outcropping flow in the annulus with linearly sloping bottom.*

both layers, $u_j(r_1) = 0$. The boundary conditions at the front in the upper layer, at $r = L_R(\theta)$ are

$$H_1(r) + h_1(r, \theta, t) = 0, \quad \frac{d_1}{dt} L_R = v, \qquad (11.22)$$

where $r = r_1 + L$ is the location of the free streamline in the unperturbed mean state, $L_R(\theta, t)$ is the position of the perturbed free streamline, and $\frac{d_1}{dt}$ is the Lagrangian derivative in layer 1. Physically, these conditions correspond to the fluid terminating at the boundary, which is a material line, cf. (4.53). The linearisation of the boundary conditions gives the relation between the perturbation of the position of the free streamline and the value of the height perturbation, $L_R = -\frac{h_1}{H_{1r}}\Big|_{r=r_1+L}$, and the continuity equation evaluated at $r = r_1 + L$, cf. (4.54). Therefore, the only constraint to be imposed at the front for the upper layer is the regularity of $(u_1, v_1, h_1 + h_2)$, as in Section 10.7.3.

We also have to ensure the continuity of pressure of the lower layer across the front. In the region $r > r_1 + L$ with no upper layer, the lower layer obeys the one-layer rotating shallow-water equations with (hydrostatic) pressure proportional to the height of the fluid column. In what follows, we consider an outer cylinder to be far enough from the front ($r_2 \gg r_0$) so that its influence is negligible. Moreover, we limit ourselves, for technical simplicity, only by the *balanced* component of π_2, which, in the leading order, satisfies the equation, which is a cylindrical version of the condition already used in studies of stability of outcropping coastal currents and double density fronts in Sections 10.7 and 10.8:

$$\frac{1}{r}\partial_r(r\partial_r\pi_2) + \left(k^2 - \frac{1}{R_{d2}^2} - \frac{k^2}{r^2}\right)\pi_2 = 0, \qquad (11.23)$$

where $R_{d2} = \sqrt{gH_2}/f$ is the Rossby deformation radius of the lower layer. We thus impose the continuity of the full solution for π_2 in the inner region $r < r_1 + L$ and of the decaying balanced solution in the outer region $r > r_1 + L$ at $r = r_1 + L$. By this choice, an unbalanced part of the one-layer flow out of the front, consisting of freely propagating surface inertia–gravity waves, is discarded. We thus lose possible resonances of the eigenmodes of the inner flow with outer inertia–gravity wave field and related radiative instabilities. For small to moderate Rossby numbers, which is the case in existing experiments as well as our case below, these instabilities are weak. As we will see later, the stability analysis under these assumptions is in a good agreement with experiments, which gives an *a posteriori* justification of our assumption.

Following the same lines as in the preceding subsection, we obtain an eigenvalue problem of order 6 which is solved by the pseudo-spectral collocation method. As in Section 10.7, only the case of the bottom layer initially at rest ($V_2 = 0$) is considered, and the upper-layer flow is supposed to have a constant rotation rate, that is $V_1 = \alpha r$.

Results of the linear stability analysis in the case of flat bottom

As in Section 10.7, the instabilities in the outcropping case originate from resonances between the wave species, which are Poincaré (inertia–gravity) modes, Rossby modes,

Kelvin modes trapped near the inner boundary, and frontal modes trapped in the vicinity of the outcropping line. Crossings of dispersion curves indicate appearance of corresponding unstable hybrid modes. We show in Figure 11.37 the dispersion diagram, and corresponding growth rates, of the eigenmodes of the outcropping coastal flow with a depth ratio $\delta_H = H_1(r_1)/(H_1(r_1) + H_2(r_1)) = 0.1$ and a density ratio $s = \rho_1/\rho_2 = 0.99$, for increasing values of vertical shear: $Ro = 0.02$, $Ro = 0.2$, and $Ro = 0.6$.

At low Rossby and Burger numbers (Figure 11.37, upper frame), Rossby modes in the lower layer resonate with Rossby modes in the upper layer. This is the standard mechanism of the baroclinic instability, as already explained, which occurs for non-dimensional wave numbers $k.R_d < 1$. The corresponding pressure and velocity fields in both layers are plotted in the panel (a) of Figure 11.38 and are typical of a Rossby mode. Rossby modes in the lower layer can also resonate with the frontal mode in the upper layer (RF interaction). The frontal mode has the characteristics of a Rossby wave for low wave numbers, and the unstable mode under consideration is therefore very similar to the classical baroclinic instability. The corresponding pressure and velocity fields in both layers are visualised in the panel (b) of Figure 11.38.

At higher Rossby and Burger numbers (Figure 11.37, middle and lower frames), the Rossby–Rossby interaction is not permitted, as the horizontal extension of the surface current is too small compared to the Rossby deformation radius. The RF mode, thus, is the leading unstable mode with wave numbers $k \cdot R_d \approx 0.5 \div 1$. The second instability in the middle panel of Figure 11.37 corresponds to the resonance of the first Poincaré mode in the upper layer with a Rossby wave in the lower layer. The pressure and velocity fields for this mode can be visualised in the panel (c) of Figure 11.38, and confirm this interpretation. Note that the same instability appears at higher k for Poincaré modes of higher order with decreasing growth rates (Figure 11.37, lower frame).

A new dispersion curve, $kc = 1$ denoted by I, appears in the middle and lower frames of Figure 11.37. It corresponds to inertial motion in the lower layer, with the quiescent upper layer, as in the case of coastal currents of Section 10.7. In spite of intersections of this curve with other branches of the dispersion diagram, no resonances and hence no instabilities between the inertial motion and other modes arise, because of its pressure-less character. It should be emphasised that resonances involving the Kelvin mode at the inner boundary, in contradistinction with the results in Section 11.4 and of Section 10.7, play no significant role in the present configuration. The reason is that the vertical shear at the inner cylinder is small, compared to the shear at the location of the front.

Figure 11.39 shows the dispersion diagram and corresponding growth rates for a larger depth ratio $\delta_H = 0.5$ and a density ratio $s = \rho_1/\rho_2 = 0.99$. Comparison of Figures 11.37 and 11.39 shows that the unstable modes become more and more vigorous when the depth ratio increases. At high Rossby numbers, a new zone of instabilities with high growth rates appears at very high wave numbers. They are due to the interaction of the frontal mode in the upper layer with various Poincaré modes in the lower layer (FP_1, Figure 11.39, lower frame). This short-wave instability is analogous to the one described in Section 10.7 for the constant-PV case in the planar geometry. The frontal mode having the characteristics of a gravity wave for high wave numbers, this instability is therefore very similar to the vertical shear instabilities, which we have seen previously (KK, KP, or

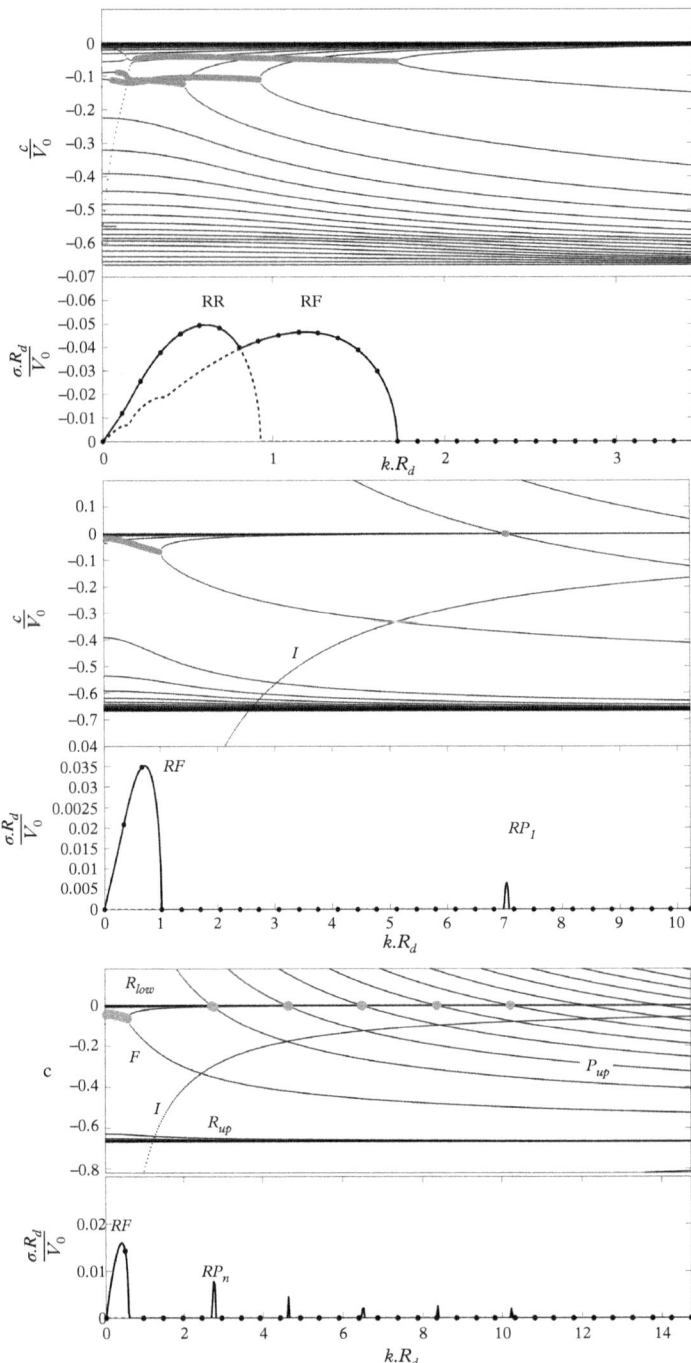

Figure 11.37 *Phase velocity (top panels) and growth rate (bottom panels) of the eigenmodes of the outcropping configuration with $\delta_H = 0.1$. Frames, from top to bottom: $Ro = 0.02$, $Ro = 0.2$, $Ro = 0.6$. Thick grey lines on the upper panels correspond to unstable modes. The phase velocity c is scaled with velocity at the front $V_1(r_0)$ and the wave number is scaled with inverse deformation radius R_d.*

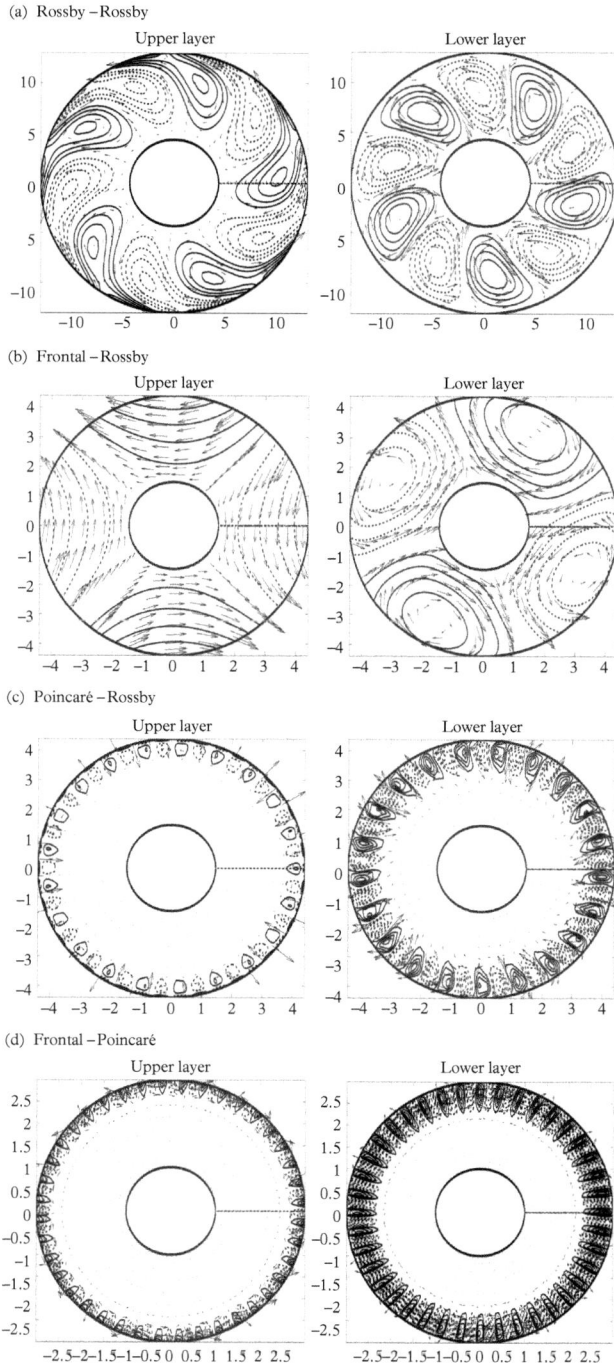

Figure 11.38 *Pressure (contours) and velocity (arrows) of the upper (left panel) and lower layer (right panel), of (a) the unstable Rossby–Rossby (RR) mode for Ro = 0.02 and k = 5/Rd, (b) the unstable Rossby–Frontal (RF) mode for Ro = 0.2 and k = 2/Rd, (c) the unstable Rossby–Poincaré mode at k = 21/Rd, and (d) the unstable Frontal–Poincaré mode at k = 40/Rd. Solid (dashed) lines: positive (negative) anomaly.*

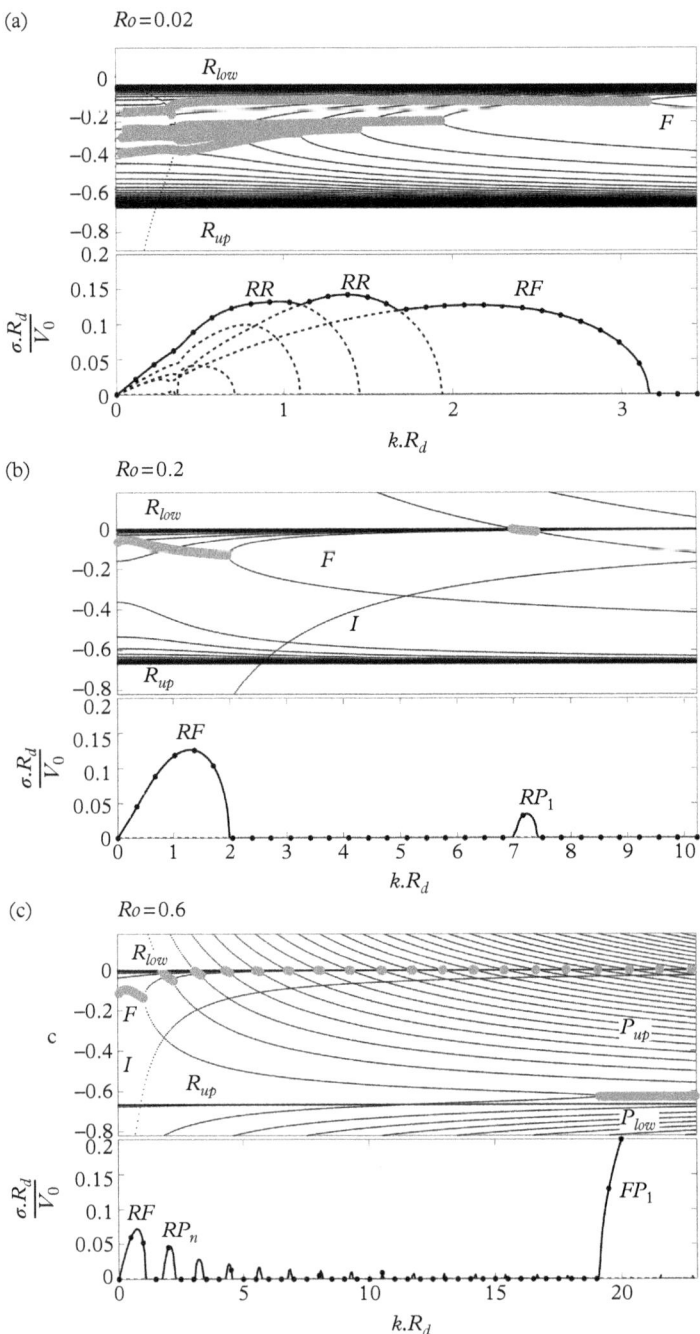

Figure 11.39 *Same as in Figure* 11.37, *but with* $\delta_H = 0.5$.

PP). The pressure and velocity fields for the first Poincaré mode in the lower layer and the frontal mode in the upper layer are plotted in the panel (*d*) in Figure 11.38.

Barotropic interactions in the upper layer, which were seen in Section 10.7 in the rectilinear case, do not manifest themselves in the present analysis due to the absence of either horizontal PV gradient, or of the current reversal in the upper layer, which would allow either Rossby–frontal or Kelvin–frontal barotropic interactions. These interactions are usually absent in experimental studies, too.

Comparison with laboratory experiments

Classical experiments use a circular tank mounted on a rotating turntable, as in Figure 11.36, filled with a solution of density ρ_2. The boundary current is created by injecting a lighter solution of density ρ_1 between the inner cylinder and a bottomless cylinder of radius r_c, such as $r_1 < r_c < r_2$. This cylinder is then drawn upward, allowing the upper layer with height h_0 and width $L_0 = r_c - r_1$ to move under the influence of buoyancy, Coriolis and centrifugal forces, and thus adjusting to a balanced state. The upper layer is initially stationary, and therefore has constant PV. If mixing during the adjustment can be neglected, PV in the upper layer Q_1 is conserved, and is the same in the final balanced state, as in the initial state. The PV conservation is then written as

$$Q_1 = \frac{f + \partial_r V_1 + \frac{V_1}{r}}{H_1} = \frac{f}{h_0},$$ (11.24)

where we restored dimensions. Steady states in cyclo-geostrophic balance in each layer are given by (11.21). If the bottom layer is initially at rest in the rotating frame, the mean velocity in the upper layer resulting from the above-described adjustment process is given by the solution of the following ordinary differential equation:

$$\partial_{rr}^2 V_1 + \frac{\partial_r V_1}{r} - \frac{V_1}{r^2} - \frac{f}{gh_0(1-S)}(fV_1 + \frac{V_1^2}{r}) = 0,$$ (11.25)

satisfying the boundary conditions $V_{1r}(r_1) = 0$, $H_1(r_1 + L) = 0$. Equation (11.25) can be easily solved numerically. The result of this calculation for a typical experimental sets of parameters, Froude number $F_0 = \frac{f^2 L_0^2}{g' h_0}$ and initial depth ratio $\delta_0 = \frac{h_0}{H}$, is shown in Figure 11.40. The structure of the most unstable mode of the adjusted state of Figure 11.40 is displayed in Figure 11.41. The most unstable mode is a hybrid RF wave. As was already discussed in the Subsection 11.5.2, the frontal mode has the characteristics of a Rossby wave at low wave numbers, and therefore the RF mode in the present case is very similar to the classical baroclinic instability, although the uniform PV in the upper layer does not have inversion of the PV gradient between the layers. Comparison of the interface deviation corresponding to the upper row of Figure 11.41 with experimental results shows a good agreement.

The influence of bottom topography

Coastal currents normally evolve over a shelf. The influence of bottom topography (bathymetry) on the instabilities of density currents is, therefore, important, and is being

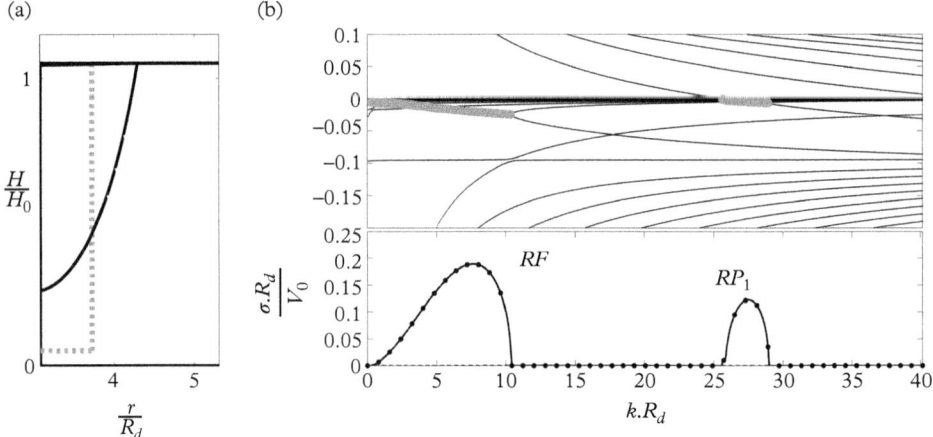

Figure 11.40 *(a) Initial state (dashed) and the corresponding state of cyclo-geostrophic equilibrium (solid) for an experimental set-up with $F_0 = 1.4$ and $\delta_0 = 0.95$, (b) Phase velocity (upper panel), and growth rate (lower panel) of the eigenmodes of perturbations of the balanced state. Thick grey lines in the upper panel correspond to the RF and RP resonances and respective unstable modes.*

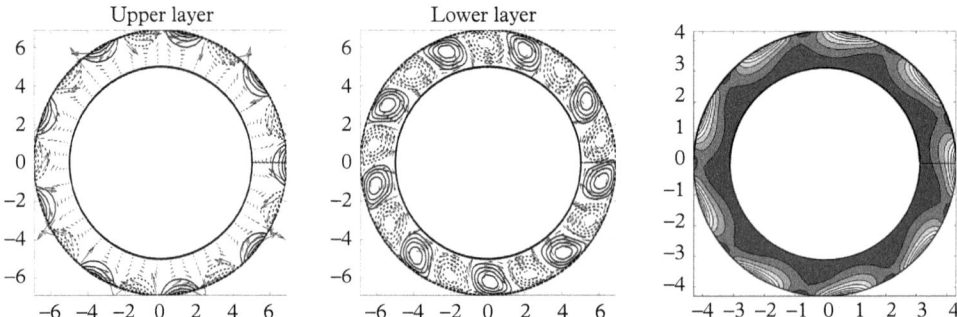

Figure 11.41 *Pressure and velocity fields of the upper (left panel) and lower (middle panel) layers of the baroclinically unstable mode at $k = 9$ of Fig. 11.40. Solid (dashed) lines: positive (negative) anomaly. Interface height perturbation of the most unstable mode with $k = 9$ of Fig. 11.40 (right panel).*

studied experimentally. We add bathymetry with a constant-slope shelf, cf. Figure 11.36, and perform stability analysis along the same lines as earlier. We use the same set of equations and parameters as in the previous subsection with the addition of a bottom topography with height $H_t(r)$ in the mass conservation equation for the lower layer. We define a topography parameter *To* as the ratio of the slope of the shelf γ to the slope of interface α,

$$To = \frac{\gamma}{\alpha} \tag{11.26}$$

where α is measured at the location of the front. Positive values of To correspond to isopycnal and shelf slopes in the same direction, as in Figure 11.36, and are typical for upwelling events along the coast of western boundary currents in the ocean. Negative values of To correspond to isopycnal and shelf slopes in the opposite directions and are typical for buoyant coastal currents. In order to study the influence of the topography on the stability of the current and to vary the parameter To without changing other parameters, we will define the aspect ratio between the two layers as $d = H_1(r_1)/H_2(r_1 + L)$ and keep it constant while varying the height of topography.

As we know from Chapters 4 and 10, the novelty in configurations with nontrivial topography is that the latter allows for topographic waves, in addition to frontal, Kelvin, Rossby, and Poincaré waves discussed earlier. Hence, as in Section 10.8, new resonances, and new instabilities are expected. On the other hand, topography changes the propagation speed of previously considered waves and may, thus, 'de-tune' the resonances, leading to stabilisation of the flow. We will show that both effects take place, depending on To.

Figure 11.42 shows the growth rates of the most unstable modes as a function of To for the set of parameters used in the dispersion diagrams of Figures 11.37 and 11.39, (c). In all cases a strong stabilisation of the flow is observed for positive To, with growth rates vanishing for To close to 1 (topography almost parallel to the interface). To explain the attenuation of instabilities, it is sufficient to look at the expressions for the PV gradient in the lower layer at rest:

$$\partial_r Q_2 = -f \frac{\partial_r H_2}{H_2^2}, \tag{11.27}$$

and for the topography parameter

$$To = \left. \frac{\partial_r H_t}{\partial_r H_2 + \partial_r H_t} \right|_{r_1 + L}. \tag{11.28}$$

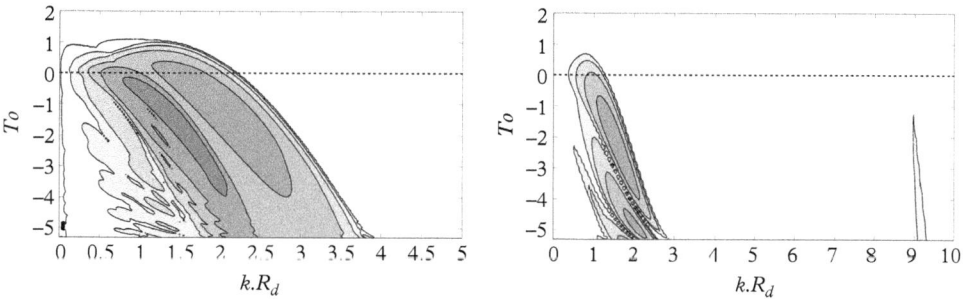

Figure 11.42 *Growth rate (levels of grey) of the most unstable modes as a function of the topography parameter To for: Ro = 0.02 (left panel) and Ro = 0.2 (right panel); d = 0.1.*

As is clear from these expressions, $To \to 1$ implies $\partial_r Q_2 \to 0$ and, hence, the disappearance of Rossby waves and related unstable hybrid modes.

Figure 11.43 displays the stability diagrams at two different negative values of To and $Ro = 0.2$, showing that the most unstable modes in such configurations are due to the resonances of a frontal wave with a topographic wave (either the first T_1 or the second T_2 radial mode), although the resonances of the first topographic mode with a Rossby wave are also observed. Thus, topography plays a crucial role in the destabilisation of the flow in this regime. Figure 11.44 displays the corresponding unstable modes.

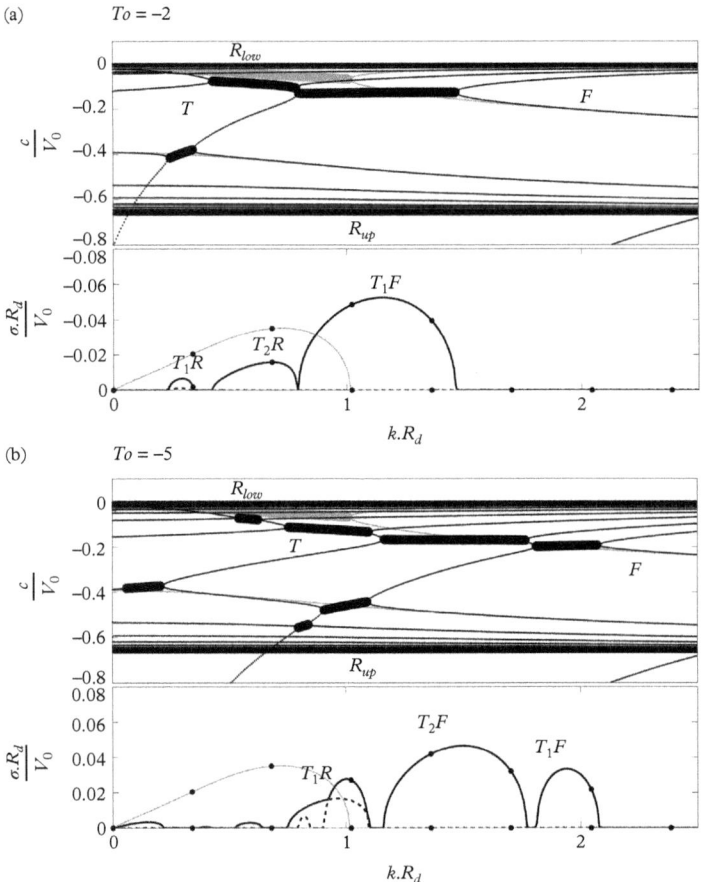

Figure 11.43 *Phase velocity (upper panel) and growth rate (lower panel) of the eigenmodes for: (a) Ro = 0.2 and To = −2 and (b) To = −5. Grey lines in the lower panels correspond to the case with no topography (cf. Fig. 11.37, middle frame).*

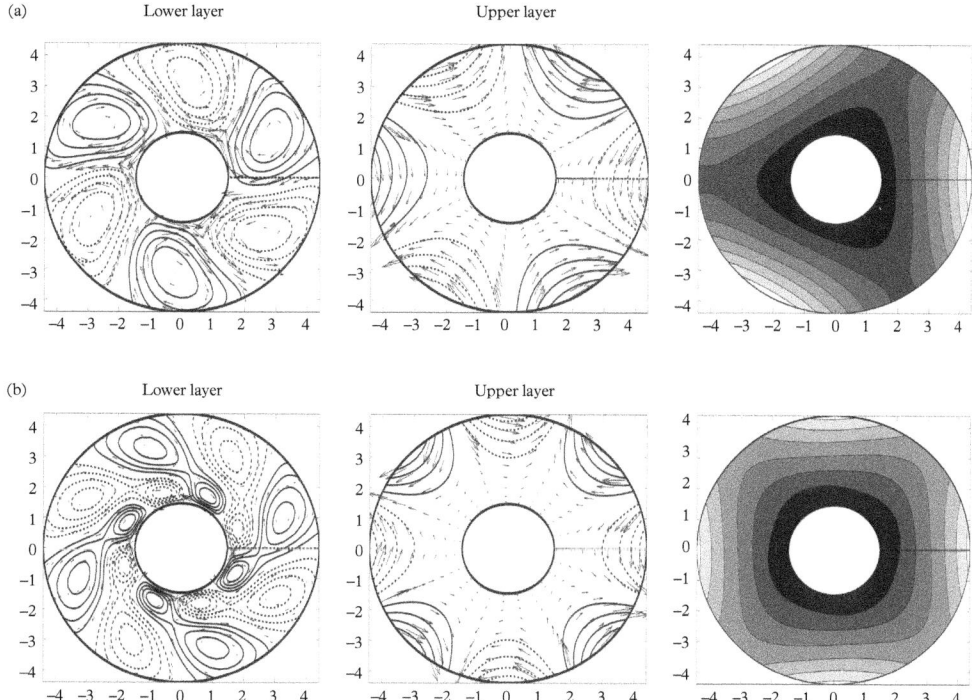

Figure 11.44 *Pressure and velocity fields of the upper (left column) and lower layer (middle column), and interface height (right column) of the most unstable mode for Ro = 0.2 and (a) To = −2 (cf. Fig. 11.43, (a)), and (b) for To = −5 (cf. Fig. 11.43, (b)). Solid (dashed) lines: positive (negative) anomaly.*

11.5.3 A brief summary of the results of analysis of instabilities in the rotating annulus

Thus, we have a detailed characterisation of the instabilities of two-layer shear flows in the rotating annulus under the rigid lid, which can be used for interpreting a class of laboratory experiments. We have seen that aside from the classical baroclinic instability, the ageostrophic instabilities, in particular the Rossby–Kelvin one, are realisable already at moderate Rossby numbers. We tracked different instabilities successively emerging with changes of the key parameters of the experiment. In the context of another class of experiments with density currents, we showed that stability analysis of the flow resulting from the initial adjustment of a fluid released from a container correctly describes the flow structure observed in experiments. We have also shown how the presence of bottom topography modifies the results, by stabilising the flows or introducing new instabilities, depending on its slope.

11.6 Summary, comments, and bibliographic remarks

We have seen that finite-energy shielded vortices at small Rossby numbers are subject to classical barotropic and baroclinic instabilities, which can be understood from the point of view of Rossby eigenmodes propagating in opposite directions around the vortex, due to the change of sign of PV gradient. If their Doppler-shifted phase velocities match, they phase-lock and form an unstable mode. The azimuthal structure of the most unstable modes is sensitive to the vortex parameters: stratification, aspect ratio, and, especially, vertical and horizontal shears. The instabilities saturate following two main scenarios, depending on the azimuthal structure of the most unstable modes: dipolar splitting and tripole formation. As small changes of vortex parameters change the structure of the most unstable mode, and hence the scenario of evolution of the perturbed vortex, accurate predictions of the behaviour of the flow are difficult without knowledge of fine details of the vortex structure.

With increasing Rossby number, the character of the barotropic and baroclinic instabilities change. While the characteristics of the eigenmodes of ageostrophic barotropic instability remain close to their low Rossby-number counterparts, the ageostrophic baroclinic instability becomes the centrifugal instability. In addition to the classical rotationally symmetric mode of the latter, unstable modes of similar type, but with non-zero azimuthal wave numbers appear, and can be dominant in some ranges of vortex parameters, although in a large part of parameter space symmetric centrifugal instability prevails. Scenarios of saturation of symmetric and asymmetric centrifugal and ageostrophic barotropic instability are different, and lead to formation of secondary vortex dipoles or tripoles. Symmetric centrifugal instability undergoes spontaneous breaking of rotational symmetry due to the instability of a concentric front that it forms. It is to be stressed that interpretation of the centrifugal instability in terms of trapped modes, as it was the case for the inertial instability in Chapter 10, complements the standard semi-heuristic explanation, and allows for better comprehension of the nature of the instability, as well as understanding of its asymmetric counterpart.

At very high Rossby numbers barotropic hurricane-like vortices exhibit instabilities of mixed type which couple counter-propagating Rossby modes inside the vortex with inertia–gravity wave field outside. The azimuthal wave number of the most unstable modes is, again, highly sensitive to fine details of the vortex structure. Evolution of these instabilities leads, first, to formation of secondary vortices in the vicinity of the hurricane's core, and then to axisymmetrisation and enhancement of vorticity and pressure anomaly in the core.

Instabilities in cylindrical geometry naturally arise in laboratory experiments with fluids in rotating annuli, which are easy to realise. In the set-up with a rigid lid, different instabilities in different regions of parameter space succeedingly emerge with changes of parameters. Their identification helps to interpret experimental results. In another widespread set-up, which models density (coastal) currents, a simple adjustment theory of constant-PV initial states, and linear stability analysis of the adjusted state, allows us to explain what is observed at initial stages of experiments. Depending on mutual orientation of the slopes of topography and outcropping, a constant-slope shelf attenuates the instabilities, or generate new ones involving topographic waves.

As with jet instabilities in Chapter 10, instabilities of vortices is a vast area, and an exhaustive review of the literature is not possible within the confines of this book. One of the main motivations for studying vortex instabilities in oceanography is enigmatic longevity of oceanic meso-scale and sub-meso-scale vortices (cf. McWilliams 1985 for review). We mention here only some key references, and references on which the presentation was based. More complete literature can be found in the cited papers.

Studies of stability of vortices in the framework of QG multi-layered models are numerous (e.g. Benilov 2004). Studies in the framework of layered shallow-water models can be found in Baey and Carton (2002), and Katsman *et al.* (2003). Material in Section 11.1 is standard, α-Gaussian vortices were used in Kloosterziel, Carnevale, and Orlandi (2007), their presentation here follows Lahaye and Zeitlin (2015). The adaptation of the pseudo-spectral collocation method to cylindrical geometry was done in Lahaye and Zeitlin (2015), following Boyd (1987). Instabilities of shielded vortices of a slightly different shape in two-layer RSW were studied in detail in Baey and Carton (2002), where the dipolar splitting scenario was discussed. The tripolar scenario was exposed in Carton, Flierl, and Polvani (1989). The presentation in Section 11.2 follows Rostami and Zeitlin (2017). Presentation in Section 11.3 is based on Lahaye and Zeitlin (2015). Centrifugal instability is a fluid dynamics classics, the criterion of centrifugal instability is due to Lord Rayleigh (1917). Explanation of the centrifugal instability in terms of trapped modes follows Lahaye and Zeitlin (2015). Presentation of Section 11.4 follows (Lahaye and Zeitlin 2016). For parameterisations of hurricane velocity profiles cf. Willoughby, Darling, and Rahn (2006). For radiative instabilities of vortices cf. Le Dizès and Billant (2009). For the observation of polygonal structures in the core, and their role in intensification of hurricanes cf. for example Menelaou, Yau, and Martinez (2013). Presentation of Section 11.5 follows Gula, Zeitlin, and Plougonven (2009) and Gula and Zeitlin (2015). Laboratory experiments on baroclinic instability in rotating cylindrical tanks have a long history (cf. Hide 1958, Hart 1972). Ageostrophic phenomena in such experiments were observed in Lovegrove, Read, and Richards (2000) and Williams, Haine, and Read (2005). For a discussion of what is captured, and what is not by shallow-water modelling in rigid-lid experiments, see Flor, Scolan, and Gula (2011). Classical experiments on gravity currents referred to in Section 11.5.2 were performed in the pioneering work by Griffiths and Linden (1982), and since repeated, with variations, by many authors. For a laboratory and numerical study of instabilities of density currents in cylindrical geometry which were discussed in Section 11.5.2 cf. Pennel, Stegner, and Beranger (2012). A general discussion of shallow-water modelling in the context of laboratory experiments can be found in Stegner (2007).

Figures 11.1 to 11.9 are reproduced from Rostami and Zeitlin (2017) with permissions. Figures 11.10 to 11.25 are reproduced from Lahaye and Zeitlin (2015) with permissions. Figures 11.26 to 11.26, and 11.30 are reproduced from Lahaye and Zeitlin (2016). Figures 11.31 to 11.35 are reproduced from Gula, Zeitlin, and Plougonven (2009), with permissions. Figures 11.36 to 11.44 are reproduced from Gula and Zeitlin (2015).

12

Resonant Wave Interactions and Resonant Excitation of Wave-guide Modes

Atmospheric and oceanic waves were repeatedly discussed in previous chapters. On almost all occasions they were harmonic waves obtained by linearisation procedure from the equations of motion, although nonlinear phenomena of breaking and solitary wave formation were evoked in Chapters 5, 7, and 8. Even when wave solutions were found to grow in amplitude due to instabilities considered in Chapters 10 and 11, the theoretical analysis stopped at the linear stage, and the evolution of the system was followed by direct numerical simulations. Yet, linear waves with infinitesimal amplitudes are a theoretical abstraction, and waves observed in the ocean and atmosphere always have finite amplitudes. The primitive equations, as well as their rotating shallow-water emanations, are nonlinear, and this nonlinearity leads to interactions and self-interactions of waves. The nonlinearity of the equations of hydrodynamics is typically quadratic. As we have already seen in Chapter 6, this nonlinearity leads to *triadic interactions* among the Fourier modes of hydrodynamic variables. Triadic interactions of waves engender a variety of specific dynamical phenomena. In this chapter, after introducing general ideas of resonant wave interactions, we will consider their applications to resonant excitation of wave-guide modes in geophysical flows.

12.1 Resonant wave triads: first acquaintance

12.1.1 Perturbation theory for Rossby waves

In order to make acquaintance with nonlinear wave interactions, let us consider the technically simplest case where waves appear in a system described by a single scalar equation. A geophysically relevant example of this kind is QG equation on the β plane which, if linearised, has barotropic Rossby-wave solutions. In non-dimensional form the one-layer QG equation on the β plane reads

Geophysical Fluid Dynamics. Vladimir Zeitlin,
Oxford University Press (2018). © Vladimir Zeitlin. DOI: 10.1093/oso/9780198804338.001.0001

$$\nabla^2 \psi_t - \psi_t + \epsilon \mathcal{J}(\psi, \nabla^2 \psi) + \psi_x = 0, \tag{12.1}$$

where we introduced a nonlinearity parameter ϵ, and suppose that nonlinearity is weak $\epsilon \to 0$. Solution of (12.1) will be sought in the form of asymptotic expansion in ϵ, that is as an asymptotic series:

$$\psi = \psi^{(0)} + \epsilon \psi^{(1)} + \epsilon^2 \psi^{(2)} \dots .$$

Obviously, to order zero in ϵ we get the linearised equation

$$\nabla^2 \psi_t^{(0)} - \psi_t^{(0)} + \psi_x^{(0)} = 0,$$

and its solution is a combination of linear Rossby waves with wave numbers **k** and wave frequencies obeying the corresponding dispersion relation, cf. Chapter 5:

$$\psi^{(0)}(\mathbf{x}, t) = \sum_i a_i e^{i(\mathbf{k}_i \cdot \mathbf{x} - \omega(\mathbf{k}_i)t)} + c.c., \quad \omega(\mathbf{k}) = -\frac{k}{\mathbf{k}^2 + 1}, \quad \mathbf{k} = (k, l). \tag{12.2}$$

The wave numbers entering the sum in (12.2) are defined by initial and boundary conditions, which we will not prescribe at the moment.

To the order one in ϵ, for the first nonlinear correction we get

$$\nabla^2 \psi_t^{(1)} - \psi_t^{(1)} + \psi_x^{(1)} = -\mathcal{J}(\psi^{(0)}, \nabla^2 \psi^{(0)}). \tag{12.3}$$

The explicit form of the term on the right-hand side is

$$\sum_{i,j} a_i a_j \left[(k_i l_j - k_j l_i) \mathbf{k}_j^2 \right] e^{i\left[(\mathbf{k}_i + \mathbf{k}_j) \cdot \mathbf{x} - (\omega(\mathbf{k}_i) + \omega(\mathbf{k}_j)) t \right]}$$
$$- \sum_{i,j} a_i a_j^* \left[(k_i l_j - k_j l_i) \mathbf{k}_j^2 \right] e^{i\left[(\mathbf{k}_i - \mathbf{k}_j) \cdot \mathbf{x} - (\omega(\mathbf{k}_i) - \omega(\mathbf{k}_j)) t \right]} + c.c. \tag{12.4}$$

We have already seen examples of inhomogeneous problems of the type (12.3) in Chapter 8, and we know that integrability conditions should be verified by the right-hand side in order to have an everywhere bounded in space and time solution $\psi^{(1)}$. If $\hat{\psi}$ is a solution of the homogeneous equation $\nabla^2 \hat{\psi}_t - \hat{\psi}_t + \hat{\psi}_x = 0$, then

$$\int_{-\infty}^{\infty} dt \int_{-\infty}^{\infty} dx \int_{-\infty}^{\infty} dy \hat{\psi}^* \left(\nabla^2 \psi_t^{(1)} - \psi_t^{(1)} + \psi_x^{(1)} \right) = 0, \tag{12.5}$$

and therefore

$$\int_{-\infty}^{\infty} dt \int_{-\infty}^{\infty} dx \int_{-\infty}^{\infty} dy \hat{\psi}^* \left(\mathcal{J}(\psi^{(0)}, \nabla^2 \psi^{(0)}) \right) = 0, \tag{12.6}$$

which means orthogonality of the right-hand side of (12.3) to the eigenvectors of the zero-order linear operator on the left-hand side. Necessarily, $\hat{\psi} \propto e^{i(\hat{\mathbf{k}} \cdot \mathbf{x} - \omega(\hat{\mathbf{k}})t)}$, with some $\hat{\mathbf{k}}$ and (12.6) becomes

$$\sum_{i,j} a_i a_j \left[\left(k_i l_j - k_j l_i \right) \mathbf{k}_j^2 \right] \cdot \int_{-\infty}^{\infty} dt \, dx \, dy e^{i\left[(\mathbf{k}_i + \mathbf{k}_j - \hat{\mathbf{k}}) \cdot \mathbf{x} - (\omega(\mathbf{k}_i) + \omega(\mathbf{k}_j) - \omega(\hat{\mathbf{k}}))t \right]} -$$
$$\sum_{i,j} a_i a_j^* \left[\left(k_i l_j - k_j l_i \right) \mathbf{k}_j^2 \right] \cdot \int_{-\infty}^{\infty} dt \, dx \, dy e^{i\left[(\mathbf{k}_i - \mathbf{k}_j - \hat{\mathbf{k}}) \cdot \mathbf{x} - (\omega(\mathbf{k}_i) - \omega(\mathbf{k}_j) - \omega(\hat{\mathbf{k}}))t \right]} + c.c. = 0.$$
(12.7)

12.1.2   Wave resonances and wave modulation

Resonant triads

Let us recall that $\int_{-\infty}^{\infty} dx e^{ikx} = \delta(k)$, the Dirac delta function. Therefore, non-zero contributions to the sums over wave numbers on the left-hand side of (12.7) come from the waves with wave-numbers obeying the *resonance conditions*

$$\mathbf{k}_i \pm \mathbf{k}_j = \hat{\mathbf{k}}, \quad \omega(\mathbf{k}_i) \pm \omega(\mathbf{k}_j) = \omega(\hat{\mathbf{k}}),$$
(12.8)

and thus forming *resonant triads*, if non trivial solutions of (12.8) exist, which can be shown to be the case for the Rossby waves. Note that if resonant triad conditions do not admit non trivial solutions, as in the case of inertia–gravity waves in RSW, in the next order of the asymptotic expansion the quartets of waves will appear and will lead to resonances in the equation for the second nonlinear correction $\psi^{(2)}$ to the wave field.

Elimination of resonances

If there exist such $\hat{\mathbf{k}}$ that the components of the wave packet (12.2) can form resonant triads with it, the first nonlinear correction to the solution of (12.1) is not bounded and, hence, the asymptotic procedure is not self-consistent. Therefore, the resonances should be 'killed'. This can be done by introducing, as in Chapter 8, a slow evolution of the wave amplitudes, and thus replacing $\partial_t \rightarrow \partial_t + \epsilon \partial_T$ in (12.1), which leads to appearance of new terms in (12.3):

$$\nabla^2 \psi_t^{(1)} - \psi_t^{(1)} + \psi_x^{(1)} = -\nabla^2 \psi_T^{(0)} + \psi_T^{(0)} - \mathcal{J}(\psi^{(0)}, \nabla^2 \psi^{(0)}).$$
(12.9)

We will be using both $(\ldots)_T$ and $\partial_T(\ldots)$ notations for slow-time derivatives in the following. Correspondingly, the new contribution in (12.7) is $\sum_i \left(\mathbf{k}^2 + 1 \right) a_{i_T} e^{i(\mathbf{k}_i \cdot \mathbf{x} - \omega(\mathbf{k}_i)t)} + c.c.$, which provides a possibility to compensate resonant terms by slow evolution (modulation) of wave amplitudes. For a single resonant triad $\mathbf{k}_1 + \mathbf{k}_2 = \mathbf{k}_3$, $\omega(\mathbf{k}_1) + \omega(\mathbf{k}_2) = \omega(\mathbf{k}_3)$ we thus get

$$\begin{cases} \left(\mathbf{k}_3^2 + 1 \right) \partial_T a_3 = c(\mathbf{k}_1, \mathbf{k}_2) a_1 a_2, \\ \left(\mathbf{k}_2^2 + 1 \right) \partial_T a_2 = c(\mathbf{k}_3, -\mathbf{k}_1) a_3 a_1^*, \\ \left(\mathbf{k}_1^2 + 1 \right) \partial_T a_1 = c(\mathbf{k}_3, -\mathbf{k}_2) a_3 a_2^*, \end{cases}$$
(12.10)

where $c(\mathbf{k}_1, \mathbf{k}_2) = \hat{\mathbf{z}} \cdot (\mathbf{k}_1 \wedge \mathbf{k}_2)\mathbf{k}_2^2$ are interaction coefficients. This is an integrable system, with general solution expressed in terms of elliptic functions. The total energy of the triad is conserved and redistributed among three waves through periodic exchanges.

Thus, we have seen on the technically simplest example of barotropic Rossby waves in the QG model that if wave amplitudes are small, but finite, nonlinear interactions between elementary harmonic waves forming a solution of the linearised problem lead to slow modulation of wave amplitudes, and related energy exchanges among waves. It is easy to understand that the above-described construction, although exposed here for Rossby waves, is general and applies to any kind of waves, provided conditions of resonance are satisfied, which is usually the case for waves with sufficiently strong dispersion. A particular case of such exchange is a decay of the energy of a wave propagating in a medium (carrier wave) due to its transfer to triad partners picked up from the noise (a so-called parametric instability of the wave). Another case, which will be illustrated on several examples shortly, are resonant triads formed by freely propagating waves and waves trapped in wave guides of various nature.

12.2 Resonant excitation of trapped coastal waves by free inertia–gravity waves

12.2.1 Resonant excitation of wave-guide modes: general philosophy

We have seen in Chapter 4 several examples of wave-guide modes in geophysical flows. On smaller scales, the wave guides arise due to the presence of coasts and/or topography, with trapped shelf and edge waves, and topographic Rossby waves, respectively, on the one hand, or, on the other hand, due to the presence of outcropping, with trapped frontal waves. On larger scales, the equator is a wave guide with trapped Kelvin, Rossby, Yanai, and inertia–gravity waves. Other examples of wave-guide modes could be evoked, like the surface waves trapped by the oceanic currents, which are of even smaller scale and are beyond the scope of this book. The common feature of all these wave guides is that they are *semi-transparent* in the sense that, together with wave-guide modes which cannot leave the wave guide, there are other waves which can propagate freely everywhere and cross the wave guide. In the case of coastal and topographic wave guides considered in one- or two-layer RSW, cf. Section 4.1, or density fronts in the two-layer model considered in Sections 10.7 and 10.8, these are inertia–gravity waves which are freely propagating, and in the case of the equatorial wave guide in the framework of the two-layer RSW with a rigid lid, cf. Section 4.4.2, these are barotropic Rossby waves. Free and wave-guide modes are coupled through nonlinear terms in the equations of motion. If there exist resonant triads including these two different types of waves, their interactions are efficient, and there is a possibility for exchanges of energy between the wave guide and its surrounding, leading to a whole new range of dynamical phenomena. It is to be emphasised that one of the wave-guide modes could have infinite wavelength, that is it could be a mean current. If corresponding resonant triads exist, this leads to a specific type of wave–mean-flow interaction. It is also to be stressed that, mathematically speaking, wave-guide modes and free waves belong to different, discrete and continuous, respectively, parts of the spectrum of the corresponding linearised problem in the across-wave-guide direction. As a consequence, they have different normalisations, and the integrability conditions (conditions of 'killing' the resonances) are formulated differently for these two types of waves. In this, and the subsequent sections, we will see that resonant interactions between trapped and free waves exist. In particular, we will consider *resonant excitation* of trapped (T) waves by free (F) waves. Two excitation mechanisms are, in principle, possible:

1. Free wave $\rightarrow$ Trapped wave + Trapped wave ($F \rightarrow T + T$),
2. Free wave + Free wave $\rightarrow$ Trapped wave ($F + F \rightarrow T$).

As just stated, a particular case of the first possibility is a resonance of a free wave with an infinitely long trapped wave, in other words, a coastal current: $F + C \rightarrow T$. In this section, as in Section 4.1, we will distinguish the cases of shelves with abrupt and gentle slope. In the first case there is only one kind of trapped wave, the non-dispersive Kelvin wave. As we will see, the only possible resonances are $F + F \rightarrow T$, and $F + C \rightarrow T$, the latter realisable only in the baroclinic two-layer set-up. In the second case trapped waves are dispersive shelf and edge waves, and all types of resonance are possible.

12.2.2 Resonant excitation of Kelvin waves by free inertia–gravity waves at abrupt shelf: barotropic model

Reminder of the linear wave spectrum

Let us recall the non-dimensional RSW equations with idealised straight coast and no bathymetry:

$$
\begin{cases}
u_t - v + h_x = -\epsilon(uu_x + vu_y), \\
v_t + u + h_y = -\epsilon(uv_x + vv_y), \\
h_t + u_x + v_y = -\epsilon\left((hu)_x + (hv)_y\right),
\end{cases}
\tag{12.11}
$$

which are considered with the boundary conditions: $x \geq 0$, $u|_{x=0} = 0$. The following scaling is used above: time $\propto f^{-1}$, space $\propto L \sim R_d = \frac{\sqrt{gH}}{f}$, and velocity $\propto U$, leading to the appearance of the Rossby number $Ro = \frac{U}{fL} \equiv \epsilon$ in the equations.

Linear wave solutions $(u, v, h) \propto e^{i(\sigma t - kx - ly)}$ of (12.11) in the limit $\epsilon \rightarrow 0$ are (σ in this chapter denotes the real wave frequency, not to be confused with growth rate in preceding chapters)

- Free inertia–gravity waves (IGW) with real k and dispersion relation

$$
\sigma = \sqrt{1 + k^2 + l^2},
$$

- Trapped non-dispersive Kelvin wave (KW) with $k = -i$ and dispersion relation

$$\sigma = -l.$$

Their polarisation relations were given in (4.10), (4.11) and (4.12) in Chapter 4.

Conditions of inertia–gravity wave (IGW)–Kelvin wave (KW) resonance

It is easy to see that a pair of IGW with frequencies $\sigma_{1,2}$ and along-coast wave numbers $l_{1,2}$ is in resonance with a KW with wave number l only when

$$\sigma_1 - \sigma_2 = \sigma_K = -l, \quad l_1 - l_2 = l, \ l \neq 0 \tag{12.12}$$

(the 'difference' resonance). In this case the interactions between IGW produce resonant-forcing $\propto e^{\pm il(y+t)}$ for KW. We choose $l < 0$ to have positive σ_K, therefore

$$|l| = \sqrt{1 + k_1^2 + l_1^2} - \sqrt{1 + k_2^2 + l_2^2}, \quad l_2 = l_1 + |l|, \tag{12.13}$$

and

$$\sqrt{1 + k_1^2 + l_1^2} - |l| = \sqrt{1 + k_2^2 + (l_1 + |l|)^2}. \tag{12.14}$$

Hence,

$$k_2^2 = k_1^2 - 2|l| \left(\sqrt{1 + k_1^2 + l_1^2} + l_1 \right), \tag{12.15}$$

and the resonance is possible only if

$$k_1^2 - 2|l| \left(\sqrt{1 + k_1^2 + l_1^2} + l_1 \right) \geq 0, \tag{12.16}$$

so

$$k_1^2 - 2|l| \, l_1 \geq 2|l| \sqrt{1 + k_1^2 + l_1^2} > 0, \tag{12.17}$$

and

$$k_1^2 \geq 2|l| \left(\sqrt{1 + l_2^2} + l_2 \right) > 0, \quad l_2 = l_1 + |l|. \tag{12.18}$$

Therefore, in order to find a pair of IGW in resonance with the KW, it is sufficient to 1) take any l and l_1, then $l_2 = l_1 + |l|$, 2) take *arbitrary* k_1 satisfying $k_1^2 \geq 2|l| \left(\sqrt{1 + l_2^2} + l_2 \right)$, and 3) define k_2 from

$$k_2^2 = k_1^2 - 2|l| \left(\sqrt{1 + k_1^2 + l_1^2} + l_1 \right).$$

Thus, a KW with wave number l may be resonantly excited by *a continuum* of incident IGW with wave numbers l_1 and $|k_1| > \sqrt{2|l| \left(\sqrt{1 + (l_1 + |l|)^2} + l_1 + |l| \right)}$ interacting with another incident wave with k_2, l_2:

$$k_2^2 = k_1^2 - 2|l| \left(\sqrt{1 + k_1^2 + l_1^2} + l_1 \right), \quad l_2 = l_1 + |l|. \tag{12.19}$$

Nonlinear expansions

We look for solutions of (12.11) in the form

$$(u, v, h) = (u^{(0)}, v^{(0)}, h^{(0)})(x, y, t, T) + \epsilon(u^{(1)}, v^{(1)}, h^{(1)})(x, y, t, T) + ..., \tag{12.20}$$

where slow time $T = \epsilon t$ is introduced. To the lowest order in ϵ we get the linearised system, which was analysed in Section 4.1.1. We consider a solution in the form

$$(u^{(0)}, v^{(0)}, h^{(0)}) = (u_K^{(0)}, v_K^{(0)}, h_K^{(0)}) + \sum_{i=1,2} (u_{i_{IG}}^{(0)}, v_{i_{IG}}^{(0)}, h_{i_{IG}}^{(0)}), \tag{12.21}$$

where the KW:

$$(u_K^{(0)}, v_K^{(0)}, h_K^{(0)}) = (0, K(y + t), -K(y + t)) e^{-x},$$

and a pair of IGW,

$$(u_{i_{IG}}^{(0)}, v_{i_{IG}}^{(0)}, h_{i_{IG}}^{(0)}) = (U_i(x), V_i(x), H_i(x)) e^{i(l_i y - \sigma_i t)} + c.c, \ i = 1, 2,$$

are supposed to be in resonance: $\sigma_1 - \sigma_2 = -(l_1 - l_2) = -l$.
    To the first order in ϵ we get for the first nonlinear correction:

$$\begin{cases} u_t^{(1)} - v^{(1)} + h_x^{(1)} = -u_T^{(0)} - u^{(0)} u_x^{(0)} + v^{(0)} u_y^{(0)} = R_u, \\ v_t^{(1)} + u^{(1)} + h_y^{(1)} = -v_T^{(0)} - u^{(0)} v_x^{(0)} + v^{(0)} v_y^{(0)} = R_v, \\ h_t^{(1)} + u_x^{(1)} + v_y^{(1)} = -h_T^{(0)} - (h^{(0)} u^{(0)})_x + (h^{(0)} v^{(0)})_y = R_h, \end{cases} \tag{12.22}$$

with boundary conditions $u^{(1)}\big|_{x=0} = 0$.

Removal of resonances in the first-order equations

As is easy to check, the resonant contributions to nonlinear terms in (12.22) are exclusively of the form $M(x)F(y + t)$, with various M and F. The resonant terms come from: 1) IGW–IGW interaction, 2) KW self-interaction, and 3) slow time dependence of the KW field, which is introduced to compensate the two previous ones. In order to understand how to treat such resonances, let us consider a general problem:

$$\begin{cases} u_t - v + h_x = M^{(u)}(x)F^{(u)}(y+t), \\ v_t + u + h_y = M^{(v)}(x)F^{(v)}(y+t), \\ h_t + u_x + v_y = M^{(h)}(x)F^{(h)}(y+t), \end{cases} \tag{12.23}$$

with the same boundary conditions: $x \geq 0$, $u|_{x=0} = 0$. By combining the v and h equations, we get

$$(v-h)_t - (v-h)_y = u_x - u + M^{(v)}(x)F^{(v)}(y+t) - M^{(h)}(x)F^{(h)}(y+t). \tag{12.24}$$

Multiplying this equation by e^{-x} and integrating in x from 0 to ∞ we obtain

$$I_t - I_y = \hat{M}^{(v)}F^{(v)}(y+t) - \hat{M}^{(h)}F^{(h)}(y+t), \tag{12.25}$$

where we defined

$$I = \int_0^\infty dx\,(v-h)e^{-x}, \quad \hat{M}^{(v,h)} = \int_0^\infty dx\,M^{(v,h)}e^{-x}. \tag{12.26}$$

Up to a constant, the left-hand side of (12.25) is a derivative of I with respect to the variable $\xi_- = y - t$, while the right-hand side is the function of $\xi_+ = y + t$ only. Hence, a linear resonant growth of I in $y - t$ results, if the right-hand side is not identically zero, which constitutes the *neccessary and sufficient condition* for existence of bounded solutions:

$$\hat{M}^{(v)}F^{(v)}(y+t) - \hat{M}^{(h)}F^{(h)}(y+t) = 0. \tag{12.27}$$

Slow evolution of the KW

The expressions for the resonant contributions to the right-hand side of the v and h equations in (12.22), to the first order are, respectively,

$$R_v = -K_T e^{-x} - KK_{\xi_+}e^{-2x} - \left[(U_1 V_{2x}^* + V_{1x}U_2 + ilV_1 V_2^*)e^{il\xi_+} + c.c.\right], \tag{12.28}$$

$$R_h = K_T e^{-x} + KK_{\xi_+}e^{-2x} - \left[((H_1 U_2^* + U_1 H_2^*)_x + il(H_1 V_2^* + V_1 H_2^*))\,e^{il\xi_+} + c.c.\right]. \tag{12.29}$$

The condition of absence of resonances (12.27) takes the form

$$\int_0^\infty dx\,e^{-x}(R_h - R_v) = 0, \tag{12.30}$$

and hence

$$K_T + KK_{\xi_+} = Se^{il\xi_+} + S^*e^{-il\xi_+},$$ (12.31)

where

$$S = \int_0^\infty dx\, e^{-x}\left[(H_1 U_2^* + U_1 H_2^*)_x - U_1 V_{2x}^* - V_{1x} U_2\right.$$

$$\left. + il(H_1 V_2^* + V_1 H_2^* - V_1 V_2^*)\right].$$ (12.32)

Using the polarisation relation for IGW we get $S = iA_1 A_2 s$, where $A_{1,2}$ are complex amplitudes of the two IGW, and

$$s = \frac{4l}{(k_1^2 + 1)(k_2^2 + 1)[1 + (k_1 + k_2)^2][1 + (k_1 - k_2)^2]}\cdot$$

$$\left[(\sigma_1 l_2 + \sigma_2 l_1 - l_1 l_2)(1 + k_1^2 + k_2^2) + \sigma_2 l_1 k_1(1 + k_1^2 - k_2^2)\right.$$

$$\left. + \sigma_1 l_2 k_2(1 + k_2^2 - k_1^2) + 2k_1 k_2(l_1 l_2 - (1 + k_1^2)(1 + k_2^2))\right].$$ (12.33)

By virtue of (12.19), the interaction coefficient s depends only on l, k_1, l_1. The isopleths of this coefficient in the (k_1, l_1) plane are presented in Figure 12.1, which, at the same time shows the domain where the resonances of the considered type are possible in the phase space.

Analysis of the evolution equation for KW

The slow evolution equation for KW, therefore, is

$$K_T + KK_{\xi_+} = -2s\,|A_1|\,|A_2|\sin l\xi_+ + \arg A_1 - \arg A_2.$$ (12.34)

After renormalisations and change of independent variables, this equation becomes a forced simple-wave (Hopf) equation:

$$K_\tau + KK_\chi = -\sin \chi,$$ (12.35)

and can be integrated by using Lagrangian variables $\mathcal{X}$: $K = \dot{\mathcal{X}}$, where the dot denotes the Lagrangian derivative $\partial_\tau + K\partial_\chi(\ldots)$:

$$\ddot{\mathcal{X}} + \sin \mathcal{X} = 0.$$ (12.36)

This is the well-known pendulum equation and is integrable, giving Lagrangian trajectories (characteristics). As in the case of the unforced simple wave equation considered in Section 7.1.1, intersection of characteristics corresponds to shock (front) formation, and it is easy to check that trajectories starting from different initial conditions do intersect.

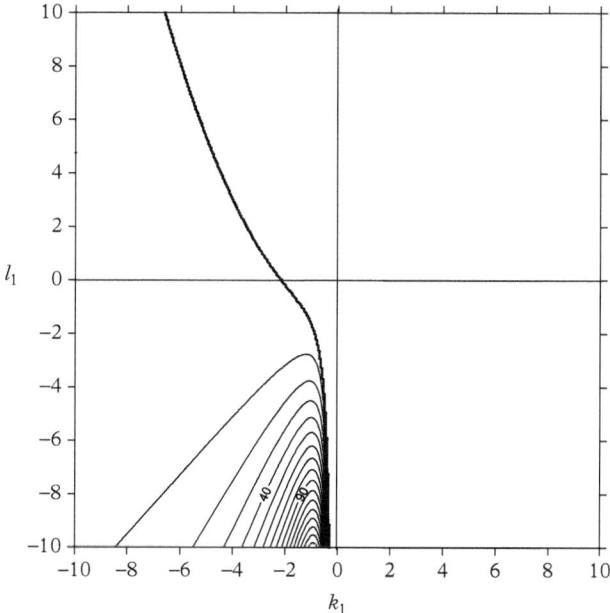

Figure 12.1 *Isopleths of the interaction coefficient $s(l, k_1, l_1)$ for $l = -1$, at interval 10. Solid line indicates the boundary of the resonance domain.*

As equation (12.35) is forced, a breaking Kelvin wave would eventually appear even if its amplitude was negligible initially. Implications of this process for transport and mixing near the coast are obvious, as Kelvin fronts are zones of enhanced localised mixing.

12.2.3   Resonant excitation of Kelvin waves by free inertia–gravity waves at abrupt shelf: baroclinic model

In this section we will extend the analysis of Section 12.2.2 to the baroclinic case. The main observation below is that including baroclinic effects, again in the simplest two-layer RSW model, allows for new resonances leading to excitation of Kelvin waves by inertia–gravity waves impinging on the coast. Probably the most interesting of these new resonances is the one including an inertia–gravity wave, a Kelvin wave, and a coastal current.

Brief reminder of the model and wave spectrum

We work with the same two-layer RSW model with a free surface, and use the same notation as in Section 3.3.2. Barotropic and baroclinic components of velocity $\mathbf{v}^{\pm}$ and surface and interface displacements $\eta^{\pm}$ will be used, instead of their layer-wise counterparts. In terms of the baroclinic and the barotropic components the non-dimensional equations of the model become

$$\partial_t \mathbf{v}^+ + \hat{z} \wedge \mathbf{v}^+ + \frac{1 + \sqrt{r}}{2} \nabla \eta^+ = -\frac{\epsilon}{4} \left[\left(\frac{1}{\sqrt{r}} + 1 \right) \left(\mathbf{v}^+ \cdot \nabla \mathbf{v}^+ + \mathbf{v}^- \cdot \nabla \mathbf{v}^- \right) \right.$$

$$\left. + \left(\frac{1}{\sqrt{r}} - 1 \right) \left(\mathbf{v}^+ \cdot \nabla \mathbf{v}^- + \mathbf{v}^- \cdot \nabla \mathbf{v}^+ \right) \right], \tag{12.37}$$

$$\partial_t \eta^+ + \nabla \cdot \mathbf{v}^+ = -\frac{\epsilon}{4} \left[\left(\frac{1}{\sqrt{r}} + 1 \right) \nabla \cdot \left(\eta^+ \mathbf{v}^+ + \eta^- \mathbf{v}^- \right) \right.$$

$$\left. + \left(\frac{1}{\sqrt{r}} - 1 \right) \nabla \cdot \left(\eta^+ \mathbf{v}^- + \eta^- \mathbf{v}^+ \right) \right], \tag{12.38}$$

$$\partial_t \mathbf{v}^- + \hat{z} \wedge \mathbf{v}^- + \frac{1 - \sqrt{r}}{2} \nabla \eta^- = -\frac{\epsilon}{4} \left[\left(\frac{1}{\sqrt{r}} - 1 \right) \left(\mathbf{v}^+ \cdot \nabla \mathbf{v}^+ + \mathbf{v}^- \cdot \nabla \mathbf{v}^- \right) \right.$$

$$\left. + \left(\frac{1}{\sqrt{r}} + 1 \right) \left(\mathbf{v}^+ \cdot \nabla \mathbf{v}^- + \mathbf{v}^- \cdot \nabla \mathbf{v}^+ \right) \right], \tag{12.39}$$

$$\partial_t \eta^- + \nabla \cdot \mathbf{v}^- = -\frac{\epsilon}{4} \left[\left(\frac{1}{\sqrt{r}} - 1 \right) \nabla \cdot \left(\eta^+ \mathbf{v}^+ + \eta^- \mathbf{v}^- \right) \right.$$

$$\left. + \left(\frac{1}{\sqrt{r}} + 1 \right) \nabla \cdot \left(\eta^+ \mathbf{v}^- + \eta^- \mathbf{v}^+ \right) \right]. \tag{12.40}$$

We place ourselves in the same set-up as in Section 12.2.2, that is in the semi-infinite plane with an idealised straight coast. The y axis of the coordinate system is directed along the coast, and the x axis points out of the coast. We recall that in the linear limit the system splits in two decoupled rotating shallow-water subsystems. The wave spectrum consists of incident and reflected by the coast barotropic (upper sign) or baroclinic (lower sign) IGW and KW propagating along the y axis with the non-dimensional phase velocity $c_\pm$ and decaying with x as $e^{-x/c_\pm}$. Their dispersion relations are, respectively,

$$\sigma_{IG\pm}^2 = 1 + c_\pm^2 \mathbf{k}^2, \quad \sigma_{K\pm}^2 = c_\pm^2 l^2, \quad c_\pm = \sqrt{\frac{1 \pm \sqrt{r}}{2}}, \tag{12.41}$$

where σ is the wave frequency and $\mathbf{k} = (k, l)$ is the wave vector. Velocity and thickness fields for IGW and KW are given by

$$(u_{IG}^\pm, v_{IG}^\pm, \eta_{IG}^\pm) = \left(U^\pm(x), V^\pm(x), N^\pm(x) \right) e^{i(ly - \sigma t)} + c.c., \tag{12.42}$$

and

$$(u_K^\pm, v_K^\pm, \eta_K^\pm) = (0, c_\pm K^\pm(y + c_\pm t), -K^\pm(y + c_\pm t)) e^{-x/c_\pm}, \tag{12.43}$$

where $K^{\pm}$ are arbitrary functions of arguments $\xi^{\pm} = y + c_{\pm}t$. The boundary condition of vanishing normal velocity at the coast imposes

$$U^{\pm} = A\left(e^{ikx} - e^{-ikx}\right), \tag{12.44}$$

where A is the complex wave amplitude, and the expressions for V and N follow from the linearised equations (12.37) to (12.40), assuming $k < 0$:

$$V^{\pm} = A(1 + c_{\pm}^2 l^2)\left(\frac{e^{ikx}}{i\sigma^{\pm} + lkc_{\pm}^2} + \frac{e^{-ikx}}{-i\sigma^{\pm} + lkc_{\pm}^2}\right),$$

$$\tag{12.45}$$

$$N^{\pm} = A\left(\frac{ik + \sigma_{\pm}l}{c_{\pm}^2 lk + i\sigma_{\pm}}e^{ikx} - \frac{ik - \sigma_{\pm}l}{c_{\pm}^2 lk - i\sigma_{\pm}}e^{-ikx}\right).$$

Resonant excitation of Kelvin waves in the two-layer system. Wave–wave resonances

We recall that non linear terms in (12.37) to (12.40) are resonant if the frequencies and wave vectors of a pair of waves obey the relations

$$\mathbf{k}_1 \pm \mathbf{k}_2 = \mathbf{k}, \quad \sigma(\mathbf{k}_1) \pm \sigma(\mathbf{k}_2) = \sigma(\mathbf{k}),$$

where the indices $1, 2$ (not to be confused with layer numbers) refer to resonating waves, whatever their nature (baroclinic or barotropic). As in the previous section, we are interested in resonant excitation of Kelvin wave by free inertia–gravity waves which are coming from infinity and are reflected by the coast. Kelvin wave are non-propagative in the x direction perpendicular to the coast, hence the resonance conditions include only y component of the wave-vector in this case, as in Section 12.2.2,

$$l_{IG_1} \pm l_{IG_2} = l_K, \quad \sigma_{IG}(\mathbf{k}_1) \pm \sigma_{IG}(\mathbf{k}_2) = \sigma_K(l), \tag{12.46}$$

where, again, each wave may be either barotropic or baroclinic. We refer to the upper- and lower-sign cases in (12.46) as the 'sum' and the 'difference' resonances, respectively.

Let us suppose that a pair of IGW, whatever their character, is in a resonance (12.46) with either a barotropic or a baroclinic KW. These three waves constitute a zero-order solution which will be considered to be dependent on slow time $T = \epsilon t$:

$$(u^{(0)}, v^{(0)}, \eta^{(0)}) = (u_K^{(0)}, v_K^{(0)}, \eta_K^{(0)})(x, y, t, T) + \sum_{i=1,2}(u_{i_{IG}}^{(0)}, v_{i_{IG}}^{(0)}, \eta_{i_{IG}}^{(0)})(x, y, t, T), \tag{12.47}$$

where the appropriate components, barotropic or baroclinic, are understood for each wave, depending on the concrete nature of the resonance. For the first-order correction

$(u^{(1)\pm}, v^{(1)\pm}, \eta^{(1)\pm})(x, y, t, T)$ we get

$$
\begin{cases}
u_t^{(1)\pm} - v^{(1)\pm} + c_\pm^2 \eta_x^{(1)\pm} = -u_T^{(0)\pm} + (NL)_u^\pm \equiv R_u^\pm, \\
v_t^{(1)\pm} + u^{(1)\pm} + c_\pm^2 \eta_y^{(1)\pm} = -v_T^{(0)\pm} + (NL)_v^\pm \equiv R_v^\pm, \\
\eta_t^{(1)\pm} + u_x^{(1)\pm} + v_y^{(1)\pm} = -\eta_T^{(0)\pm} + (NL)_\eta^\pm \equiv R_\eta^\pm,
\end{cases}
\tag{12.48}
$$

where $(NL)_{u,v,\eta}^\pm$ stand for corresponding non linear terms in (12.37) to (12.40) built from the zero-order fields (12.47). As was shown in Section 12.2.2, all essential information for Kelvin-wave excitation is contained in the expressions $R_v^\pm$ and $R_\eta^\pm$. Up to non-resonant terms, they have the form

$$
R_v^\pm = -c_\pm K_T^\pm e^{-x/c_\pm} - a_\pm c_\pm^2 K^\pm K_{\xi^\pm}^\pm e^{-2x/c_\pm} - \left[(NL)_v^{\pm IG} e^{il\xi^\pm} + c.c. \right],
\tag{12.49}
$$

$$
R_\eta^\pm = K_T^\pm e^{-x/c_\pm} + 2a_\pm c_\pm K^\pm K_{\xi^\pm}^\pm e^{-2x/c_\pm} - \left[(NL)_\eta^{\pm IG} e^{il\xi^\pm} + c.c. \right],
\tag{12.50}
$$

where we introduced a shorthand notation, $a_\pm = \frac{1}{4} \left(\frac{1}{\sqrt{r}} \pm 1 \right)$, and $(NL)_{v,\eta}^{\pm IG}$ denote bilinear combinations proportional to $|A|^2$ and resulting from insertion in the nonlinear terms the zero-order IGW fields with polarisation relations (12.44) to (12.45). To obtain the evolution equation for the Kelvin wave $K^\pm(\xi^\pm, T)$ we follow the lines of Section 12.2.2 and subtract the third equation from the second one, multiplied by $c_\pm^{-1}$ in (12.48). After multiplying by $e^{-x/c_\pm}$ and integrating in x we get

$$
B_t^\pm - c_\pm B_y^\pm = -\int_0^\infty dx\, e^{-x/c_\pm} (R_\eta^\pm - c_\pm^{-1} R_v^\pm), \quad B = \int_0^\infty dx\, e^{-x/c_\pm} (c_\pm^{-1} v^{(1)\pm} - \eta^{(1)\pm}). \tag{12.51}
$$

As follows from (12.49) and (12.50), the right-hand side of (12.51) is a function of $\xi^\pm$ only and, by the same reasoning as in Section 12.2.2, in order to guarantee the boundedness of $B^\pm$ it should be equal to zero:

$$
\int_0^\infty dx\, e^{-x/c_\pm} (R_\eta^\pm - c_\pm^{-1} R_v^\pm) = 0,
\tag{12.52}
$$

which leads to the evolution equation for K:

$$
K_T^\pm + a_\pm c_\pm K^\pm K_{\xi^\pm}^\pm = \mathcal{S}.
\tag{12.53}
$$

As in Section 12.2.2, $\mathcal{S}$ is proportional to the product of amplitudes of the two IGW, and is harmonic in $\xi^\pm$. The detailed expression for $\mathcal{S}$ in terms of the wave numbers of the resonating IGW depends on the precise nature of the resonance. As compared to the one-layer case, where only the 'difference' resonances were operational, the 'sum' resonances between the waves of the same and different kinds (barotropic and baroclinic) can also lead to the resonance excitation of the KW in the two-layer model.

Thus, barotropic KW can be excited by the 'difference' resonances of two barotropic IGW, as in the one-layer case, and by the 'sum' resonances including either a baroclinic and a barotropic IGW, or a pair of baroclinic IGW. The baroclinic KW can be only excited by 'difference' resonances of all kinds (baroclinic–baroclinic, barotropic–baroclinic, and barotropic–barotropic). The 'difference' resonances can be analysed as in Section 12.2.2, so we sketch below only the mechanism of the sum resonance of two baroclinic IGW generating a barotropic KW.

The corresponding resonance conditions are, cf (12.46),

$$l_{IG_1} + l_{IG_2} = l_K, \quad \sigma_{IG}(\mathbf{k}_1) + \sigma_{IG}(\mathbf{k}_2) = \sigma_K(l), \tag{12.54}$$

and give (we will omit the subscripts *IG* and *K* from now on)

$$l_1 + l_2 = l = -|l|, \quad \sqrt{1 + c_-^2(k_1^2 + l_1^2)} + \sqrt{1 + c_-^2(k_2^2 + l_2^2)} = c_+|l|, \tag{12.55}$$

where, as before, we adopt a convention of positive wave frequencies, and thus the wave-vector l of the Kelvin wave is negative. The resonance condition on frequencies, after renormalising the wave vectors with a factor c_-^{-1} is

$$\sqrt{1 + k_1^2 + l_1^2} + \sqrt{1 + k_2^2 + l_2^2} = -\frac{c_+}{c_-}(l_1 + l_2). \tag{12.56}$$

Suppose that $l_{1,2}$ are both negative $l_{1,2} = -|l_{1,2}|$. Equation (12.56) then takes the form

$$\sqrt{1 + k_1^2 + |l_1|^2} + \sqrt{1 + k_2^2 + |l_2|^2} = +\frac{c_+}{c_-}(|l_1| + |l_2|), \tag{12.57}$$

and admits a simple geometric interpretation presented in Figure 12.2. At fixed $l_{1,2}$ the minimum value of the left-hand side of (12.57) is equal to $\sqrt{1 + |l_1|^2} + \sqrt{1 + |l_2|^2} > |l_1| + |l_2|$ (sum of the lengths of the dashed intervals in Figure 12.2 vs the length of the horizontal interval). Thus if the ratio $\frac{c_+}{c_-}$ is insufficiently large: $\frac{c_+}{c_-} < \frac{\sqrt{1+|l_1|^2}+\sqrt{1+|l_2|^2}}{|l_1|+|l_2|}$, that is the stratification is sufficiently strong, the resonance is not possible. Otherwise, as is easy to see from Figure 12.2, we can always find $k_{1,2}$ to satisfy the resonance condition. Inversely, at fixed $\frac{c_+}{c_-}$, that is at a given stratification, the condition on $l_{1,2}$ for the existence of resonance is

$$\frac{c_+}{c_-}(|l_1| + |l_2|) > \sqrt{1 + |l_1|^2} + \sqrt{1 + |l_2|^2}. \tag{12.58}$$

Due to the $\sqrt{r}$ dependence, the resonance does not exist only for unrealistically small values of r. Analysis of (12.58) shows that, except for the values of $\frac{c_+}{c_-}$ close to 1 (strong stratification), the resonance is always possible for high enough $|l|_{1,2}$.

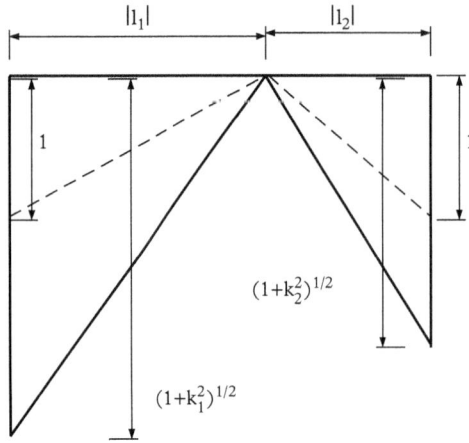

Figure 12.2 *Geometric representation of the resonance condition (12.57). Hypotenuses of the right-angle triangles (solid lines) represent the two entries on the left-hand side of (12.57), with their minimal values indicated by dashed lines.*

Resonant excitation of Kelvin waves in the two-layer system. Wave–mean-flow resonance

As already mentioned, the two-layer system admits a new type of resonance leading to the excitation of Kelvin waves: IGW–mean-flow resonance. Indeed, a mean-flow solution, which is a solution of not only linearised, but also of the full system, should be added to the above-described wave solutions of the linear system. In non-dimensional terms it is given by

$$u_{1_M} = u_{2_M} = 0, \quad -v_{1_M} + \partial_x\left(\eta_{1_M} + \eta_{2_M}\right) = 0, \quad -v_{2_M} + \partial_x\left(r\eta_{1_M} + \eta_{2_M}\right) = 0, \quad (12.59)$$

and corresponds to an along-coast current in geostrophic equilibrium. The corresponding expressions in terms of baroclinic and barotropic components are

$$u_M^\pm = 0, \quad v_M^\pm = \frac{1 \pm \sqrt{r}}{2}\partial_x\eta_M^\pm. \quad (12.60)$$

We will consider the functions $v_{1,2}$ rapidly decaying out of the coast (at $x = 0$), and thus representing coastal currents which are ubiquitous in the ocean. In the case of surface currents, which is probably the most relevant configuration, the lower layer is motionless:

$$v_{2_M} = 0, \quad r\eta_{1_M} + \eta_{2_M} = 0. \quad (12.61)$$

Note that, unlike Chapters 4 and 10, we do not introduce here currents with outcroppings. A weak mean current along the coast may be considered as a wave with zero frequency and zero wave number in the direction along the coast. For a resonance of an IGW with the current and with a KW to exist, the wave frequency and the wave number in the direction along the coast of the incoming free wave should coincide with those of the KW. From the intersection of the dispersion surfaces of the baroclinic IGW and the barotropic KW in Figure 5.4 of Chapter 4 it is obvious that the resonance of a baroclinic IGW, a mean current, and a barotropic KW and, hence, resonant excitation of barotropic Kelvin wave via this resonance, is possible. We, thus, should have

$$\sigma_{IG^-} = \sigma_{K^+}, \quad l_{IG^-} = l_{K^+} \equiv l, \text{ and } c_+^2 l^2 = 1 + c_-^2(k^2 + l^2), \tag{12.62}$$

which leads to the following expression for the x component k of the wave number of the incoming IGW:

$$k^2 = \left(\frac{c_+^2}{c_-^2} - 1 \right) l^2 - \frac{1}{c_-^2}. \tag{12.63}$$

Therefore, for a given stratification, that is a given ratio $\frac{c_+^2}{c_-^2} > 1$, the resonance is always possible for sufficiently large negative l. With respect to the wave–wave resonance, the conditions of the present one are less restrictive, as a single wave with a sufficiently short wavelength, impinging on the coastal current is enough. We thus have a universal mechanism of Kelvin-wave excitation, as such situations are ubiquitous in the ocean. We should stress that this mechanism is impossible in the barotropic model, because the frequencies of KW are always smaller than the frequencies of IGW of the same type.

Let us now see how the mechanism works. The zero-order flow has the form

$$(u^{(0)}, v^{(0)}, \eta^{(0)}) = (u_K^{(0)}, v_K^{(0)}, \eta_K^{(0)})(x, y, t, T) + (u_{IG}^{(0)}, v_{IG}^{(0)}, \eta_{IG}^{(0)})(x, y, t, T)$$

$$+ (0, v_M^{(0)}, \eta_M^{(0)})(x, T), \tag{12.64}$$

and we look for the first-order corrections; solutions of the system (12.48). According to our scaling the amplitude of the mean current is of the same order as that of impinging wave, that is it is small. We will discuss later how this constraint may be relaxed. The expressions for the resonant contributions to R_v^+ and R_η^+ in the case of wave–meanflow resonance are (cf. (12.49) and (12.50))

$$R_v^{WM^+} = -c_\pm K_T^+ e^{-x/c_+} - a_+ c_+^2 K^+ K_{\xi^+}^+ e^{-2x/c_+} - \left[(NL)_v^{WM^+} e^{il\xi^+} + c.c. \right]$$

$$- c_+ \left(a_+ v_M^+ + a_- v_M^- \right) K_{\xi^+}^+ e^{-x/c_+}, \tag{12.65}$$

$$R_\eta^{WM^+} = K_T^+ e^{-x/c_+} + 2a_+ c_+ K^+ K_{\xi^+}^+ e^{-2x/c_+} - \left[(NL)_\eta^{WM^+} e^{il\xi^+} + c.c. \right]$$

$$- \left(a_+ \left(c_+ \eta_M^+ - v_M^+ \right) + a_- \left(c_+ \eta_M^- - v_M^- \right) \right) K_{\xi^+}^+ e^{-x/c_+}, \tag{12.66}$$

with nonlinear terms $(NL)_v^{WM^+}$ and $(NL)_\eta^{WM^+}$ resulting from the wave–mean-flow interaction. Contributions from the Kelvin wave–mean-flow interactions containing $\partial_{\xi^+} K^+$, which were absent in the case of wave–wave resonance, appear in (12.65) and (12.66).

As above, in order to obtain the evolution equation for the Kelvin wave K^+, we construct the combination

$$\mathcal{K} = R_\eta^{WM^+} - c_+^{-1} R_v^{WM^+} = 2K_T^{\dagger} e^{-x/c_+} + 3c_+ K^{\dagger} K_{\xi+}^{\dagger} e^{-2x/c_+}$$

$$+ \left[2 \left(a_+ v_M^+ + a_- v_M^- \right) - c_+ \left(a_+ \eta_M^+ + a_- \eta_M^- \right) \right] K_{\xi+}^+ e^{-x/c_+} + \mathcal{R}^{WM}, \tag{12.67}$$

and require that

$$\int_0^\infty dx\, e^{-x/c_+} \mathcal{R} = 0. \tag{12.68}$$

Taking into account the relations (12.60) between velocities and thicknesses for the mean currents, the expression for $\mathcal{R}^{WM}$ in the relevant case of upper-layer current is

$$\mathcal{R}_{(upper)}^{WM} = \frac{1-r}{4c_+} \left(\bar{u}_{IG} \partial_{xx}^2 \eta_{1M} + \partial_y \bar{v}_{IG} \partial_x \eta_{1M} \right) - (1+r) \partial_x \left(\bar{u}_{IG} \eta_{1M} \right)$$

$$- \partial_y \left[(1+r) \bar{v}_{IG} \eta_{1IG} + \frac{1-r}{4} \bar{\eta}_{IG} \partial_x \eta_{1M} \right] + c.c. \tag{12.69}$$

Injecting the expression for $\mathcal{R}^{WM}$ into (12.68) we obtain the evolution equation for K:

$$K_T^+ + CK_{\xi+}^+ + a_+ c_+ K^+ K_{\xi+}^+ = \mathcal{S}. \tag{12.70}$$

Here C is the non-dimensional phase velocity induced by the mean current and depending on the profile of the mean flow, cf. (16.40), while $\mathcal{S}$ is the source term harmonic in ξ^+ proportional to the amplitude of the IGW. The equation (12.70) is of the same form as (12.53), apart from the linear term $CK_{\xi+}^+$ which represents a Doppler shift in the phase velocity of the Kelvin wave due to the presence of the mean current. Note that by integration by parts the integral in x of the second term on the right-hand side of (12.67) may be transformed into an integral over of the mean-flow velocity minus the value of the thickness at the coast (up to a factor). Positive coastal currents will thus have lesser C, and vice versa, as it should be, because Kelvin waves propagate in the negative direction in y. As to the efficiency of the forcing $\mathcal{S}$, it depends both on the characteristics of the mean flow, and of the parameters of impinging IGW. For the coastal currents of finite width D, the integration over x in (12.68) is reduced to the finite interval $[0, D]$. From the standard steepest descent estimates applied to such integrals, one can deduce that the forcing is optimal for IGW with $kD \leq 1$, that is long waves, and negligible for $kD \gg 1$. The Doppler-shift term in (12.70) can be removed by the change of the dependent variable $K \to K - C$ and essentially does not change the properties of the equation (12.70), as compared to (12.53). Thus, the same conclusions as to the behaviour of its solutions hold; namely, the amplitue of the barotropic Kelvin wave starts growing due to the resonant forcing produced by the interaction of free baroclinic

IGW and the coastal current. Nonlinear evolution of the excited Kelvin wave results, in general, in breaking and formation of characteristic mixing zones, the Kelvin fronts.

12.2.4   Resonant excitation of coastal waves by free inertia–gravity waves at the shelf with gentle slope

In order to study how the process of resonant (parametric) excitation of coastal waves by free inertia–gravity waves works in the presence of nontrivial shelf bathymetry we place ourselves in the idealised configuration of section 4.1.2. As in that section all calculations can be done explicitly by using the exponential shelf profile of the Ball's model, but the results remain valid for any monotonous profile of the shelf. The calculations are much more cumbersome, as compared to the case without bathymetry treated in Section 12.2.2 so we will omit many technical details and will concentrate mostly on the results.

Reminder on the linear wave spectrum of the RSW model with a gentle shelf

We work with one-layer model with a straight coast and bathymetry. The equations of the model are RSW equations with topography

$$\begin{cases} u_t + uu_x + vu_y - fv + g\eta_x = 0, \\ v_t + uv_x + vv_y + fu + g\eta_y = 0, \\ \eta_t + ((\eta + h)u)_x + ((\eta + h)v)_y = 0, \end{cases} \tag{12.71}$$

where the coast is situated at $x = 0$, $h(x)$ is water depth at rest, η is perturbation of the free surface, and bathymetry h is one-dimensional, $h|_{x=0} = 0$, $h|_{x\to\infty} \to H = $ const, with all derivatives of $h(x)$ tending to zero at infinity. The boundary conditions are

$$\eta|_{x\to 0} \text{ is regular}, \quad \eta u|_{x\to 0} \to 0, \tag{12.72}$$

and the linearised non-dimensional equations are

$$\begin{cases} u_t - \hat{f}v + \eta_x = 0, \\ v_t + \hat{f}u + \eta_y = 0, \\ \eta_t + (hu)_x + (hv)_y = 0, \end{cases} \tag{12.73}$$

where $\hat{f} = \frac{fL}{\sqrt{gH}}$ is the Froude number (inverse square root of the Burger number, as used in section 4.2.2) constructed from the typical velocity of inertia–gravity (Poincaré) waves $\sqrt{gH}$, and typical scale of topography L, the hat over f will be omitted from now on. As follows from Section 4.1.2, the spectrum of (12.73) consists of *trapped* waves with $(\eta, u, v)|_{x\to\infty} \to 0$, and *free (incident + reflected)* waves: $(\eta, u, v)|_{x\to\infty} \propto e^{i(ly-\sigma t)} Re\left(A^+ e^{ikx} + A^- e^{-ikx}\right)$. If solutions of (12.73) are sough in the form

$(u, v, \eta) = (iU, V, Z)(x)e^{i(ly-\sigma t)}$, then

$$U = \frac{flZ - \sigma Z'}{\sigma^2 - f^2}, \quad V = \frac{\sigma lZ - fZ'}{\sigma^2 - f^2}, \tag{12.74}$$

where prime denotes the x derivative, and (12.73) can be reduced to a single equation

$$\left(hZ'\right)' + \left(\sigma^2 - f^2 - l^2h - \frac{fl}{\sigma}h'\right)Z = 0. \tag{12.75}$$

This gives, at fixed l, an eigenproblem for eigenfunctions Z and eigenfrequencies σ, which can be solved analytically in the Ball's model with $h = 1 - e^{-x}$, cf. Section 4.1.2. It is worth noting that the spectrum of the eigenfrequencies is symmetric in the along-coast wave number l in the absence of rotation, and becomes asymmetric when rotation is present, cf. Figure 12.3.

With the chosen shape of h, (12.75) has one regular singular point at $x = 0$ and an irregular singular point at $x = \infty$. If we assume for simplicity that $h'(0) \neq 0$, then the application of the standard methods of ordinary differential equations theory shows that in the vicinity of $x = 0$ the fundamental system of solutions of (12.75) can be written as

$$\phi_1 = \sum_{m=0}^{\infty} a_m x^m, \quad \phi_2 = C\log x \phi_1 + \bar{\phi}, \quad \bar{\phi} = \sum_{m=0}^{\infty} b_m x^m, \tag{12.76}$$

where C is some function of b_k. At $x \to \infty$, (12.75) takes the form

$$Z'' + \left(\sigma^2 - f^2 - l^2\right)Z = 0. \tag{12.77}$$

Hence, a trapped wave solution of (12.75) exists only if $\sigma^2 - f^2 - l^2 = -p^2 < 0$, while the Poincaré wave solution exists only if $\sigma^2 - f^2 - l^2 = k^2 > 0$.

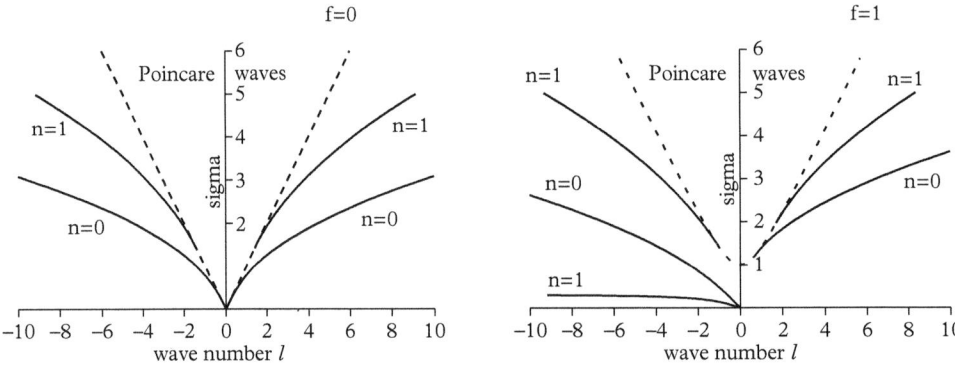

Figure 12.3 *Comparison of the spectra of the Ball's model in the absence (left panel) and in the presence (right panel) of rotation.*

General conditions of resonance of a pair of trapped modes with a free wave

Let us start with a qualitative analysis of interactions of trapped modes with Poincaré waves. Suppose that trapped modes have the phases $\theta_{1,2} = l_{1,2}y - \sigma_{1,2}t$. Due to their interaction, two mixed trapped harmonics $\sim e^{i(\theta_1 \pm \theta_2)}$ arise. When the frequencies and the wave numbers of the mixed harmonics $\sigma_\pm = \sigma_1 \pm \sigma_2$, and $l_\pm = l_1 \pm l_2$ do not belong to the Poincaré continuum in the plane σ, l, cf. Figure 12.3, that is $\sigma_\pm^2 < l_\pm^2 + f^2$, the interaction leads only to a modification of the trapped wave field. If, however, at least one of the pairs $\sigma_\pm$, $l_\pm$, for example σ_+, l_+, is inside the Poincaré continuum, that is $\sigma_+^2 > l_+^2 + f^2$, then the combinational harmonic $\sim e^{i(\theta_1 + \theta_2)}$ produces a forced Poincaré wave with σ_+, l_+ and $k_+ = \sqrt{\sigma_+^2 - l_+^2 - f^2}$, propagating away from the wave guide, and similar for the beat harmonic with (σ_-, L). The radiation of the forced Poincaré wave causes energy leakage from the wave guide, which can be compensated by the incoming free Poincaré wave with the same parameters (σ_+, l_+, k_+). The interactions between such free Poincaré wave and trapped modes would result in the exponential growth of the trapped modes, that is in energy input into the wave guide. Competition between the mechanisms of radiation and compensation determines the joint evolution of the system of interacting waves. We confirm these qualitative conclusions by asymptotic analysis below.

Multi-scale asymptotic expansions

The non-dimensional nonlinear equations of the model are

$$\begin{cases} u_t - fv + \eta_x = -\epsilon(uu_x + vu_y), \\ v_t + fu + \eta_y = -\epsilon(uv_x + vv_y), \\ \eta_t + (hu)_x + (hv)_y = -\epsilon\left((\eta u)_x + (\eta v)_y\right), \end{cases} \tag{12.78}$$

where the nonlinearity parameter ϵ is assumed to be small. Solutions are sought in the following form:

$$(u, v, \eta) = (u^{(0)}, v^{(0)}, \eta^{(0)})(x, y, t, T_1, T_2, \ldots) + \epsilon(u^{(1)}, v^{(1)}, \eta^{(1)})(x, y, t, T_1, T_2, \ldots) + \ldots,$$

where $T_n = \epsilon^n t, n = 1, 2, \ldots$ are slow times. In the zeroth order in ϵ the linear system (12.73) arises, and in the first and second orders, we get, respectively

$$\begin{cases} u_t^{(1)} - fv^{(1)} + \eta_x^{(1)} = -u_{T_1}^{(0)} - (u^{(0)}u_x^{(0)} + v^{(0)}u_y^{(0)}), \\ v_t^{(1)} + fu^{(1)} + \eta_y^{(1)} = -v_{T_1}^{(0)} - (u^{(0)}v_x^{(0)} + v^{(0)}v_y^{(0)}), \\ \eta_t^{(1)} + \left(hu^{(1)}\right)_x + \left(hv^{(1)}\right)_y = -\eta_{T_1}^{(0)} - ((\eta^{(0)}u^{(0)})_x + (\eta^{(0)}v^{(0)})_y), \end{cases} \tag{12.79}$$

$$\begin{cases} u_t^{(2)} - fv^{(2)} + \eta_x^{(2)} = -u_{T_2}^{(0)} - u_{T_1}^{(1)} - \left((u^{(0)}u^{(1)})_x + v^{(1)}u_y^{(0)} + v^{(0)}u_y^{(1)}\right), \\ v_t^{(2)} + fu^{(2)} + \eta_y^{(2)} = -v_{T_2}^{(0)} - v_{T_1}^{(1)} - \left(u^{(1)}v_x^{(0)} + u^{(0)}v_x^{(1)} + (v^{(0)}v^{(1)})_y\right), \\ \eta_t^{(2)} + \left(hu^{(2)}\right)_x + \left(hv^{(2)}\right)_y = -\eta_{T_2}^{(0)} - \eta_{T_1}^{(1)} - ((\eta^{(1)}u^{(0)} + \eta^{(0)}u^{(1)})_x + (\eta^{(1)}v^{(0)} + \eta^{(0)}v^{(1)})_y), \end{cases}$$
$$\tag{12.80}$$

These equations constitute a basis of calculations of all possible interactions, which differ only by choices of the input zero-order solution.

Solutions of the forced linear RSW system with a shelf and conditions of their regularity

Prior to studying nonlinear interactions of the waves in the system, let us consider the linear system (12.73) with forcing:

$$
\begin{cases}
u_t - fv + \eta_x = R_u(x)e^{i\theta}, \\
v_t + fu + \eta_y = R_v(x)e^{i\theta}, \\
\eta_t + (hu)_x + (hv)_y = R_\eta(x)e^{i\theta},
\end{cases}
\tag{12.81}
$$

where $\theta = ly - \sigma t$, and the wave number l and the frequency σ are arbitrary. The solution is sought in the form $(u, v, \eta) = (iU, V, Z)(x)e^{i\theta}$, and after elimination of U and V we obtain the equation for Z:

$$
(hZ')' + \left(\sigma^2 - f^2 - l^2 h - \frac{fl}{\sigma} h' \right) Z = R_z,
\tag{12.82}
$$

with

$$
R_z = \frac{i}{\sigma}(\sigma^2 - f^2)R_\eta + (hR_u)' + \frac{fl}{\sigma}hR_u + i\left[\frac{f}{\sigma}(hR_v)' + lhR_v \right].
\tag{12.83}
$$

Two cases are possible, depending on the sign of $\sigma^2 - f^2 - l^2$. In both cases we use a system of fundamental solutions $Z_{1,2}$ of the homogeneous problem to solve the inhomogeneous problem, according to standard recipes of the theory of ordinary differential equations.

If $\sigma^2 - f^2 - l^2 = -p^2 < 0$, and σ and l *do not satisfy the dispersion relation for trapped waves*, there is no regular solution of the homogeneous equation. We then define

$$
\begin{aligned}
Z_1|_{x \to 0} = C\phi_1 \log x + \bar{\phi}, \quad Z_1|_{x \to \infty} \sim e^{-px}, \\
Z_2|_{x \to 0} = \phi_1, \quad Z_2|_{x \to \infty} \sim e^{px},
\end{aligned}
\tag{12.84}
$$

with the Wronskian $W = Z_1 Z_2' - Z_1' Z_2 = \frac{2p}{h}$. According to the general rules, the solution of (12.82) decaying at infinity is written as

$$
Z = -\frac{1}{2p}\left(Z_1 \int_0^x dx \, R_z Z_2 - Z_2 \int_x^\infty dx R_z Z_1 \right),
\tag{12.85}
$$

and can be easily shown to be regular. Hence, no special regularity condition is needed in this case.

If $\sigma^2 - f^2 - l^2 = -p^2 < 0$, and σ and l *do satisfy the dispersion relation for trapped waves*, then one of the fundamental solutions, say Z_1, is regular, and Z_2 is singular both at $x = 0$ and $x \to \infty$:

$$Z_1|_{x\to 0} = \phi_1, \quad Z_1|_{x\to\infty} \sim e^{-px},$$

$$Z_2|_{x\to 0} = C\phi_1 \log x + \bar\phi, \quad Z_2|_{x\to\infty} \sim e^{px}. \tag{12.86}$$

The solution of (12.82) decaying at infinity takes the form

$$Z = -\frac{1}{2p}\left(Z_1 \int_0^x dx\, R_z Z_2 - Z_2 \int_x^\infty dx R_z Z_1\right) + cZ_1, \tag{12.87}$$

where c is an arbitrary constant. This solution cannot be regular at $x = 0$, unless the orthogonality condition,

$$\int_0^\infty dx R_z Z_1 = 0 \tag{12.88}$$

is satisfied.

Example: self-interactons of a single trapped mode and its excitation by a free wave

We illustrate the application of the technique of the previous subsection in the case of self-interaction of a single trapped mode, when zero-order solution is

$$(u^{(0)}, v^{(0)}, \eta^{(0)}) = A(T_1, \ldots)(iU, V, Z)(x)e^{i\theta} + c.c., \quad \theta = ly - \sigma t. \tag{12.89}$$

We suppose, for simplicity, that the double harmonic is not resonant, that is the double frequency 2σ and the wave number $2l$ do not belong to any dispersion curve for trapped waves. This means that there is no dependence on the first slow time in zero-order fields, and (12.79) takes the form

$$\begin{cases} u_t^{(1)} - fv^{(1)} + \eta_x^{(1)} = \bar R(x) + R_u(x)e^{i2\theta} + c.c., \\ v_t^{(1)} + fu^{(1)} + \eta_y^{(1)} = R_v(x)e^{i2\theta}, \\ \eta_t^{(1)} + \left(hu^{(1)}\right)_x + \left(hv^{(1)}\right)_y = R_\eta(x)e^{i2\theta}, \end{cases} \tag{12.90}$$

where all functions R on the right-hands side contain A^2. Hence, solution for the first nonlinear correction consists of a zero-mode, that is a mean current, and a double harmonic. Expressions for both are found by reducing the corresponding system (12.90) to a single equations for the free-surface elevation, and solving the latter along the lines of the previous subsection. The equation for the free-surface elevation of the double-harmonic part of the solution is

$$\left(hZ'\right)' + \left(4\sigma^2 - f^2 - 4l^2 h - \frac{fl}{\sigma}h'\right)Z = R_z^{11}, \tag{12.91}$$

where R_z^{11} is obtained from (12.83) by the replacement $\sigma \rightarrow 2\sigma$, $l \rightarrow 2l$. As in the previous subsection, the form of solution depends on the sign of $4\sigma^2 - f^2 - 4l^2$. If it is negative, the solution is trapped near the shore. In the opposite case $4\sigma^2 - f^2 - 4l^2 = k_{11}^2 > 0$ the solution belongs to the free-wave spectrum. In the first case, the self-interaction of the trapped mode does not lead to a leakage of the energy from the wave guide and the double harmonic amplitudes are localised in x and are real, as well as the zero mode. In the second case, self-interaction of the trapped mode results in the offshore radiation of the inertia–gravity wave with frequency 2σ, and wave numbers k_{11} and $2l$. The corresponding solution for the double harmonic is complex, non-localised in space and obeys the radiation condition, giving an outgoing wave $\sim e^{ik_{11}x}$ at large x. It is to be emphasised that without rotation the condition of far-field radiation is impossible to satisfy, and the double harmonic is always trapped near the shore. In the presence of rotation the resonance condition is verified for the trapped modes in narrow bands in the dispersion diagram corresponding to the pieces of dispersion curves between the curve $\sigma = \sqrt{l^2 + f^2}$ and the cone $\sigma = \pm l$, as shown in Figure 12.4. In both cases, the obtained corrections are regular, and in order to determine the slow-dependence of the trapped-wave amplitude on time, we have to go to the next order. The resonant terms in the second order are precisely of the form used in (12.81), with forcing terms proportional to A^3, plus the terms with derivatives in second slow time, cf. (12.80). Imposing the condition of elimination of resonances (12.88), we get the modulation equation

$$\partial_{T_2}A + iM|A|^2A = 0, \tag{12.92}$$

where the coeficient M is determined from the zero- and first-order solutions, and is given by a rather cumbersome expression which we do not present. The most important role is played by the imaginary part of M, as it determines the evolution of

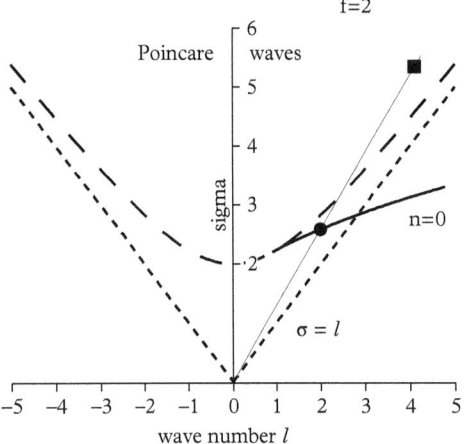

Figure 12.4 *Self-interaction of the trapped mode with frequency and wave number indicated by the circle on the dispersion curve $n = 0$ (thick solid line), giving a double harmonic belonging to the free-wave part of the spectrum, and indicated by the square.*

energy of the trapped mode $E \propto |A|^2$. Indeed, it is easy to get from (12.92) that $E = E_0 [1 - 2E_0 \, ImM \, T_2]^{-1}$, where $E_0 = E|_{t=0}$. As could be expected, if the double harmonic does not belong to the Poincaré continuum, $ImM = 0$, as in no-rotation case, the energy of the trapped mode is conserved, and does not leave the coastal wave guide. On the contrary, if the double harmonic does belong to the Poincaré continuum, M is complex, and the energy E changes with time. Obviously, the sign of ImM is of great importance. Expression for the coefficient M is rather complicated, and difficult to analyse analytically. Yet its numerical computations with the Ball's model always give negative values, which is natural, as otherwise the energy in the wave guide would blow up in finite time.

The decrease of energy of the trapped mode because of the free-wave radiation can be compensated by the incoming free wave with an amplitude of the order of ϵ, the frequency 2σ, and the wave numbers $(2l, -k_{11})$. To show this, it is sufficient to add to the first-order solution, which consisted of zero-mode and double-harmonic corrections denoted by overbar and index 11, respectively, a free wave:

$$(u^{(1)}, v^{(1)}, \eta^{(1)}) = (\bar{u}^{(1)}, \bar{v}^{(1)}, \bar{\eta}^{(1)}) + (u_{11}, v_{11}, \eta_{11}) + (u_{fr}, v_{fr}, \eta_{fr}),$$

denoted by the subscript *fr*. The Poincaré wave is proportional to $e^{2i\theta}$; therefore, its presence results in some additional terms in the amplitude equation (12.92). The modified amplitude equation becomes

$$\partial_{T_2} A + iM|A|^2 A + iM_{fr}A_{fr}A^* = 0. \tag{12.93}$$

The new term $iM_{fr}A_{fr}A^*$ appearing in this equation due to the incoming free wave is of great importance, as it provides exponential growth $\propto exp(|M_{fr}A_{fr}|T_2)$ of the trapped mode amplitude A, when A is small and the cubic term can be neglected. With increasing amplitude, the cubic term comes into play and gradually stops the growth. As a result, the amplitude stabilises at one of the two constant limit values:

$$A_{lim} = \frac{\sqrt{|M_{fr}||A_{fr}|}}{\sqrt{|M|}} e^{i\mu_s}, \quad \mu_s = \frac{1}{2}(\arg M_{fr} + \arg A_{fr} - \arg M) + \left(s - \frac{1}{2}\pi\right), \, s = 1, 2. \tag{12.94}$$

We therefore get a generation of the trapped mode by incident-reflected inertia–gravity wave (Poincaré wave) with double frequency and double alongshore wave number. An important point is that the amplitude of the Poincaré wave is small, of the order ϵ, while that of the excited trapped wave is much larger, of the order 1. Thus, an effect of amplification takes place: a relatively weak free wave resonantly excites much stronger trapped waves. If the amplitude of the generator free wave is considered to be of the order of 1, the asymptotic expansion of the solution should be reorganised, and start with terms of the order $\epsilon^{-\frac{1}{2}}$ for input trapped waves. A systematic expansion of this kind leads to the same equation (12.93). We will see examples of such modified asymptotic expansions in the next section.

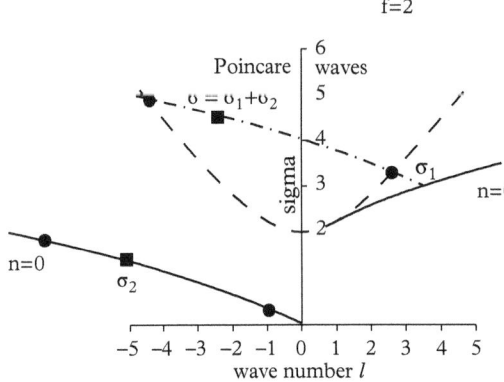

Figure 12.5 *Graphical representation of the synchronism condition for two trapped modes and the combinational free wave. The interaction of a given mode with σ_1, $n = 0$ (circle) with a mode with σ_2, $n = 0$ (lower square) gives a combination harmonic, in resonance with a free wave (upper square) with $\sigma = \sigma_1 + \sigma_2$, $l = l_1 + l_2$.*

Interactions between a pair of trapped modes and its resonant excitation by the free waves

Let us finally sketch what happens when we take a pair of different trapped waves as an input, and repeat the same reasoning. The solution of resonance conditions can be understood graphically, as in the previous subsection, and is illustrated in Figure 12.5. Elimination of resonances gives a system of coupled modulation equations for the amplitudes $A_{1,2}$ of the trapped waves:

$$\begin{cases} \partial_{T_2} A_1 + iM_{11} |A_1|^2 A_1 + iM_{12} |A_2|^2 A_1 + iM_{fr2} A_{fr} A_2^* = 0 \\ \partial_{T_2} A_2 + iM_{21} |A_1|^2 A_2 + iM_{22} |A_2|^2 A_2 + iM_{fr1} A_{fr} A_1^* = 0, \end{cases} \tag{12.95}$$

where the coefficients M_{ij} are given by convolutions of zeroth and first orders for waves i and j, and $M_{fr2} M_{fr1}^* > 0$, which provides a condition for resonant growth. A general property of (12.95) is that nonlinear saturation of the growing amplitudes of trapped waves consists in reaching either an *attracting fixed point* in the space of A_1, A_2, or of an *attracting limit cycle*. We will see in more details these patterns of saturation in the next Section. Finally, if slow spatial modulation of the wave-amplitudes is taken into account, the resulting modulation equations exhibit spatio-temporal organisation. Again, we will see in more detail how it happens in Section 12.3.

Resonant excitation of a trapped mode by pairs of free waves

As in the case of the abrupt shelf, resonant excitation of trapped waves by pairs of inertia–gravity waves is possible over the shelf. The difference, however, is that all waves, including Kelvin waves, are dispersive now, and the integrability conditions leading to modulation equations differ from the abrupt shelf case. Application of the 'naive' perturbation theory still leads to (12.79) and (12.80), which should be solved with initial fields of the form

$$(u^{(0)}, v^{(0)}, \eta^{(0)}) = \sum_{m=1,2} A_m (iU_m, V_m, Z_m)(x) e^{i\theta_m} + A(T_1, \ldots)(iU, V, Z)(x) e^{i\theta} c.c, \quad (12.96)$$

where the subscript m denotes inertia–gravity waves with fixed amplitudes, and $A(T_1)$ is the amplitude of a trapped wave, with modulation in slow time to be found. By applying the same procedure as above, we arrive at a very simple modulation equation:

$$\partial_{T_1} A = \bar{c} A_1 A_2 = \text{const}, \quad (12.97)$$

where $\bar{c}$ is some constant determined by U_m, V_m, Z_m. Thus, the interaction of the Poincaré waves results in a linear growth of the trapped mode. As in the cases considered earlier, this growth is to be bounded by the cubic nonlinearity induced by the self-interaction of the trapped mode. Taking into account these cubic terms in the amplitude equation requires a modification of the 'naive' expansions. Let the amplitude of the trapped mode be of the order of ϵ^β. In this case, self-interaction of the trapped mode produces a double harmonic $e^{2i\theta}$ of the order $\epsilon^{1+2\beta}$. The cubic terms result from the interaction of this double harmonic with the initial trapped mode and are of the order $\epsilon^{2+3\beta}$. In turn, the interaction of the Poincaré waves produces the terms of the order of ϵ. Obviously, for these terms to be of the order of the cubic terms, it is necessary that $\beta = -1/3$. In turn, this means that the modified expansions should have the form

$$
\begin{aligned}
(u, v, \eta) &= \epsilon^{-\frac{1}{3}} (u^{(0)}, v^{(0)}, \eta^{(0)})(x, y, t, T_1, T_2, \ldots) \\
&\quad + (u^{(1)}, v^{(1)}, \eta^{(1)})(x, y, t, T_1, T_2, \ldots) \\
&\quad + \epsilon^{+\frac{1}{3}} (u^{(2)}, v^{(2)}, \eta^{(2)})(x, y, t, T_1, T_2, \ldots), \quad T_n = \epsilon^{\frac{n}{3}}. \quad (12.98)
\end{aligned}
$$

A sequential analysis of solutions of increasing order is straightforward but cumbersome, and we omit the details. We only note that the lowest-order solution $(u^{(0)}, v^{(0)}, \eta^{(0)})$ now describes the trapped mode and the first-order solution $(u^{(1)}, v^{(1)}, \eta^{(1)})$ is the sum of two Poincaré waves. Then, the second- and the third-order solutions are determined by nonlinear interactions between the free waves and the trapped mode. Elimination of secular terms in the fourth-order equations for $(u^{(4)}, v^{(4)}, \eta^{(4)})$ gives the expected amplitude equation

$$\partial_{T_1} A + iM|A|^2 A = \bar{c} A_1 A_2, \quad (12.99)$$

where $\bar{c}$ is a constant, and the interaction coefficient M is the same as in (12.92). The steady solution of (12.99) is

$$A = c_0/|c_0|^{2/3}, \quad c_0 = -ic/M, \quad (12.100)$$

which is stable and attractive if $ImM < 0$. In the case $ImM = 0$, solutions of (12.99) are periodic, with maximal amplitude $A_{max} = \sqrt[3]{4|c_0|}$.

12.3   Resonant excitation of baroclinic Rossby and Yanai waves in the equatorial wave guide

12.3.1   Reminder on two-layer equatorial RSW and general conditions of removal of resonances

The same idea of parametric generation of wave-guide modes by free waves can be applied to the equatorial wave gude. We consider it in the framework of the two-layer RSW model with a rigid lid on the equatorial beta plane of Section 4.4.2. The model, in terms of barotropic–baroclinic decomposition, is given by (4.89) and (4.90), and we use it here with the same conventions and notation. We recall that the wave spectrum of the model consists of barotropic Rossby waves expressed in terms of the barotropic streamfunction: $\psi_0 = A_\psi e^{i(kx-\sigma t+ly)} + c.c.$, with non-dimensional dispersion relation $\sigma = -k/(k^2 + l^2)$, and baroclinic Kelvin, Yanai, Rossby, and inertia–gravity waves, expressed in terms of the components of the baroclinic velocity, and the position of the interface between the layers: $(u, v, h) = (iU_n(y), V_n(y), iH_n(y)) A e^{i(kx-\sigma_n t)} + c.c.$, with non-dimensional dispersion relation $\sigma_n^3 - (k^2 + 2n + 1)\sigma_n - k = 0; \; n = -1, 0, 1, 2, \ldots$. If a linear combination of these linear wave solutions is considered, the nonlineairity in (4.89) to (4.91) provides a forcing for the first nonlinear correction which can be resonant. As was already mentioned, the orthogonality and normalisation conditions are different for free and wave-guide eigenmodes of the linear spectrum. In order to understand how to treat resonances in the present case, let us consider the linearised system (4.92) with a forcing

$$\nabla^2 \psi_t + \psi_x = Q_\psi \tag{12.101}$$

$$u_t - yv + h_x = Q_u, \; v_t + yu + h_y = Q_v, \; h_t + u_x + v_y = Q_h, \tag{12.102}$$

where $Q_{u,v,h}$ are some forcing functions. The solution of (12.101) and (12.102) is bounded provided the following orthogonality conditions are satisfied:

$$\langle \hat{\psi} Q_\psi \rangle_{x,y,t} = 0, \tag{12.103}$$

$$\int_{-\infty}^{\infty} dy \, \langle \hat{u} Q_u + \hat{v} Q_v + \hat{h} Q_h \rangle_{x,t} = 0. \tag{12.104}$$

Here, $\hat{\psi}, \hat{u}, \hat{v}, \hat{h}$ represent an arbitrary bounded solution of the homogeneous equations, and angles denote averaging, for example:

$$\langle \ldots \rangle_x = \lim_{L_x \to \infty} \frac{1}{2L_x} \int_{-L_x}^{L_x} dx \ldots .$$

In what follows, the source terms are of the form

$$Q_{\psi,u,v,h}(y) = \sum_q Q^q_{\psi,u,v,h}(y)e^{i(k_q x - \sigma_q t)}$$

with $Q^q_{\psi,u,v,h}(y)$ rapidly decaying at $y \to \pm\infty$. It can be shown that for such $Q_{\psi,...,h}$ the conditions (12.103) and (12.104) are not only necessary but also sufficient for existence of bounded solutions, if $k_q \neq 0$, $\sigma_q \neq 0$. These conditions will be applied to remove resonances, together with introduction of slow modulation of wave amplitudes, as above.

12.3.2   Wave–wave resonances

Synchronism conditions

Because of special structure of nonlinear terms in (4.89), (4.90), only the interaction between one barotropic and two baroclinic waves can be resonant as barotropic–barotropic interaction is absent in (4.90). The synchronism conditions for a pair of trapped waves with zonal wave numbers $k_{1,2}$, and frequencies $\sigma_{1,2}$, and a barotropic free wave with k, σ are standard:

$$k_1 \pm k_2 = k; \quad \sigma_1 \pm \sigma_2 = \sigma, \tag{12.105}$$

where trapped waves have some meridional structure characterised by indices of Gauss–Hermite functions $n_{1,2}$, and the barotropic wave has a meridional wave number l. We are considering Rossby and Yanai waves, for which zonal wave numbers are negative, at positive frequencies. Effective energy transfer to the equatorial wave guide turns to be possible only for the upper sign in (12.105), which will be the only one considered hereafter. Finding the domains of the phase space, which are allowed for each type of resonance results in a series of algebraic problems. Although their solution is technically easier than in the case of coastal waves, we will not present technical details, and display the results for allowed domains in the phase space in Figure 12.6.

Resonant interactions between a pair of baroclinic Rossby/Yanai waves and a barotropic Rossby wave

To analyse resonant interactions we start from a wave triad obeying the synchronism conditions (12.105):

$$\psi^{(0)} = A_\psi(T)e^{i(\theta + ly)} + c.c., \quad (u^{(0)}, v^{(0)}, h^{(0)}) = \sum_{\alpha=1,2}(u^{(0)}_\alpha, v^{(0)}_\alpha, h^{(0)}_\alpha), \tag{12.106}$$

$$(u^{(0)}_\alpha, v^{(0)}_\alpha, h^{(0)}_\alpha) \propto (iU_\alpha(y), V_\alpha(y), iH_\alpha(y)) A_\alpha(T)e^{i\theta_\alpha} + c.c. \tag{12.107}$$

Here $\alpha = 1, 2$ distinguishes between two baroclinic waves, $\theta_{1,2} = k_{1,2}x - \sigma_{1,2}t$, $\theta = kx - \sigma t$, and we introduce modulation of wave amplitudes in slow time $T = \epsilon t$, and omit a normalisation factor in (12.107). We inject the straightforward asymptotic expansion

$$(\psi, u, v, h) = (\psi^{(0)}, u^{(0)}, v^{(0)}, h^{(0)})(x, y, t, T) + \epsilon(\psi^{(1)}, u^{(1)}, v^{(1)}, h^{(1)})(x, y, t, T) + ..., \tag{12.108}$$

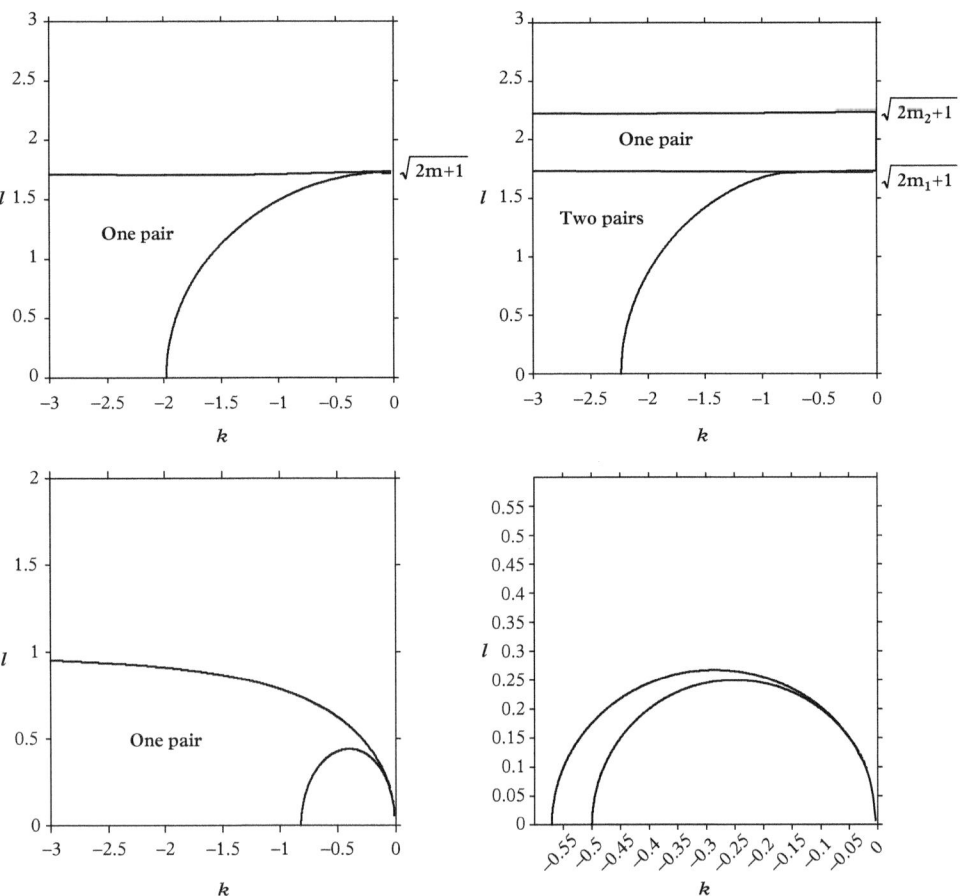

Figure 12.6 *Domains in the space of zonal k and meridional l wave numbers of an incoming barotropic Rossby wave allowed for resonance with trapped equatorial waves. One or two pairs of resonant wave guide modes exist in the domains, as indicated. From top to bottom and from left to right: two equatorial Rossby waves with identical meridional structure ($n_1 = n_2$), two equatorial Rossby waves with different meridional structure ($n_1 \neq n_2$), one equatorial Rossby and one Yanai wave, two Yanai waves (with only one resonant pair possible).*

in the equations of motion, and consider them order by order in ϵ. The equations for the first correction are

$$
\begin{cases}
\nabla^2 \psi_t^{(1)} + \psi_x^{(1)} = -\nabla^2 \psi_T^{(0)} + (NL)_\psi, \\
u_t^{(1)} - y v^{(1)} + h_x^{(1)} = -u_T^{(0)} + (NL)_u, \\
v_t^{(1)} + y u^{(1)} + h_y^{(1)} = -v_T^{(0)} + (NL)_v, \\
h_t^{(1)} + u_x^{(1)} + v_y^{(1)} = -h_T^{(0)} + (NL)_h.
\end{cases}
$$

The nonlinear interaction terms are

$$
\begin{cases}
(NL)_\psi = -\mathcal{J}(\psi^{(0)}, \nabla^2 \psi^{(0)}) - s(\partial_{xx} - \partial_{yy})(u^{(0)} v^{(0)}) + s\partial_{xy}(u^{(0)2} - v^{(0)2}), \\[2mm]
(NL)_u = \underline{-\mathcal{J}(\psi^{(0)}, u^{(0)})} + \mathbf{u}^{(0)} \cdot \nabla \psi_y^{(0)} - q\mathbf{u}^{(0)} \cdot \nabla u^{(0)} \\[2mm]
(NL)_v = \underline{-\mathcal{J}(\psi^{(0)}, v^{(0)})} - \mathbf{u}^{(0)} \cdot \nabla \psi_x^{(0)} - q\mathbf{u}^{(0)} \cdot \nabla v^{(0)} \\[2mm]
(NL)_h = \underline{-\mathcal{J}(\psi^{(0)}, h^{(0)})} - q\nabla \cdot (\mathbf{u}^{(0)} h^{(0)}),
\end{cases}
\tag{12.109}
$$

where only underlined contributions may contain resonances. Applying the resonance removal condition (12.104) with $(\hat{u}, \hat{v}, \hat{h}) = (\overline{u^{(0)}}_\alpha, \overline{v^{(0)}}_\alpha, \overline{h^{(0)}}_\alpha)$ results in the following equations for the slow evolution of the amplitudes of the baroclinic waves:

$$
A_{1_T} = L_1 A_\psi \bar{A}_2, \quad \bar{A}_{2_T} = \bar{L}_2 \bar{A}_\psi A_1,
\tag{12.110}
$$

which can be reduced to a single one for each amplitude (the overbars here and below denote complex conjugation):

$$
\partial_{TT}^2 A_{1(2)} = C |A_\psi|^2 A_{1(2)}, \quad C = L_1 \bar{L}_2.
\tag{12.111}
$$

It can be shown that $Im(C) = 0$. An important property of the 'sum' resonance in (12.105) is that $C > 0$, which means *exponential growth* of amplitudes in slow time. This result seems paradoxical, in comparison to triad interactions of Section 12.1, as it means that energy is not conserved within the resonant triad. The energy balance, therefore, should be clarified.

The energy is conserved in the system: $E = E_{bt} + E_{bc} = const$, where E_{bt}, E_{bc} are the barotropic and the baroclinic energies:

$$
E_{bt} = \int_{-\infty}^{\infty} dy \, \langle (\nabla \psi)^2 \rangle_x, \quad E_{bc} = \frac{s}{2} \int_{-\infty}^{\infty} dy \, \langle (1 + \epsilon q h)(u^2 + v^2 + h^2) \rangle_x.
$$

Expanding these expressions in ϵ, and omitting the constant energy of the initial barotropic wave, we see that to the lowest order the conserved energy is

$$
E_0 = \frac{s}{2} \int_{-\infty}^{\infty} dy \, \langle (u^{(0)2} + v^{(0)2} + h^{(0)2}) \rangle_x + \int_{-\infty}^{\infty} dy \, \langle \nabla \psi^{(0)} \cdot \nabla \psi^{(1)} \rangle_x.
\tag{12.112}
$$

The second term in this expression is interaction between primary and secondary barotropic modes, the latter appearing as a result of self-interaction of the baroclinic modes. Hence, the first barotropic correction (the barotropic response of the equator) is crucial. It is to be determined from the first equation in (12.109).

Nonlinear saturation of resonantly excited baroclinic waves

The first-order correction to the barotropic wave $\psi^{(1)}$ is quadratic in baroclinic wave amplitudes and, thus, is exponentially growing in slow time as the latter, becoming comparable to $\psi^{(0)}$ when the baroclinic amplitudes become of the order $\epsilon^{-1/2}$. It is the interaction between the secondary barotropic mode and the baroclinic waves which arrests the growth of the baroclinic amplitudes and leads to their saturation. To study this process, the asymptotic expansion should be rearranged in anticipation of resonant growth, as before:

$$\psi = \psi^{(0)}(x, y, t, T_1, T_2, ...) + \epsilon^{\frac{1}{2}} \psi^{(1)}(x, y, t, T_1, T_2, ...) + ...,$$

$$(u, v, h) = \epsilon^{-\frac{1}{2}} (u^{(0)}, v^{(0)}, h^{(0)})(x, y, t, T_1, T_2, ...) \tag{12.113}$$

$$+ (u^{(1)}, v^{(1)}, h^{(1)})(x, y, t, T_1, T_2, ...) + ...,$$

with introduction of a sequence of slow times $T_n = \epsilon^{n/2}t$. We do not present the details of order-by-order calculations, which now need to continue to the second order of the perturbation theory. Let us simply stress that assigning different orders of magnitude to barotropic and baroclinic fields in (12.113) renders the lowest-order equation for ψ inhomogeneous, with a source due to baroclinic self-interactions. So the solution for the barotropic component is sought from the very beginning as a sum of incoming wave and outgoing correction produced by this source. In order to determine such barotropic correction, which plays a crucial role as we have seen, the correct boundary condition at $y \to \pm\infty$ should be chosen. This is the *radiation boundary condition* with energy flux going out of the equator, according to the physical nature of this correction.

The calculations are simpler in the case of a pure parametric resonance, when the barotropic wave resonates with a trapped baroclinic wave of twice less frequency: $\sigma = 2\hat{\sigma}$, $k = 2\hat{k}$. In this case, the following modulation equation for the amplitude A of the baroclinic wave results:

$$\partial_{T_2} A + LA_\psi \bar{A} + (P + iQ) |A|^2 A = 0. \tag{12.114}$$

Here, $Q = Q_0 + qQ_1 + sQ_2$, L, P, Q are real, and $P \geq 0$. This is the classical Landau equation. The term $\propto L$ is due to the interaction of the primary harmonic barotropic wave with the zero-order baroclinic one, the term $\propto P + iQ_0$ is due to the interaction between the secondary barotropic mode and the zero-order baroclinic mode, the term containing Q_1 is due to the interaction between zero- and first-order baroclinic fields, and the term containing Q_2 is due to the cubic self-interaction of the zero-order baroclinic mode.

Stationary solutions of (12.114) are

$$A_0 = 0, \ A_\pm^2 = -\frac{LA_\psi}{P + iQ}.$$

It can be easily shown that A_0 is unstable, and $A_\pm$ are stable if $P > 0$ (neutrally stable if $P = 0$). Hence *nonlinear saturation* always takes place, as confirmed by Figure 12.7, where typical behavior of solutions in time is displayed. The increment and the level

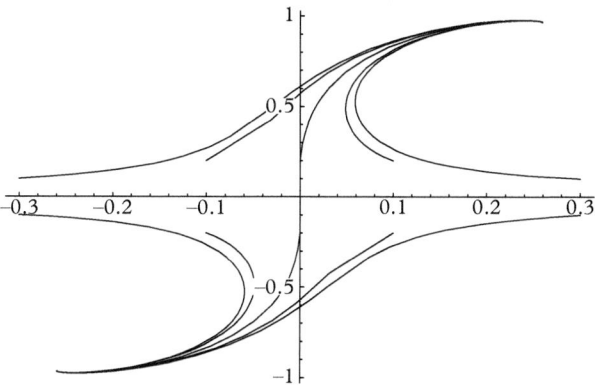

Figure 12.7 *Typical behaviour of the trajectories of solutions of the Landau equation in the phase space ReA – ImA. The trajectories in the upper (lower) half-plane are attracted to $A_{-(+)}$.*

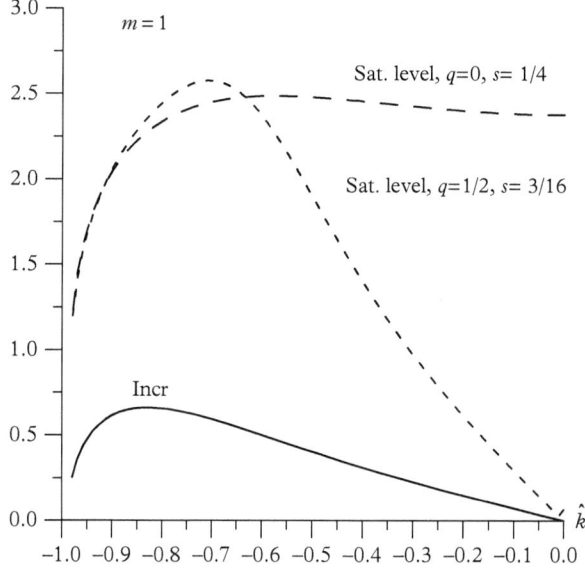

Figure 12.8 *Increment and saturation amplitude of the baroclinic Rossby wave with $n = 1$ as functions of zonal wave number $\hat{k}$ in the case of parametric resonance.*

of saturation of the baroclinic mode with a given meridional structure are sensible to its zonal wave number, and to the values of parameters of the model, as follows from Figure 12.8, both having maxima at some zonal wave number.

In the case of *two different baroclinic waves* excited by the barotropic wave, by applying the same procedure we arrive at a pair of coupled Landau equations for the amplitudes of the baroclinic waves:

$$\begin{cases} \partial_{T_2} A_1 + \alpha_1 \bar{A}_2 + \beta_1 |A_2|^2 A_1 + \gamma_1 |A_1|^2 A_1 = 0 \\ \partial_{T_2} A_2 + \alpha_2 \bar{A}_1 + \beta_2 |A_1|^2 A_2 + \gamma_2 |A_2|^2 A_2 = 0. \end{cases} \tag{12.115}$$

The following properties of the coefficients in these equations can be established:

$$\alpha_1 \bar{\alpha}_2 > 0, \ Re(\beta_1 + \beta_2) \leq 0, \ Re(\gamma_1) \leq 0, \ Re(\gamma_2) \leq 0, \tag{12.116}$$

with $\alpha_{1,2}$ either both real or both imaginary, and at least one of $Re(\gamma_1), Re(\gamma_2)$ negative. With these constraints nonlinear saturation of the solutions is guaranteed. However, in general, a stationary solution is an *attracting limit cycle*, with a frequency ω determined by the values of coefficients in (12.115):

$$A_{1,2} \propto e^{\pm i\omega T_2}, \tag{12.117}$$

and not a fixed point, which means that two resonantly excited equatorial waves undergo slow oscillations.

The effects of spatial modulation

Up to now we were considering slow modulation of wave amplitudes in time. Slow spatial dependence can be added along the same lines, by using multiple scales both in space and time in the asymptotic expansion:

$$\psi = \psi^{(0)}(\mathbf{x}, X_1, X_2, ..., t, T_1, T_2, ...)$$

$$+ \epsilon^{\frac{1}{2}} \psi^{(1)}(\mathbf{x}, X_1, X_2, ..., t, T_1, T_2, ...) + ...,$$

$$(u, v, h) = \epsilon^{-\frac{1}{2}}(u^{(0)}, v^{(0)}, h^{(0)})(\mathbf{x}, X_1, X_2, ..., t, T_1, T_2, ...)$$

$$+ (u^{(1)}, v^{(1)}, h^{(1)})(\mathbf{x}, X_1, X_2, ..., t, T_1, T_2, ...) + ... \tag{12.118}$$

Again, calculations are simpler in the case of pure parametric resonance with a single baroclinic wave excited by a barotropic one. The amplitude of the baroclinic wave is now a function of slow time and space variables: $A = A(T_1, T_2, ..., X_1, X_2, ...)$. In this case, in the first order of the perturbation theory which was giving a trivial result $A_{T_1} = 0$ without spatial modulation, we get

$$\partial_{T_1} A + c_g A_{X_1} = 0, \tag{12.119}$$

where $c_g = \hat{\sigma}'(\hat{k})$ is the group velocity of the baroclinic wave. Thus, as could be expected, the first effect of spatial dependence of the amplitude of the baroclinic wave is propagation of the modulations with the group velocity along the equator. Therefore, in the reference frame moving with the group velocity, the envelope A does not change its

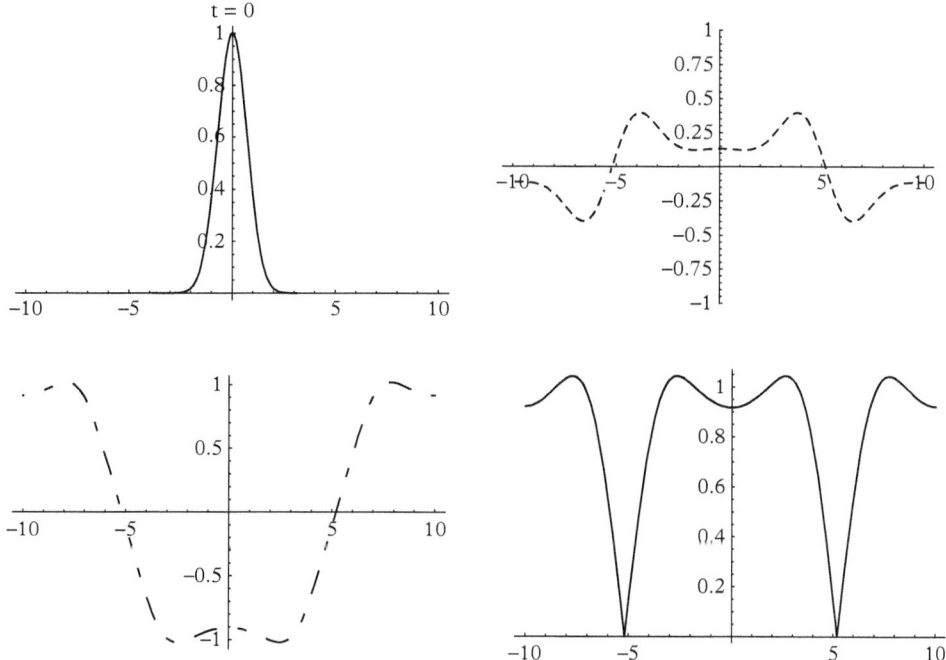

Figure 12.9 *Initial (top left), and final profiles of the real (dashed) and imaginary (dash-dotted) parts of A, and its amplitude (solid), as functions of X, as follows from numerical simulations of (12.120).*

form. In the next order of the perturbation theory, in the reference frame moving with the group velocity of the trapped wave, we get

$$\partial_{T_2}A - \frac{i}{2}\hat{\sigma}''(\hat{k})A_{X_1X_1} - LA_\psi\bar{A} + (P + iQ)\,|A|^2\,A = 0, \tag{12.120}$$

where $\hat{\sigma}''(\hat{k})$ is the derivative of the group velocity with respect to the zonal wave number $\hat{k}$. Equation (12.120) is of the Ginzburg–Landau (GL) type, and is well-known in non-linear optics or solid-state physics, for example. The spatially homogeneous solutions, given by (12.114), are still solutions, and are stable for $P > 0$. However, there also exist domain-wall type solutions, corresponding to two homogeneous solutions separated by a transition zone. A pair of such solutions can form a 'bubble'. Formation of such a bubble from a localised initial distribution $A(0, X)$ is shown in Figure 12.9.

In the case of a pair of baroclinic waves, the amplitude equations to the lowest order are

$$\partial_{T_1}A_i + c_{g_i}A_{iX_1} = 0, \quad i = 1, 2, \tag{12.121}$$

where c_{g_i} are the group velocities of the respective waves. To the next order:

$$\begin{cases} \partial_{T_2}A_1 + c_{g_1}A_{1X_2} - \frac{i}{2}\sigma_1''(k_1)A_{1X_1X_1} + \alpha_1\bar{A}_2 + \beta_1\,|A_2|^2\,A_1 + \gamma_1\,|A_1|^2\,A_1 = 0, \\ \partial_{T_2}A_2 + c_{g_2}A_{2X_2} - \frac{i}{2}\sigma_2''(k_2)A_{2X_1X_1} + \alpha_2\bar{A}_1 + \beta_2\,|A_1|^2\,A_2 + \gamma_2\,|A_2|^2\,A_2 = 0. \end{cases} \tag{12.122}$$

Equations (12.121) and (12.122) can be combined in the following 'synthetic' equations:

$$A_{1_{T_1}} + c_{g_1} A_{1_{X_1}} +$$

$$\epsilon^{\frac{1}{2}} \left(-\frac{i}{2}\sigma_1''(k_1)A_{1_{X_1 X_1}} + \alpha_1\bar{A}_2 + \beta_1 |A_2|^2 A_1 + \gamma_1 |A_1|^2 A_1 \right) = 0$$

$$A_{2_{T_1}} + c_{g_2} A_{2_{X_1}} + \qquad (12.123)$$

$$\epsilon^{\frac{1}{2}} \left(-\frac{i}{2}\sigma_2''(k_2)A_{2_{X_1 X_1}} + \alpha_2\bar{A}_1 + \beta_2 |A_1|^2 A_2 + \gamma_2 |A_2|^2 A_2 \right) = 0.$$

As generally $c_{g_1} \neq c_{g_2}$, it is impossible to remove the propagative effects in these equations by a change of reference frame. A number of interesting dynamical phenomena, like front formation and propagation, are described by these equations, which are, however, beyond the scope of the present chapter.

12.3.3 Wave mean current resonances

Synchronism conditions

As in the case of coastal waves, a mean current in the equatorial wave guide can be considered as an infinite-wavelength wave. Any zonal flow with baroclinic and barotropic components

$$\bar{u} = \bar{u}(y), \quad \bar{h} = \bar{h}(y), \quad y\bar{u} + \bar{h}_y = 0; \quad \bar{\psi} = \bar{\psi}(y). \qquad (12.124)$$

is an exact solution of the equations of the model. We take the same model as in Section 12.3.2, but for simplicity consider layers of equal depth, i.e. $q \equiv 0$. In order to have a resonance of a free Rossby wave with a trapped equatorial wave and a mean current, the barotropic and the baroclinic waves should have the same zonal wave number and frequency k, σ. As follows from respective dispersion relations, this is possible if

$$l^2 = 2n + 1 - \sigma_n^2, \qquad (12.125)$$

where n indicates the order of the Gauss–Hermite function describing the trapped wave. For baroclinic Rossby and Yanai waves $\sigma_n < 1$ and, hence, for a given baroclinic mode n a matching barotropic mode exists. In the case of Yanai wave, when $n = 0$ and $\sigma_0 = k/2 + \sqrt{1 + k^2/4}$ we have

$$l^2 = 1 - \sigma_0^2 = -k\sigma_0, \qquad (12.126)$$

and (12.125) is satisfied for any $k < 0$. In the case of Rossby wave, when $m \geq 1$ and, with high accuracy $\sigma_n \simeq -\frac{k}{k^2+2n+1}$, we obtain

$$l^2 \simeq 2n + 1 - \frac{k^2}{(k^2 + 2n + 1)^2} \approx 2n + 1. \qquad (12.127)$$

The relation (12.125) defines a curve in the (k, l) plane and thus, unlike the synchronism conditions in the wave-triad case of Section 12.3.2, only a discrete spectrum of barotropic Rossby waves will resonate with equatorial wave-guide modes in the presence of the equatorial current.

Resonant interactions and removal of resonances

As in Section 12.3.2, we start with a straightforward asymptotic expansion in ϵ, with $T = \epsilon t$:

$$(\psi, u, v, h) = (\psi^{(0)}, u^{(0)}, v^{(0)}, h^{(0)})(x, y, t, T, ...) + \epsilon(\psi^{(1)}, u^{(1)}, v^{(1)}, h^{(1)})(x, y, t, T, ...) + ...$$

and

$$(\psi^{(0)}, u^{(0)}, v^{(0)}, h^{(0)}) = (\tilde{\psi}^{(0)}, \tilde{u}^{(0)} + \bar{u}_0, \tilde{v}^{(0)}, \tilde{h}_0 + \bar{h}_0), \tag{12.128}$$

where tilde denotes the wave part comprising a free barotropic and a trapped baroclinic waves, and overbar denotes a zonal flow (12.124), which, for simplicity will be considered to be purely baroclinic, $\bar{\psi}(y) = 0$. By applying the conditions of absence of resonances established in Section 12.3.2, to the lowest order in ϵ we get that: 1) the barotropic wave remains unchanged and A_ψ does not depend on T, 2) the zonal flow $\bar{u}_0, \bar{h}_0$ does not change in time, and 3) the slow-time evolution of the amplitude of the baroclinic wave obeys the following equation:

$$A_T = -kL_\psi A_\psi, \tag{12.129}$$

where L_ψ is a (complex) interaction coefficient. As the resonance is possible only for negative k, (12.129) leads to *linear growth* of the amplitude of the baroclinic wave. (It can be shown that this conclusion is general and holds for layers of arbitrary depths, and in the presence of the barotropic component of the mean current as well). Note that the growth is linear, and not exponential as in the wave-triad case of Section 12.3.2. Considerations similar to those in Section 12.3.2 then apply, showing that the first barotropic wave correction is important to maintain the energy conservation, and that the perturbation series should be reorganised in order to account properly for this correction, which will be the source of saturation of the baroclinic wave amplitude. Solutions of the equations of motion are then sought in the form of generalised asymptotic expansion:

$$\psi^{(0)} = A_\psi e^{i(\theta + ly)} + c.c. + \epsilon^\gamma \bar{\psi}(y) + \psi^{(1)}(x, y, t, T_{\beta'}, \epsilon),$$

$$(u^{(0)}, v^{(0)}, h^{(0)}) = \epsilon^\alpha(\bar{u}_0, 0, \bar{h}_0)(y, T_{\alpha'}) + \epsilon^\beta(iU_m, \phi_m, iH_m)A(T_{\beta'})e^{i\theta}$$

$$+ (u^{(1)}, v^{(1)}, h^{(1)})(x, y, t, T_{\alpha'}, T_{\beta'}, \epsilon) + c.c, \tag{12.130}$$

where $T_{\alpha', \beta'} = \epsilon^{\alpha', \beta'} t$, $\alpha', \beta' > 0$, and it is supposed that the mean flow is sufficiently intense: $1 < \alpha \le 0$, $\gamma \le 0$. Eliminating resonances while finding the baroclinic correction $(u^{(1)}, v^{(1)}, h^{(1)})$ leads to a generalised modulation equation for A:

$$\epsilon^{\beta'} \partial_{T_{\beta'}} A + \epsilon^{2+2\alpha}(R+iS)A + \epsilon^{2+2\beta}(P+iQ)|A|^2 A = -\epsilon^{1+\alpha-\beta} kL_\psi. \tag{12.131}$$

Here R, S arise from wave–mean–mean interactions, and P, Q arise from three-wave interaction. It should be emphasised that this equation is not a result of a rigorous asymptotic procedure, but of an heuristic approach based on balancing different contributions. In order to confirm its validity, the parameters α, β, γ should be fixed and ordered, and asymptotic procedure performed, which can be always done. The 'optimal' for the excitation case is that of a strong mean flow, when $\alpha = \beta = -\frac{1}{2}$, which gives the modulation equation

$$A_{T_2} + (R+iS)A + (P+iQ)|A|^2 A = -kL_\psi A_\psi. \tag{12.132}$$

Two important points are to be stressed: it can be shown that 1) $R > 0$, and $P \geq 0$, which ensures the saturation of the amplitude of the baroclinic wave; 2) $P = 0$, if $l^2 - 3k^2 < 0$, and $P \neq 0$, if $l^2 - 3k^2 > 0$, which gives different properties of saturated solutions depending on the angle of incidence of the incoming barotropic wave.

Analysis of saturated solutions

By renormalising A and T the number of relevant parameters in (12.132) can be reduced:

$$\partial_T A + e^{i\xi} A + e^{i\eta}|A|^2 A = c_0 |A_\psi| \equiv c, \quad Im\, c_0 = 0. \tag{12.133}$$

For time-independent solutions, a cubic equation for $|A|^2$ results:

$$|A|^6 + 2\cos\chi\, |A|^4 + |A|^2 - c^2 = 0, \quad \chi = \xi - \eta. \tag{12.134}$$

It has either three positive roots, or a single positive root. Analysis of these roots show that 1) in the case of a single root, it is always stable and attractive, and 2) in the case of three different roots, the largest and the smallest are stable and attractive, while the intermediate one is unstable. We show in Figure 12.10 the phase portraits of (12.133) in the two cases. It is to be emphasised that in the second case the saturated amplitude is sensitive to initial conditions, because the phase space is non-trivially divided in domains of attraction of two stable solutions.

Effects of spatial modulation

We consider the case of strong baroclinic zonal current $\sim \epsilon^{-\frac{1}{2}}$, with $\alpha = \beta = -\frac{1}{2}$, and introduce slow spatial modulation in the zonal direction of the baroclinic *and* barotropic waves with the scale $X = \epsilon^{\frac{1}{2}} x$. The 'synthetic' modulation equations for A and A_ψ follow:

$$\left(\partial_{T_1} + c_g^{bt} \partial_X\right) A_\psi - \epsilon^{\frac{1}{2}} \frac{i}{2} \left(\sigma^{bt}\right)'' \partial_{XX}^2 A_\psi = 0, \tag{12.135}$$

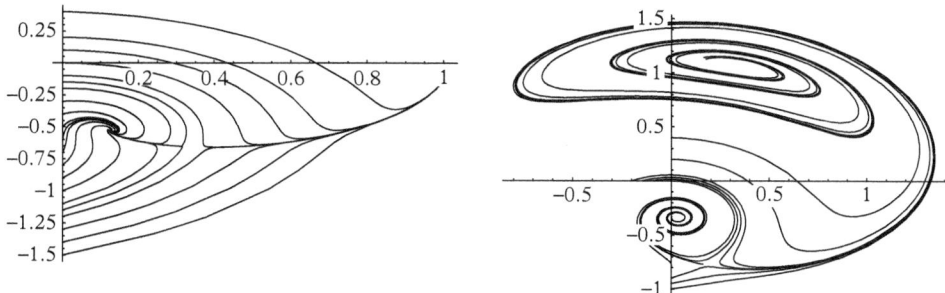

Figure 12.10 *Phase portraits of the system (12.133) in the plane ReA − ImA.* Left panel:
$\eta = -.4\pi$, $\xi = 19\pi/20$, $c = .4$; Right panel: $\eta = -\pi/2$, $\xi = 19\pi/20$, $c = .3$.

$$\left(\partial_{T_1} + c_g^{bc}\partial_X\right)A + \epsilon^{\frac{1}{2}}\left[-\frac{i}{2}\left(\sigma^{bc}\right)''\partial_{XX}^2 A + (R + iS)A\right.$$

$$\left. + (P + iQ)|A|^2 A\right] = -\epsilon^{\frac{1}{2}}c_0 A_\psi. \tag{12.136}$$

The group velocity of the Yanai wave may differ significantly from the group velocity of the barotropic Rossby wave of the same frequency. For long waves, $k \ll 1$, $c_g^{bc} \approx \frac{1}{2} \ll c_g^{bt} \approx -\frac{1}{k}$. On the contrary, the group velocities of the baroclinic and the barotropic Rossby waves of the same frequency are practically the same. In the former case, the only situation where barotropic and baroclinic waves have possibility to interact is that of 'gentle' modulation when the fields depend on $X_1 = \epsilon x$, and not on X, and on $T_2 = \epsilon t$, and not on T_1. In this case dispersion effects are weak, and

$$\partial_{T_2}A_\psi + c_g^{bt}\partial_{X_1}A_\psi = 0, \tag{12.137}$$

$$\partial_{T_2}A + c_g^{bc}\partial_{X_1}A + (R + iS)A + (P + iQ)|A|^2 A = -c_0 A_\psi. \tag{12.138}$$

In the latter case, by choosing the reference frame moving with the common group velocity, we obtain

$$\partial_{T_1}A_\psi - \frac{i}{2}\left(\sigma^{bt}\right)''\partial_{XX}^2 A_\psi = 0, \tag{12.139}$$

$$\partial_{T_1}A - \frac{i}{2}\left(\sigma^{bc}\right)''\partial_{XX}^2 A + (R + iS)A + (P + iQ)|A|^2 A = -c_0 A_\psi. \tag{12.140}$$

This is a Ginzburg - Landau equation for A forced by the wave packet of barotropic waves which, in turn, is subject to dispersion. If there is no spatial modulation of the barotropic wave (i.e. a plane barotropic wave occupies the whole equatorial plane) we get the equation (12.140) with constant A_ψ. We should stress that in the case where two stationary independent solutions of the modulation equation without spatial modulation exist, the domain-wall-like structures, presented in the Section 12.3.2, and related spatio-temporal organisation, will appear.

12.4 Summary, comments, and bibliographic remarks

The main message of this chapter is that classical resonant wave interactions recalled in Section 12.1 lead to highly nontrivial phenomena, if applied to triads composed of free and wave-guide modes. This kind of interaction leads to energy exchanges between atmospheric and oceanic wave guides and far wave fields, and to specific slow-time large-scale variability in the wave guides, with characteristic time and spatial scales determined by intrinsic properties of the system.

We described the mechanisms of resonant excitation of trapped coastal waves by inertia–gravity waves coming from the large. In the case of an abrupt shelf, excitation of dispersionless Kelvin waves was described both in barotropic and baroclinic configurations, including excitation of a barotropic Kelvin wave by a free inertia–gravity wave impinging upon a coastal current. Such situations are ubiquitous in the ocean. As we have shown, the wavelength of the incoming inertia–gravity wave should be comparable with the width of the coastal current; that is the wave should be long enough for efficient generation. This mechanism provides a route of dissipation of energy in the ocean, as the energy transported by inertia–gravity waves is transferred to the coastal wave guide, and dissipated there due to the breaking of Kelvin waves and formation of Kelvin fronts (cf. Fedorov and Melville 1995). Although dissipation was not explicitly included in our considerations, it is easy to anticipate what would happen if it was. Harmonically forced Burgers equation would result in this case instead of a simple-wave equation. It is the same as (12.34), with an additional term $\nu\nabla^2 K$, and can be also integrated. The sharp shocks then would be replaced by smoothed profiles with localised zones of strong gradients. Burgers equation is known for clustering properties of related transport of matter (cf. Bec and Khanin 2007) for a review on Burgers equation and its properties. Thus, the described parametric excitation mechanism of Kelvin waves leads to specific transport and mixing properties in the coastal zones, with obvious practical implications. It is worth emphasising that the realisation of the excitation mechanism in the baroclinic case, where a barotropic Kelvin wave can be excited by a pair of baroclinic waves, or a baroclinic wave and coastal current, amounts to appearance of a Kelvin wave 'from nothing', as the baroclinic waves, which can be associated with internal waves in the ocean with weak signature at the surface, give rise to a barotropic wave with a clear signature at the surface. If the two-layer model is applied to model the real ocean, the density ratio r should be taken to be close to unity. At the same time, the upper layer should be shallower than the lower one. Corresponding changes may be done in the above-derived expressions, yet the realistic situation of continuously stratified ocean is much more interesting for applications. As is well known, cf. for example, Gill (1982), by separation of variables in the vertical and horizontal directions in the continuously stratified primitive equations one arrives to a solution of the linearised equations in the form of a series of baroclinic modes. For each baroclinic mode the dispersion relation is analogous to the rotating shallow-water one, with the phase velocity depending on the vertical structure. Therefore, it is possible to satisfy the resonance conditions of the discussed types for frequency and the horizontal wave number, and the mechanism works. However, compatibility with vertical structure implies additional constraints, to be analysed.

In view of practical importance of the described phenomena, a question of efficiency of the excitation mechanism naturally arises. Above, the inertia–gravity wave impinging on the coast were considered to be harmonic, and only one pair of impinging waves was considered. Needless to say that a given Kelvin wave may be excited by different pairs of inertia–gravity waves, if their parameters belong to the resonance region, and contributions of all such pairs should be summed up on the right-hand side of (12.34). In reality, the inertia–gravity waves arrive at the coast in wave packets, being produced, for example, by distant storms. The efficiency of energy transfer would depend then on the characteristics of these wave packets. Although the calculations in this chapter were displayed in the case when the amplitude of the wave and the magnitude of the mean current are both small, following the argument of Section 12.3.3 on resonant wave excitation in the presence of the mean equatorial current, we can deduce that if the magnitude of the current is much larger than that of the wave, but not too large, of the order $\epsilon^{-\frac{1}{2}}$, the same evolution equation (12.70) results, but with a 'faster' slow time $T_{\frac{1}{2}} = \epsilon^{\frac{1}{2}} t$.

We also showed how the resonant, or parametric, excitation mechanisms work for dispersive coastal waves evolving over a shelf. Although we did not gave much of technical details, the full analysis of all possible cases can be found in Reznik and Zeitlin (2011). Generation of coastal waves by this mechanism leads to slow variability and structure formation in the coastal zone. We thus see that several mechanisms of resonant excitation of trapped coastal waves exist, all leading to formation of coherent and/or dissipative structures in the coastal zones. Resonant pairs of free inertia–gravity waves, or resonances of free inertia–gravity waves with coastal currents can produce coastal waves of significant amplitude from a small-amplitude noise.

The analysis of excitation of the equatorial waves of Section 12.3 is related to the question of teleconnections between mid-latitudes and tropics in the context of global weather and climate. As we show, this mechanism is ubiquitous, however its efficiency, like in the case of coastal waves, depends on intensity and duration of the trains of Rossby waves arriving to the tropics. As in the case of the coastal waves, the resulting modulation equations for wave-guide modes exhibit specific patterns of spatio-temporal variability, which are important to bear in mind while analysing observational and computational data. It should be stressed that these modulation equations are of the types arising in other areas of nonlinear physics, which allows us to use the corpus of knowledge and methods of this field in geophysics. It is also to be emphasised that in the case of excitation of equatorial waves by free wave–trapped wave–mean current resonances, the mean equatorial current acts like a resonator, amplifying eigenmodes with selected frequency, unlike the wave-triad excitation, where the spectrum of excited modes can be continuous.

It is worth repeating that the phenomena considered in this chapter make a link of geophysical fluid dynamics with other branches of nonlinear physics such as nonlinear optics. Although the scales and the nature of waves are completely different, similar evolution equations for wave envelopes arise, and are treated by the same methods.

The presentation in Section 12.1 is standard. The first paper considering resonant triads of Rossby waves was probably Kenyon (1964). The idea of resonant, or parametric, excitation of the coastal waves by incoming gravity waves, the topic of Section 12.2, has a long history, although only the resonance of a single free wave with a pair of coastal waves was traditionally considered by Guza and Bowen (1975), Minzoni and Whitham (1977), Akylas (1983), and Miles (1990). The full analysis of this case was given in Reznik and Zeitlin (2011), where details of various particular cases, estimates of efficiency of the processes, and their dependence on parameters can be found. The elements of the ordinary differential equations theory used in this chapter can be found in textbooks (e.g. Morse and Feshbach (1953)). Resonant excitation of Kelvin waves by pairs of inertia–gravity waves in the barotropic case was described in Reznik and Zeitlin (2009a), and generalised to the baroclinic case in Zeitlin (2013). The presentation in Section 12.3 follows Reznik and Zeitlin (2007) and Reznik and Zeitlin (2009b). Again, the details of (often cumbersome) calculations can be found in these papers.

Figure 12.1 is reproduced from Reznik and Zeitlin (2009a), with the permission, and Figure 12.2 is reproduced from Zeitlin (2013), with permission. Figures 12.3 to 12.5 are reproduced from Reznik and Zeitlin (2011), with the permission. Figures 12.6 to 12.9 are reproduced from Reznik and Zeitlin (2007), and Figure 12.10 from Reznik and Zeitlin (2009b), with permissions.

13

Wave Turbulence

As we have already seen in Chapter 12, triadic interactions of waves lead to specific dynamical phenomena, if conditions of wave resonance are verified. In this chapter, unlike Chapter 12 where we were investigating separate triads, wave interactions will be considered within large ensembles of waves with a wide spectrum corresponding to so-called wave turbulence, where a statistical approach is applicable. We will first sketch the standard wave-turbulence theory and methods, and then give applications to rotating shallow water model.

13.1 The main hypotheses and ideas of the wave turbulence approach

A systematic way to describe non-dissipative wave systems is to use the Hamiltonian description. A universal general scheme arises in this framework, which can be then applied to any concrete system. It should be emphasised that a direct approach with the help of perturbation theory in small nonlinearity applied to an ensemble of waves, and ensemble-averaging under the random-phase approximation leads to the same results in any concrete case.

13.1.1 A reminder on Hamiltonian description of wave systems

Let us consider an ensemble of large number of weakly nonlinear, small-amplitude, harmonic waves, each of which has the form

$$\hat{a}_{\mathbf{k}}^{(0)} e^{i(\mathbf{k} \cdot \mathbf{x} - \omega(\mathbf{k})t)} + c.c., \tag{13.1}$$

where $\hat{a}_{\mathbf{k}}^{(0)}$ is a complex wave amplitude, $\Phi(\mathbf{k}) = \mathbf{k} \cdot \mathbf{x} - \omega(\mathbf{k})t$ is the phase, $\mathbf{k}$ is wave number, and $\omega = \omega(\mathbf{k})$ is the dispersion relation. The latter gives a spectrum of infinitesimal perturbations of linearised over some stationary state equations of motion of the underlying model. The eigenfrequencies $\omega(\mathbf{k})$ are assumed to be positive. In the simplest case when there is only one species of waves in the model, such linearised equations can be cast in the form

Geophysical Fluid Dynamics. Vladimir Zeitlin,
Oxford University Press (2018). © Vladimir Zeitlin. DOI: 10.1093/oso/9780198804338.001.0001

$$\dot{a}_{\mathbf{k}} + i\omega(\mathbf{k})a_{\mathbf{k}} = 0; \quad \text{with} \quad a_{\mathbf{k}} = \hat{a}_{\mathbf{k}}^{(0)} e^{-i\omega(\mathbf{k})t}, \tag{13.2}$$

where the dot notation is used for time derivatives. If there are several wave species, complex Fourier amplitude should be introduced for each of them (see the examples below). Note that if $a_{\mathbf{k}}$ is the Fourier transform of some real space-time field, then $a_{\mathbf{k}}^* = a_{-\mathbf{k}}$, where the asterisk denotes complex conjugation. Equation (13.2) can be written in Hamiltonian form:

$$\dot{a}_{\mathbf{k}} = \{a_{\mathbf{k}}, H_0\} = -i\frac{\delta H_0}{\delta a_{\mathbf{k}}^*}, \tag{13.3}$$

with the Hamiltonian

$$H_0 = \int d\mathbf{k}\omega(\mathbf{k})\, a_{\mathbf{k}}a_{\mathbf{k}}^*, \tag{13.4}$$

and the Poisson bracket

$$\{A, B\} = \frac{\delta A}{\delta a_{\mathbf{k}}}\frac{\delta B}{\delta a_{\mathbf{k}}^*} - \frac{\delta B}{\delta a_{\mathbf{k}}}\frac{\delta A}{\delta a_{\mathbf{k}}^*}, \tag{13.5}$$

obeying the Jacobi identity. The identity $\frac{\delta a_{\mathbf{k}}}{\delta a_{\mathbf{k}'}} = \delta(\mathbf{k}-\mathbf{k}')$, where $\delta(\mathbf{k})$ is Dirac's delta function, is to be used while calculating variational derivatives. Note that, as compared to the brief description of the Hamiltonian formalism in Section 2.2.1, the phase space of the variables $a_{\mathbf{k}}$, $a_{\mathbf{k}}^*$ is complex, and these variables can be thought of as complex combinations of canonical coordinates $\mathcal{Q}$ and momenta $\mathcal{P}$: $a_{\mathbf{k}} = \mathcal{P}_{\mathbf{k}} + i\mathcal{Q}_{\mathbf{k}}$. The Hamiltonian of free waves H_0 is thus the energy of an ensemble of non-interacting harmonic oscillators with frequencies $\omega(\mathbf{k})$, and the quantities $a_{\mathbf{k}}a_{\mathbf{k}}^*$ give intensity of oscillations with a given wave number $\mathbf{k}$.

The full equations of any model are, generally, nonlinear, and nonlinear terms engender interactions of harmonic waves. The full Hamiltonian contains an interaction term, $H = H_0 + H_{int}$, with H_{int} which may be expanded in powers of $a_{\mathbf{k}}$ and their complex conjugates $a_{\mathbf{k}}^*$: $H_{int} = H_3 + H_4 + \dots$ where

$$H_3 = \frac{1}{2}\int d\mathbf{k}d\mathbf{k}_1 d\mathbf{k}_2\, \delta(\mathbf{k} - \mathbf{k}_1 - \mathbf{k}_2)\, V_{\mathbf{k},\mathbf{k}_1,\mathbf{k}_2} a_{\mathbf{k}}^* a_{\mathbf{k}_1} a_{\mathbf{k}_2} + c.c., \tag{13.6}$$

and

$$H_4 = \frac{1}{2}\int d\mathbf{k}d\mathbf{k}_1 d\mathbf{k}_2 d\mathbf{k}_3\, \delta(\mathbf{k} + \mathbf{k}_1 - \mathbf{k}_2 - \mathbf{k}_3)\, W_{\mathbf{k},\mathbf{k}_1,\mathbf{k}_2,\mathbf{k}_3} a_{\mathbf{k}}^* a_{\mathbf{k}_1}^* a_{\mathbf{k}_2} a_{\mathbf{k}_3}, \tag{13.7}$$

for statically stable homogeneous media, where a decay of vacuum in three or more waves is impossible. The full equations of motion are, correspondingly,

$$\dot{a}_{\mathbf{k}} = \{a_{\mathbf{k}}, H_0 + H_{int}\} = -i\omega(\mathbf{k})a_{\mathbf{k}} - i\frac{\delta H_{int}}{\delta a_{\mathbf{k}}^*}, \tag{13.8}$$

and we recall that wave amplitudes are always supposed to be small and, hence, the interaction Hamiltonian gives only small corrections to the linear solutions. It is clear from (13.8) that a cubic Hamiltonian H_3 produces quadratic nonlinearity in the equations of motion, the quartic one H_4 produces cubic nonlinearity, etc.

Let us consider as an example a system with cubic Hamiltonian, with higher-order interactions being absent or neglected. Equations of motion (13.8) take the form

$$\dot{a}_{\mathbf{k}} = -i\omega(\mathbf{k})a_{\mathbf{k}} - i\int d\mathbf{k}_1 d\mathbf{k}_2\, \delta(\mathbf{k} - \mathbf{k}_1 - \mathbf{k}_2)\left[V_{\mathbf{k},\mathbf{k}_1,\mathbf{k}_2}a_{\mathbf{k}_1}a_{\mathbf{k}_2} + 2V^*_{\mathbf{k}_1,\mathbf{k},-\mathbf{k}_2}a_{\mathbf{k}_1}a^*_{-\mathbf{k}_2}\right] \tag{13.9}$$

where relabelling of integration variables was used, together with the symmetry of the interaction coefficients $V_{\mathbf{k},\mathbf{k}_1,\mathbf{k}_2}$ with respect to permutations of its arguments, to simplify the resulting expression. Solution of (13.9) can be represented as $a_{\mathbf{k}}^{(0)}(t) + a_{\mathbf{k}}^{(1)}(t)$, where $a^{(0)}$ is a solution of the linearised equations (13.2). Obviously, $a_{\mathbf{k}}^{(0)}(t) = \hat{a}_{\mathbf{k}}^{(0)}e^{-i\omega(\mathbf{k})t}$, and if the first nonlinear correction $a_{\mathbf{k}}^{(1)}(t)$ is sought in the form $a_{\mathbf{k}}^{(1)}(t) = \int d\omega\, e^{-i\omega t}\hat{a}_{\mathbf{k}}^{(1)}(\omega)$, we obtain

$$\hat{a}_{\mathbf{k}}^{(1)}(\omega) = \frac{1}{\omega - \omega(\mathbf{k})}\int d\mathbf{k}_1 d\mathbf{k}_2\, \delta(\mathbf{k} - \mathbf{k}_1 - \mathbf{k}_2)\left[V_{\mathbf{k},\mathbf{k}_1,\mathbf{k}_2}\hat{a}_{\mathbf{k}_1}^{(0)}\hat{a}_{\mathbf{k}_2}^{(0)}e^{-i(\omega(\mathbf{k}_1)+\omega(\mathbf{k}_2))}\right.$$

$$\left. + 2V^*_{\mathbf{k}_1,\mathbf{k},-\mathbf{k}_2}\hat{a}_{\mathbf{k}_1}^{(0)}\hat{a}_{-\mathbf{k}_2}^{(0)*}e^{-i(\omega(\mathbf{k}_1)-\omega(\mathbf{k}_2))}\right]. \tag{13.10}$$

In order to get $a_{\mathbf{k}}^{(1)}(t)$, an inverse Fourier transformation should be applied to (13.10), and the singularity of the resulting integral in ω at $\omega = \omega(\mathbf{k})$ should be properly treated. Shifting the singularity in the upper complex half-plane by infinitesimal amount $i\epsilon$, and taking the integral by the method of residues, results in a retarded reaction of the wave system to the perturbation, while the opposite choice results in an advanced reaction, which is physically unreasonable. With a choice of retarded reaction, the Fourier transformation leads to the expression

$$a_{\mathbf{k}}^{(1)}(t) \propto \int d\mathbf{k}_1 d\mathbf{k}_2\, \delta(\mathbf{k} - \mathbf{k}_1 - \mathbf{k}_2)\left[V_{\mathbf{k},\mathbf{k}_1,\mathbf{k}_2}\frac{a_{\mathbf{k}_1}^{(0)}a_{\mathbf{k}_2}^{(0)}}{\omega(\mathbf{k}) - \omega(\mathbf{k}_1) - \omega(\mathbf{k}_2) + i\epsilon}\right.$$

$$\left. + 2V^*_{\mathbf{k}_1,\mathbf{k},-\mathbf{k}_2}\frac{a_{\mathbf{k}_1}^{(0)}a_{-\mathbf{k}_2}^{(0)*}}{\omega(\mathbf{k}) - \omega(\mathbf{k}_1) + \omega(\mathbf{k}_2) + i\epsilon}\right]. \tag{13.11}$$

We see that *resonant triads* obeying the relations

$$\omega(\mathbf{k}) - \omega(\mathbf{k}_1) \pm \omega(\mathbf{k}_2) = 0, \quad \mathbf{k} - \mathbf{k}_1 - \mathbf{k}_2 = 0, \tag{13.12}$$

give dominant contributions to the first nonlinear correction to the wave field, if the dispersion relation allows for nontrivial solutions of (13.12). The spectrum allowing for resonant triads is called *decay spectrum*; in the opposite case it is a *non-decay spectrum*. It should be emphasised that in the case of the non-decay spectrum *resonant quartets* always exist, obeying

$$\omega(\mathbf{k}) + \omega(\mathbf{k}_1) = \omega(\mathbf{k}_2) + \omega(\mathbf{k}_2), \quad \mathbf{k} + \mathbf{k}_1 = \mathbf{k}_2 + \mathbf{k}_3, \tag{13.13}$$

with, for example, $\mathbf{k} = \mathbf{k}_2$, $\mathbf{k}_1 = \mathbf{k}_3$ for any dispersion law. In the case of non-decay spectrum, a canonical transformation, that is a transformation produced by Poisson brackets of the phase-space variables with some generating function, exists, allowing us to remove the cubic part of the Hamiltonian:

$$a_\mathbf{k} \rightarrow a_\mathbf{k} - \int d\mathbf{k}_1 d\mathbf{k}_2 \, \delta(\mathbf{k} - \mathbf{k}_1 - \mathbf{k}_2) \frac{V_{\mathbf{k}\mathbf{k}_1\mathbf{k}_2}}{\omega(\mathbf{k}) - \omega(\mathbf{k}_1) - \omega(\mathbf{k}_2)} a_{\mathbf{k}_1} a_{\mathbf{k}_2}$$

$$- 2 \int d\mathbf{k}_1 d\mathbf{k}_2 \, \delta(\mathbf{k} + \mathbf{k}_1 - \mathbf{k}_2) \frac{V^*_{\mathbf{k}\mathbf{k}_1\mathbf{k}_2}}{-\omega(\mathbf{k}) + \omega(\mathbf{k}_1) - \omega(\mathbf{k}_2)} a^*_{\mathbf{k}_1} a_{\mathbf{k}_2}. \tag{13.14}$$

A normal form of quartic Hamiltonian thus follows:

$$H = \int d\mathbf{k}\,\omega(\mathbf{k})\, a_\mathbf{k} a^*_\mathbf{k}$$
$$+ \tfrac{1}{2} \int d\mathbf{k} d\mathbf{k}_1 d\mathbf{k}_2 d\mathbf{k}_3 \, \delta(\mathbf{k} + \mathbf{k}_1 - \mathbf{k}_2 - \mathbf{k}_3) \, T_{\mathbf{k}\mathbf{k}_1\mathbf{k}_2\mathbf{k}_3} \, a^*_\mathbf{k} a^*_{\mathbf{k}_1} a_{\mathbf{k}_2} a_{\mathbf{k}_3}, \tag{13.15}$$

where

$$T_{\mathbf{k}\mathbf{k}_1\mathbf{k}_2\mathbf{k}_3} = -2 \frac{V_{\mathbf{k}+\mathbf{k}_1,\mathbf{k},\mathbf{k}_1} V^*_{\mathbf{k}_2+\mathbf{k}_3,\mathbf{k}_2,\mathbf{k}_3}}{\omega(\mathbf{k}+\mathbf{k}_1) - \omega(\mathbf{k}) - \omega(\mathbf{k}_1)} - 2 \frac{V_{\mathbf{k},\mathbf{k}_2,\mathbf{k}-\mathbf{k}_2} V^*_{\mathbf{k}_3,\mathbf{k}_1,\mathbf{k}_3-\mathbf{k}_1}}{\omega(\mathbf{k}_3 - \mathbf{k}_1) + \omega(\mathbf{k}_1) - \omega(\mathbf{k}_3)}$$

$$-2 \frac{V_{\mathbf{k},\mathbf{k}_3,\mathbf{k}-\mathbf{k}_3} V^*_{\mathbf{k}_2,\mathbf{k}_1,\mathbf{k}_2-\mathbf{k}_1}}{\omega(\mathbf{k}_2 - \mathbf{k}_1) + \omega(\mathbf{k}_1) - \omega(\mathbf{k}_2)} - 2 \frac{V_{\mathbf{k}_1,\mathbf{k}_3,\mathbf{k}_1-\mathbf{k}_3} V^*_{\mathbf{k}_2,\mathbf{k},\mathbf{k}_2-\mathbf{k}}}{\omega(\mathbf{k}_3 - \mathbf{k}_1) + \omega(\mathbf{k}_1) - \omega(\mathbf{k}_3)}$$

$$- 2 \frac{V_{\mathbf{k}_1,\mathbf{k}_2,\mathbf{k}_1-\mathbf{k}_2} V^*_{\mathbf{k}_3,\mathbf{k},\mathbf{k}_3-\mathbf{k}}}{\omega(\mathbf{k}_2 - \mathbf{k}_1) + \omega(\mathbf{k}_1) - \omega(\mathbf{k}_2)} + W_{\mathbf{k}\mathbf{k}_1\mathbf{k}_2\mathbf{k}_3}. \tag{13.16}$$

By the same reasoning as the one used above in the decay spectrum case, we can show that dominant contributions into the first nonlinear correction to the wave field in the case of non-decay spectrum are given by resonant quartets.

13.1.2 The principal idea of wave turbulence approach

The main hypothesis of the weak turbulence theory is that weak interactions of a large number of waves lead to *phase randomisation* and *Gaussian statistics* of the wave field. As usual in statistical description, the ergodic hypothesis is supposed to be valid, meaning that time averages and ensemble averages are equivalent. Instead of the full dynamical description in the framework of (13.8) a statistical description is thus used, where the system is described by a set of correlation functions of complex amplitudes defined with the help of ensemble averaging $\langle...\rangle$. Gaussian statistics in a spatially uniform medium means that all odd-order correlation functions of wave amplitudes a_k vanish and that all even-order correlation functions are expressed in terms of the (real) quadratic one:

$$\langle a_k a_{k'} \rangle = 0, \quad \langle a_k a_{k'}^* \rangle = N(\mathbf{k})\delta(\mathbf{k} - \mathbf{k}'),$$

$$\langle a_{k_1}^* a_{k_2}^* a_{k_3} a_{k_4} \rangle = N(\mathbf{k_1})N(\mathbf{k_2})\left(\delta(\mathbf{k_1} - \mathbf{k_3})\delta(\mathbf{k_2} - \mathbf{k_4}) + \delta(\mathbf{k_1} - \mathbf{k_4})\delta(\mathbf{k_2} - \mathbf{k_3})\right). \quad (13.17)$$

The quantity $N(\mathbf{k})$ is the mean intensity of waves with wave number $\mathbf{k}$. The mean energy of waves in the leading approximation is thus $\langle H_0 \rangle = \int d\mathbf{k}\omega(\mathbf{k})N(\mathbf{k})$. The above-described statistical hypothesis, by definition, is applied to harmonic waves so, in fact, a_k should be read as $a_k^{(0)}$, and we omitted the superscript (0) in order not to overcharge the formulae. As all correlation functions are expressed in terms of the quadratic one, that is in terms of $N(\mathbf{k})$, the main goal of the statistical theory of waves is to determine the evolution of this quantity in time, which is given by the *kinetic equation*

$$\dot{N}(\mathbf{k}) = \mathcal{I}[N(\mathbf{k})], \quad (13.18)$$

where expression on the right-hand side is called a *collision integral*.

13.1.3 Kinetic equations for decay and non-decay dispersion laws

Let us consider first the case of decay spectrum. In order to obtain the expression for collision integral, we multiply the left-hand and the right-hand sides of (13.8) by a_k^*, add the complex conjugates of the expressions in both sides and average. In this way we get

$$\dot{N}(\mathbf{k}) = -i \int d\mathbf{k_1} d\mathbf{k_2} \, \delta(\mathbf{k} - \mathbf{k_1} - \mathbf{k_2}) \left[V_{k,k_1,k_2} \langle a_k^* a_{k_1} a_{k_2} \rangle + 2V_{k_1,k,-k_2}^* \langle a_k^* a_{k_1} a_{-k_2}^* \rangle \right] + c.c.. \quad (13.19)$$

The correlation functions on the right-hand side are cubic, and seem to disappear under the hypothesis of Gaussian statistics adopted earlier. However, this statistics apply to zero-order wave amplitudes, as already mentioned. So, to calculate the right-hand side of (13.19) we use the perturbative expansion of a_k up to the first order and the expression for the first nonlinear correction (13.11). We will not dwell on the details of the calculation but only stress two key points: 1) it is easy to see that, to the leading order in nonlinearity, the correlation functions on the right-hand side become quartic, and split in pair correlation functions according to (13.17); 2) the singular denominators appearing on the right-hand side due to (13.11) are treated with the help of Sokhotsky formula which reads for an arbitrary real A:

$$\frac{1}{A \pm i\epsilon} = \mathcal{P}\frac{1}{A} \mp i\pi\delta(A), \tag{13.20}$$

where $\mathcal{P}$ denotes the principal value, and $\delta(A)$ is Dirac's delta-function. This formula should be understood in integral sense, that is it is for the integrands of integrals. The principal value contributions vanish after integration, by symmetry, and only delta-function contributions remain. Using the symmetries of interaction coefficients in this way we obtain the following expression for the collision integral in the case of decay spectrum:

$$\mathcal{I}^{(3)}\left[N(\mathbf{k})\right] = \int d\mathbf{k}_1 d\mathbf{k}_2 \left[W_{\mathbf{k}\mathbf{k}_1\mathbf{k}_2}f_{\mathbf{k}\mathbf{k}_1\mathbf{k}_2} - W_{\mathbf{k}_1\mathbf{k}_2\mathbf{k}}f_{\mathbf{k}_1\mathbf{k}_2\mathbf{k}} - W_{\mathbf{k}_2\mathbf{k}\mathbf{k}_1}f_{\mathbf{k}_2\mathbf{k}\mathbf{k}_1}\right], \tag{13.21}$$

where

$$W_{\mathbf{k}\mathbf{k}_1\mathbf{k}_2} = 2\pi \left|V_{\mathbf{k}\mathbf{k}_1\mathbf{k}_2}\right|^2 \delta(\mathbf{k} - \mathbf{k}_1 - \mathbf{k}_2)\delta(\omega(\mathbf{k}) - \omega(\mathbf{k}_1) - \omega(\mathbf{k}_2)), \tag{13.22}$$

and

$$f_{\mathbf{k}\mathbf{k}_1\mathbf{k}_2} = N(\mathbf{k}_1)N(\mathbf{k}_2) - N(\mathbf{k})\left(N(\mathbf{k}_1) + N(\mathbf{k}_2)\right). \tag{13.23}$$

The case of the non-decay spectrum is treated similarly, on the basis of (13.15) and (13.16), and gives, for the collision integral,

$$\mathcal{I}^{(4)}\left[N(\mathbf{k})\right] = \pi \int d\mathbf{k}_1 d\mathbf{k}_2 d\mathbf{k}_3 \, W_{\mathbf{k}\mathbf{k}_1\mathbf{k}_2\mathbf{k}_3} \, f_{\mathbf{k}\mathbf{k}_1\mathbf{k}_2\mathbf{k}_3}, \tag{13.24}$$

where

$$W_{\mathbf{k}\mathbf{k}_1\mathbf{k}_2\mathbf{k}_3} = \left|T_{\mathbf{k}\mathbf{k}_1\mathbf{k}_2\mathbf{k}_3}\right|^2 \delta(\mathbf{k} + \mathbf{k}_1 - \mathbf{k}_2 - \mathbf{k}_3)\delta(\omega(\mathbf{k}) + \omega(\mathbf{k}_1) - \omega(\mathbf{k}_2) - \omega(\mathbf{k}_3)), \tag{13.25}$$

and

$$f_{\mathbf{k}\mathbf{k}_1\mathbf{k}_2\mathbf{k}_3} = N(\mathbf{k}_1)N(\mathbf{k}_2)N(\mathbf{k}_3) + N(\mathbf{k})N(\mathbf{k}_2)N(\mathbf{k}_3)$$

$$- N(\mathbf{k})N(\mathbf{k}_1)N(\mathbf{k}_2) - N(\mathbf{k})N(\mathbf{k}_1)N(\mathbf{k}_3). \tag{13.26}$$

13.1.4   Exact solutions of kinetic equations

The beauty of the weak turbulence theory is that for wide classes of dispersion laws it allows us to obtain exact stationary solutions of the kinetic equations in the form of power law Kolmogorov-like distributions. These solutions make the collision integral vanish identically.

Let us first consider the *decay dispersion* law. The first observation is that energy equipartition, that is the Rayleigh–Jeans distribution according to the standard physical terminology,

$$N^{RJ}(\mathbf{k}) \propto \omega(\mathbf{k})^{-1}, \qquad (13.27)$$

is always a solution. It is sufficient to rewrite $f_{\mathbf{k},\mathbf{k}_1,\mathbf{k}_2}$ in the form

$$f_{\mathbf{k},\mathbf{k}_1,\mathbf{k}_2} = N(\mathbf{k})N(\mathbf{k}_1)N(\mathbf{k}_2)\left(N(\mathbf{k})^{-1} - N(\mathbf{k}_1)^{-1} - N(\mathbf{k}_2)^{-1}\right), \qquad (13.28)$$

to see that due to the delta function in ω in the collision integral $N^{RJ}(\mathbf{k})$ makes it vanish, as $f_{\mathbf{k},\mathbf{k}_1,\mathbf{k}_2}$ becomes proportional to $\omega(\mathbf{k}) - \omega(\mathbf{k}_1) - \omega(\mathbf{k}_2)$. Let us recall that the Rayleigh - Jeans distribution $N = \frac{T}{\omega}$ corresponds to thermodynamic equilibrium with temperature T. It is equipartition of energy because in zeroth order the mean energy density of each mode is $\epsilon(\mathbf{k}) = \omega(\mathbf{k})N(\mathbf{k}) = \text{const}$.

Along with the equilibrium solutions, there are other, non-equilibrium ones with $f_{\mathbf{k},\mathbf{k}_1,\mathbf{k}_2} \neq 0$, at least for certain classes of dispersion laws. Let us suppose that dispersion is isotropic and has the form of power law with some exponent β:

$$\omega(\mathbf{k}) = |\mathbf{k}|^\beta \qquad (13.29)$$

(this is the case of gravity waves in non-rotating shallow water, for example). It is easy to check that $\beta > 1$ corresponds to a decay dispersion law. The existence of non-equilibrium solutions is based on symmetries of the collision integral. The dispersion law (13.29) is invariant with respect to rotations in $\mathbf{k}$ space. This means that $\omega(\hat{R}\mathbf{k}) = \omega(\mathbf{k})$, where the operator of rotation is denoted by $\hat{R}$. The interaction coefficient $V_{\mathbf{k},\mathbf{k}_1\mathbf{k}_2}$ is a scalar function of the vector arguments $\mathbf{k}, \mathbf{k}_1, \mathbf{k}_2$. It is, therefore, a function of scalar products of these vectors and is also invariant with respect to rotations. The delta functions are invariant with respect to rotations, too. Hence, $W_{\hat{R}\mathbf{k},\hat{R}\mathbf{k}_1\hat{R}\mathbf{k}_2} = W_{\mathbf{k},\mathbf{k}_1\mathbf{k}_2}$.

Another symmetry of the dispersion law (13.29) is scale invariance. The eigen-frequencies scale under dilatations as $\omega(\lambda\mathbf{k}) = \lambda^\beta \omega(\mathbf{k})$. Hence, the delta function of frequencies scales with a factor $\lambda^{-\beta}$, while the delta function of wave numbers scales as λ^{-d}, where d is space and $\mathbf{k}$-space dimension (this scaling follows from the fact that the integral of delta function is unity). If the squares of interaction coefficients $|V|^2$ entering the collision integral are homogeneous functions with an index m, then

$$W_{\lambda\mathbf{k},\lambda\mathbf{k}_1\lambda\mathbf{k}_2} = \lambda^{m-\beta-d} W_{\mathbf{k},\mathbf{k}_1\mathbf{k}_2}. \qquad (13.30)$$

The triads of wave vectors enter each of three terms in the collision integral (13.21) in the form of three triangles, with one of the sides fixed by the external argument $\mathbf{k}$, while $\mathbf{k}_1, \mathbf{k}_2$ are integration variables. A crucial observation is that by appropriate rotations and dilatations the second and the third triangles may be transformed into the first one. However, while integration measure is invariant under rotations, dilatations provide an extra Jacobian factor λ^{3d} which will add up in the rescaling of W, cf. (13.30). In this way the collision integral may be cast in the following form:

$$\mathcal{I}^{(3)}\left[N(\mathbf{k})\right] = \int d\mathbf{k}_1 d\mathbf{k}_2\, \delta(\mathbf{k}-\mathbf{k}_1-\mathbf{k}_2)\delta(\omega(\mathbf{k})-\omega(\mathbf{k}_1)-\omega(\mathbf{k}_2))\, W_{\mathbf{k},\mathbf{k}_1\mathbf{k}_2} \cdot$$

$$\left(f_{\mathbf{k},\mathbf{k}_1,\mathbf{k}_2} - \lambda_1^\alpha f_{\hat{G}_1^2\mathbf{k}_1,\mathbf{k},\hat{G}_1\mathbf{k}_2} - \lambda_2^\alpha f_{\hat{G}_2^2\mathbf{k}_2,\hat{G}_2\mathbf{k}_1,\mathbf{k}}\right). \tag{13.31}$$

Here the scaling factors are $\lambda_{1,2} = \frac{|\mathbf{k}|}{|\mathbf{k}_{1,2}|}$, and the transformations $\hat{G}_{1,2}$ are defined by rotations and dilatations with $\lambda_{1,2}$ such that $\hat{G}_{1,2}\mathbf{k}_{1,2} = \mathbf{k}$, with the scaling factor $\alpha = m + 2d - \beta$.

A tentative power-law solutions $N(\mathbf{k}) \propto \omega^s(\mathbf{k})$, behave under rescaling as $N(\lambda\mathbf{k}) = \lambda^{\beta s}$, and hence $f_{\lambda\mathbf{k},\lambda\mathbf{k}_1,\lambda\mathbf{k}_2} = \lambda^{2\beta s}f_{\mathbf{k},\mathbf{k}_1,\mathbf{k}_2}$. In addition, f is invariant with respect to rotations in this case. Inverting the dispersion relation $|\mathbf{k}| = \omega^{\frac{1}{\beta}}$ and introducing this expression into the definitions of $\lambda_{1,2}$ we reduce the collision integral to the following expression:

$$\mathcal{I}^{(3)}\left[N(\mathbf{k})\right] = \omega^\nu(\mathbf{k})\int d\mathbf{k}_1 d\mathbf{k}_2\, W_{\mathbf{k},\mathbf{k}_1\mathbf{k}_2} f_{\mathbf{k},\mathbf{k}_1,\mathbf{k}_2}\left(\omega^{-\nu}(\mathbf{k})-\omega^{-\nu}(\mathbf{k}_1)-\omega^{-\nu}(\mathbf{k}_2)\right), \tag{13.32}$$

where

$$\nu = 2s - 1 + \frac{2d+m}{\beta}, \tag{13.33}$$

and

$$f_{\mathbf{k},\mathbf{k}_1,\mathbf{k}_2} = (\omega(\mathbf{k})\omega(\mathbf{k}_1)\omega(\mathbf{k}_2))^s\left(\omega^{-s}(\mathbf{k})-\omega^{-s}(\mathbf{k}_1)-\omega^{-s}(\mathbf{k}_2)\right). \tag{13.34}$$

There are two values of s which annihilate this collision integral and, hence, give stationary distributions of $N(\mathbf{k})$. At $s = -1$ $f_{\mathbf{k},\mathbf{k}_1,\mathbf{k}_2}$ vanishes: this is a thermodynamic equilibrium solution. However there is another solution: $\nu = -1$ and $s = -\frac{2d+m}{2\beta}$, which gives

$$N(\mathbf{k}) = |\mathbf{k}|^{-\frac{2d+m}{2}}. \tag{13.35}$$

Contrary to the equilibrium spectrum, this one is model-dependent, and is determined by the scaling exponent of the interaction coefficients in the Hamiltonian.

The same approach may be applied to the *non-decay case*. The resonant wave quadrangle $(\mathbf{k}, \mathbf{k}_1, \mathbf{k}_2, \mathbf{k}_3)$ defining the integration measure in the collision integral (13.24) can be transformed, with the help of symmetry transformations $\hat{G}_i : \hat{G}_i\mathbf{k}_i = \mathbf{k}$ $i = 1, 2, 3$ including rotations and dilatations, into another resonant quadrangle $(\mathbf{k}, \mathbf{q}_1, \mathbf{q}_2, \mathbf{q}_3)$ in three different ways:

$$\begin{cases} \mathbf{q}_1 = \hat{G}_1^{\,2}\mathbf{k}_1, \; \mathbf{q}_2 = \hat{G}_1\mathbf{k}_2, \; \mathbf{q}_3 = \hat{G}_1\mathbf{k}_3; \\[2mm] \mathbf{q}_1 = \hat{G}_2\mathbf{k}_3, \; \mathbf{q}_2 = \hat{G}_2^{\,2}\mathbf{k}_2, \; \mathbf{q}_3 = \hat{G}_2\mathbf{k}_1; \\[2mm] \mathbf{q}_1 = \hat{G}_3\mathbf{k}_2, \; \mathbf{q}_2 = \hat{G}_3\mathbf{k}_1, \; \mathbf{q}_3 = \hat{G}_3^{\,2}\mathbf{k}_3. \end{cases} \tag{13.36}$$

For power-law dispersion $\omega(\mathbf{k}) = |\mathbf{k}|^\beta$ the interaction coefficients and delta function scale in a way similar to the decay-spectrum case, and the collision integral may be represented as a sum of four replicas of itself in the form

$$\mathcal{I}^{(4)}\,[N(\mathbf{k})] = \frac{1}{4}\int d\mathbf{k}_1 d\mathbf{k}_2 d\mathbf{k}_3 \; W_{\mathbf{k}\mathbf{k}_1\mathbf{k}_2\mathbf{k}_3} \left(f_{\mathbf{k},\mathbf{k}_1,\mathbf{k}_2,\mathbf{k}_3} + \lambda_1^\alpha f_{\hat{G}_1\mathbf{k},\mathbf{k}_1,\mathbf{k}_2,\mathbf{k}_3} \right.$$

$$\left. + \lambda_2^\alpha f_{\hat{G}_2\mathbf{k},\mathbf{k}_1,\mathbf{k}_2,\mathbf{k}_3} + \lambda_3^\alpha f_{\hat{G}_3\mathbf{k},\mathbf{k}_1,\mathbf{k}_2,\mathbf{k}_3} \right), \tag{13.37}$$

where the scale factors are $\lambda_i = \frac{|\mathbf{k}|}{|\mathbf{k}_i|}$, $i = 1, 2, 3$, and $\alpha = m + 3d - \beta$. As a result, the collision integral for isotropic and homogeneous distributions $N(\mathbf{k}) \propto \omega^s(\mathbf{k})$, is factorised:

$$\mathcal{I}^{(4)}\,[N(\mathbf{k})] = \frac{\omega^\nu(\mathbf{k})}{4}\int d\mathbf{k}_1 d\mathbf{k}_2 \; d\mathbf{k}_3 \; W_{\mathbf{k},\mathbf{k}_1\mathbf{k}_2,\mathbf{k}_3} f_{\mathbf{k},\mathbf{k}_1,\mathbf{k}_2,\mathbf{k}_3} \cdot$$

$$\left(\omega^{-\nu}(\mathbf{k}) + \omega^{-\nu}(\mathbf{k}_1) - \omega^{-\nu}(\mathbf{k}_2) - \omega^{-\nu}(\mathbf{k}_3)\right), \tag{13.38}$$

where

$$\nu = 3s - 1 + \frac{3d + m}{\beta} \tag{13.39}$$

and

$$f_{\mathbf{k},\mathbf{k}_1,\mathbf{k}_2,\mathbf{k}_3} = (\omega(\mathbf{k})\omega(\mathbf{k}_1)\omega(\mathbf{k}_2)\omega(\mathbf{k}_2))^s \, (\omega^{-s}(\mathbf{k}) + \omega^{-s}(\mathbf{k}_1)$$

$$- \omega^{-s}(\mathbf{k}_2) - \omega^{-s}(\mathbf{k}_2)). \tag{13.40}$$

The collision integral vanishes, giving stationary distributions of $N(\mathbf{k})$, either when $f_{\mathbf{k},\mathbf{k}_1,\mathbf{k}_2,\mathbf{k}_3}$ vanishes, or when the last factor in (13.38) vanishes. The first possibility corresponds to one of the two limiting cases $\mu \gg \omega$ and $\mu \ll \omega$ of the equilibrium distribution with 'chemical potential' μ:

$$N^{RJ}(\mathbf{k}) \propto (\omega(\mathbf{k}) - \mu)^{-1}. \tag{13.41}$$

The appearance of a chemical potential is related to an additional conservation law (see the next section) of the number of 'particles' in non-decay kinetic equations, as in classical thermodynamics.

The non-equilibrium stationary distributions correspond to $\nu = 0, -1$ and are related to conservation of number of waves and energy, respectively, in an elementary collision proccess. Correspondingly, $s = \frac{1}{3} - \frac{3d+m}{3\beta}$, and $s = -\frac{3d+m}{3\beta}$ which gives two solutions for N:

$$N(\mathbf{k}) = |\mathbf{k}|^{\frac{\beta - 3d + m}{3}}, \quad N(\mathbf{k}) = |\mathbf{k}|^{-\frac{3d + m}{3}} \tag{13.42}$$

13.1.5   Conservation laws and dimensional estimates

The *mean energy* of the medium in the lowest order (cf. (13.4)) is $E = \int d\mathbf{k}\,\omega(\mathbf{k})N(\mathbf{k})$, whence $\epsilon = \omega(\mathbf{k})N(\mathbf{k})$ is the mean energy density. Multiplying both sides of (13.18) by $\omega(\mathbf{k})$, integrating by $\mathbf{k}$, and using the symmetry properties of interaction coefficients we get in the decay spectrum case

$$\dot{E} = \int d\mathbf{k}\,d\mathbf{k}_1\,d\mathbf{k}_2\,(\omega(\mathbf{k}) - \omega(\mathbf{k}_1) - \omega(\mathbf{k}_2)) \cdot$$

$$2\pi \left| V_{\mathbf{k}\mathbf{k}_1\mathbf{k}_2} \right|^2 \delta(\mathbf{k} - \mathbf{k}_1 - \mathbf{k}_2)\delta(\omega(\mathbf{k}) - \omega(\mathbf{k}_1) - \omega(\mathbf{k}_2)) \equiv 0, \tag{13.43}$$

and similarly in the non-decay case. Thus, the total energy is conserved by the kinetic equation. It is clear from the above expression that total energy conservation is ensured by energy conservation in each elementary interaction act. Energy conservation can be rewritten in (pseudo-) local form:

$$\frac{\partial \epsilon}{\partial t} + \frac{\partial \mathbf{P}}{\partial \mathbf{k}} = 0, \tag{13.44}$$

where $\mathbf{P}$ is (the density of) the energy flux in the $\mathbf{k}$ space: $\frac{\partial \mathbf{P}}{\partial \mathbf{k}} = -\omega(\mathbf{k})\mathcal{I}\,[N(\mathbf{k})]$. The true locality is ensured if integrals defining P are convergent (the formal *locality criterion*). It may be shown in an analogous way that the three components of the *total momentum* of the medium $\mathbf{K} = \int d\mathbf{k}\,\mathbf{k}N(\mathbf{k})$ are conserved by the kinetic equation, as they are conserved in each elementary interaction. Moreover, in the non-decay case the total *wave action*, or the total number of waves $N = \int d\mathbf{k}N(\mathbf{k})$ is also conserved, as it is conserved in elementary interactions. It may be again rewritten in the (pseudo-) local form:

$$\frac{\partial N(\mathbf{k})}{\partial t} + \frac{\partial \mathbf{Q}}{\partial \mathbf{k}} = 0, \tag{13.45}$$

where $\mathbf{Q}$ is (the density of) the wave-action flux in the $\mathbf{k}$ space: $\frac{\partial \mathbf{Q}}{\partial \mathbf{k}} = -\mathcal{I}^{(4)}\,[N(\mathbf{k})]$, the locality being ensured if integrals converge.

Like the famous Kolmogorov–Obukhov spectrum of (strong) hydrodynamical turbulence, which we have seen in Chapter 6, the energy spectra corresponding to constant fluxes of conserved quantities can be constructed by dimensional considerations. The difference between isotropic and homogeneous developed turbulence and wave turbulence is that in the last case an additional parameter, the frequency of the waves, is present. Hence, pure dimensional considerations are insufficient and some dynamical information should be added. It is provided by the very structure of the kinetic equation with either $\mathcal{I}^{(3)} \propto N^2$, or $\mathcal{I}^{(4)} \propto N^3$. The collision integral defines the fluxes of conserved quantities, as just explained. The dimensions of N, P, and Q are

$$[N] = L^5 T^{-1}, \quad [P] = L^{5-d} T^{-3}, \quad [Q] = L^{5-d} T^{-2}. \tag{13.46}$$

In the decay-spectrum case $N \sim P^{\frac{1}{2}}$ and looking for the spectrum of the form $N = P^{\frac{1}{2}} \omega^a |\mathbf{k}|^b$ we get by comparing dimensions

$$N \propto P^{\frac{1}{2}} \omega^{-\frac{1}{2}} |\mathbf{k}|^{-\frac{d+5}{2}}. \tag{13.47}$$

In the non-decay spectrum case, $N \sim P^{\frac{1}{3}}$ or $N \sim Q^{\frac{1}{3}}$ and we get a spectrum with a constant energy flux:

$$N \propto P^{\frac{1}{3}} |\mathbf{k}|^{-\frac{d+10}{3}}, \tag{13.48}$$

and a spectrum with a constant wave-action flux:

$$N \propto Q^{\frac{1}{3}} \omega^{\frac{1}{3}} |\mathbf{k}|^{-\frac{d+10}{3}}. \tag{13.49}$$

In the case of self-similar dispersion laws these spectra coincide with those found in Section 13.1.4.

13.2 Applications of the wave turbulence theory to waves in rotating shallow water

13.2.1 Wave turbulence of inertia–gravity waves on the *f* plane

Method of factorising collision integrals for near-acoustic dispersion laws

The dispersion relation of inertia–gravity waves (IGW) in the f - plane approximation, $\omega(\mathbf{k}) = \sqrt{gh_0 \mathbf{k}^2 + f^2}$, where h_0 is the mean thickness, is isotropic but does not have a power-law form in the presence of rotation. Yet, for the *short IGW* (SIGW) with $|\mathbf{k}| \to \infty$ it has a form of an isotropic in $\mathbf{k}$ power law, which corresponds to non-dispersive gravity waves (or acoustic waves, in virtue of the acoustic analogy discussed in Chapter 3), plus a small dispersive correction, which is also isotropic and self-similar:

$$\omega(\mathbf{k}) \approx \sqrt{gh_0} |\mathbf{k}| \left(1 + \frac{1}{2 \left(|\mathbf{k}| R_d \right)^2} \right), \quad R_d = \frac{\sqrt{gh_0}}{f}. \tag{13.50}$$

The method of obtaining stationary solutions of the kinetic equation of wave turbulence described earlier can be generalised to such dispersion laws. Let us consider a dispersion law of a general form of this type,

$$\omega(\mathbf{k}) = c|\mathbf{k}| + \gamma(\mathbf{k}), \tag{13.51}$$

such that $|\gamma(\mathbf{k})| \ll c|\mathbf{k}|$, like in (13.50). The leading contribution to the spectrum (13.51) is of non-decay type. Thus, collision integral is of the form (13.24). The smallness of dispersive correction means that nonlinear wave scattering is dominated by interaction of waves with almost collinear wave numbers, as in the absence of dispersion only collinear wave vectors produce resonances. For angles between the wave vectors such that

$$\theta_i = (\widehat{\mathbf{k}, \mathbf{k}_i}) \leq \sqrt{\frac{\gamma}{c|\mathbf{k}|}} \tag{13.52}$$

the following estimate holds for the four-wave interaction coefficient in (13.16): $T \sim \dfrac{|V|^2}{c|\mathbf{k}| \left(\theta^2 + |\gamma/c|\mathbf{k}||\right)}$, and everywhere, except for resonant denominators, we may consider that $\mathbf{k}_i$ is parallel to $\mathbf{k}$. Introducing the variables $\mathbf{s}_i = \dfrac{\mathbf{k}_{i\perp}}{|\mathbf{k}_{i\perp}|}$, where the subscript $\perp$ means perpendicular to $\mathbf{k}$ component, the product of delta functions in the collision integral (13.24), (13.25) can be then rewritten as

$$\delta(|\mathbf{k}| + |\mathbf{k}_1| - |\mathbf{k}_2| - |\mathbf{k}_3|) \, \delta(\mathbf{s}_1|\mathbf{k}|_1\theta_1 - \mathbf{s}_2|\mathbf{k}|_2\theta_2 - \mathbf{s}_3|\mathbf{k}|_3\theta_3)$$

$$\delta\left(\gamma + \gamma_1 - \gamma_2 - \gamma_3 + \frac{c}{2}\left(|\mathbf{k}|_1\theta_1^2 - |\mathbf{k}|_2\theta_2^2 - |\mathbf{k}|_3\theta_3^2\right)\right). \tag{13.53}$$

The collision integral is not conformal-invariant. Nevertheless, if

$$\gamma(\lambda\mathbf{k}) = \lambda^\beta \gamma(\mathbf{k}), \; V_{\lambda\mathbf{k},\lambda\mathbf{k}_1,\lambda\mathbf{k}_2} = \lambda^\mu V_{\mathbf{k},\mathbf{k}_1,\mathbf{k}_2}, \tag{13.54}$$

then it can be factorised using rescaling of angles and conformal transformations. Indeed, in this case the conservation laws and resonant denominators are mapped onto themselves under the transformation $\mathbf{k}_i \to \lambda\mathbf{k}_i$, $\theta_i \to \lambda^{\frac{\beta-1}{2}}\theta_i$. Correspondingly,

$$W_{\mathbf{k},\mathbf{k}_1,\mathbf{k}_2,\mathbf{k}_3} \to \lambda^w W_{\mathbf{k},\mathbf{k}_1,\mathbf{k}_2,\mathbf{k}_3}, \; w = 4\mu - 3\beta - 1 - \frac{1}{2}(\beta+1)(d-1). \tag{13.55}$$

If we suppose that $N(\mathbf{k}) \propto |\mathbf{k}|^s$, with possible weak dependence on the angle, the following transformations of wave vectors,

$$k_2' = \lambda_2 k = \lambda_2^2 k_2, \; k_1' = \lambda_2 k_3, \; k_3' = \lambda_2 k_1, \tag{13.56}$$

and angles,

$$\theta_2' = -\lambda_2^{\frac{\beta-1}{2}}\theta_2, \; \theta_1' = \lambda_2^{\frac{\beta-1}{2}}(\theta_3 - \theta_2), \; \theta_3' = \lambda_2^{\frac{\beta-1}{2}}(\theta_1 - \theta_2), \tag{13.57}$$

lead to factorisation of the collision integral

$$\mathcal{I}^{(4)}\left[N(\mathbf{k})\right] = \frac{k^{\nu}}{4} \int dk_1\, dk_2\, dk_3\, W_{\mathbf{k},\mathbf{k}_1\mathbf{k}_2,\mathbf{k}_3}\, f_{\mathbf{k},\mathbf{k}_1,\mathbf{k}_2,\mathbf{k}_3}\,\cdot$$

$$\left(k^{-\nu} + k_1^{-\nu} - k_2^{-\nu} - k_3^{-\nu}\right), \tag{13.58}$$

where

$$\nu = 3s + w + 4d + \frac{3}{2}(d-1)\,(\beta - 1). \tag{13.59}$$

Hence, there are two non-equilibrium power-law spectra with $\nu = 0$ (constant wave-action flux solution), and with $\nu = -1$ (constant energy flux solution). It should be stressed that we thus get an exact solution of an approximate kinetic equation, obtained under hypothesis that exchanges between short-wave and long-wave parts of the wave spectrum are slow. Such hypothesis is supported by direct numerical simulations presented at the end of this Chapter.

Turbulence of short inertia–gravity waves

As was discussed on several occasions in previous chapters, the RSW equations describe both slow vortex (zero frequency in the linear approximation) and fast inertia–gravity wave (IGW) motions. An invariant criterion of their separation, at least at weak nonlinearities, is provided by the conservation of PV. Indeed, as was shown in Chapter 3, IGW do not bear PV anomaly, that is deviation of PV from the background value $\frac{f}{h_0}$. The motions with zero PV anomaly—with constant PV $q = \frac{f}{h_0}$—are described by a pair of canonical Hamiltonian variables: velocity potential Φ and h. Any two-dimensional velocity field can be represented as a sum of a gradient of a potential and curl of a stream function. Using Φ and h for this purpose, and imposing the constant value of PV, we get for the two components of velocity (u, v):

$$u = \frac{\partial \Phi}{\partial x} - \frac{f}{h_0}\frac{\partial}{\partial y}\nabla^{-2}(h - h_0), \quad v = \frac{\partial \Phi}{\partial y} + \frac{f}{h_0}\frac{\partial}{\partial x}\nabla^{-2}(h - h_0), \tag{13.60}$$

where ∇^{-2} denotes inverse Laplacian. The RSW system, thus restrained to have zero PV anomaly, is Hamiltonian:

$$\dot{h} = \frac{\delta H}{\delta \Phi}, \quad \dot{\Phi} = -\frac{\delta H}{\delta h}, \tag{13.61}$$

with a Hamiltonian given by the full (kinetic + potential) energy:

$$H = \frac{1}{2} \int dx\, dy\, \left[h(u^2 + v^2) + gh^2\right], \tag{13.62}$$

and a canonical Poisson bracket. Introducing normalised wave amplitudes b_k, such that

$$h_k = \sqrt{\frac{k^2 h_0}{2\omega(k)}}(b^*_{-k} + b_{-k}), \quad \Phi_k = i\sqrt{\frac{\omega(k)}{2k^2 h_0}}(b^*_{-k} - b_{-k}), \tag{13.63}$$

we get the interaction Hamiltonian of the form

$$H_3 = \frac{1}{2}\int dk dk_1 dk_2\, V_{kk_1k_2} b^*_k b_{k_1} b_{k_2} + \frac{1}{3}\int dk dk_1 dk_2\, U_{kk_1k_2} b_k b_{k_1} b_{k_2} + c.c. \tag{13.64}$$

The interaction coefficients are

$$V_{kk_1k_2} = 2U_{kk_1k_2} = \sqrt{18}\frac{k_2 \cdot k_3\, \omega(k_1) + k_1 \cdot k_3\, \omega(k_2) + k_1 \cdot k_2\, \omega(k_3)}{\sqrt{\omega(k_1)\omega(k_2)\omega(k_3)}}. \tag{13.65}$$

The canonical transformation

$$b_k \to b_k - \int dk_1 dk_2\, \delta(k - k_1 - k_2)\frac{V_{kk_1k_2}}{\omega(k) - \omega(k_1) - \omega(k_2)}\, b_{k_1} b_{k_2}$$

$$-2\int dk_1 dk_2\, \delta(k + k_1 - k_2)\frac{V_{kk_1k_2}}{-\omega(k) + \omega(k_1) - \omega(k_2)}\, b^*_{k_1} b_{k_2}$$

$$-\int dk_1 dk_2\, \delta(k + k_1 + k_2)\frac{U_{kk_1k_2}}{\omega(k) + \omega(k_1) + \omega(k_2)}\, b_{k_1} b_{k_2} \tag{13.66}$$

casts the Hamiltonian in the normal form, cf. Section 13.1.1:

$$H = \int dk\omega(k)\, a_k a^*_k$$

$$+ \frac{1}{2}\int dk dk_1 dk_2 dk_3\, \delta(k + k_1 - k_2 - k_3)\, T_{kk_1k_2k_3}\, a^*_k a^*_{k_1} a_{k_2} a_{k_3}. \tag{13.67}$$

There was no loss of generality up until now. However, the full dispersion law of IGW is not scale-invariant and does not allow us to obtain analytically stationary solutions annihilating the collision integral. We consider therefore an ensemble of SIGW with $|k| \to \infty$ and retain only the leading term in the expansion of the dispersion law in powers of $|k|^{-1}$, which leads to (13.50). We also suppose that the time needed to transfer the energy to the long-wave part of the spectrum is very long. Then we can limit ourselves by short-wave contributions to the collision integral, and the above-described method of factorisation in the case of the dispersion law (13.51) applies, giving two power-law spectra,

$$N(k) \propto |k|^{-\frac{14}{3}}, \quad N(k) \propto |k|^{-\frac{13}{3}}, \tag{13.68}$$

corresponding to constant energy and wave-action fluxes through the spectrum, respectively. By construction, these are approximate solutions of the kinetic equations and their relevance should be checked by comparing with observation and/or experiment. The related energy spectra, in the leading approximations are, correspondingly, $\epsilon(\mathbf{k}) \propto |\mathbf{k}|^{-\frac{11}{3}}$, $\epsilon(\mathbf{k}) \propto |\mathbf{k}|^{-\frac{10}{3}}$. The spectra of the energy density integrated over orientations of the wave vector, i.e. in the space of $k = |\mathbf{k}|$, are $\epsilon(k) \propto |\mathbf{k}|^{-\frac{8}{3}}$, $\epsilon(k) \propto |\mathbf{k}|^{-\frac{7}{3}}$, taking into account that integration measure is $k\,dk$ in the space of scalar k.

13.2.2   Weak turbulence of short inertia–gravity waves on the equatorial β plane

Normal form of the RSW equations on the equatorial tangent plane

Another example where approximate power-law solutions of the kinetic equations can be analytically obtained is turbulence of short inertia–gravity waves (SIGW) on the equatorial β plane. The non-dimensional RSW equations for two components of velocity u, v, and normalised deviation of the thickness field from the rest value $z = \frac{h-h_0}{h_0}$ are

$$
\begin{cases}
u_t + uu_x + vu_y - \beta yv + z_x = 0, \\
v_t + uv_x + vv_y + \beta yu + z_y = 0, \\
z_t + u_x + v_y + (uz)_x + (vz)_y = 0.
\end{cases}
\tag{13.69}
$$

We will consider motions with a characteristic scale which is small with respect to the equatorial deformation radius $R_e = \frac{(gh_0)^{\frac{1}{4}}}{\sqrt{\beta}}$, meaning that non-dimensional β is small: $\beta \sim \epsilon \ll 1$, and assume weak nonlinearity of the same order of magnitude: $u, v, z \sim \epsilon$. Under these hypotheses the leading-order part of the system (13.69) has constant coefficients:

$$
\begin{cases}
u_t + z_x = 0, \\
v_t + z_y = 0, \\
z_t + u_x + v_y = 0.
\end{cases}
\tag{13.70}
$$

We make a Fourier transformation $(u(\mathbf{x}), v(\mathbf{x}), z(\mathbf{x})) = \int (u_\mathbf{k}, v_\mathbf{k}, z_\mathbf{k})\exp(i\mathbf{kx})d\mathbf{k}$, where $\mathbf{x} = (x, y)$ and $\mathbf{k} = (k_1, k_2)$, and write down the equations for the Fourier-transformed variables, following from (13.69) in symmetric form:

$$
\begin{pmatrix} \partial u_\mathbf{k}/\partial t \\ \partial v_\mathbf{k}/\partial t \\ \partial z_\mathbf{k}/\partial t \end{pmatrix} + \begin{pmatrix} 0 & -i\beta\frac{\partial}{\partial k_2} & ik_1 \\ i\beta\frac{\partial}{\partial k_2} & 0 & ik_2 \\ ik_1 & ik_2 & 0 \end{pmatrix} \begin{pmatrix} u_\mathbf{k} \\ v_\mathbf{k} \\ z_\mathbf{k} \end{pmatrix} +
\tag{13.71}
$$

$$
+ \begin{pmatrix} ik_1/2 \int (u_1 u_m + v_1 v_m)d\lambda - \int \Omega_1 v_m d\lambda \\ ik_2/2 \int (u_1 u_m + v_1 v_m)d\lambda + \int \Omega_1 u_m d\lambda \\ ik_1 \int z_1 u_m d\lambda + ik_2 \int z_1 v_m d\lambda \end{pmatrix} = 0,
$$

where $\Omega = v_x - u_y$, $\Omega_1 = il_1 v_1 - il_2 u_1$, and we use the shorthand notation $d\lambda = \delta(\mathbf{k} - \mathbf{1} - \mathbf{m}) d\mathbf{l}\, d\mathbf{m}$. The linear term in (13.71) is diagonalised by the following change of variables:

$$\begin{pmatrix} u_k \\ v_k \\ z_k \end{pmatrix} = \begin{pmatrix} \dfrac{-ih_2}{|k|} & \dfrac{h_1}{\sqrt{2}|k|} & \dfrac{-h_1}{\sqrt{2}|k|} \\[2mm] \dfrac{ik_1}{|k|} & \dfrac{k_2}{\sqrt{2}|k|} & \dfrac{-k_2}{\sqrt{2}|k|} \\[2mm] 0 & \dfrac{1}{\sqrt{2}} & \dfrac{1}{\sqrt{2}} \end{pmatrix} \begin{pmatrix} a_k \\ b_k \\ c_k \end{pmatrix}, \tag{13.72}$$

and we get

$$\begin{pmatrix} \partial a_k/\partial t \\ \partial b_k/\partial t \\ \partial c_k/\partial t \end{pmatrix} + \begin{pmatrix} 0 & 0 & 0 \\ 0 & i\,|\,\mathbf{k}\,| & 0 \\ 0 & 0 & -i\,|\,\mathbf{k}\,| \end{pmatrix} \begin{pmatrix} a_k \\ b_k \\ c_k \end{pmatrix} +$$

$$+ \beta \begin{pmatrix} -\dfrac{ik_1}{|k|^2} & \dfrac{1}{\sqrt{2}}\dfrac{\partial}{\partial k_2} & -\dfrac{1}{\sqrt{2}}\dfrac{\partial}{\partial k_2} \\[2mm] \dfrac{1}{\sqrt{2}}\dfrac{\partial}{\partial k_2} & -\dfrac{ik_1}{2|k|^2} & \dfrac{ik_1}{2|k|^2} \\[2mm] -\dfrac{1}{\sqrt{2}}\dfrac{\partial}{\partial k_2} & \dfrac{ik_1}{2|k|^2} & -\dfrac{ik_1}{2|k|^2} \end{pmatrix} \begin{pmatrix} a_k \\ b_k \\ c_k \end{pmatrix} + (NL) = 0 . \tag{13.73}$$

Here, (NL) denotes the nonlinear terms:

$$\begin{pmatrix} \int (U_{klm}^{(0)} a_l a_m + U_{klm}^{(1)} a_l b_m + U_{klm}^{(2)} a_l c_m)\, d\lambda \\[2mm] \int (V_{klm}^{(0)} a_l b_m + V_{klm}^{(1)} a_l a_m + V_{klm}^{(2)} a_l c_m)\, d\lambda \\[2mm] \int (W_{klm}^{(0)} a_l c_m + W_{klm}^{(1)} a_l a_m + W_{klm}^{(2)} a_l b_m)\, d\lambda \end{pmatrix} +$$

$$+ \begin{pmatrix} 0 \\[2mm] \int (V_{klm}^{(3)} b_l b_m + V_{klm}^{(4)} c_l c_m + V_{klm}^{(5)} b_l c_m)\, d\lambda \\[2mm] \int (W_{klm}^{(3)} c_l c_m + W_{klm}^{(4)} b_l b_m + W_{klm}^{(5)} c_l b_m)\, d\lambda \end{pmatrix}. \tag{13.74}$$

We do not write down explicit expressions for interaction coefficients in this formula, which are lengthy but can be straightforwardly found with the help of (13.71), (13.72).

The linear part of (13.73) may be fully diagonalised by further transforming the dependent variables:

$$\begin{pmatrix} a_k \\ b_k \\ c_k \end{pmatrix} \rightarrow \begin{pmatrix} a_k \\ b_k \\ c_k \end{pmatrix} + \beta \begin{pmatrix} 0 & -\dfrac{\partial}{\partial k_2}\dfrac{1}{\sqrt{2}|k|} & -\dfrac{\partial}{\partial k_2}\dfrac{1}{\sqrt{2}|k|} \\[2mm] \dfrac{i}{\sqrt{2}|k|}\dfrac{\partial}{\partial k_2} & 0 & -\dfrac{k_1}{4|k|^3} \\[2mm] \dfrac{i}{\sqrt{2}|k|}\dfrac{\partial}{\partial k_2} & \dfrac{k_1}{4|k|^3} & 0 \end{pmatrix} \begin{pmatrix} a_k \\ b_k \\ c_k \end{pmatrix}, \tag{13.75}$$

which gives, to the leading order in ϵ,

$$
\begin{pmatrix} \partial a_{\mathrm{k}}/\partial t \\ \partial b_{\mathrm{k}}/\partial t \\ \partial c_{\mathrm{k}}/\partial t \end{pmatrix} + \begin{pmatrix} i\Omega_{\mathrm{k}} & 0 & 0 \\ 0 & i\omega_{\mathrm{k}} & 0 \\ 0 & 0 & -i\omega_{-\mathrm{k}} \end{pmatrix} \begin{pmatrix} a_{\mathrm{k}} \\ b_{\mathrm{k}} \\ c_{\mathrm{k}} \end{pmatrix} +
$$

$$
+ \begin{pmatrix} \int (U_{klm}^{(0)} a_1 a_{\mathrm{m}} + U_{klm}^{(1)} a_1 b_{\mathrm{m}} + U_{klm}^{(2)} a_1 c_{\mathrm{m}}) d\lambda \\ \int (V_{klm}^{(0)} a_1 b_{\mathrm{m}} + V_{klm}^{(1)} a_1 a_{\mathrm{m}} + V_{klm}^{(2)} a_1 c_{\mathrm{m}}) d\lambda \\ \int (W_{klm}^{(0)} a_1 c_{\mathrm{m}} + W_{klm}^{(1)} a_1 a_{\mathrm{m}} + W_{klm}^{(2)} a_1 b_{\mathrm{m}}) d\lambda \end{pmatrix} +
$$

$$
+ \begin{pmatrix} 0 \\ \int (V_{klm}^{(3)} b_1 b_{\mathrm{m}} + V_{klm}^{(4)} c_1 c_{\mathrm{m}} + V_{klm}^{(5)} b_1 c_{\mathrm{m}}) d\lambda \\ \int (W_{klm}^{(3)} c_1 c_{\mathrm{m}} + W_{klm}^{(4)} b_1 b_{\mathrm{m}} + W_{klm}^{(5)} c_1 b_{\mathrm{m}}) d\lambda \end{pmatrix} = 0. \tag{13.76}
$$

Here,

$$
\Omega_{\mathrm{k}} = -\frac{\beta k_1}{|\mathbf{k}|^2}, \qquad \omega_{\mathrm{k}} = |\mathbf{k}| - \frac{\beta k_1}{2 |\mathbf{k}|^2}, \tag{13.77}
$$

and, thus, the variable a_{k} describes the short equatorial Rossby waves, while the variables b_{k}, c_{k} describe the short inertia–gravity waves.

Dynamical splitting of short inertia–gravity and Rossby waves

The Rossby waves split out from the inertia–gravity waves, as follows from (13.76); that is if $a_{\mathrm{k}} = 0$ initially, then $a_{\mathrm{k}} = 0$ for all times, in the domain of validity of these equations. Hence, a separate system for SIGW results:

$$
\begin{pmatrix} \partial b_{\mathrm{k}}/\partial t \\ \partial c_{\mathrm{k}}/\partial t \end{pmatrix} + \begin{pmatrix} i\omega_{\mathrm{k}} & 0. \\ 0 & -i\omega_{-\mathrm{k}} \end{pmatrix} \begin{pmatrix} b_{\mathrm{k}} \\ c_{\mathrm{k}} \end{pmatrix} +
$$

$$
+ \begin{pmatrix} \int (V_{klm}^{(3)} b_1 b_{\mathrm{m}} + V_{klm}^{(4)} c_1 c_{\mathrm{m}} + V_{klm}^{(5)} b_1 c_{\mathrm{m}}) d\lambda \\ \int (W_{klm}^{(3)} c_1 c_{\mathrm{m}} + W_{klm}^{(4)} b_1 b_{\mathrm{m}} + W_{klm}^{(5)} c_1 b_{\mathrm{m}}) d\lambda \end{pmatrix} = 0. \tag{13.78}
$$

We introduce a pair of new variables $\phi(\mathbf{x}, t)$, $\zeta(\mathbf{x}, t)$ with Fourier transforms,

$$
\begin{pmatrix} \varphi_{\mathrm{k}} \\ \zeta_{\mathrm{k}} \end{pmatrix} = \begin{pmatrix} \frac{i}{\sqrt{2}k} & -\frac{i}{\sqrt{2}k} \\ \frac{1}{\sqrt{2}} & \frac{1}{\sqrt{2}} \end{pmatrix} \begin{pmatrix} b_{\mathrm{k}} \\ c_{\mathrm{k}} \end{pmatrix}, \tag{13.79}
$$

where $c_{\mathbf{k}} = b^*_{-\mathbf{k}}$, as the initial variables are real. In terms of these new variables the equations (13.78) take the canonical Hamiltonian form:

$$\begin{pmatrix} \dot{\varphi} \\ \dot{\zeta} \end{pmatrix} + \begin{pmatrix} 0 & 1 \\ -1 & 0 \end{pmatrix} \begin{pmatrix} \delta H/\delta\varphi \\ \delta H/\delta\zeta \end{pmatrix} = 0, \tag{13.80}$$

with the Hamiltonian

$$H = \int \left(\frac{1}{2}(1+\zeta)(\varphi_x^2 + \varphi_y^2) + \frac{1}{2}\zeta^2 + \beta\varphi\nabla^{-2}\zeta_x \right) dxdy. \tag{13.81}$$

In the Fourier space (13.80) become

$$\frac{\partial b_{\mathbf{k}}}{\partial t} + i\frac{\delta H}{\delta b^*_{\mathbf{k}}} = 0, \tag{13.82}$$

with the Hamiltonian $H = H_2 + H_3$, where

$$H_2 = \int \omega_{\mathbf{k}} \mid b_{\mathbf{k}} \mid^2 d\mathbf{k}, \quad \omega_{\mathbf{k}} = k - \frac{\beta}{2}\frac{k_x}{k^2}, \quad k = \sqrt{k_x^2 + k_y^2}. \tag{13.83}$$

The frequency $\omega_{\mathbf{k}}$ is positive, and

$$H_3 = \frac{1}{2}\int V_{123}\left(b_1 b_2^* b_3^* + b_1^* b_2 b_3 \right) \delta(\mathbf{k}_1 - \mathbf{k}_2 - \mathbf{k}_3)d\mathbf{k}_1 d\mathbf{k}_2 d\mathbf{k}_3 +$$

$$+ \quad \frac{1}{3}\int U_{123}\left(b_1 b_2 b_3 + b_1^* b_2^* b_3^* \right) \delta(\mathbf{k}_1 + \mathbf{k}_2 + \mathbf{k}_3)d\mathbf{k}_1 d\mathbf{k}_2 d\mathbf{k}_3. \tag{13.84}$$

To the leading order in β

$$2U_{123} = V_{123} = \sqrt{18}\frac{k_1(\mathbf{k}_2, \mathbf{k}_3) + k_2(\mathbf{k}_3, \mathbf{k}_1) + k_3(\mathbf{k}_1, \mathbf{k}_2)}{\sqrt{k_1 k_2 k_3}}. \tag{13.85}$$

We then apply the weak turbulence approach to equatorial SIGW. The results, however, depend on the decay or non-decay character of the dispersion law (13.83).

Analysis of resonance conditions

The following analysis shows that the dispersion law changes its type depending on orientation of the wave vectors of interacting waves. We consider the three-wave synchronism conditions

$$\mathbf{k} = \mathbf{k}_1 + \mathbf{k}_2, \quad \omega_{\mathbf{k}} = \omega_{\mathbf{k}_1} + \omega_{\mathbf{k}_2},$$

and reorient the coordinate system using the vector **k** as an axis, and introducing the polar coordinate θ. Then

$$k = k_1 \cos\theta_1 + k_2 \cos\theta_2, \quad 0 = k_1 \sin\theta_1 + k_2 \sin\theta_2, \tag{13.86}$$

$$k_1 + k_2 - k - \frac{\beta}{2}\left(\frac{\cos(\alpha + \theta_1)}{k_1} + \frac{\cos(\alpha + \theta_2)}{k_2} - \frac{\cos\alpha}{k}\right) = 0, \tag{13.87}$$

where α is the angle between **k** and the direction $\hat{x}$. The equation for frequencies is rewritten as

$$\left(2k_1 \sin^2\frac{\theta_1}{2} + 2k_2 \sin^2\frac{\theta_2}{2}\right) -$$

$$- \frac{\beta\cos\alpha}{2}\left(\frac{\cos\theta_1}{k_1} + \frac{\cos\theta_2}{k_2} - \frac{1}{k}\right) - \frac{\beta\sin\alpha}{2}\left(\frac{\sin\theta_1}{k_1} + \frac{\sin\theta_2}{k_2}\right) = 0. \tag{13.88}$$

The characteristic scales are $k_1 \sim k_2 \sim 1 \gg \beta > 0$, and hence $\theta_1, \theta_2 \ll 1$. As follows from this equation, two situations are possible.

Let us assume first that the third bracket is small: $\cos\alpha \gg \theta_j \sin\alpha$. Then the first and the second ones should balance each other: $\theta_j^2 \sim \beta\cos\alpha$. From these relations it is easy to find that such situation is possible if

$$\frac{\cos\alpha}{\sin^2\alpha} \gg \beta. \tag{13.89}$$

Let us assume then that the second bracket in (13.88) is small: $\cos\alpha \ll \theta_j \sin\alpha$. Then $\theta_j \sim \beta\sin\alpha$ and therefore

$$\frac{\cos\alpha}{\sin^2\alpha} \ll \beta. \tag{13.90}$$

In the first case we get the following balance, to the leading order:

$$2k_1 \sin^2\frac{\theta_1}{2} + 2k_2 \sin^2\frac{\theta_2}{2} = \frac{\beta\cos\alpha}{2}\left(\frac{1}{k_1} + \frac{1}{k_2} - \frac{1}{k}\right) \tag{13.91}$$

and since

$$\frac{1}{k_1} + \frac{1}{k_2} - \frac{1}{k_1 + k_2} = \frac{(k_1 + k_2/2)^2 + 3k_2^2/4}{k_1 k_2 (k_1 + k_2)} > 0,$$

solutions exist only for $\cos\alpha > 0$.

In the second case we have

$$2k_1 \sin^2\frac{\theta_1}{2} + 2k_2 \sin^2\frac{\theta_2}{2} = \frac{\beta\sin\alpha}{2}\left(\frac{\sin\theta_1}{k_1} + \frac{\sin\theta_2}{k_2}\right) \tag{13.92}$$

and $\theta_1, \theta_2 \sim \beta \ll 1$. Solutions exist at any sign of $\sin\alpha$, because there always exist a trivial solution $\theta_1 = \theta_2 = 0$.

As a result of this analysis, we deduce that resonant triads exist:

1. in a relatively wide segment around the x axis in the right half-plane in the $\mathbf{k}$ space;
2. in two narrow segments in the upper and lower half-plane around the y axis in the $\mathbf{k}$ space. Apart from the narrow regions around the y axis, there are no resonant triads in the left half-plane in the $\mathbf{k}$ space.

Kinetic equations and energy spectra for short equatorial IGW

The wave interactions are, therefore, different in the regions of the phase space with allowed and forbidden resonant triads. As a consequence, supposing that exchanges of energy between these two regions are slow, and waves in each are randomised rapidly, we can formulate approximate kinetic equations for each region. In the first case the standard three-wave kinetic equation holds:

$$
\frac{\partial N(\mathbf{k}, t)}{\partial t} = \pi \int [\ | V_{k12} |^2 f_{k12} \delta(\mathbf{k} - \mathbf{k}_1 - \mathbf{k}_2) \delta(\omega_k - \omega_1 - \omega_2)
$$

$$
- | V_{12k} |^2 f_{12k} \delta(\mathbf{k}_1 - \mathbf{k}_2 - \mathbf{k}) \delta(\omega_1 - \omega_2 - \omega_k) \tag{13.93}
$$

$$
- | V_{2k1} |^2 f_{2k1} \delta(\mathbf{k}_2 - \mathbf{k}_k - \mathbf{k}_1) \delta(\omega_2 - \omega_k - \omega_1) \ ] \, d\mathbf{k}_1 d\mathbf{k}_2,
$$

where we introduced the shorthand notation $f_{\mathbf{k}\mathbf{k}_1\mathbf{k}_2} = f_{k12} = N_1 N_2 - N N_1 - N N_2$, and $N_j = N(\mathbf{k}_j, t)$. Dispersion is weak, and using techniques similar to those used in Section 13.2.1, the three-wave collision integral in the case $\cos\alpha \gg \beta \sin\alpha$, may be factorised assuming that $N(\mathbf{k}) = k^s$:

$$
I(\mathbf{k}) = \pi k^r \int | V_{k12} |^2 f_{k12} \delta(k - k_1 - k_2) \delta(k_1 \theta_1 + k_2 \theta_2)
$$

$$
\delta\left(k_1 \theta_1^2 / 2 + k_2 \theta_2^2 / 2 - \frac{\beta \cos\alpha}{2} \left(\frac{1}{k_1} + \frac{1}{k_2} - \frac{1}{k} \right) \right) \tag{13.94}
$$

$$
\left(k^{-r} - k_1^{-r} - k_2^{-r} \right) d\mathbf{k}_1 d\mathbf{k}_2.
$$

Here $r = 2s + 7$ is a sum of powers coming from: f_{klm} which give $2s$, the interaction coefficient $| V_{klm} |^2$ which gives 3, the delta functions which give $-1 + 0 + 1 = 0$ each, and Jacobians which give 4. We obtain a non-equilibrium power-law solution $N \propto k^{-4}$ at $r = -1$, with the spectral energy density $\varepsilon_k = \omega_k k N_k \propto k^{-2}$. It should be noted that the case of resonant triads with almost meridional wave vectors, at $\cos\alpha \ll \beta \sin\alpha$, does not allow such treatment.

The four-wave collision integral in the non-decay regions of the phase space can be factorised as well by similar techniques, giving

$$I(\mathbf{k}) = k^r \int |T_{k123}|^2 \, \delta(k + k_1 - k_2 - k_3)$$

$$\delta\left(k_1\theta_1^2/2 - k_2\theta_2^2/2 - k_3\theta_3^2/2 - \frac{\beta\cos\alpha}{2}\left(k^{-1} + k_1^{-1} - k_2^{-1} - k_3^{-1}\right)\right)$$

$$\left(N_k^{-1} + N_1^{-1} - N_2^{-1} - N_3^{-1}\right)\left(k^{-r} + k_1^{-r} - k_2^{-r} - k_3^{-r}\right) d\mathbf{k}_1 d\mathbf{k}_2 d\mathbf{k}_3,$$

with $r = 3s + 13$. We therefore get two stationary solutions: one with a constant energy flux: $r = -1$ and $N \propto k^{-14/3}$, and one with a constant wave-action flux: $r = 0$ and $N \propto k^{-13/3}$.

13.2.3 Weak turbulence of the Rossby waves on the β plane

The non-dimensional quasi-geostrophic equation for the geostrophic stream function (pressure) h on the beta plane was already considered in Chapters 5, 6, and 12:

$$\partial_t \left(\nabla^2 h - h\right) + \beta\partial_x h + \epsilon \mathcal{J}\left(h, \nabla^2 h\right) = 0, \tag{13.95}$$

where we introduced the non-dimensional β and rescaled the amplitude of h to make appear the nonlinearity parameter ϵ. The specificity of this equation is that it is Hamiltonian, but the corresponding Hamiltonian structure is not canonical. This is reflected, in particular, in the fact that it possesses an infinity of integrals of motion, the Casimirs, which were discussed in Chapter 6. Indeed, as we have seen in Chapter 5 the essence of the QG dynamics, either on f- or on β plane, is conservation of PV. The quasi-geostrophic absolute PV on the β plane is $q_a = \nabla^2 h - h + \beta y$, and any function of q_a integrated over the domain of the flow obeying (13.95) is conserved. The Hamiltonian structure of the flow is given by the non-canonical Poisson bracket defined for any pair A and B of functionals of the quasi-geostrophic relative PV $q = \nabla^2 h - h$:

$$\{A[q], B[q]\} = -\int dxdy \, (q + \beta y)\mathcal{J}\left(\frac{\delta A}{\delta q}, \frac{\delta B}{\delta q}\right). \tag{13.96}$$

It is easy to check that with the quadratic Hamiltonian:

$$H = \frac{1}{2}\int dxdy \left((\nabla h)^2 + h^2\right), \tag{13.97}$$

which represents the sum of kinetic energy calculated with geostrophic velocity, and of the available potential energy due to elevations of the free surface, the Poisson bracket (13.96) reproduces (13.95) written in terms of q.

It should be emphasised at this point that the topology of isopleths of the quasi-geostrophic PV on the $x - y$ plane allows us to distinguish between wave and vortex flows, which can be both described by (13.95). The regions with closed isopleths of q_a correspond to *vortices* while small-amplitude *waves* have isopleths of $q_a = q + \beta y$ only slightly deviating from the straight lines $y = const$. For configurations with no closed isopleths of $q + \beta y$ there exist a change of independent variables which lets us transform the non-canonical Poisson bracket (13.96) into the canonical one. (This transformation, in fact, straightens the isolines of q_a.) The transformation adds nonlinear terms to the initial quadratic Hamiltonian (13.97). The standard approach described in preceding sections can be then applied to the turbulence of Rossby waves. Yet, a direct approach, which we describe below, results in the same three-wave kinetic equation and we proceed along these (simpler) lines.

Following Section 13.1, we introduce the Fourier transformation of the geostrophic stream function $h_{\mathbf{k}}$, make the Fourier transformation of (13.95), multiply the result by the complex conjugate of $h_{\mathbf{k}}$, average by supposing Gaussian statistics for an ensemble of linear Rossby waves, and use the expansion of h in ϵ. We get in this way the following kinetic equation for the spectral density $N_{\mathbf{k}} = h_{\mathbf{k}} h_{\mathbf{k}}^*$ of h:

$$\dot{N}_{\mathbf{k}} = 4\pi \int d\mathbf{k}_1 d\mathbf{k}_2 \, \delta(\mathbf{k} + \mathbf{k}_1 + \mathbf{k}_2) \delta(\omega(\mathbf{k}) + \omega(\mathbf{k}_1) + \omega(\mathbf{k}_2)) \cdot$$

$$\frac{D_{\mathbf{k}_1 \mathbf{k}_2}}{(k^2 + 1)(k_1^2 + 1)(k_2^2 + 1)} \left(D_{\mathbf{k}_1 \mathbf{k}_2} N_{\mathbf{k}_1} N_{\mathbf{k}_2} + D_{\mathbf{k}\mathbf{k}_2} N_{\mathbf{k}} N_{\mathbf{k}_1} + D_{\mathbf{k}\mathbf{k}_2} N_{\mathbf{k}} N_{\mathbf{k}_2} \right),$$

$$(13.98)$$

where the interaction coefficients are

$$D_{\mathbf{k}_1 \mathbf{k}_2} = \frac{1}{4\pi} \left(\mathbf{k}_1 \wedge \mathbf{k}_2 \right) \left(\mathbf{k}_2^2 - \mathbf{k}_1^2 \right), \qquad (13.99)$$

and $\omega = -\beta \frac{k_x}{k^2 + 1}$ is the standard dispersion relation for Rossby waves.

It may be checked by direct substitution into (13.98) that generalised thermodynamic equilibria of the form

$$N(\mathbf{k}) = \frac{1}{a + b\mathbf{k}^2} + \Phi(k_y)\delta(k_x), \qquad (13.100)$$

where a, b are arbitrary constants, and Φ is an arbitrary function of meridional wave vector, annihilate the collision integral. The first term of this solution corresponds to energy equipartition, while the second corresponds to an arbitrary zonal current, if transformed back to the physical space. This is a peculiar property of the Rossby waves dynamics, which accounts, at least partially, for alternating zonal currents frequently produced in numerical simulations of the QG turbulence on the β plane. There are no other analytic solutions annihilating the collision integral in (13.98). However, we notice that for short Rossby waves the dispersion relation to the leading order becomes $\omega = -\beta \frac{k_x}{|\mathbf{k}|^2}$, and for waves with $k_x \ll k_y$ the eigenfrequency scales independently with k_x and k_y: $k_x \to \lambda_x k_x$, $k_y \to \lambda_y k_y$, $\Rightarrow \omega \to \lambda_x \lambda_y^{-2} \omega$. Using this scale invariance, the collision integral in the domain of short in the meridional direction waves can be represented as a sum of three integrals and the integration domains may be transformed one into another using the new integration variables $s_j = \frac{k_{jx}}{k_x}$ and $t_j = \frac{\omega(k_j)}{\omega(k)}$, along the lines of Section 13.1.4. Under hypothesis of self-similar solutions $N(k_x, k_y) \propto k_x^\alpha k_y^\beta$, the whole collision integral then is factorised, and the following stationary solutions are obtained:

$$N_k \propto k_x^{-\frac{3}{2}}, \quad N_k \propto k_x^{-\frac{3}{2}} k_y^{-1}. \tag{13.101}$$

As in the previous examples, these are exact solutions of an approximate kinetic equation, obtained under *ad hoc* assumption that collision integral is saturated by the waves of the chosen kind.

13.3 Turbulence of inertia–gravity waves in rotating shallow water: theory vs numerical experiment

As we have seen in Section 13.2.1, four-wave resonances of inertia–gravity waves lead, in the limit of large wave numbers (i.e. weak dispersion) to the power energy spectra of the form $k^{-8/3}$, corresponding to a constant energy flux, and of the form $k^{-7/3}$, corresponding to a constant wave-action flux, predicted by the wave turbulence theory. These spectra are given by approximate solutions of the kinetic equation established under assumption of small wave amplitudes and quasi-Gaussian statistics. At higher wave amplitudes wave breaking and shock formation inevitably take place in the RSW model, as explained in Chapter 7. Gentler spectra in k^{-2} are expected for ensembles of random shocks. The predictions of wave-turbulence theory, and of random shocks theory, can be checked by direct numerical simulations of the evolution of a random Gaussian field of inertia–gravity waves with weak amplitude in the RSW model. In the simulations presented below, an ensemble of waves was initialised with random phases, and amplitudes corresponding to a power-law energy spectrum. The highest initial amplitude of the longest wave was $3.5 \cdot 10^{-3}$, and the mean absolute value of the non-dimensional surface elevation was 10^{-3}, in non-dimensional terms. Typical evolution of the wave spectrum from $t = 0$ to $t = 20/f$ in such simulations is presented in Figure 13.1. As follows from the figure, the distinguishable slopes exhibit a power law close to k^{-6}, far steeper than theoretical predictions. Moreover, we see that the long-wave component hardly evolves throughout the simulation, whereas the short-wave one quickly adjusts to the -6 slope.

The explanation of such discrepancy between theoretical predictions and simulations is, probably, the finite size of the computational domain. If, due to finite-size effects, which lead to sparser resonant quartets, the exchange of energy of some waves with the rest of the spectrum is suppressed, these waves will eventually break down and form shocks in RSW. This scenario is confirmed in simulations. The divergence field of a typical simulation is displayed in Figure 13.2 at $t = 10/f$ and exhibits coexistence of long breaking waves, which are represented by the straight convergence fronts (breaking

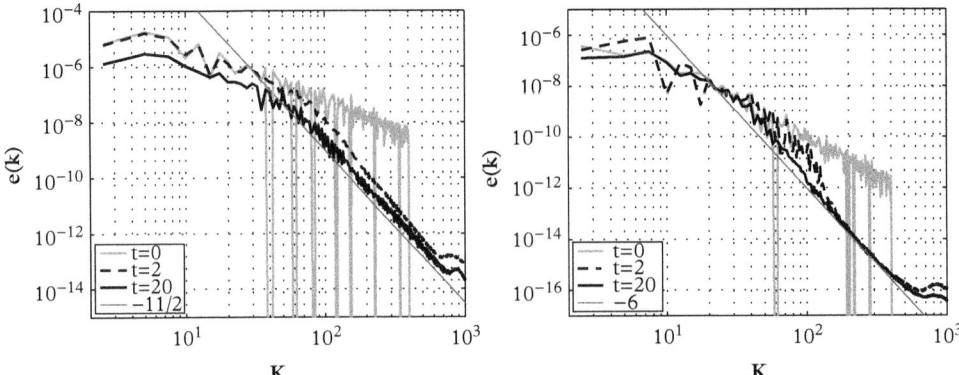

Figure 13.1 *Evolution of the energy spectra for the simulations of wave turbulence with initial energy spectra k^{-2} (left panel) and k^{-3} (right panel).*

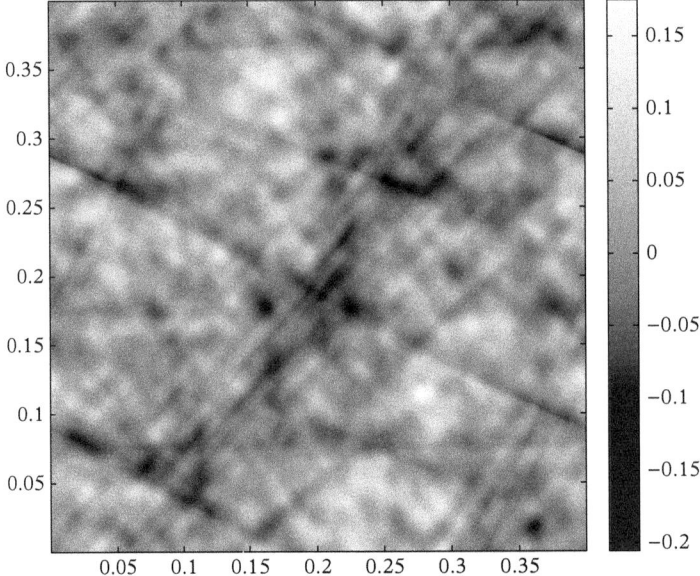

Figure 13.2 *Snapshot of the divergence field at $t = 10 f^{-1}$ in a simulation with a random inertia–gravity wave initialisation. Straight dark lines: convergence fronts.*

zone is characterised by high and concentrated values of convergence), and a 'soup' of shorter non-breaking waves at this stage of evolution. Wave profiles clearly show sharp gradients of all fields and wave breaking, as shown in Figure 13.3. Such behaviour is observed in all shock-resolving wave-turbulence simulations. Thus, the longest waves of

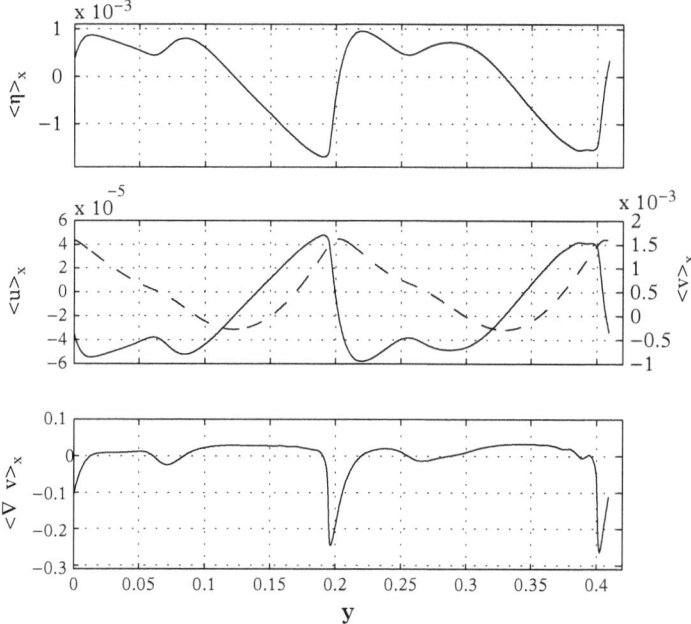

Figure 13.3 *Height (upper panel), and parallel and perpendicular to the wave-front velocities (middle panel, respectively, dashed and solid), and divergence (lower panel) averaged along a moving line of strong convergence.*

the initial spectrum do not exchange enough energy with other parts of the spectrum, and they break. However, they are not numerous enough to produce shock turbulence with its characteristic spectrum. Nevertheless, as coherent structures they contribute to the non-Gaussianity of the wave field, which is confirmed by Figure 13.4.

13.4 Historical comments, summary, and bibliographic remarks

The concept of weak turbulence appeared in plasma physics (e. g. Sagdeev and Galeev 1969, Zaslavsky and Sagdeev 1981). These ideas were first applied to hydrodynamical waves in Zakharov (1984). Earlier, a statistical approach to water waves was developed in Hasselmann (1967). Kinetic equations were derived in these works. At the same time, random-wave closures in the hydrodynamical context were proposed in Benney and Saffman (1966) and Benney and Newell (1969). The wave-turbulence approach has a long history in oceanography, where it was repeatedly evoked to explain the observed spectra of internal inertia–gravity waves, in particular the famous Garret–Munk spectrum (e.g. the reviews of Muller *et al.* (1976) and Polzin and Lvov (2011), and

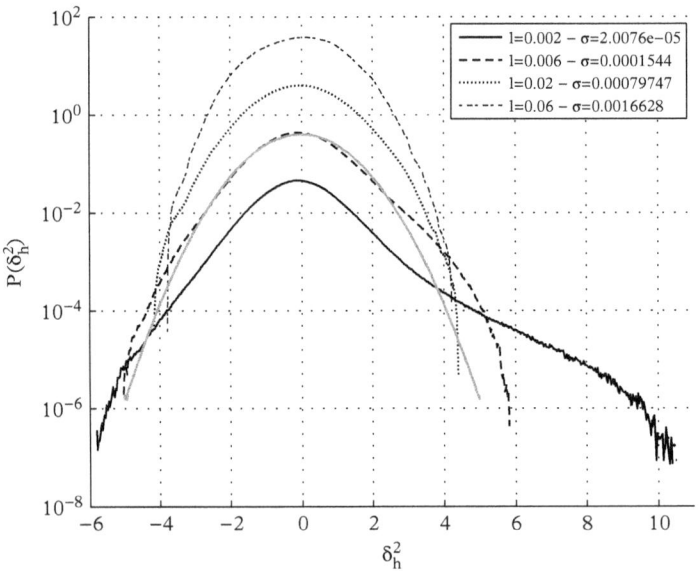

Figure 13.4 *Normalised probability distribution function of the second order increment of the height deviation at different separations averaged over one inertial period. Gaussian distribution is superimposed in grey. Curves are shifted vertically at equal intervals for clarity.*

references therein). Stationary energy spectra of inertia–gravity waves in the framework of the primitive equations were obtained, along the same lines as above, in Caillol and Zeitlin (2000). We should stress that, as the previous Chapter 12, this one connects geophysical fluid dynamics to other branches of physics.

Apart from the universal energy spectra corresponding to equipartition, all wave turbulence spectra found above are, in spite of the elegance of the derivation, intermediate asymptotics, as they were obtained by using asymptotic expressions for dispersion laws, valid only in some subspace of the full wave-number space. Yet, the transformations of the integration domains in the collision integral, which are necessary to obtain factorisation, are in the whole wave-number space. This means that, implicitly, a hypothesis that the contributions of the considered type of waves saturate the collision integral is made. Such a hypothesis is difficult to justify *a priori*. As we have seen, an *a posteriori* verification by numerical simulations has its own difficulties, mainly due to the boundary effects, which tend to supress energy transfers among waves. So the question about the relevance of obtained spectra, in spite of their physical consistence as spectra related to constant fluxes of energy or wave action, remains open, although the wave-turbulence spectra are invoked in relation to observations (cf. Polzin and Lvov 2011). Nevertheless, the kinetic equations themselves are solid, and can be used for numerical simulations of evolution of ensembles of various types of waves, as it is being done for surface waves in the ocean (e.g. Annenkov and Shrira 2013).

The presentation of the main ideas of wave turbulence in this chapter is standard. There is a number of books and reviews on the subject: (Zakharov 1984, Zakharov, Lvov, and Falkovich 1992, Nazarenko 2011). For advanced and retarded solutions for nonlinear corrections to linear waves, and Sokhotsky formulas cf. Morse and Feshbach (1953) for example. In describing factorisation of the collision integral we followed the approach developed in Kats and Kontorovich (1973). Discussion of applicability criteria of kinetic equations can be found in Zakharov, Lvov, and Falkovich (1992) and Nazarenko (2011). Factorisation of the collision integral in the non-decay case close to the acoustic law follows a virtually unknown paper (Volotsky, Kats, and Kontorovich 1980) which was published in a local journal. Kinetic equation and weak turbulence spectra of short inertia–gravity waves in the f plane were obtained in Falkovich and Medvedev (1992), and we follow this paper, exposition of results on turbulence of short equatorial waves follows Medvedev and Zeitlin (2005).

Weak turbulence of the Rossby waves has a long history. Kinetic equation was obtained by Kenyon (1964) and Longuet-Higgins and Gill (1967). Equilibrium spectra were derived and investigated in Reznik and Soomere (1983) and Reznik (1984). Canonical variables for the Rossby-waves equation were obtained by Zakharov and Piterbarg (1987) and used to construct the kinetic equation. The relation between canonical and non-canonical Hamiltonian structure for nonlinear Rossby waves was established in Zeitlin (1992). Equivalence of the thus obtained kinetic equation and the old one of Longuet-Higgins and Gill (1967) was established in Monin and Piterbarg (1987) and the nonequilibrium Kolmogorov-type spectra were found following the method of Kuznetsov (1972). Stability of non-equilibrium power-law solutions of the kinetic equation for Rossby waves was studied in Balk and Nazarenko (1990).

The presentation in Section 13.3 follows Lahaye and Zeitlin (2012c). Energy spectra for an ensemble of random shock waves were obtained in Kuznetsov (2004). For the role of finite size effects in steepening of the wave spectra see Nazarenko (2011), yet no theoretical prediction along these lines for the RSW model is available, to our knowledge.

Figures 13.1 to 13.4 are reproduced from Lahaye and Zeitlin (2012c) with the permission of AIP publishing.

Part III

Generalisations of Standard Rotating Shallow-water Model, and their Applications

14

Rotating Shallow-Water model with Horizontal Density and/or Temperature Gradients

As we have seen in the preceding chapters, the rotating shallow-water model allows us to understand the nature and the properties of fundamental dynamical processes. However, in the models we were considering up to now an important ingredient was missing. Indeed, the RSW models, as they were constructed in Chapter 3, do not allow for horizontal density or temperature gradients. Chapter 3 mentioned that there exists a possibility of modifying the derivation of the RSW models, in order to include this effect. This will be done here and will lead to the so-called thermal shallow-water model.

14.1 Derivation of the thermal rotating shallow-water model and its properties

14.1.1 Derivation of the model

Let us recall the 'master' equation, which was obtained in Chapter 3 by applying vertical mass-weighted averaging between a pair of material surfaces at z_1, z_2, and the mean field hypothesis to the horizontal momentum equations in general form in the presence of the Coriolis force:

$$\bar{\rho}(z_2 - z_1)(\partial_t \boldsymbol{v} + \boldsymbol{v} \cdot \nabla \boldsymbol{v} + f\hat{z} \wedge \boldsymbol{v}) = \tag{14.1}$$

$$-\nabla \left(-g\bar{\rho}\frac{(z_2 - z_1)^2}{2} + (z_2 - z_1)\,p|_{z_1} \right) - \nabla z_1\,p|_{z_1} + \nabla z_2\,p|_{z_2}.$$

Recall that $\bar{\rho}$ is the vertically averaged density of the layer, and $\boldsymbol{v}(x, y, t)$ is vertically averaged horizontal velocity. The hydrostatic relation

$$p(x, y, z, t) = -g\bar{\rho}(z - z_1) + p|_{z_1} \tag{14.2}$$

Geophysical Fluid Dynamics. Vladimir Zeitlin,
Oxford University Press (2018). © Vladimir Zeitlin. DOI: 10.1093/oso/9780198804338.001.0001

was used to express pressure in terms of other variables. The continuity equation, averaged in the same way, is

$$\partial_t \left(\bar{\rho}(z_2 - z_1) \right) + \nabla \cdot \left(\bar{\rho}(z_2 - z_1)\boldsymbol{v} \right) = 0. \tag{14.3}$$

Note that if the starting point is the horizontal momentum part of the primitive equations, which does not explicitly contain density, mass-weighted averaging is replaced by simple geometric averaging, and the master equation is modified:

$$(z_2 - z_1)(\partial_t \boldsymbol{v} + \boldsymbol{v} \cdot \nabla \boldsymbol{v} + f\hat{z} \wedge \boldsymbol{v}) = \tag{14.4}$$

$$-\nabla \left(-g\frac{\bar{\rho}}{\rho_0} \frac{(z_2 - z_1)^2}{2} + (z_2 - z_1)\, \phi|_{z_1} \right) - \nabla z_1\, \phi|_{z_1} + \nabla z_2\, \phi|_{z_2}.$$

Here, $\bar{\rho}$ is the vertically averaged variable part of density, to be replaced by potential temperature in the atmospheric context, and ρ_0 is some reference value. The hydrostatic relation is to be used to eliminate geopotential $\phi(x, y, z, t) = -g\frac{\bar{\rho}}{\rho_0}(z - z_1) + \phi|_{z_1}$. Averaging of the incompressibility condition, together with the mean field hypothesis, gives

$$\partial_t(z_2 - z_1) + \nabla \cdot ((z_2 - z_1)\boldsymbol{v}) = 0. \tag{14.5}$$

At the same time, averaging and applying the mean field assumption to the equation of advection of density $\frac{d\rho}{dt} = 0$ results in

$$\partial_t \bar{\rho} + \boldsymbol{v} \cdot \nabla \bar{\rho} = 0. \tag{14.6}$$

We notice at this point that relaxing the additional hypothesis $\bar{\rho} = \text{const}$, which was made in Chapter 3 and led to the standard RSW equations, is straightforward. In the simplest one-layer configuration with flat bottom the system (14.4) to (14.6) with variable $\bar{\rho} = \bar{\rho}(x, y, t)$ becomes

$$\begin{cases} \partial_t \boldsymbol{v} + \boldsymbol{v} \cdot \nabla \boldsymbol{v} + f\hat{z} \wedge \boldsymbol{v} = -g\frac{\bar{\rho}}{\rho_0}\nabla h - g\frac{h}{2}\nabla\frac{\bar{\rho}}{\rho_0}, \\ \partial_t h + \nabla \cdot (\boldsymbol{v}h) = 0, \, \partial_t \bar{\rho} + \boldsymbol{v} \cdot \nabla \bar{\rho} = 0. \end{cases} \tag{14.7}$$

Here $h = z_2 - z_1$, as usual, and we will omit the bar over ρ from now on. The buoyancy variable $b = g\bar{\rho}(x, y, t)/\rho_0$ is natural to use in these equations, which constitute the thermal rotating shallow-water model (TRSW). We will be considering it in the f-plane approximation, although it is straightforward to reinstitute the beta effect.

If the starting point is the system (14.1) to (14.3), with non-constant $\bar{\rho}$ and without incompressibility constraint it is not closed for a non-constant mean density, and an additional hypothesis is needed in order to get three independent equations for three dynamical variables $\boldsymbol{v}, h, \bar{\rho}$, for example, incompressibility. In this case we get the same second and third equations (14.7), but the first equation changes:

$$\partial_t \boldsymbol{v} + \boldsymbol{v} \cdot \nabla \boldsymbol{v} + f\hat{z} \wedge \boldsymbol{v} = -g\nabla h - g\frac{h}{2}\frac{\nabla\bar{\rho}}{\bar{\rho}}, \tag{14.8}$$

where it is natural to introduce a new variable $\sigma = \log \bar{\rho}$. We will call these equations modified TRSW. If the Boussinesq approximation is applied to (14.8), that is $\bar{\rho}$ is represented by a sum of a constant part, and small variable part, negligible in denominator, we recover (14.7).

14.1.2 Gas dynamics analogy

TRSW equations (14.7) in terms of buoyancy can be rewritten as

$$\begin{cases} \partial_t \boldsymbol{v} + \boldsymbol{v} \cdot \nabla \boldsymbol{v} + f\hat{z} \wedge \boldsymbol{v} = -\dfrac{1}{h}\nabla\dfrac{bh^2}{2}, \\ \partial_t h + \nabla \cdot (\boldsymbol{v}h) = 0, \\ \partial_t b + \boldsymbol{v} \cdot \nabla b = 0. \end{cases} \tag{14.9}$$

We recognise in these equations adiabatic (i.e. without heat exchanges among fluid parcels) dynamics of an ideal gas with density h, entropy b, and pressure $P = \frac{bh^2}{2}$ in a rotating frame. It is curious that while 'normal' RSW is equivalent to dynamics of a barotropic (isentropic) gas in two dimensions in the presence of rotation, the TRSW is equivalent to dynamics of a non-isentropic gas.

14.1.3 Waves and vortices

Linearisation of the TRSW equations in the f-plane approximation about a state of rest with constant buoyancy $b = B$ and constant $h = H$ immediately shows that buoyancy perturbation is time-independent and that the spectrum of small perturbations consists of usual inertia–gravity waves with the dispersion relation $\omega = \pm\sqrt{BH\mathbf{k} + f^2}$, as in the 'normal' RSW. On the contrary, the PV $q = \frac{\partial_x v - \partial_y u + f}{h}$ is not a Lagrangian invariant in TRSW. A direct calculation shows that

$$\frac{dq}{dt} = \partial_t q + \boldsymbol{v} \cdot \nabla q = \frac{1}{2h}\mathcal{J}(h, b). \tag{14.10}$$

This appears to be a drawback, although the following integral conservation law for PV can be still established:

$$\frac{d}{dt}\int_{S_B} dxdy\, hq = 0, \tag{14.11}$$

where S_B is an area inside a closed 'isentrope' $b = $ const. Indeed, the analogue of the Kelvin theorem for the absolute velocity $\mathbf{v}_a = \mathbf{v} + \frac{f}{2}\hat{\mathbf{z}} \wedge \mathbf{x}$ can be obtained by direct calculation, as in Chapter 2:

$$\frac{d}{dt}\int_{\Gamma} \mathbf{v}_a \cdot d\mathbf{l} = \int_{\Gamma} \frac{h}{2}\nabla b \cdot d\mathbf{l}, \tag{14.12}$$

where Γ is a material contour in the plane. If it is chosen to coincide with a close isentrope $b = \text{const}$, which is consistent in view of the Lagrangian conservation of b, then the right-hand side of (14.12) vanishes and the circulation of absolute velocity is conserved. The integral on the left-hand side of (14.12) is then transformed with the help of Green's formula to the integral on the left-hand side of (14.11), which ends the proof.

14.1.4 Quasi-geostrophic TRSW

Quasi-geostrophic version of TRSW equations can be obtained along the same lines as for 'ordinary' one-layer RSW in Chapter 5; namely, we consider fluid motions with a single spatial and a single velocity scales L and U, respectively, introduce unperturbed thickness H_0 and reference buoyancy B_0, and define in a standard way the Rossby and Burger numbers which characterise a solution:

$$Ro = U / fL, \quad Bu = B_0 H_0 / f^2 L^2.$$

It should be noted that Bu is a sort of baroclinic Burger number, as it is built with buoyancy which plays the role of reduced gravity. Keeping in mind the results of Section 5.2, we apply the quasi-geostrophic scaling $(u, v) \sim U$, $h \sim H_0 (1 + (Ro/Bu)\eta)$, $b \sim B_0 (1 + 2(Ro/Bu)\beta)$ (the perturbation of the reference buoyancy β is not to be confused with the meridional gradient of the Coriolis parameter, which is neglected in this chapter), and suppose that $Bu = \mathcal{O}(1)$ and $Ro = \epsilon = \ll 1$. The non-dimensional TRSW equations then take the form

$$\begin{cases} \epsilon \left(\partial_t \mathbf{v} + \mathbf{v} \cdot \nabla \mathbf{v}\right) + (1 + \epsilon y)\hat{\mathbf{z}} \wedge \mathbf{v} = -(1 + 2\epsilon\beta)\nabla\eta - (1 + \epsilon\eta)\nabla\beta, \\ \epsilon\partial_t \eta + \nabla \cdot (\mathbf{v}(1 + \epsilon\eta)) = 0, \\ \partial_t \beta + \mathbf{v} \cdot \nabla\beta = 0. \end{cases} \tag{14.13}$$

By expanding all variables in asymptotic series in ϵ, we get in the leading order

$$\hat{\mathbf{z}} \wedge \boldsymbol{v}_0 = -\nabla(\eta_0 + \beta_0), \quad \Rightarrow \nabla \cdot \boldsymbol{v}_0 = 0, \tag{14.14}$$

where the subscript 0 refers to the zero order in ϵ. This is an expression, within the given scaling, of *thermo-geostrophic equilibrium*:

$$f\hat{\mathbf{z}} \wedge \boldsymbol{v} = -\frac{1}{h}\nabla P \leftrightarrow f\hat{\mathbf{z}} \wedge \boldsymbol{v} = -b\nabla h - \frac{h}{2}\nabla b, \tag{14.15}$$

which replaces in TRSW the standard geostrophic equilibrium.

The expressions for the first-order components of velocity acquire corrections with respect to the standard QG derivation in Section 5.2.1:

$$\begin{cases} u^{(1)} = -\dfrac{d^{(0)}}{dt} v^{(0)} - 2\beta_0 \partial_y \eta_0 - \eta_0 \partial_y \beta_0, \\[2mm] v^{(1)} = \dfrac{d^{(0)}}{dt} u^{(0)} + 2\beta_0 \partial_x \eta_0 + \eta_0 \partial_x \beta_0, \end{cases} \tag{14.16}$$

where the advective derivative is now $\frac{d^{(0)}}{dt} \cdots = \partial_t \cdots + \mathcal{J}(\eta_0 + \beta_0, ...)$, and $\psi = \eta_0 + \beta_0$ is the geostrophic stream function in TRSW, which has, as in the standard QG, the meaning of quasigestrophic pressure, although the pressure here is understood in the sense of gas dynamics analogy introduced earlier.

By injecting the expressions (14.16) into (14.13) we get

$$\begin{cases} \partial_t \left(\nabla^2 \psi - \psi + \beta_0 \right) + \mathcal{J}(\psi, \nabla^2 \psi) = 0, \\[2mm] \partial_t \beta_0 + \mathcal{J}(\psi, \beta_0) = 0. \end{cases} \tag{14.17}$$

These are the QG equations in TRSW. Obviously, they can be rewritten in terms of η_0 and β_0. Some interesting properties of the TRSW system can be deduced from the QG equations (14.17). It is easy to see that stationary solutions of this system are given by any pair of functionally dependent stream function and buoyancy. For example, any buoyancy front $\beta_0 = B(y)$ is a stationary solution provided ψ is zero, meaning that buoyancy and thickness gradients compensate each other, such that the right-hand side of the thermo-geostrophic balance equation (14.15) is zero. A straightforward linearisation about a state with constant buoyancy gradient leads to wave solutions with the dispersion relation of Rossby waves, where the role of the gradient of planetary vorticity is played by the gradient of buoyancy.

14.1.5   Variational principle for TRSW

The Hamilton's principle with the action

$$S = \int dt\, dx\, dy\, \mathcal{L} = \int dx\, dy \left(\frac{\dot{X}^2 + \dot{Y}^2}{2} - \frac{b(X, Y) h(X, Y)}{2} - f Y \dot{X} \right) \tag{14.18}$$

in terms of Lagrangian variables X, Y, where as in Section 3.4.1 we supposed that initial distribution of thickness $h_I(x, y)$ is 'straightened': $h_I = h_0 = 1$, gives equations (14.9) in the Lagrangian form:

$$\begin{cases} \ddot{X} - f\dot{Y} = -b(X, Y)\partial_X h(X, Y) - \dfrac{h(X, Y)}{2} \partial_X b(X, Y), \\[2mm] \ddot{Y} + f\dot{X} = -b(X, Y)\partial_Y h(X, Y) - \dfrac{h(X, Y)}{2} \partial_Y b(X, Y), \end{cases} \tag{14.19}$$

if we recall that buoyancy is a Lagrangian invariant: $b_I(x, y) = b(X, Y)$. The buoyancy thus replaces in TRSW the neutral buoyancy, which is simply g, the gravity acceleration, in RSW. This seemingly minor change, however, is sufficient to break the invariance of the action with respect to relabelling, which leads to the loss of PV conservation.

14.2   Instabilities of jets and vortices in thermal rotating shallow water

14.2.1   New instabilities in TRSW: first example

We will see in this section that adding horizontal density/temperature gradients in RSW radically changes stability properties of stationary solutions. To show this, let us take an almost trivial solution of QG TRSW equations (14.17) corresponding to spatially uniform flow with constant velocity $u = U$, $v = 0$ which is in thermo-geostrophic equilibrium with buoyancy, without thickness perturbation:

$$\beta_0 = - Uy, \ \eta_0 = 0, \ U = \text{const} \tag{14.20}$$

in non-dimensional terms. This configuration is, obviously, a solution of thermal QG equations, and we can study its stability by linearising (14.17) about it, looking for harmonic solutions $\propto \exp i(\mathbf{k} \cdot \mathbf{x} - \omega t)$, with $\mathbf{k} = (k, l)$, and finding the eigenfrequencies ω. This procedure is straightforward, as linearised equations have constant coefficients, and leads to the following dispersion relation for $c = \omega/k$

$$c = U \left(\frac{\mathbf{k}^2}{\mathbf{k}^2 + 1} \pm \sqrt{\left(\frac{\mathbf{k}^2}{\mathbf{k}^2 + 1} \right)^2 - \frac{\mathbf{k}^2}{\mathbf{k}^2 + 1}} \right). \tag{14.21}$$

As $\mathbf{k}^2 < \mathbf{k}^2 + 1$, the flow is unstable at any $\mathbf{k}$, a situation resembling the classical convective instability.

14.2.2   Instabilities of thermal vortices in TRSW

In order to show that a whole realm of new instabilities opens up in TRSW, as was suggested by the example in Section 14.2.1, we will concentrate on vortices, but a similar picture arises also for jets. We will be following the general scheme already abundantly used in Chapter 11; namely, we find stationary vortex solutions, perform their detailed linear stability analysis, find unstable modes, and make high-resolution numerical simulations of the saturation of instabilities using the found unstable modes for initialisations. For linear stability analysis we are using, as in Chapter 11 the pseudo-spectral collocation method adapted to polar coordinates with grid stretching. For numerical simulations we are applying a second-order centred finite-difference scheme adapted to TRSW by adding corresponding terms in the momentum equations, and an upwind-biased finite-volume buoyancy transport scheme.

14.2.3   Stationary vortex solutions

Let us recall that axisymmetric stationary vortex solutions with azimuthal velocity profile $V(r)$ in one-layer RSW obey the cyclo-geostrophic equilibrium

$$\frac{V^2}{r} + fV = g\partial_r H,$$

and for a given $V(r)$, the corresponding thickness profile $H(r)$ can be found by integration. In TRSW the cyclo-geostrophic equilibrium is replaced by *thermo-cyclo-geostrophic equilibrium*

$$\frac{V^2}{r} + fV = B\partial_r H + \frac{H}{2}\partial_r B, \tag{14.22}$$

where both H and B are present with nontrivial radial profiles, and thus some additional hypotheses are needed to determine $B(r)$ and $H(r)$. We first non-dimensionalise (14.22) by using the same quasi-geostrophic scaling as above $h \sim H_0\,(1 + (Ro/Bu)H)$, $b \sim b_0\,(1 + 2(Ro/Bu)B)$. The equation of thermo-cyclo-geostrophic balance thus becomes

$$\left(1 + Ro\frac{V}{r}\right)V = \left(1 + 2\frac{Ro}{Bu}B\right)H' + \left(1 + \frac{Ro}{Bu}H\right)B', \tag{14.23}$$

where prime denotes the derivative with respect to r. We work with α-Gaussian velocity profiles, already used in Chapter 11: $V(r) = re^{-\frac{r^\alpha - 1}{\alpha}}$ with $\alpha = 3$, and consider a one-parameter family of solutions:

$$B(r) = (\kappa - 1)\int_r^\infty (1 + RoV/r')V dr', \tag{14.24}$$

with buoyancy proportional to the thickness of the corresponding solution in the standard non-thermal RSW. Once $B(r)$ is fixed, $H(r)$ can be found from (14.23). The character of such solution is determined by the parameter κ, which will be called thermal. At $\kappa = 1$ we recover a usual isentropic, or 'isothermal', RSW vortex with flat B, and at $\kappa = 0$ we have a purely 'thermal' vortex with flat H. An example of a vortex with $\kappa = 0.5$ is presented in Figure 14.1.

14.2.4   Results of the linear stability analysis of a thermal cyclone

We present in Figure 14.2 a comparison of the eigen-frequency spectrum resulting from the linear stability analysis of 'isothermal' and purely 'thermal' vortices with the same

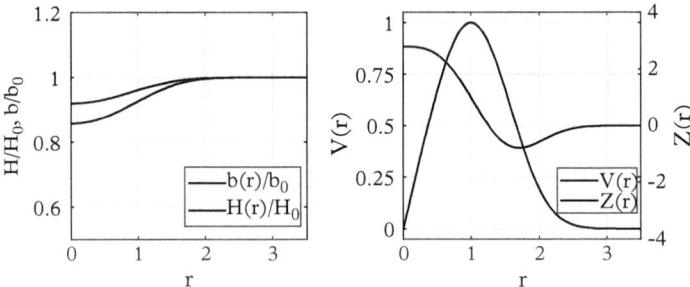

Figure 14.1 *Radial profiles of thickness and buoyancy (left panel), and azimuthal velocity and relative vorticity (right panel) for a cyclonic vortex with κ = 0.5 in thermo-cyclo-geostrophic equilibrium.*

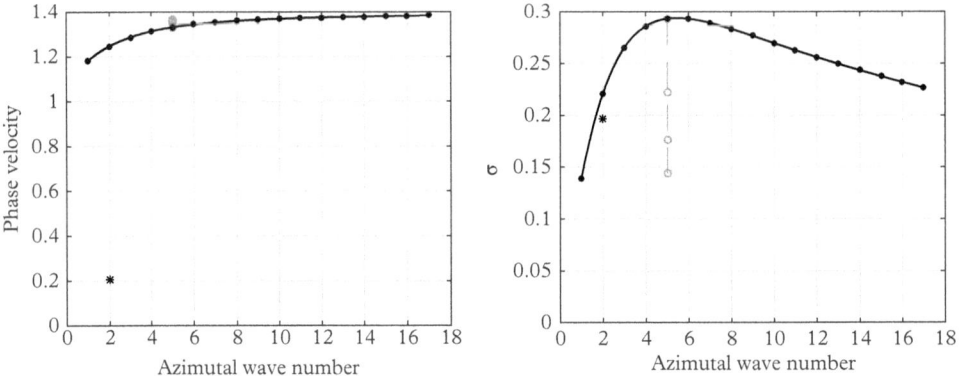

Figure 14.2 *Phase velocities (left panel), and growth rates (right panel) of unstable modes of the thermal vortex with κ = 0, Ro = 0.1, Bu = 1, as functions of azimuthal wave number. Asterisk: barotropic instability. Dots: thermal instability. Grey circles: unstable modes with more nodes in r and lesser growth rates, which are shown only for the most unstable mode with l = 5.*

velocity profile. The difference between the two is striking: there exists only one unstable mode in the case of the 'isothermal' vortex, and a whole series of them in the case of the 'thermal' vortex with the same velocity profile. The unstable mode marked by asterisk in Figure 14.2 is the standard barotropic instability mode. This eigenmode is presented in Figure 14.3, in comparison with the 'thermal' instability mode with the same azimuthal structure corresponding to the dot at $l = 2$ in Figure 14.2. As follows from the left panel of Figure 14.3, 'thermal' instability modes are strongly localised in the radial direction. This is confirmed by Figure 14.4, which presents the radial

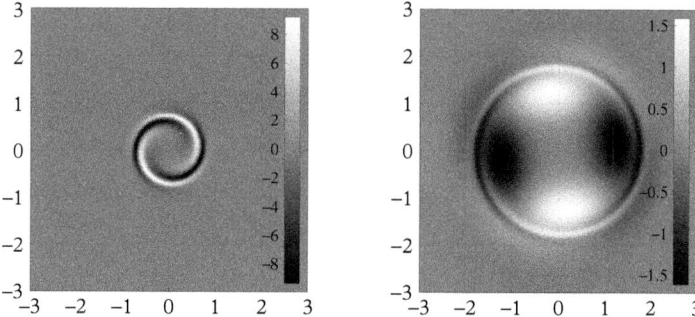

Figure 14.3 *Vorticity distribution in the x − y plane of the unstable mode with l = 2 of the thermal (left) and standard barotropic (right) instabilities of a purely 'thermal vortex' with κ = 0.*

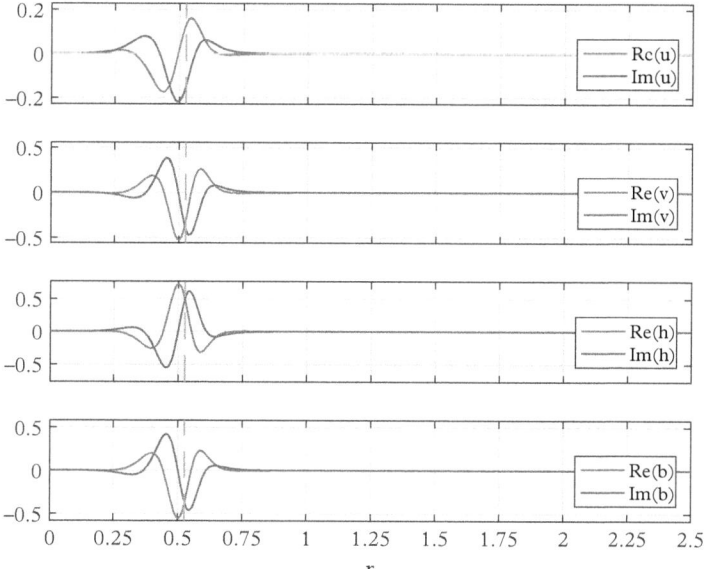

Figure 14.4 *Radial structure of the most unstable mode of the thermal instability with azimuthal wave number l = 5. Position of the critical level is indicated by a dashed line.*

structure of an unstable mode. Note that the mode is localised in the vicinity of the critical level, the position where its phase velocity coincides with the local velocity of the background vortex. In this respect, the unstable modes of the thermal instability are in striking resemblance with the modes of asymmetric centrifugal instability studied in Chapter 11.

14.2.5   Nonlinear saturation of the instability

The change of acoustic analogy from an isentropic gas for RSW to a non-isentropic gas for TRSW renders the numerical method of Chapter 7 inapplicable, so a direct finite difference approach, with an addition of finite-volume scheme for passive scalar advection, is used in numerical simulations presented in the following. A small Newtonian dissipation was added to stabilise the numerical scheme during development of fine-scale structures due to the instability. We present in Figure 14.5 the results of nonlinear evolution of the most unstable mode of Figure 14.2 with azimuthal wave number $l = 5$ and small amplitude, which was superimposed onto the background vortex to initialise the simulation. As seen in the figure, nonlinear saturation produces typically convective 'mushroom' patterns in the buoyancy field, followed by a strong mixing. This pattern is totally different as compared to nonlinear saturation of the barotropic instability, which was analysed in Chapter 11, where it was shown that it produces a regular secondary dipolar structure at small Rossby numbers.

Figure 14.5 *Buoyancy field during nonlinear evolution of instability with $l = 5$ of purely 'thermal' vortex in the $x - y$ plane. Colours correspond to the values of b from 0.95 (brown) to 0.74 (blue). Snapshots at times $t \approx 40$, $t \approx 70$, $t \approx 110$, $t \approx 150 f^{-1}$.*

14.2.6 Discussion of the results

Although we presented here only the analysis of vortex instabilities, and only for cyclonic vortices, similar results can be obtained for instabilities of anticyclonic vortices and of jets. Namely, 'isothermal' jets in geostrophic equilibrium in TRSW, with a flat profile of buoyancy/temperature, are subject only to the standard barotropic instability at small Rossby numbers, as in Chapter 11, while 'thermal' jets in thermo-geostrophic equilibrium exhibit a whole spectrum of new 'thermal' instabilities with higher than the barotropic instability growth rates and convective patterns of nonlinear evolution. It is important to stress that, for both vortices and jets, the new instabilities tend to disappear when gradients of buoyancy (potential temperature) and pressure (thickness) become opposite. These instabilities are, therefore, operational for cold-core cyclones and warm-core anticyclones, in the atmospheric case, and for cyclonic vortices with negative density anomaly, and anticyclonic vortices with positive density anomaly, in the oceanic case, within the limits of applicability of the model.

14.3 Summary, comments, and bibliographic remarks

We have thus seen that inclusion of horizontal buoyancy/potential temperature gradients in RSW is possible, with only a slight modification of the derivation of the RSW model by vertical averaging. Although we presented here only the one-layer case, the procedure can be extended to two-layer models. The resulting TRSW model is equivalent to two dimensional gas dynamics with a specific equation of state, with variable buoyancy/potential temperature playing a role of entropy. The model can be obtained from a physically transparent Hamilton's principle in Lagrangian terms. Yet, the TRSW model does not conserve PV, although a weak integral version of PV conservation can be established. This could seem strange, in view of the above-mentioned gas-dynamics analogy, and the Ertel theorem stating that gas dynamics does have PV conservation. The explanation comes from the fact that Ertel's PV is constructed by projecting (absolute) vorticity onto the gradient of density, or entropy, cf. Chapter 2. Unlike the ordinary RSW, there is an explicit density gradient in TRSW and, being horizontal, it is *orthogonal* to the vertical vorticity. Therefore, conserved Ertel's PV does exist in the model, but is identically zero. The second unusual feature of TRSW model is that stationary jet and vortex solutions exhibit a whole spectrum of instabilities of convective type, if buoyancy/potential temperature gradients are aligned with thickness gradients. A possible link of such instabilities to symmetric-type instabilities of primitive equations was discussed in the literature, but no definitive conclusion was reached. It should be stressed that although the derivation of the TRSW model above is consistent, the stationary states in TRSW do not respect the thermal wind balance, which relates horizontal density/potential temperature gradients to vertical shear of the horizontal velocity in geostrophically and hydrostatically balanced states, cf. (5.5). This means that TRSW can be relevant only for thin enough layers, or weak enough density/temperature gradients.

The TRSW model was independently rediscovered by different authors many times, both in the atmospheric (Lavoie 1972, Salby 1989), and oceanic (McCreary *et al.* 1993,

Ripa 1995) contexts. Its main applications in the literature are to well-mixed atmospheric (Dempsey and Rotunno 1988), and oceanic (McCreary *et al.* 1993), layers, although it was also applied to modelling of planetary atmospheres (Cho *et al.* 2008, Warneford and Dellar 2014, in a modified version in the first paper). The instability of density fronts, even with a uniform velocity, in TRSW was discovered by Young and Chen (1995) and Fukamachi, McCreary, and Proehl (1995). The QG version of the model, with additional salinity field, not introduced above, was derived in Young (1994) and Young and Chen (1995), and corrections due to the vertical shear of the horizontal velocity in the mixed layer were considered in QG approximation. The question of relation of this instability to the classical symmetric instability (Stone 1966), was raised in Fukamachi, McCreary, and Proehl (1995). It is worth emphasising that the patterns of the instabilities of vortices in TRSW presented above resemble much the patterns of vortex instabilities observed in direct numerical simulations of the oceanic mixed layer (cf. Brannigan 2016). Presentation of vortex instabilities in TRSW above follows Gouzien *et al.* (2017). The derivation of thermal QG equations by asymptotic expansions, as well as derivation of integral conservation law for PV, were given in Warneford and Dellar (2013), and we followed this paper. Figures 14.1 to 14.5 are reproduced from Gouzien *et al.* (2017) with permissions.

15

Rotating Shallow-Water Models with Moist Convection

15.1 Constructing moist-convective shallow-water models

15.1.1 General context and philosophy of the approach

The importance of moisture and related moist convection in atmospheric dynamics is obvious. In spite of significant progress and massive efforts, water vapour condensation and precipitations remain a weak point of weather forecasts, especially long-term ones, while predictions of climate models are notoriously divergent in what concerns humidity and precipitations. Physics of moist air including phase transitions is too complicated to be fully reproduced in the general atmospheric circulation models, and condensation processes are usually represented through simplified parameterisations, yet the essentially non-linear, switch character of condensation and related latent heat release poses specific problems in modelling the effects of humidity

Moisture influences large-scale dynamics, which is the main subject of this book, via the latent heat release, due to condensation. As this latter has no linear limit, thinking in linear terms, like modal decomposition, linear stability analysis, etc., is not helpful any more, and one has to resort to direct numerical simulations. Yet, modelling dynamics with inclusion of full thermodynamics and microphysics of the moist air in all its complexity is a tremendous task. If, in continuity with the philosophy of building vertically averaged models adopted in this book, one wants to proceed by a direct averaging of the complete system of equations, a series of specific *ad hoc* hypotheses should be made, which is not satisfactory. Instead, in what follows we will adopt a hybrid approach, which allows us to circumvent the above-mentioned difficulties, while retaining the essentials for large-scale motions part of the moist-air thermodynamics. In order to do this we will combine the vertical averaging of primitive equations between the isobaric surfaces, used in Chapter 3, with that of Lagrangian conservation of the equivalent potential temperature, which is an essential quantity in moist atmosphere and is directly related to the moist static energy frequently used in describing the moist air. We will allow for convective fluxes (i.e. an extra vertical velocity) across the material surfaces

Geophysical Fluid Dynamics. Vladimir Zeitlin,
Oxford University Press (2018). © Vladimir Zeitlin. DOI: 10.1093/oso/9780198804338.001.0001

defining shallow-water layers of the model, and will link these fluxes to condensation. Finally, for the condensation process itself we will use a relaxation parameterisation, of the type applied in general circulation models, in terms of the bulk moisture content of the air columns. We will see that the resulting moist-convective rotating shallow-water models (mcRSW) combine simplicity and fidelity of reproduction of the moist phenomena at large scales, allowing us to use efficient numerical tools available for rotating shallow-water equations, and tending, in various limits, to the previously known simplified models of moist dynamics with condensation.

15.1.2   Introducing moisture in primitive equations

Let us recall the primitive equations in pseudo-height coordinates, cf. Chapter 2:

$$\frac{d}{dt}\boldsymbol{v} + f\hat{\boldsymbol{z}} \wedge \boldsymbol{v} = -\nabla\phi, \tag{15.1}$$

$$\nabla \cdot \boldsymbol{v} + \partial_z w = 0, \quad \partial_z \phi = g\frac{\theta}{\theta_0}, \tag{15.2}$$

$$\frac{d}{dt}\theta = 0, \tag{15.3}$$

where $\boldsymbol{v} = (u, v)$ and w are horizontal and vertical velocities, $\frac{d}{dt} = \partial_t + \boldsymbol{v} \cdot \nabla + w\partial_z$, f - Coriolis parameter, θ - potential temperature, and ϕ - geopotential. If the condensation of water vapour is turned off, the specific humidity q of any air parcel, which measures the water vapour content, is conserved along its trajectory:

$$\frac{d}{dt}q = 0. \tag{15.4}$$

If the condensation and related latent heat release are allowed, θ and q equations (15.3) and (15.4) acquire source and sink, respectively. Yet the equivalent potential temperature $\theta + \frac{L}{c_p}q$, where L is the latent heat of condensation, c_p is specific heat at constant pressure, is conserved for any air parcel in moist-adiabatic processes:

$$\frac{d}{dt}\left(\theta + \frac{L}{c_p}q\right) = 0. \tag{15.5}$$

15.1.3   Vertical averaging with convective fluxes

Let us now reconsider the procedure of vertical averaging of the primitive equations which led us to the RSW model in Chapter 3. By anticipating applications to convection, we allow for extra fluxes across material surfaces delimiting fluid layers. In the case of two layers, that is of three material surfaces, the vertical velocities at the interfaces thus become

$$w_0 = \frac{dz_0}{dt}, \quad w_1 = \frac{dz_1}{dt} + W_1, \quad w_2 = \frac{dz_2}{dt} + W_2, \tag{15.6}$$

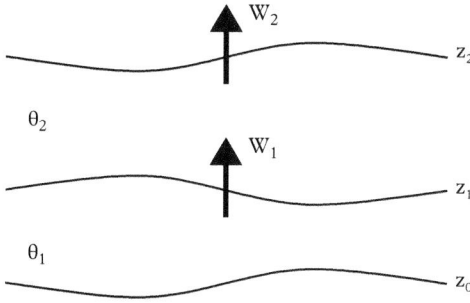

Figure 15.1 *Sketch of the atmospheric two-layer RSW model with convective exchanges between the layers through additional vertical velocities W_1 and W_2.*

where $W_{1,2}$ are contributions from the extra fluxes, whatever their origin, cf. Figure 15.1. It is easy to see that such modification of vertical velocities does not prevent averaging from being carried through. The modified expressions for the vertical velocity (15.6) are introduced in vertical fluxes of mass and momentum in the continuity and horizontal momentum equations, respectively. Modifications of the resulting continuity equations are straightforward, giving physically transparent source and sink terms

$$\begin{cases} \partial_t h_1 + \nabla \cdot (h_1 \boldsymbol{v}_1) = -W_1, \\ \partial_t h_2 + \nabla \cdot (h_2 \boldsymbol{v}_2) = +W_1 - W_2, \end{cases} \tag{15.7}$$

where $h_1 = z_1 - z_0$, $h_2 = z_2 - z_1$ are the thicknesses of the layers. Modifications of the momentum equations contain the terms of the form $W_i \boldsymbol{v}$, which should be calculated at the layers' boundaries z_i. Such terms do not appear in the 'conventional' derivation of Chapter 3. An additional hypothesis is, therefore, needed, as to the value of the horizontal velocity at the interface. The layered models, by construction, have the overall horizontal velocity as a function of z of the form $\boldsymbol{v}(z) = \sum_{i=1}^{N} \boldsymbol{v}_i H(z_i - z) H(z - z_{i-1})$, where $H(z)$ denotes the Heaviside (step-) function. Assigning a value to velocity at z_i is equivalent to assigning a value to the Heaviside function at zero, where this function is not defined. This is a well-known modelling ambiguity, and any value between zero and one can be chosen, depending on the physics of the underlying system. In the case of the shallow-water model this choice would reflect the processes in a buffer layer interpolating between the main layers, and replacing the sharp interface, if a more vertically refined model is used. In the illustrations that follow the assignment $H(0) = 1$ is chosen. A centred assignment $H(0) = 1/2$ can be also made, leading to qualitatively the same results. In any case, extra vertical fluxes lead to exchange of momentum and Rayleigh-type drag, but it should be stressed that, with any assignment, the total momentum of the closed system, that is the system without flux to/from the exterior, is conserved.

The vertically averaged momentum equations become

$$\begin{cases} \partial_t \boldsymbol{v}_1 + (\boldsymbol{v}_1 \cdot \nabla)\boldsymbol{v}_1 + f\hat{\boldsymbol{z}} \wedge \boldsymbol{v}_1 = -\nabla \phi(z_1) + g\frac{\theta_1}{\theta_0}\nabla z_1, \\ \partial_t \boldsymbol{v}_2 + (\boldsymbol{v}_2 \cdot \nabla)\boldsymbol{v}_2 + f\hat{\boldsymbol{z}} \wedge \boldsymbol{v}_2 = -\nabla \phi(z_2) + g\frac{\theta_2}{\theta_0}\nabla z_2 + \frac{\boldsymbol{v}_1 - \boldsymbol{v}_2}{h_2}W_1. \end{cases} \tag{15.8}$$

To close the system, the hydrostatic relation between geopotential and potential temperature is used to express the geopotential at the upper levels in terms of the lower-level one:

$$\phi(z) = \begin{cases} \phi(z_0) + g\frac{\theta_1}{\theta_0}(z - z_0), & \text{if } z_0 \leq z \leq z_1, \\[2ex] \phi(z_0) + g\frac{\theta_1}{\theta_0}(z_1 - z_0) + g\frac{\theta_2}{\theta_0}(z - z_1), & \text{if } z_1 \leq z \leq z_2. \end{cases} \tag{15.9}$$

15.1.4 Linking convection and condensation

Vertical averaging suggests the use of the bulk humidity layer-wise $Q_i = \int_{z_{i-1}}^{z_i} q\,dz$, where q is specific humidity. Q_i thus measure the total water vapour content of the air columns. Condensation introduces a humidity sink in the equation of local conservation of moisture:

$$\partial_t Q_i + \nabla \cdot (Q_i \boldsymbol{v}_i) = -P_i, \quad i = 1, 2. \tag{15.10}$$

In the regions of condensation ($P_i > 0$), specific humidity q is saturated $q(z_i) = q^s(z_i)$ and the temperature of an elementary air mass $W_i\,dt\,dx\,dy$, which moves upward due to released heat $\theta(z_i) + \frac{L}{c_p}q^s(z_i)$, is the one of the upper layer: θ_{i+1}. If a stable background stratification

$$\theta_{i+1} = \theta(z_i) + \frac{L}{c_p}q(z_i) \approx \theta_i + \frac{L}{c_p}q(z_i) > \theta_i, \tag{15.11}$$

with constant $\theta(z_i)$ and constant $q(z_i)$, is assumed, by integrating equation (15.5) we obtain

$$W_i = \beta_i P_i, \quad \beta_i = \frac{L}{c_p(\theta_{i+1} - \theta_i)} \approx \frac{1}{q(z_i)} > 0. \tag{15.12}$$

In this way the extra vertical fluxes in (15.7) and (15.8) are linked to condensation. In order to close the system the last step remains, as condensation should be linked, in turn, to the moisture content. We use the relaxation parameterisation, where the moisture, once beyond the saturation value Q^s, is relaxing towards this value with characteristic time τ:

$$P_i = \frac{Q_i - Q_i^s}{\tau} H(Q_i - Q_i^s). \tag{15.13}$$

The examples treated in what follows use the simplest parameterisation with a constant value of Q^s, yet, the pressure of the saturated vapour depends on temperature, according to Clausius–Clapeyron law, and therefore Q^s too. By construction, temperature in the standard RSW models is considered to be horizontally constant. (Horizontal inhomogeneity of density/temperature can be introduced as in Chapter 14, giving another variant of the model, which we will not discuss.) However, small variations of temperature and pressure in the atmosphere are related linearly, and hence a parameterisation which crudely takes into account this fact consists in replacing constant Q^s by $Q^s = Q_0 - \alpha\eta$, where η is the deviation of the thickness h of the fluid layer from its equilibrium value, and $\alpha = \text{const}$.

We present here the simplest version of moist-convective shallow-water model, with emphasis on condensation and related latent heat release. Therefore, the condensed water is dropped off in the dynamics, and we do not consider the inverse phase transition of liquid water to water vapour (vapourisation). In this sense there is no difference between condensation and precipitation in the model. We will briefly discuss how to include the water cycle in a more comprehensive manner at the end of the chapter.

15.1.5 Surface evaporation and its parameterisations

The above-described construction links convective fluxes to latent heat release in physically and mathematically consistent way. In order to render the model more realistic, some parameterisation of exchanges of the lower layer of the model with the boundary layer is to be considered. We will limit ourselves by only introducing the moisture input from the boundary layer through surface evaporation, which is important for maintaining the moist convection. (See discussion of other improvements at the end of this chapter.) Surface evaporation E provides a source in the bulk moisture equation, which thus becomes

$$\partial_t Q + \nabla \cdot (Q\boldsymbol{v}_1) = E - P \tag{15.14}$$

As already stated, the evaporation is due to external factors related to physical processes at the lower boundary of the model. The simplest parameterisations being used in the literature are:

- Relaxational: $E = \frac{\hat{Q}-Q}{\tau_E} H\left(\hat{Q}-Q\right)$, where $\hat{Q}$ is an equilibrium value.
- Dynamical: $E = \delta|\mathbf{v}|$

The first choice roughly corresponds to a situation where the lower layer of the model evolves over a humid soil, with a buffer layer of constant humidity (not represented in the model) which provides a humidity source tending to bring the humidity of the air column to its own value. The second choice roughly corresponds to a situation where the lower layer of the model evolves over the ocean, and the wind leads to evaporation from the ocean surface. A disadvantage of the dynamical parameterisation is that evaporation continues even if Q reaches saturation. An improved combined parameterisation

$$E \propto |\mathbf{v}| \frac{Q^s - Q}{\tau_E} H\left(\hat{Q}-Q\right) \tag{15.15}$$

can therefore be used. The evaporation coefficients $\gamma = \tau_E^{-1}$ and δ are adjustable parameters. The typical evaporation relaxation time τ_E is about one day in the atmosphere, to be compared with τ, which is about one hour, and can be even less. Thus $\tau_E \gg \tau$. The choice of $\hat{Q}$, as compared to Q^s, is not unique.

15.1.6 Two-layer model with a dry upper layer and its one-layer limit

We will consider a configuration which mimics a situation frequently encountered in the atmosphere where humidity and condensation are concentrated in the lower layer, while the upper layer is dry: $Q_2 = 0$, $Q_1 = Q$. The vertical boundary conditions are: $z_2 = $ const, the isobaric upper surface, and ground pressure (i.e. z_0) free, cf. Figure 15.2. The equations of the model in the absence of evaporation are, cf. (3.25)

$$
\begin{cases}
\partial_t \boldsymbol{v}_1 + (\boldsymbol{v}_1 \cdot \nabla)\boldsymbol{v}_1 + f\hat{z} \wedge \boldsymbol{v}_1 = -g\nabla(h_1 + h_2), \\[4pt]
\partial_t \boldsymbol{v}_2 + (\boldsymbol{v}_2 \cdot \nabla)\boldsymbol{v}_2 + f\hat{z} \wedge \boldsymbol{v}_2 = -g\nabla(h_1 + \alpha h_2) + \frac{\boldsymbol{v}_1 - \boldsymbol{v}_2}{h_2}\beta P, \\[4pt]
\partial_t h_1 + \nabla \cdot (h_1 \boldsymbol{v}_1) = -\beta P, \\[4pt]
\partial_t h_2 + \nabla \cdot (h_2 \boldsymbol{v}_2) = +\beta P, \\[4pt]
\partial_t Q + \nabla \cdot (Q\boldsymbol{v}_1) = -P, \quad P = \frac{Q - Q^s}{\tau} H(Q - Q^s)
\end{cases}
\tag{15.16}
$$

where $\alpha = \frac{\theta_2}{\theta_1}$ is stratification parameter.

The limit of very deep upper layer $h_2 \gg h_1$ gives a simplified one-layer version of the model:

$$
\begin{cases}
\partial_t \boldsymbol{v}_1 + (\boldsymbol{v}_1 \cdot \nabla)\boldsymbol{v}_1 + f\hat{z} \wedge \boldsymbol{v}_1 = -g\nabla h_1, \\[4pt]
\partial_t h_1 + \nabla \cdot (\boldsymbol{v}_1 h_1) = -\beta P, \\[4pt]
\partial_t Q + \nabla \cdot (Q\boldsymbol{v}_1) = -P.
\end{cases}
\tag{15.17}
$$

The models (15.16) and (15.17) will be called moist-convective rotating shallow water (mcRSW). We should stress that the addition of relaxation terms does not pose numerical difficulties, and numerical schemes of Chapter 7 can be applied.

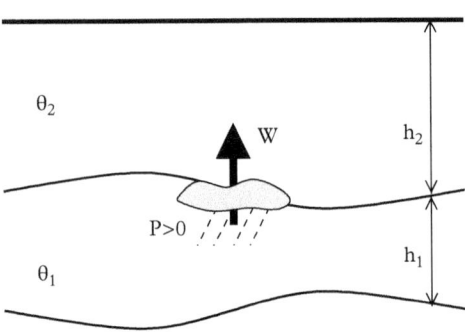

θ_2

θ_1

W

h_2

h_1

P>0

Figure 15.2 *Sketch of the moist-convective two-layer RSW model with moist lower layer.*

15.2   Properties of moist-convective RSW models

15.2.1   Limiting cases

Immediate relaxation limit

As already mentioned, the relaxation time τ in the atmosphere is short. In the limit $\tau \to 0$, the deviation of humidity from the saturation value $\tilde{q} = Q - Q^s$ remains small in the region of condensation, and the dominant terms in the humidity equation (15.10) with $E = 0$ are

$$\partial_t \tilde{q} + Q^s \nabla \cdot \boldsymbol{v} = -\frac{\tilde{q}}{\tau}. \tag{15.18}$$

Supposing that at very small τ, variations of q are much faster than variations of velocity, then the approximate solution of this equation is

$$\tilde{q}|_{\tau \to 0} \approx -\tau Q^s \nabla \cdot \boldsymbol{v}. \tag{15.19}$$

Replacing q by this expression in the definition of P gives

$$P = -Q^s \nabla \cdot \boldsymbol{v}, \tag{15.20}$$

and the equations of the two-layer model with dry upper layer, in the regions of condensation, take the following form:

$$\begin{cases} \partial_t \boldsymbol{v}_1 + (\boldsymbol{v}_1 \cdot \nabla)\boldsymbol{v}_1 + f\hat{\boldsymbol{z}} \wedge \boldsymbol{v}_1 = -g\nabla(h_1 + h_2), \\ \partial_t \boldsymbol{v}_2 + (\boldsymbol{v}_2 \cdot \nabla)\boldsymbol{v}_2 + f\hat{\boldsymbol{z}} \wedge \boldsymbol{v}_2 = -g\nabla(h_1 + \alpha h_2) - \frac{v_1 - v_2}{h_2}\beta Q^s \nabla \cdot \boldsymbol{v}_1, \\ \partial_t h_1 + \nabla \cdot (h_1 \boldsymbol{v}_1) = +\beta Q^s \nabla \cdot \boldsymbol{v}_1, \\ \partial_t h_2 + \nabla \cdot (h_2 \boldsymbol{v}_2) = -\beta Q^s \nabla \cdot \boldsymbol{v}_1, \end{cases} \tag{15.21}$$

with humidity always staying at the saturation value: $Q = Q^s$. This simplification is useful for understanding the numerical results presented here, as τ is very small in simulations, several time-steps of the numerical scheme. The system (15.21) does not contain step functions and can be analysed by standard means.

One-layer quasi-geostrophic model with immediate relaxation

With the standard QG scaling of Chapter 5, at $Ro \to 0$ and $Bu \sim O(1)$, and close to saturation $\psi \sim \tilde{q} << 1$, where ψ is non-dimensional geostrophic stream function in terms of deviation of the free surface, we get in the f-plane approximation

$$\begin{cases} (\partial_t + \boldsymbol{v}^{(0)} \cdot \nabla)\left(\nabla^2 \psi - \psi\right) = \beta P, \\ (\partial_t + \boldsymbol{v}^{(0)} \cdot \nabla)\left(\tilde{q} - Q^s \nabla^2 \psi\right) = -P, \end{cases} \tag{15.22}$$

(an extra term, $\partial_x \psi$, is to be added on the left-hand side of the first equation on the β-plane).

Here $v^{(0)} = (-\partial_y\psi, \partial_x\psi)$ is the geostrophic velocity. In the immediate relaxation limit $\tau \to 0$, $\tilde{q} \approx 0$, and the second equation becomes

$$Q^s(\partial_t + v^{(0)} \cdot \nabla)(\nabla^2\psi) \approx P_{\tau\to0} > 0, \qquad (15.23)$$

which means an increase of geostrophic vorticity in the regions of condensation. This result by itself is sufficient to give a rough understanding of the dynamical role of condensation: it leads to intensification of cyclonic (positive) vorticity, and depletion of anticyclonic (negative) vorticity.

Linear baroclinic reduction

Rewriting the model in terms of baroclinic and barotropic velocities,

$$v^{bt} = \frac{h_1 v_1 + h_2 v_2}{h_1 + h_2}, \quad v^{bc} = v_1 - v_2,$$

and formally linearising it in the hydrodynamic sector leads to decoupling of the barotropic velocity:

$$\begin{cases} \partial_t v^{bc} + f\hat{z} \wedge v^{bc} = -g'\nabla\eta, \\ \partial_t\eta + H_e\nabla \cdot v^{bc} = -\beta P, \\ \partial_t Q + Q_e\nabla \cdot v^{bc} = -P, \end{cases}$$

where $g' = g(\alpha - 1)$ is the reduced gravity, $Q_e = \frac{H_e}{H_1}Q^s$, η is the perturbation of the interface, and H_e is the equivalent height.

Two-layer quasigeostrophic limit

In the small Rossby number limit on the β plane, the following non-dimensional equations result:

$$\begin{cases} \dfrac{d_1^{(0)}}{dt}\left(\nabla^2\psi_1 + y - \dfrac{\eta_1}{\bar{h}_1}\right) = \dfrac{\beta P}{\bar{h}_1}, \\ \dfrac{d_2^{(0)}}{dt}\left(\nabla^2\psi_2 + y - \dfrac{\eta_2}{\bar{h}_2}\right) = -\dfrac{\beta P}{\bar{h}_2}. \end{cases} \qquad (15.24)$$

Here $\frac{d_i^{(0)}}{dt} = \partial_t + (v_i^{(0)} \cdot \nabla)$, $\hat{z} \times v_i^{(0)} = -\nabla\psi_i$, $\bar{h}_i = \frac{H_i}{H_0}$, and $\psi_{1,2}$, the geostrophic stream functions, are related to the free-surface and interface perturbations, respectively $\eta_{2,1}$, as: $\psi_1 = \eta_1 + \eta_2$, $\psi_2 = \eta_1 + \alpha\eta_2$, and the standard QG scaling for the two-layer RSW model with a free surface was applied, as in Section 5.4.

15.2.2   Conservation laws

Horizontal momentum

The horizontal momentum density equations in the layers are

$$(\partial_t + v_1 \cdot \nabla)(v_1 h_1) + v_1 h_1 \nabla \cdot v_1 + f\hat{z} \wedge (v_1 h_1)$$
$$= -g\nabla \frac{h_1^2}{2} - gh_1 \nabla h_2 - \underline{v_1 \beta P}, \qquad (15.25)$$

$$(\partial_t + v_2 \cdot \nabla)(v_2 h_2) + v_2 h_2 \nabla \cdot v_2 + f\hat{z} \wedge (v_2 h_2)$$
$$= -\alpha g\nabla \frac{h_2^2}{2} - gh_2 \nabla h_1 + \underline{v_1 \beta P}. \qquad (15.26)$$

The underlined terms in these equations correspond to the moist convection drag. These terms have opposite signs, like the terms due to condensation in the mass-conservation equations in (15.16), and cancel each other in the equation for the total momentum density $v_1 h_1 + v_2 h_2$ which is, therefore, not affected by convection.

Energy

The energy densities of the layers are

$$e_1 = h_1 \frac{v_1^2}{2} + g\frac{h_1^2}{2}, \quad e_2 = h_2 \frac{v_2^2}{2} + gh_1 h_2 + \alpha g\frac{h_2^2}{2}. \qquad (15.27)$$

In a closed system with no energy fluxes at the boundaries, we get for the total energy $E = \int dx dy (e_1 + e_2)$

$$\partial_t E = -\int dx \, \beta P \left(gh_2(1-\alpha) + \frac{(v_1 - v_2)^2}{2} \right). \qquad (15.28)$$

The first term on the right-hand side corresponds to *production* of potential energy (for stable stratifications) due to upward convection fluxes; the second term corresponds to *destruction* of kinetic energy due to the Rayleigh drag.

Potential vorticity

Potential vorticity equations in each layer in the presence of condensation become

$$\begin{cases} (\partial_t + v_1 \cdot \nabla)\dfrac{\zeta_1 + f}{h_1} = \dfrac{\zeta_1 + f}{h_1^2}\beta P, \\[2ex] (\partial_t + v_2 \cdot \nabla)\dfrac{\zeta_2 + f}{h_2} = -\dfrac{\zeta_2 + f}{h_2^2}\beta P + \dfrac{\hat{z}}{h_2} \cdot \left\{ \nabla \wedge \left(\dfrac{v_1 - v_2}{h_2}\beta P \right) \right\}, \end{cases} \qquad (15.29)$$

where $\zeta_i = \hat{z} \cdot (\nabla \wedge v_i) = \partial_x v_i - \partial_y u_i$ ($i = 1, 2$) is relative vorticity . Hence, PV in each layer is not a Lagrangian invariant in convective regions.

The moist enthalpy in the lower layer is $m_1 = h_1 - \beta Q$ and is locally conserved in the absence of evaporation:

$$\partial_t m_1 + \nabla \cdot (m_1 \boldsymbol{v}_1) = 0. \tag{15.30}$$

The conservation of the moist enthalpy in the lower layer allows us to derive a *new Lagrangian invariant*, the moist PV q_m in the humid layer:

$$(\partial_t + \boldsymbol{v}_1 \cdot \nabla)q_m = 0, \quad q_m = \frac{\zeta_1 + f}{m_1}. \tag{15.31}$$

How evaporation influences the conservation laws

The moist enthalpy is directly affected by evaporation:

$$\partial_t m_1 + \nabla \cdot (m_1 \boldsymbol{v}_1) = -\beta E. \tag{15.32}$$

If no constraint is imposed on the source E, moist enthalpy can become negative, which poses a problem of consistency. Indeed, the mean content of the water vapour in the fluid column $\frac{Q}{h_1}$ evolves as follows:

$$(\partial_t + \boldsymbol{v}_1 \cdot \nabla) \frac{Q}{h_1} = -\frac{m_1}{h_1^2} + \frac{1}{h_1} E, \tag{15.33}$$

and negative m_1 would lead to an increase of mean water vapour content with condensation, which is nonsensical. Hence, the model with included (external) evaporation is valid only while m_1 remains positive. This is important to bear in mind while choosing parameters in numerical simulations with evaporation.

The moist PV is also affected by evaporation, and is not Lagrangian-invariant anymore:

$$(\partial_t + \boldsymbol{v}_1 \cdot \nabla) \frac{\zeta_1 + f}{m_1} = \frac{\zeta_1 + f}{m_1^2} \beta E. \tag{15.34}$$

15.3   Mathematics of moist-convective rotating shallow water

15.3.1   Quasilinear form and characteristic equations

In this section, as in Chapter 7, we will consider the one-dimensional reduction of the model, supposing that all variables do not depend on one of the spatial variables, $\partial_y(...) = 0$, and omitting rotation for simplicity. Again, for simplicity, we will be considering the model in the immediate relaxation limit, that is we will be working with a one-dimensional non-rotating version of (15.21). This gives a *quasilinear system* of the form

$$\partial_t f + \mathbf{A}(f)\partial_x f = \mathbf{b}(f).$$

The expression of the matrix $\mathbf{A}$ is easily recovered from the equations of the model and the characteristic equation is obtained from $\det(\mathbf{A} - c I) = 0$. The 'dry' characteristic equation in the absence of condensation is, cf. section 7.2.2

$$\mathcal{F}(c) = \left\{ (u_1 - c)^2 - g h_1 \right\} \left\{ (u_2 - c)^2 - \alpha g h_2 \right\} - g h_1 g h_2 = 0, \tag{15.35}$$

and the 'moist' characteristic equation is

$$\mathcal{F}^m(c) = \mathcal{F}(c) + ((u_1 - u_2)^2 - (\alpha - 1) g h_2) g \beta Q^s = 0. \tag{15.36}$$

Characteristic velocities about the rest state are obtained by linearisation of the characteristic equations (15.35) and (15.36). In this way we get the 'dry' characteristics, cf. (7.75)

$$C_\pm = g(H_1 + \alpha H_2) \frac{1 \pm \sqrt{\Delta}}{2}, \tag{15.37}$$

and the 'moist' characteristics

$$C_\pm^m = g(H_1 + \alpha H_2) \frac{1 \pm \sqrt{\Delta^m}}{2}. \tag{15.38}$$

Here the subscript $\pm$ refers to 'external' and 'internal' characteristic velocities, corresponding to barotropic (external) and baroclinic (internal) gravity waves in the system. We used the notation: $C = c^2$ and the same definition (7.76), as earlier

$$\Delta = 1 - \frac{4 H_1 H_2 (\alpha - 1)}{(H_1 + \alpha H_2)^2} = \frac{(H_1 - \alpha H_2)^2 + 4 H_1 H_2}{(H_1 + \alpha H_2)^2}, \tag{15.39}$$

$$\Delta^m = \Delta + \frac{4(\alpha - 1)\beta Q^s H_2}{(H_1 + \alpha H_2)^2}. \tag{15.40}$$

C^m are real if the moist enthalpy of the lower layer in the state of rest is positive: $M_1 = H_1 - \beta Q^s > 0$. They obey the following inequalities:

$$C_-^m < C_- < \frac{g(H_1 + \alpha H_2)}{2} < C_+ < C_+^m. \tag{15.41}$$

As $0 < M_1 < H_1$, the moist internal (baroclinic) mode propagates slower than the dry one, which is consistent with observations of internal waves in the moist atmosphere. A characteristic with zero speed $c_q = 0$, corresponding to the linearised conservation of moisture outside of condensation regions, should be also added.

15.3.2 Discontinuities and Rankine–Hugoniot conditions

Discontinuities in dependent variables

In the absence of rotation, Rankine–Hugoniot (RH) conditions in the immediate relaxation limit are

$$
\begin{cases}
- s[v_1 h_1 + v_2 h_2] + [u_1 v_1 h_1 + u_2 v_2 h_2] = 0, \\
- s[m_1] + [m_1 u_1] = 0, \\
- s[h_2] + [h_2 u_2 + \beta Q^s u_1] = 0.
\end{cases}
\tag{15.42}
$$

Here s is the propagation speed of the discontinuity. It should be emphasised that the standard mass conservation condition is replaced by the moist enthalpy conservation in the lower layer. It is also important to stress that, due to the fact that $\lim_{x_s \to a} \lim_{b \to x_s} \int_a^b P = 0$, P does not enter the RH conditions for u, v, h.

Discontinuities in derivatives and condensation fronts

The RH conditions linearised about the rest state are

$$
\begin{cases}
(s^2 - C_+)(s^2 - C_-)[\partial_x u_1] = -(\alpha - 1) g H_2 g \beta [P], \\
(s^2 - C_+^m)(s^2 - C_-^m)[\partial_x u_1] = - s(\alpha - 1) g H_2 g \beta [\partial_x Q].
\end{cases}
\tag{15.43}
$$

Let the condensation happen at the right-hand side of the discontinuity: $P_- = 0$, $P_+ = -Q^s \partial_x u_{1+} > 0$. Then there exist *five types* of condensation fronts:

1. the dry barotropic fronts, $\sqrt{C_+} < s < \sqrt{C_+^m}$;
2. the dry baroclinic subsonic fronts, $\sqrt{C_-^m} < s < \sqrt{C_-}$;
3. the moist baroclinic subsonic fronts, $-\sqrt{C_-^m} < s < 0$;
4. the moist baroclinic supersonic fronts, $-\sqrt{C_+} < s < -\sqrt{C_-}$;
5. the moist barotropic fronts, $s < -\sqrt{C_+^m}$.

All of them appear in numerical simulations of wave propagation in the presence of moisture and condensation.

15.3.3 Illustration: wave scattering on a moisture front

In this and the following sections we will give examples of how the condensation changes wave and vortex dynamics in the two-layer mcRSW. We will be not introducing surface evaporation for the moment. Let us consider a localised internal simple wave centred at $x_P = 2$, and moving from left to right:

$$
u_1(x, 0) =
\begin{cases}
-\sigma (x - x_P)^2 + U_0, & \text{if } \quad -\sqrt{\dfrac{U_0}{\sigma}} \le x - x_P \le \sqrt{\dfrac{U_0}{\sigma}}, \\
0, & \text{otherwise,}
\end{cases}
\tag{15.44}
$$

where $U_0 = 0.01$ and $\sigma = 1$, and a stationary moisture front situated at $x_M = 5$, with saturated air on its right, and unsaturated air on its left:

$$Q(x, 0) = Q^s\{1 + q_0 \tanh(x - x_M)H(-x + x_M)\}, \quad q_0 = 0.05. \tag{15.45}$$

We initialise the one-dimensional version of the numerical scheme for the one-layer mcRSW of Section 15.3.2 with these fields and simulate their evolution. From the strong down-flow convergence in the lower layer we can expect condensation to be triggered ($P > 0$) when the wave approaches the moisture front, which is indeed the case, as follows from Figure 15.3. The wave is partially reflected from the front, and partially penetrates in the saturated region provoking condensation. A schematic zoom of the saturation region showing the condensation fronts in the one-layer limit is presented in Figure 15.4. Hence, the presence of moisture close to saturation induces a new wave-scattering process, which is absent in the 'dry' dynamics, and thus qualitatively modifies the behavior of the system.

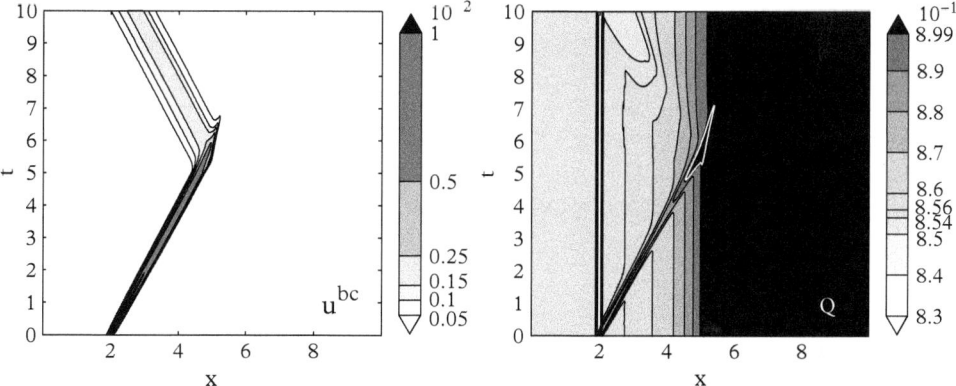

Figure 15.3 *Wave scattering on a moisture front: baroclinic velocity, left panel. Moisture (levels of grey) and condensation (white contour), right panel.*

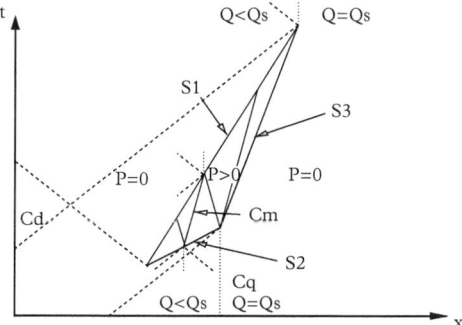

Figure 15.4 *Schematic zoom of the condensation zone forming during the wave scattering on the moisture front. Different characteristics and fronts are indicated.*

15.4   Applications to 'moist' instabilities of geostrophic jets and vortices

15.4.1   Moist instability of the baroclinic Bickley jet

As already mentioned, condensation is an essentially nonlinear process. Hence the linear stability analysis is not applicable to 'moist' instabilities. We will therefore resort to direct numerical simulations to analyse the moist baroclinic instability. For this, we will exploit the results of the linear stability analysis in the 'dry' case of chapter 10; namely, we will use the most unstable mode of the baroclinic Bickley jet of Chapter 10, which was identified by the 'dry' linear stability analysis, in order to initialise numerical simulations. We will then compare the evolution of the perturbation in 'dry' and 'moist-precipitating' cases, moisture being passively advected in the first, and subject to condensation beyond the saturation level in the second. Obviously, a reliable 'dry' simulation with such initialisation should follow the linear growth of perturbation at initial stages, which represents, in fact, a benchmark for nonlinear simulations. Nonlinear saturation of the instability then follows, cf. Chapter 10. The effects of condensation change both the growth and the saturation stages, and we display these changes below. In this and subsequent sections the initial distribution of moisture Q_0 is chosen to be uniform and close to the saturation value.

A comparison of moist and dry growth rates of the instability of the baroclinic Bickley jet is shown in Figure 15.5. The growth rate in numerical simulations is calculated with

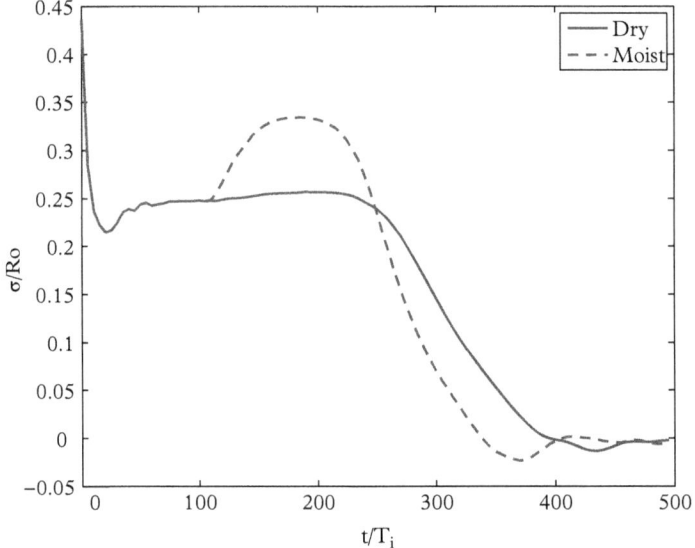

Figure 15.5 *Comparison of the growth rate of the instability of the baroclinic Bickley jet in dry and moist-precipitating cases, as follows from direct numerical simulations initialised with the most unstable mode. ©American Meteorological Society. Used with permissions.*

the help of the dry energy norm of the perturbation. After initial adjustment due to discretisation errors, the growth rate in the dry case follows the predictions of linear stability analysis (the plateau in the figure). A transient increase of the growth rate, perfectly correlated with condensation is observed in the moist-precipitating case. The evolution of moisture and condensation in the moist-precipitating simulation is presented in Figure 15.6. A comparison of dry and moist-precipitating saturations of the baroclinic instability for the relative vorticity is given in Figure 15.7. As follows from the figure, the saturation of the instability in the moist-precipitating case is faster, and leads to pronounced cyclone–anticyclone asymmetry. This asymmetry is easy to understand in the QG approximation discussed earlier and it is relevant here, as the Rossby number of the jet is small. If we take, for simplicity, the one-layer version of the mcRSW model in the immediate relaxation limit, as follows from (15.22), the PV of the air columns which pass through the condensation regions increases, which automatically leads to an increase in the absolute value of cyclonic (positive) and decrease in the absolute value of anticyclonic (negative) anomalies of vorticity. The characteristic comma-shaped regions of cyclonic vorticity and associated condensation fronts are qualitatively close to what is typically observed in nature, cf. Figure 15.9.

An important quantity is the divergence of the baroclinic velocity. Let us recall that in the shallow-water models it is a proxy for the vertical velocity. The comparison of the dry and moist-precipitating simulations presented in Figure 15.8 shows that not only the divergence anomalies related to ageostrophic motions are an order of magnitude larger in the moist case, but their spatial distribution is different, with strong enhancement in the vicinity of the condensation fronts.

Thus, the comparison between dry and moist-precipitating saturations of the baroclinic instability of the jet shows significant differences, namely:

- enhancement of the growth rate of the moist-convective instability at the condensation onset,

- significant increase in intensity of ageostrophic motions during the evolution of the moist-convective instability, and

- substantial cyclone–anticyclone asymmetry, which develops due to the effects of moist convection.

At the same time, the evolution of the moist baroclinic instability in mcRSW simulations is in qualitative agreement with observations.

15.4.2   Moist instability of geostrophic vortices

'Moist' instabilities of vortices can be analysed along the same lines. We offer in the following an example of the influence of the effects of moisture and condensation upon the instability of idealised upper-tropospheric cyclone, which was studied in the 'dry' case in Chapter 11. We follow the same strategy as in Section 5.4.1, by superposing the most unstable mode found by the linear stability analysis, with small (of the order of 2% with the respect to background values) amplitude, onto the alpha- Gaussian

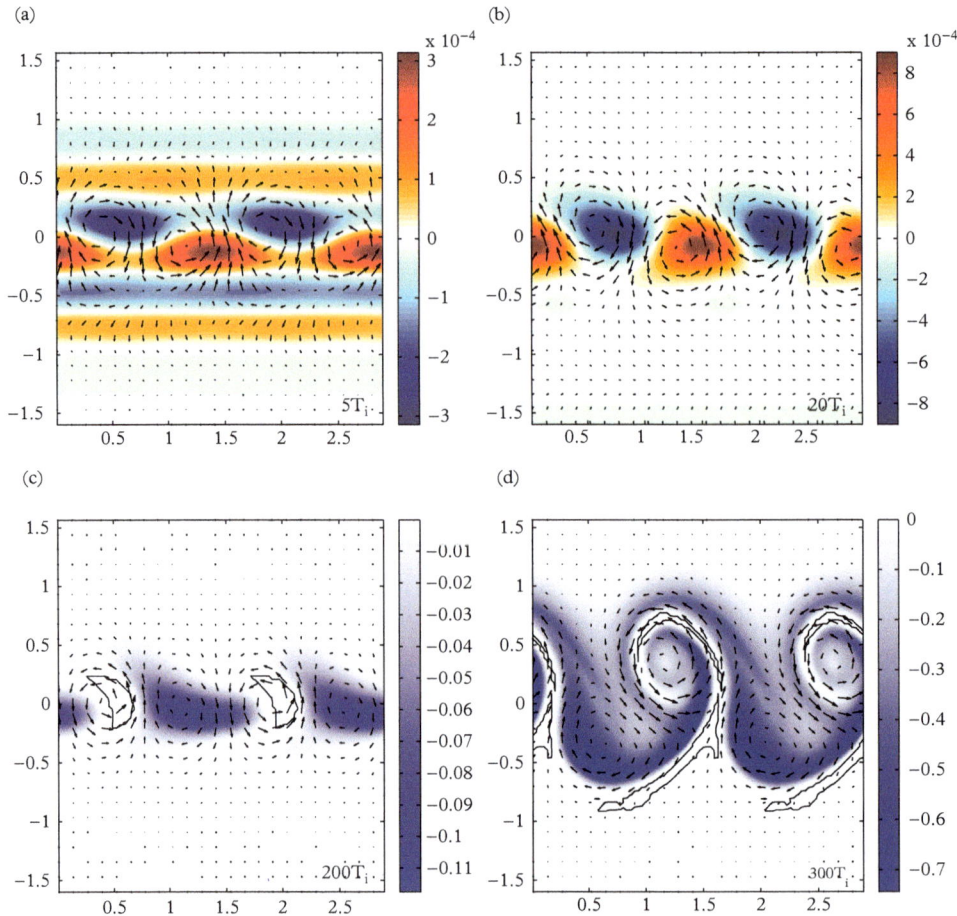

Figure 15.6 *Evolution of the moisture anomaly $Q - Q_0$ with superimposed lower-layer velocity. Snapshots at $5, 20, 200, 300f^{-1}$, from left to right, and from top to bottom. Black contours: condensation zones.* ©*American Meteorological Society. Used with permissions.*

vortex profile of Section 11.2.2, and compare the evolution of the system in 'dry' and moist-precipitating cases. The initial moisture is distributed uniformly and unsaturated: $Q_0 = 0.89$, and $Q_s = 0.9$.

Comparison of the evolution of the moisture content in 'dry' (M) and moist-precipitating (MP) simulations of the development of tripolar instability mode of an idealised upper-tropospheric cyclone, together with joint evolution of PV anomaly and condensation in the lower layer in MP case are presented in Figure 15.10. The MP simulation produces spiral condensation bands. Although the clouds (cf. e.g. Figure 15.9) are droplets of liquid water, and the latter was not included in the model (see, however, a discussion below), they are produced by the condensation, and hence the shape of the condensation regions gives an idea of clouds distribution. Thus, development of the

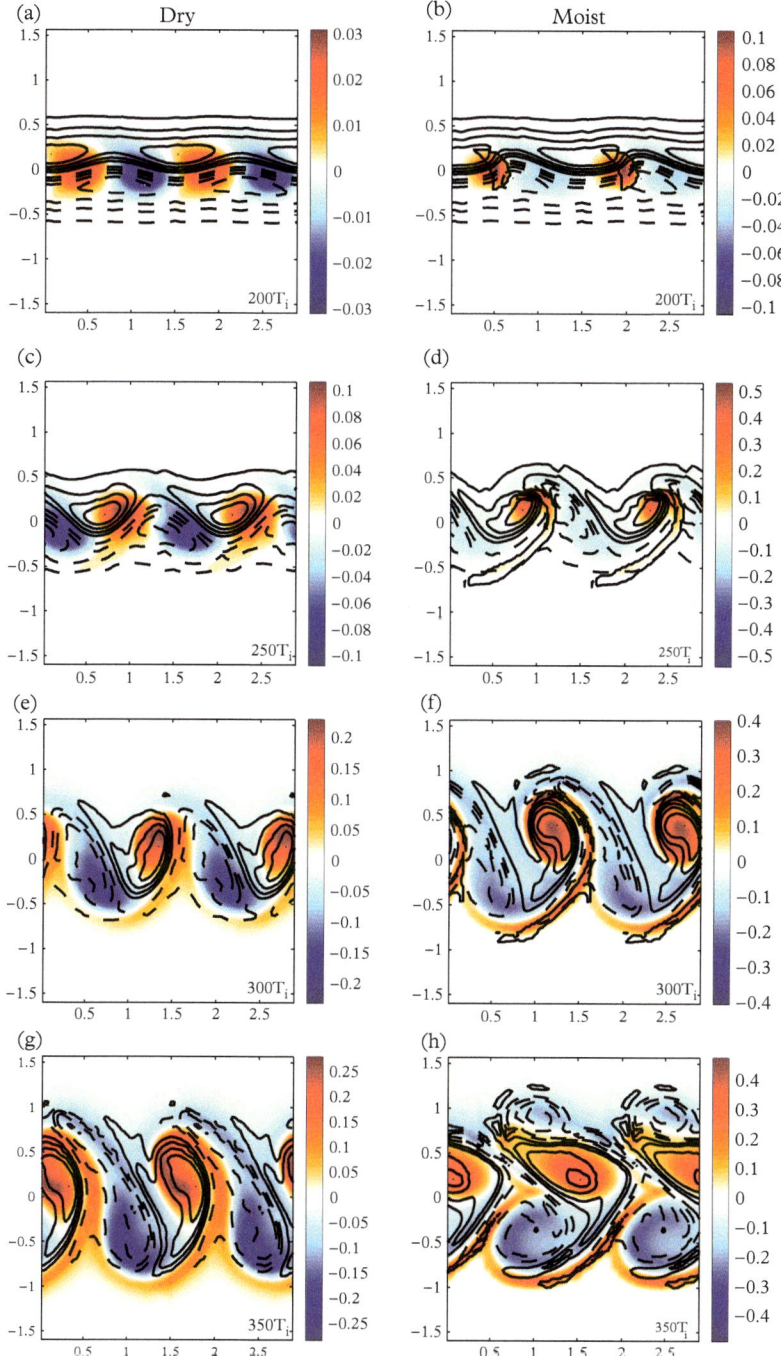

Figure 15.7 *Nonlinear evolution of the total relative vorticity in the dry (left column) and moist-convective (right column) simulations. Lower layer: colours, upper layer: contours. Condensation: solid black contours.* ©*American Meteorological Society. Used with permissions.*

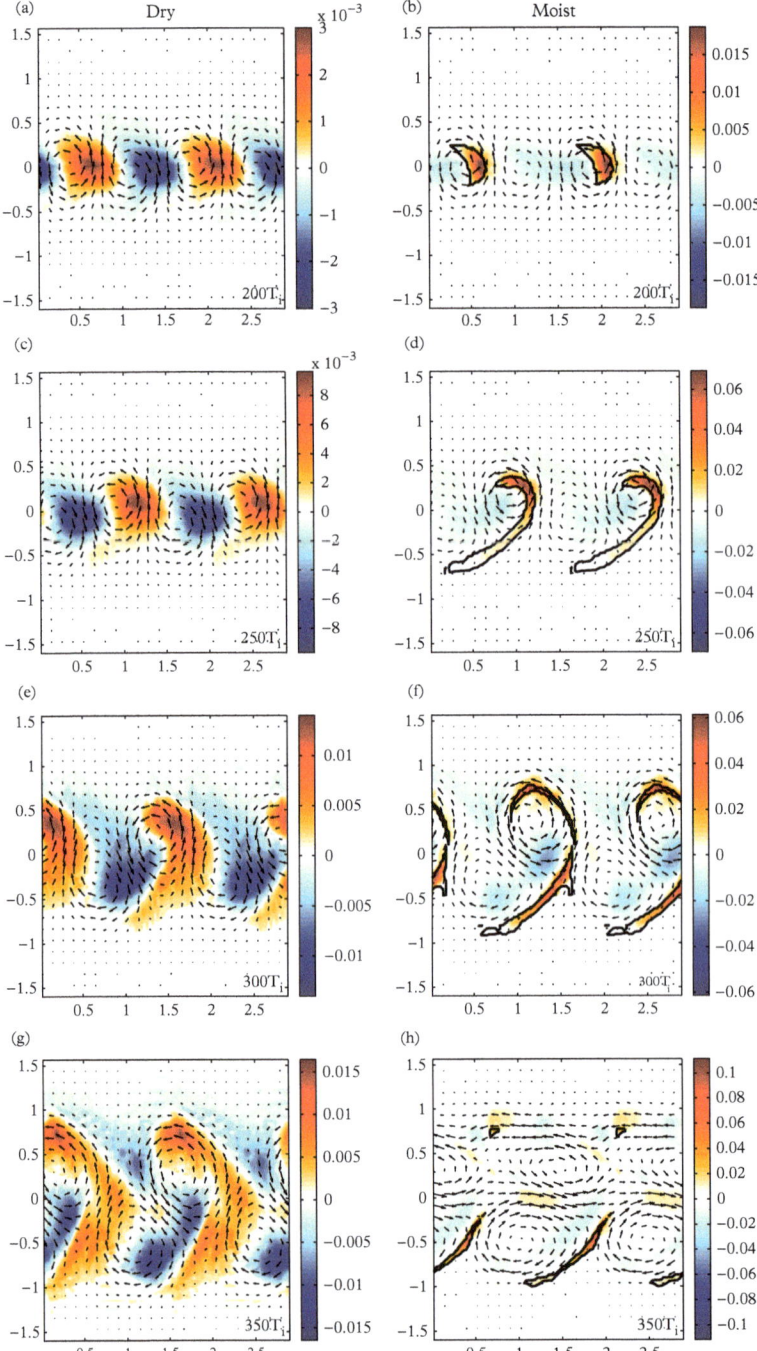

Figure 15.8 *Dry vs moist simulations of the baroclinic instability: divergence of the baroclinic velocity. Black contours: condensation zones.* ©*American Meteorological Society. Used with permissions.*

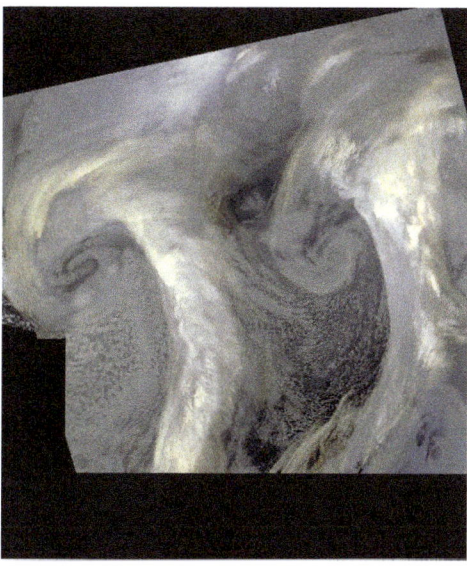

Figure 15.9 *Satellite view of a couple of synoptic disturbances over the North Atlantic. Courtesy NOAA.*

instability of the upper-layer cyclone leads to appearance of specific cloud patterns in the lower layer. Inversely, condensation affects the growth rate of the instability, as in the case of jet instability above. Evolution of the growth rate, calculated, as before, with the help of the dry energy norm is presented in Figure 15.11 and shows, again, a transient increase of the growth rate and more rapid nonlinear saturation of the instability in the presence of condensation. As discussed in Chapter 11, development of the instability is accompanied by emission of inertia–gravity waves. At the onset of condensation this emission is enhanced (not shown).

15.5   Moist dynamics of tropical cyclone-like vortices

Tropical storms, obviously, evolve in the moist-precipitating and evaporating environment, cf. Figure 11.29. Therefore, although the evolution of the instabilities of the hurricane-like vortices studied in Section 11.4 allowed us to understand dynamical origins of the life-cycle of hurricanes, including the effects of moisture is of crucial importance. Following the same approach as in Section 15.4, we use the mcRSW model to see how dynamical effects of moisture influence the evolution of hurricane-type vortices. Numerical simulations with the model are initiated with the most unstable mode, identified by linear stability analysis. The background vortex profile is the same as in Section 11.4. As in previous sections, the initial humidity was chosen to be uniform and close to saturation. We perform simulations in moist-precipitating configurations, with and without surface evaporation. The hurricanes evolve mostly over the ocean, so the evaporation proportional to the wind velocity is used.

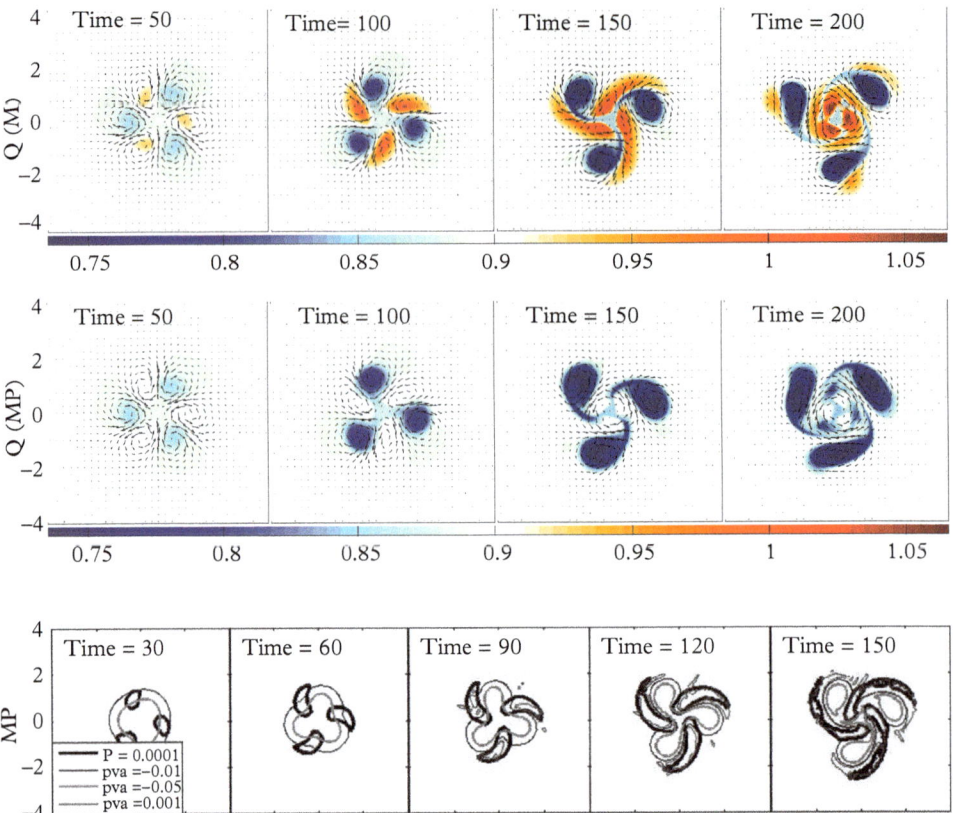

Figure 15.10 *Evolution of moisture in M (top panel) and MP (middle panel) environments, and joint evolution of PVA and condensation in MP simulation (bottom panel) during saturation of the geostrophic vortex instability with azimuthal wavenumber $l = 3$. Time is measured in units of f^{-1}.*

We present the results of a simulation without evaporation in terms of humidity and relative vorticity in Figure 15.12. Although the evolution pattern with transient formation of meso-vortices and subsequent uniformisation of the core is qualitatively similar to what was observed in 'dry' simulations of Section 11.4, there are quantitative differences which are clearly visible in the inter-comparison of simulations in 'dry' (moisture behaving as a passive scalar), moist-precipitating (moisture with condensation, but no evaporation), and moist-precipitating and evaporating configurations presented in Figure 15.13. As follows from the figure, the condensation, especially in the presence of evaporation, significantly influences the evolution of tropical cyclones: the pressure minimum deepens, and the end-state vortex is intensified with a net increase in the wind strength.

The influence of the effects of moisture is important not only in the evolution of the vortex itself but also for the gravity wave activity it generates. We show an inter-comparison of this activity at the early stages of the evolution in simulations in

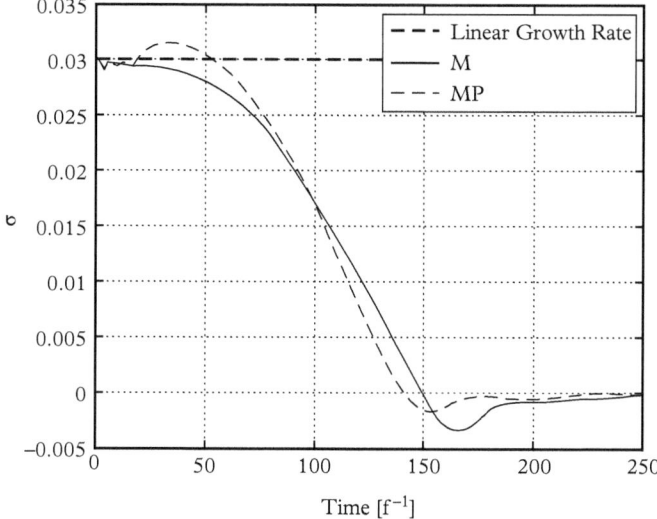

Figure 15.11 *Evolution of the growth rate of the perturbation during the tripolar evolution scenario of the baroclinic vortex instability in M and MP environments.*

different configurations in Figure 15.14. As before, wind divergence is taken as a proxy of the wave activity. In order to measure the latter, it was averaged over an annulus sufficiently far from the vortex. As is clear from the figure, the wave activity in precipitating and evaporating environment is an order of magnitude greater. Moreover, convection-coupled waves (i.e. inertia–gravity waves associated with condensation fronts) are produced during the evolution of the instability, as follows from the Figure 15.15. Their passages through the control annulus are at the origin of bursts visible in Figure 15.14.

Thus, inclusion of condensation and evaporation leads to substantial changes in the dry scenario of saturation of the instabilities of tropical cyclones, with intensification of the vortex, substantial increase in associated gravity wave activity, and appearance of convection-coupled waves.

15.6 Summary, discussion, and bibliographic remarks

We have shown in chapter that there exists a simple and consistent method to include the effects of moist convection in vertically integrated shallow-water models of the atmosphere. The moist-convective shallow-water models obtained in this way allow us, by simple technical means, to understand the influence of moisture and related heat release upon the large-scale atmospheric dynamics. A simple relaxational parameterisation of condensation, which is switched on whenever a saturation threshold is reached in the places where convergence of velocity leads to accumulation of moisture, makes

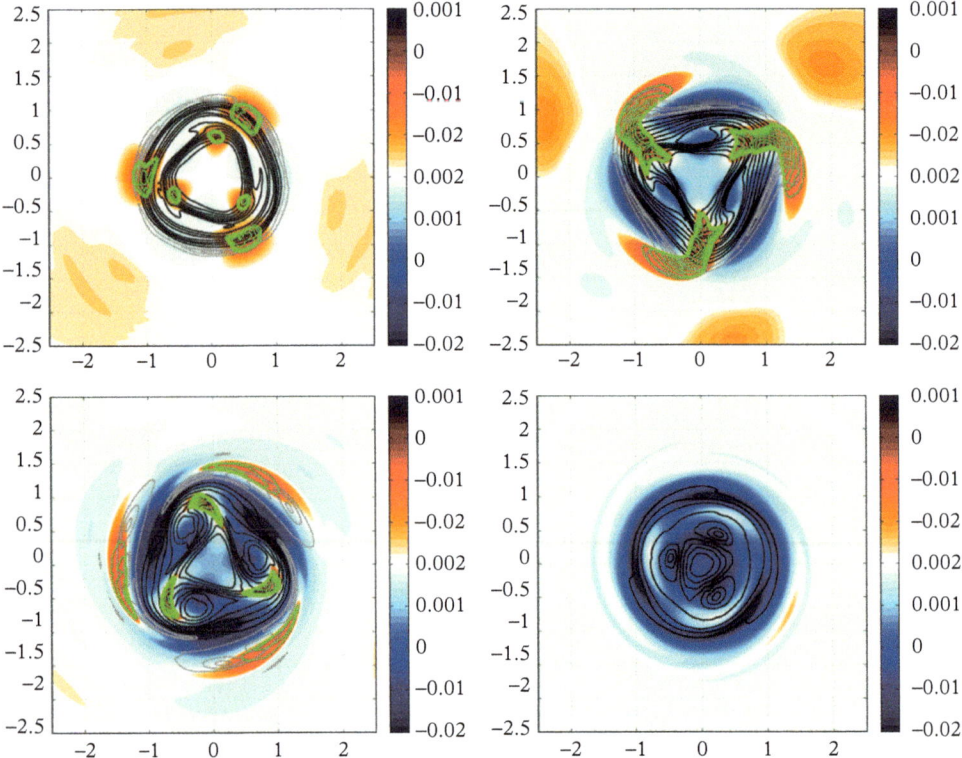

Figure 15.12 *Specific humidity (colors), relative vorticity (black contours), and condensation (green contours) at non-dimensional times $t = 2, 20, 50, 250f^{-1}$, left to right and top to bottom, during moist-precipitating evolution of the most unstable mode with small amplitude superimposed on the background vortex.* ©*American Meteorological Society. Used with permissions.*

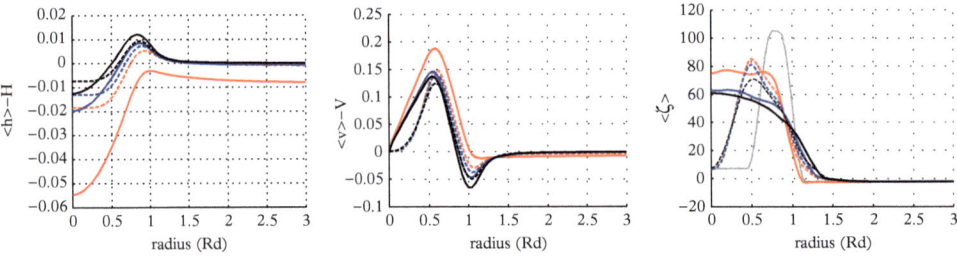

Figure 15.13 *Azimuthal averages of thickness (left), azimuthal velocity (middle), and relative vorticity (right) during the saturation of the instability. Dashed curves: $t = 1.5f^{-1}$. Solid curves: $t = 4f^{-1}$. Dry simulation: black. Moist-precipitating simulation: blue. Moist-precipitating-evaporating simulation: red. Relative vorticity of the initial vortex: grey.* ©*American Meteorological Society. Used with permissions.*

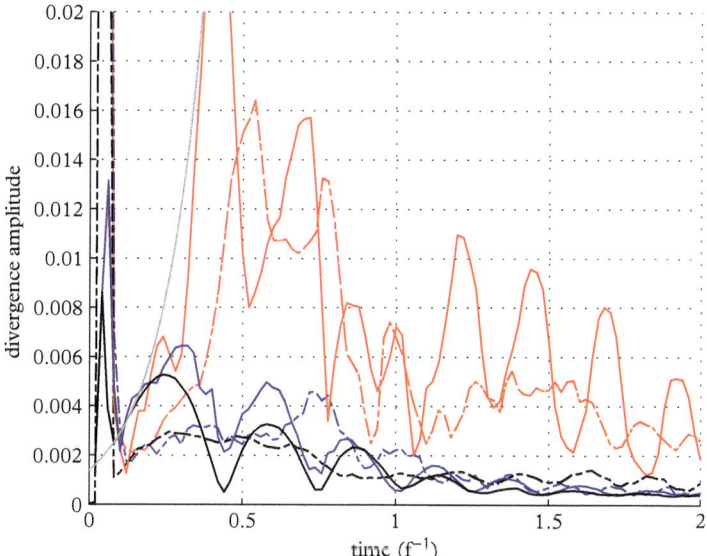

Figure 15.14 *Wind divergence averaged over an annulus around the vortex. Black: dry. Blue: moist-precipitating. Red: moist-precipitating and evaporating cases. Solid curves: initialisation with solely $l = 3$ unstable mode. Dashed curves: initialisation with an ensemble of unstable modes with different l and random phases. Thin grey: prediction of linear stability analysis. Sharp initial peak is due to initial adjustment provoked by discretisation errors.* ©*American Meteorological Society. Used with permissions.*

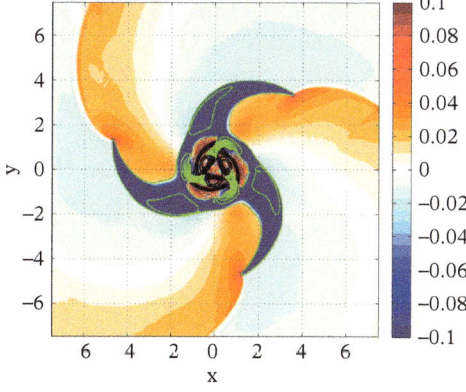

Figure 15.15 *Convection-coupled waves, as seen in divergence field at $t = 0.6f^{-1}$ in the moist-precipitating-evaporating simulation. Green lines delimit the condensation zones.* ©*American Meteorological Society. Used with permissions.*

obvious the mechanism of intensification of cyclonic motions, and leads to qualitative and quantitative (within obvious limits of the simple model) understanding of the life-cycles of moist baroclinic and barotropic instabilites of jets and vortices, including tropical cyclones.

It should be stressed that the simplest version of the model discussed allows for straightforward ameliorations. Thus, precipitable water (clouds) can be added in the model as another advected quantity, as was done for humidity (water vapour), with a source due to condensation, and a sink due to vapourisation. A (true) precipitation sink is then introduced in the equation for precipitable water, and a simple relaxational parameterisation is again possible. Vapourisation leads to cooling, so related downward convective fluxes can be introduced in the model along the same lines as upward fluxes due to the heat release by condensation. In this way the model becomes 'cloud resolving', in a sense that it allows us to trace the evolution of precipitable water. Other tracers, like aerosols, can be introduced in a similar way, and coupled, for example to precipitation by providing precipitation nuclei. Radiative cooling can be also introduced with the help of a sink term in the equation of h of the form $\propto (h - h_e)$, where h_e is an equilibrium geopotential height. Finally, horizontal temperature gradients can be introduced along the lines described in Chapter 14. In this way(s) the model can be considerably 'enhanced', while keeping its advantage of being computationally light, and opens wide range of possibilities of modelling large-scale atmospheric phenomena of various kinds. It should be stressed that the representation of the boundary layer in the model can be also ameliorated.

We should also emphasise that an approach similar to the that described above, with simpler implementation, can be applied to convection in the ocean.

Thermodynamics of convection in full complexity can be found in Emmanuel (1994). A discussion of problems of predicting clouds and precipitations in climate models is given in Stevens and Bony (2013). The exposition of mcRSW models above follows Bouchut *et al.* (2009), Lambaerts, Lapeyre, and Zeitlin (2011), and Lambaerts *et al.* (2011). Construction of these models was motivated by the pioneering ideas of Gill (1982), who proposed using shallow-water modelling of moist-convective systems and introduced the immediate relaxation limit, by works on condensation fronts (Frierson, Majda, and Pauluis 2004; Pauluis, Frierson, and Majda 2008), and by work on oceanic convection (Marchal and Nycander 2004). The linear baroclinic reduction limit discussed corresponds to the model used in Frierson, Majda, and Pauluis (2004) and Pauluis, Frierson, and Majda (2008). A model similar to the mcRSW, with much attention paid to detailed parameterisation of the marine boundary layer, was used in Schecter and Dunkerton (2009) for studying tropical cyclogenesis. The boundary-layer parameterisations of this work are easily transposable to mcRSW models. The two-layer quasi-geostrophic limit of two-layer mcRSW coincides with the model of Lapeyre and Held (2004). The relaxation parameterisation of condensation above is a 'poor man's' version of the parameterisations used in the 'big' models (Betts and Miller 1986). The simple parameterisations of evaporation we considered are of use in the literature (e.g. Neelin, Held, and Cook 1987; Goswami and Goswami 1991). The description of the moist instability of the baroclinic jet follows Lambaerts, Lapeyre, and Zeitlin (2012), that of the moist instability of vortices follows Rostami and Zeitlin (2017), and that of the moist dynamics of hurricanes, Lahaye and Zeitlin (2016). A review of multi-layer modelling of the tropical atmosphere can be found in Khouider, Majda, and Stechman (2013).

Figures 15.1 to 15.4 are reproduced from Lambaerts, Lapeyre, and Zeitlin (2011) with the permission of AIP publishing. Figures 15.5 to 15.8 are reproduced from Lambaerts, Lapeyre, and Zeitlin (2012). Figures 15.10 and 15.11 are reproduced from Rostami and Zeitlin (2017) with permissions. Figures 15.12 to 15.15 are reproduced from Lahaye and Zeitlin (2016).

16

Rotating Shallow-Water Models with Full Coriolis Force

As explained in Chapter 2, in the standard primitive equations both the vertical (in the direction of gravity) component of the Coriolis force, and the contribution to the Coriolis force due to the vertical component of velocity are neglected. This is a traditional assumption, which is then transposed to the tangent (f– or β–) plane approximation. Yet, the 'non-traditional' terms are present, as shown in Figure 16.1 and can give significant corrections under some circumstances, as we will demonstrate in this chapter.

Independently of its main subject, the 'non-traditional' corrections, the present chapter can be considered as a complement to Chapters 2, 3, and 10, as we will revisit the derivation of rotating shallow-water model from the primitive equations, the primitive equations themselves, and the inertial instability. The variational principle for the primitive equations thus will be established, which is useful for many purposes. In particular, it provides an alternative derivation of the shallow-water equations by applying the hypothesis of columnar motion directly in the action functional.

16.1 'Non-traditional' rotating shallow-water model in the tangent plane approximation

16.1.1 Vertical averaging of 'non-traditional' primitive equations

Our starting point will be full incompressible non-hydrostatic primitive equations on the 'non-traditional' f-plane, where all components of the Coriolis force with constant Coriolis parameters are taken into account:

$$
\begin{cases}
\dfrac{du}{dt} - fv + Fw + \Phi_x = 0, \\[2mm]
\dfrac{dv}{dt} + fu + \Phi_y = 0, \\[2mm]
\dfrac{dw}{dt} - Fu + b + \Phi_z = 0, \\[2mm]
\dfrac{db}{dt} = 0, \quad u_x + v_y + w_z = 0.
\end{cases}
\tag{16.1}
$$

Geophysical Fluid Dynamics. Vladimir Zeitlin,
Oxford University Press (2018). © Vladimir Zeitlin. DOI: 10.1093/oso/9780198804338.001.0001

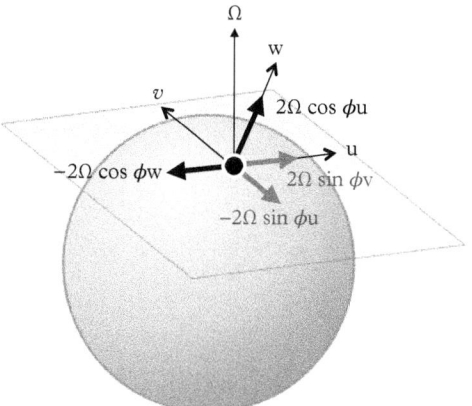

Figure 16.1 *Components of fluid velocity u, v, w, and of the Coriolis force with respect to the tangent plane at latitude ϕ. Grey: traditional; Black: non-traditional.*

Here, as usual, $\frac{d}{dt} = \partial_t + u\partial_x + v\partial_y + w\partial_z$, Φ is geopotential, $b = g\frac{\rho}{\rho_0}$ is buoyancy, f is the 'traditional' Coriolis parameter, and F is the 'non-traditional' one. If we recall their origin, $f = 2\Omega\sin\phi$, and $F = 2\Omega\cos\phi$, where Ω is the angular velocity of the planet's rotation, and Φ is the latitude of the tangent plane, cf. (2.37) to (2.39), and Figure 16.1.

Following the procedure which led to the rotating shallow-water (RSW) model in Chapter 3, we derive the averaged over the fluid layer equations and compare them with the standard RSW equations. The latter being hydrostatic, we will correspondingly neglect the vertical acceleration in (16.1):

$$-Fu + b + \Phi_z = 0. \tag{16.2}$$

Hydrostatic balance now includes the vertical component of the Coriolis force, and thus becomes *quasi-hydrostatic*. We proceed by vertical averaging of the x-momentum equation in (16.1), after rewriting it in conservative form, between a pair of material surfaces $z_{1,2}$, with $\frac{dz_{1,2}}{dt} = w|_{z_{1,2}}$ (the y-momentum equation is treated analogously):

$$\int_{z_1}^{z_2} dz \left(u_t + \left(u^2\right)_x + (vu)_y + (wu)_z - fv + Fw\right) = -\int_{z_1}^{z_2} dz\,\Phi_x. \tag{16.3}$$

By using the Leibniz formula, cf. Chapter 3, we obtain

$$\partial_t \int_{z_1}^{z_2} dz\,u + \partial_x \int_{z_1}^{z_2} dz\,u^2 + \partial_y \int_{z_1}^{z_2} dz\,uv - f\int_{z_1}^{z_2} dz\,v + F\int_{z_1}^{z_2} dz\,w = -\int_{z_1}^{z_2} dz\,\partial_x\Phi. \tag{16.4}$$

We then apply the mean-field approximation to express integrals in terms of vertical averages, denoted by a bar:

$$\int_{z_1}^{z_2} dz\,u \approx (z_2 - z_1)\,\bar{u}, \quad \int_{z_1}^{z_2} dz\,u^2 \approx (z_2 - z_1)\,\bar{u}^2, \quad \ldots, \tag{16.5}$$

hence

$$\partial_t \left[(z_2 - z_1)\bar{u} \right] + \partial_x \left[(z_2 - z_1)\bar{u}^2 \right] + \partial_y \left[(z_2 - z_1)\bar{u}\bar{v} \right] - f(z_2 - z_1)\bar{v} + F \int_{z_1}^{z_2} dz\, w = - \int_{z_1}^{z_2} dz\, \partial_x \Phi.$$

(16.6)

A new term, $\int_{z_1}^{z_2} dz\, w$, with respect to the derivation of Chapter 3 appears. It is calculated with the help of the continuity equation:

$$w = - \int_0^z dz\, (u_x + v_y) \approx -z\left(\bar{u}_x + \bar{v}_y\right) \Rightarrow \int_{z_1}^{z_2} dz\, w \approx -\left(\bar{u}_x + \bar{v}_y\right) \frac{z_2^2 - z_1^2}{2}.$$

(16.7)

It should be stressed that under the (strict) mean-field approximation the operations of taking spatial and temporal derivatives and averaging are commuting.

Another difference with the standard RSW comes from the correction to the hydrostatic relation due to the vertical component of the Coriolis force:

$$\Phi_z = -b + Fu \Rightarrow \Phi = -\bar{b}(z - z_1) + F \int_{z_1}^z dz\, u + \Phi(z_1) \approx -\bar{b}(z - z_1) + F(z - z_1)\bar{u} + \Phi(z_1).$$

(16.8)

Alternatively,

$$\Phi = -\bar{b}(z - z_2) - F \int_z^{z_2} dz\, u + \Phi(z_2) \approx -\bar{b}(z - z_2) - F(z_2 - z)\bar{u} + \Phi(z_2).$$

(16.9)

We will consider here the 'ordinary' shallow-water model with constant $\bar{b}$, although thermal non-traditional shallow-water model can be obtained along the lines of Chapter 14 as well. From the incompressibility equation in (16.1) and definition of material surfaces we similarly get the conservation equation for the volume of the fluid column:

$$(z_2 - z_1)_t + \left[(z_2 - z_1)\bar{u} \right]_x + \left[(z_2 - z_1)\bar{v} \right]_y = 0.$$

(16.10)

Combining (16.6), (16.8), and (16.10) we get, for any pair of material surfaces,

$$(z_2 - z_1) \left[\left(\partial_t + \bar{u}\partial_x + \bar{v}\partial_y \right)\bar{u} - f\bar{v} - \frac{F}{2}\left(\bar{u}_x + \bar{v}_y\right)(z_2 + z_1) \right]$$

(16.11)

$$= -\partial_x \int_{z_1}^{z_2} dz\, \Phi - z_{1_x} \Phi(z_1) + z_{2_x} \Phi(z_2),$$

where either (16.8) or (16.9) should be used for closure. Equation (16.11), together with a similar equation for $\bar{v}$, which may be obtained in the analogous manner, and (16.10), represent non-traditional master equations, a building block for constructing non-traditional multi-layer RSW models. They should be combined with appropriate boundary conditions at the material surfaces.

16.1.2 Non-traditional RSW models

Non-traditional one-layer RSW

We first consider the case of a single-layer RSW with a flat bottom $z_1 = 0$ and a free surface $z_2 = h$, $\Phi(z_2) = 0$. Therefore, $\Phi = g(h-z)-F(h-z)\bar{u}$, as the constant mean buoyancy is just the (reduced) gravity $\bar{b} = g$. The non-traditional RSW equations, thus, are:

$$
\begin{cases}
\dfrac{d}{dt}\bar{u} - f\bar{v} = -gh_x + \dfrac{F}{2}\left(\bar{u}_x + \bar{v}_y\right)h + F\left(\bar{u}h_x + \dfrac{\bar{u}_x h}{2}\right), \\[2mm]
\dfrac{d}{dt}\bar{v} + f\bar{u} = -gh_y + F\left(\bar{u}h_y + \dfrac{\bar{u}_y h}{2}\right), \\[2mm]
\dfrac{d}{dt}h + h\left(\bar{u}_x + \bar{v}_y\right) = 0,
\end{cases}
\tag{16.12}
$$

where $\frac{d}{dt} = \partial_t + \bar{u}\partial_x + \bar{v}\partial_y$. We will omit bars over u and v in what follows.

Non-traditional two-layer RSW with a rigid lid

We consider master equations between two pairs of material surfaces: (z_1, z_2) and (z_2, z_3), supposing uniform densities $\rho_{1,2}$ in respective layers, take $z_1 = 0$ and $z_3 = \text{const}$ (rigid lid), and express geopotentials inside the layers in terms of the pressure at z_3 and vertical position. We arrive in this way to the following system of equations for the average velocities $u_{1,2}$, $v_{1,2}$ and thicknesses $h_{1,2}$ of the respective layers:

$$
\begin{cases}
\dfrac{d_1 u_1}{dt} - fv_1 - F\left(\dfrac{1}{2}\partial_y v_1 h_1 - \partial_x(u_1 h_1)\right) = -\dfrac{1}{\rho_1}\partial_x \pi, \\[2mm]
\dfrac{d_1 v_1}{dt} + fu_1 - \dfrac{F}{2}\partial_y u_1 h_1 = -\dfrac{1}{\rho_1}\partial_y \pi, \\[2mm]
\dfrac{d_2 u_2}{dt} - fv_2 - \dfrac{F}{2}\left(\partial_y v_2 h_2 + \partial_x(h_1 u_2)\right) = -\dfrac{1}{\rho_2}\partial_x \pi - g'\partial_x h_2 + F\dfrac{\rho_1}{\rho_2}\partial_x(h_1 u_1), \\[2mm]
\dfrac{d_2 v_2}{dt} + fu_2 - \dfrac{F}{2}\left(\partial_y(u_2 h_2) + \partial_y h_2 u_2\right) = -\dfrac{1}{\rho_2}\partial_y \pi - g'\partial_y h_2 + F\dfrac{\rho_1}{\rho_2}\partial_y(h_1 u_1), \\[2mm]
\partial_t h_{1,2} + \partial_x\left(u_{1,2} h_{1,2}\right) + \partial_y\left(v_{1,2} h_{1,2}\right) = 0, \quad h_1 + h_2 = H_0 = \text{const}.
\end{cases}
\tag{16.13}
$$

Here, $\frac{d_{1,2}}{dt} = \partial_t + u_{1,2}\partial_x + v_{1,2}\partial_y$ are Lagrangian derivatives in respective layers counted from the top, $g' = g\frac{\rho_2-\rho_1}{\rho_2}$ is reduced gravity, and π denotes the pressure under the rigid lid at $z = z_3$.

Non-traditional two-layer RSW with a free surface

In the case when z_3 is a free surface in the two-layer configuration, we obtain in the same way

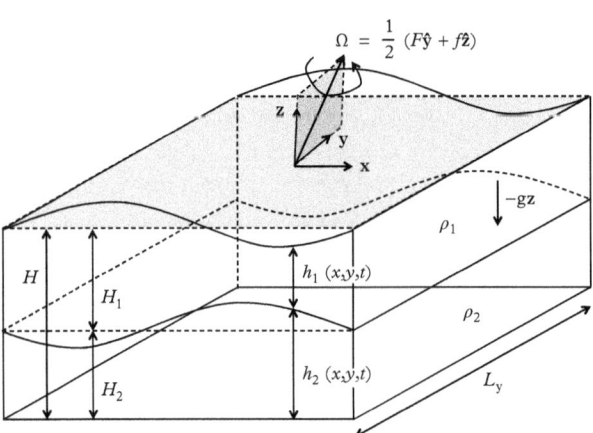

Figure 16.2 *Two-layer rotating shallow-water model with a free surface in the non-traditional f-plane approximation.*

$$\begin{cases} \dfrac{d_1}{dt}u_1 - fv_1 + g\partial_x (h_1 + h_2) - \\ F\left(-v_1\partial_y h_2 + \partial_x (h_1 u_1) + \partial_x (h_2 u_2) + \partial_y (h_2 v_2) + \dfrac{h_1\partial_y v_1}{2}\right) = 0, \\[2mm] \dfrac{d_1}{dt}v_1 + fu_1 + g\partial_y (h_1 + h_2) - F\left(u_1\partial_y (h_1 + h_2) + \dfrac{h_1\partial_y u_1}{2}\right) = 0, \\[2mm] \partial_t h_1 + \partial_x (h_1 u_1) + \partial_y (h_1 v_1) = 0, \\[2mm] \dfrac{d_2}{dt}u_2 - fv_2 + g\partial_x (h_2 + rh_1) - F\left(\partial_x (h_2 u_2) + r\partial_x (h_1 u_1) + \dfrac{h_2\partial_y v_2}{2}\right) = 0, \\[2mm] \dfrac{d_2}{dt}v_2 + fu_2 + g\partial_y (h_2 + rh_1) - \\ F\left(u_2\partial_y h_2 + \dfrac{h_2\partial_y u_2}{2} + r\partial_y (h_1 u_1)\right) = 0, \\[2mm] \partial_t h_2 + \partial_x (h_2 u_2) + \partial_y (h_2 v_2) = 0. \end{cases} \tag{16.14}$$

Here the subscripts 1 and 2 refer to the top (bottom) layers, cf. Figure 16.2. The density ratio is $r = \rho_1/\rho_2 < 1$, the height ratio is $d = H_2/H_1$ and the total thickness is $H = H_1 + H_2$.

16.2 'Non-traditional' rotating shallow-water model on the sphere

16.2.1 Including the effects of curvature in the RSW with full Coriolis force

In Section 16.1 the non-traditional (NT) corrections were considered in the f-plane approximation, where all effects of sphericity were neglected. Including the leading sphericity corrections, that is constructing consistent non-traditional beta-plane approximation, turns to be rather non-trivial. As we will see shortly, several small parameters which are present in the system should be carefully ordered, and curvature and non-traditional (NT) corrections systematically compared. In fact, it is easier to develop a consistent non-traditional shallow-water model on the whole sphere, and take the tangent-plane limit afterwards. It seems, at first glance, that a straightforward averaging along the lines of the preceding section would work. However, such a direct approach leads to a system violating conservation of the angular momentum, unless the full-power machinery of differential geometry in curvilinear coordinates is applied. An approach based on the application of the columnar motion hypothesis in the variational principle for primitive equations turns to be the most efficient. However, as we will see, the very notion of columnar motion should be reconsidered if the NT terms are present. Let us recall that the essence of the shallow-water dynamics in the plane is the columnar character of the corresponding fluid motion. Indeed, this hypothesis is crucial in the derivation of shallow-water equations by vertical averaging of the primitive equations. As we have seen in Section 16.1, the same method works for deriving non-traditional rotating shallow-water equations in the rotating plane. The traditional approximation in spherical geometry neglects the variation of the planetary component of fluid velocity with height. The motion is, thus, considered columnar in the 'flat' sense, and in this configuration, depicted in the right panel of Figure 16.3, a straightforward vertical averaging gives the standard RSW equations on the sphere of Section 3.2.4, which are consistent for negligible depths of the fluid layer. The situation is more delicate for layers of non-negligible depth, when obliqueness of the angular velocity of the overall rotation is combined with the curvature of the surface of the fluid layer. If the traditional approximation is to be relaxed, the notion of columnar motion should be revisited. Due to the curvature, the fluid 'columns' on the sphere are, in fact, solid angles. Hence, strictly speaking, the hypothesis of columnar motion (i.e. independence of radial position) should be applied to the angular, and not physical velocity, leading to the dependence of this latter on altitude, cf. the left panel of Figure 16.3. One could naively think that in the shallow-layer

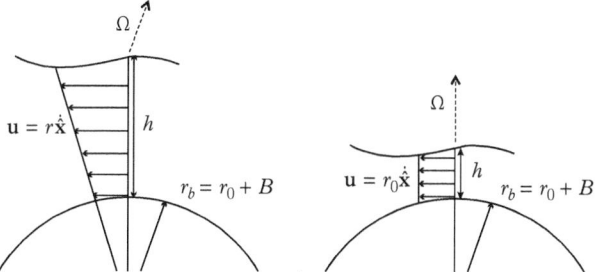

Figure 16.3 *Columnar motion on the sphere: constant angular (left panel) vs constant linear (right panel) velocity of the column.* **u** *denotes the vector of velocity tangent to the sphere.*

approximation this effect is negligible but, as already said, the 'brute-force' replacement of the radial position by the constant planet's radius leads to problems with the conservation of the axial angular momentum, if all components of the Coriolis force are retained. The best way to cope with these problems is to use the variational principle in spherical coordinates. Independently of non-traditional aspects, such an approach provides an alternative derivation of the shallow-water equations.

16.2.2 Variational principle for the primitive equations in spherical geometry

Primitive equations and their conservation laws

We start with non-hydrostatic primitive equations:

$$\begin{cases} \dfrac{d}{dt}\mathbf{v} + 2\mathbf{\Omega} \wedge \mathbf{v} + \nabla\Pi - b\mathbf{g} = 0, \\ \dfrac{d}{dt}b = 0, \quad \nabla \cdot \mathbf{v} \quad = 0. \end{cases} \tag{16.15}$$

Here $\mathbf{v}$ is the fluid velocity, $b = \frac{\rho}{\rho_0}$ is the normalized density, with a constant reference density ρ_0, Π is the geopotential with incorporated centrifugal terms, and $\frac{d}{dt} = \partial_t + \mathbf{v}\cdot\nabla$ is the advective derivative. Lagrangian invariants of the system (16.15), in addition to b, are potential vorticity, $q = \boldsymbol{\zeta}_a \cdot \nabla b$, where $\boldsymbol{\zeta}_a = \boldsymbol{\zeta} + 2\mathbf{\Omega}$ is the absolute vorticity and $\boldsymbol{\zeta} = \nabla \wedge \mathbf{v}$ is the relative vorticity, and the absolute axial angular momentum, which is an important characteristic of the general circulation of the atmosphere. The mass density of the axial momentum is obtained from the mass density of the absolute angular momentum

$$\mathbf{l}_a = \mathbf{r} \wedge (\mathbf{v} + \mathbf{\Omega} \wedge \mathbf{r}), \tag{16.16}$$

by projection onto the unit vector $\hat{\mathbf{z}}$ in the direction of $\mathbf{\Omega}$:

$$l_z = \hat{\mathbf{z}} \cdot \mathbf{l}_a, \tag{16.17}$$

where $\mathbf{r}$ is the position vector, and $\mathbf{v} + \mathbf{\Omega} \times \mathbf{r} = \mathbf{v}^a$ is the absolute velocity. The evolution of the axial momentum is given by the equation

$$\frac{d}{dt}l_z = \hat{\mathbf{z}} \cdot [\mathbf{r} \wedge (-\nabla\Pi + b\mathbf{g})] \tag{16.18}$$

Primitive equations in spherical coordinates

The system (16.15) is written in the abstract vector form independent of a concrete choice of coordinate system. As applied to a fluid on a sphere, the use of spherical coordinates (λ, ϕ, r) is natural. The equations for the three components of velocity are rewritten as follows, cf. Chapter 2:

$$\begin{cases} \dfrac{d}{dt}v_r - \dfrac{v_\lambda^2 + v_\phi^2}{r} - 2\Omega\cos\phi v_\lambda + bg = -\partial_r\Pi, \\[3mm] \dfrac{d}{dt}v_\lambda + \dfrac{v_r v_\lambda - v_\phi v_\lambda \tan\phi}{r} + 2\Omega(-\sin\phi v_\phi + \cos\phi v_r) = \\[3mm] \quad -\dfrac{1}{r\cos\phi}\partial_\lambda\Pi, \\[3mm] \dfrac{d}{dt}v_\phi + \dfrac{v_r v_\phi + v_\lambda^2 \tan\phi}{r} + 2\Omega\sin\phi v_\lambda = -\dfrac{1}{r}\partial_\phi\Pi, \end{cases} \tag{16.19}$$

The expression of the advection operator in spherical coordinates was already given in Chapter 2: $\frac{d}{dt} = \partial_t + v_r\partial_r + \frac{v_\phi}{r}\partial_\phi + \frac{v_\lambda}{r\cos\phi}\partial_\lambda$, and the incompressibility equation, together with the the buoyancy advection equation are

$$\frac{1}{r^2}\frac{\partial(r^2 v_r)}{\partial r} + \frac{1}{r\cos\phi}\left(\frac{\partial(\cos\phi v_\phi)}{\partial\phi} + \frac{\partial(v_\lambda)}{\partial\lambda}\right) = 0, \quad \frac{d}{dt}b = 0. \tag{16.20}$$

Typical for curvilinear systems of coordinates, curvature terms, proportional to r^{-1} and without derivatives, appear in (16.19) and (16.20).

Boundary conditions in λ and ϕ are periodic, if applied to the atmosphere, while for the ocean the basin boundary should be imposed. We are considering non-dissipative system, so the bottom is a material surface, and arbitrary topography is allowed. The radial position of a point at the bottom surface is given by

$$r = r_0 + B(\lambda, \phi), \tag{16.21}$$

where r_0 is the planet's radius, and $B(\lambda, \phi)$ is the bottom topography. The fluid layer is considered between the bottom and a free surface. It should be emphasised that the velocity components v_λ, v_ϕ, v_r in the above formulae are the *physical* ones, most often used in the applications. They are related to the rate of change of spherical coordinates as follows:

$$v_\lambda = r\cos\phi\dot\lambda, \quad v_\phi = r\dot\phi, \quad v_r = \dot r, \tag{16.22}$$

where we use the dot notation for the Lagrangian time derivative, as usual.

Variational principle for primitive equations

Let us recall, cf. Section 2.2.1, that the most straightforward way to establish a variational principle for fluid motion is through Lagrangian interpretation of the equations of motion. In the following we will use the Lagrangian description of the flow, where Lagrangian coordinates, denoted by capitals, are functions of Lagrangian labels and time. From the Lagrangian viewpoint, (16.15) are Newton's equations for the fluid parcels with coordinates $\mathbf{X}$, labelled by their initial positions $\chi : \mathbf{X}(\chi, t)|_{t=0} = \chi$. The fluid velocity $\mathbf{v}$ is given by $\dot{\mathbf{X}}$. An additional change of variables $\chi \mapsto \mathbf{a}$ to some other set of Lagrangian labels **a** can be made, as was discussed in Chapters 7 and 8. The non-divergence of velocity means that the elementary volumes are conserved by this mapping, and the advection of b means that the density remains the same for each fluid parcel, although densities of different parcels are different. The former condition in terms of Lagrangian coordinates means that the Jacobian of the mapping $\chi \mapsto X$ is unity, and represents a dynamical constraint, while the latter means that an initial distribution of density is transported by fluid parcels to their new locations. As was shown in Section 2.2.1, the incompressibility constraint is introduced with the help of a Lagrangian multiplier, the pressure, which becomes the geopotential in the present context. Following the rules explained in Section 2.2.1, the Lagrangian is obtained as kinetic minus potential energy. In order to obtain the variational principle in a rotating system a vector potential for the angular velocity should be introduced 'augmenting' the Lagrangian, cf. (2.36). In this Section spherical Lagrangian coordinates, and spherical Lagrangian labels are used:

$$X(\chi, t) = (\Lambda(\lambda, \phi, r, t), \Phi(\lambda, \phi, r, t), R(\lambda, \phi, r, t)).$$

Following these recipies, we arrive at the action functional for the system (16.15)

$$S = \rho_0 \int dt \int_{-\pi/2}^{+\pi/2} \cos\phi \, d\phi \int_0^{2\pi} d\lambda \int_0^{\infty} r^2 dr \mathcal{L}, \qquad (16.23)$$

with Lagrangian density

$$\mathcal{L} = \frac{1}{2} \left(R^2 \cos^2 \Phi \dot{\Lambda}^2 + R^2 \dot{\Phi}^2 + \dot{R}^2 \right) + \Omega R^2 \cos^2 \Phi \dot{\Lambda} - bgR + \Pi \left(\frac{R^2 \cos \Phi}{r^2 \cos \phi} \frac{\partial (\Lambda, \Phi, R)}{\partial (\lambda, \phi, r)} - 1 \right). \qquad (16.24)$$

Here,

$$\frac{R^2 \cos \Phi}{r^2 \cos \phi} \frac{\partial (\Lambda, \Phi, R)}{\partial (\lambda, \phi, r)} = \frac{\partial (R\Lambda \cos \Phi, R\Phi, R)}{\partial (r\lambda \cos \phi, r\phi, r)} \qquad (16.25)$$

is the Jacobian of the mapping $(\lambda, \phi, r) \mapsto (\Lambda, \Phi, R)$, and we use the standard notation

$$\frac{\partial (\Lambda, \Phi, R)}{\partial (\lambda, \phi, r)} = \begin{vmatrix} \partial_\lambda \Lambda & \partial_\lambda \Phi & \partial_\lambda R \\ \partial_\phi \Lambda & \partial_\phi \Phi & \partial_\phi R \\ \partial_r \Lambda & \partial_r \Phi & \partial_r R \end{vmatrix}. \qquad (16.26)$$

For the moment, we are considering the system in the whole three-dimensional space without boundaries, the reduction to a layer of finite depth will be made in the next section. The equations of motion follow, after applying the standard rules of the calculus of variations and integration by parts, and give a Lagrangian version of the equations (16.15), as one can easily check. We immediately apply the *quasi-hydrostatic approximation* in (16.24), which will consist in omitting the contribution of the vertical velocity $\dot{R}^2$ in the kinetic energy:

$$\mathcal{L} \to \mathcal{L}_{QH} = \frac{1}{2} \left(R^2 \cos^2 \Phi \dot{\Lambda}^2 + R^2 \dot{\Phi}^2 \right) + \Omega R^2 \cos^2 \Phi \dot{\Lambda} - bgR$$

$$+ \Pi \left(\frac{R^2 \cos \Phi}{r^2 \cos \phi} \frac{\partial (\Lambda, \Phi, R)}{\partial (\lambda, \phi, r)} - 1 \right). \tag{16.27}$$

We thus obtain the following quasi-hydrostatic equations of motion:

$$
\begin{cases}
\dfrac{d}{dt} \left[R^2 \cos^2 \Phi \left(\dot{\Lambda} + \Omega \right) \right] + \partial_\Lambda \Pi & = 0, \\[2mm]
\dfrac{d}{dt} \left[R^2 \dot{\Phi} \right] + R^2 \cos \Phi \sin \Phi \left(\dot{\Lambda}^2 + 2\Omega \dot{\Lambda} \right) + \partial_\Phi \Pi & = 0, \\[2mm]
R \left(\cos^2 \Phi \dot{\Lambda}^2 + \dot{\Phi}^2 \right) + 2\Omega R \cos^2 \Phi \dot{\Lambda} - bg - \partial_R \Pi & = 0, \\[2mm]
\dfrac{R^2 \cos \Phi}{r^2 \cos \phi} \dfrac{\partial (\Lambda, \Phi, R)}{\partial (\lambda, \phi, r)} = 1, \qquad \dfrac{d}{dt} b & = 0,
\end{cases}
\tag{16.28}
$$

where we kept d/dt instead of dots in some places, to avoid confusion. As is easy to check, upon the standard identifications of Lagrangian coordinates with Eulerian coordinates in the Eulerian description of the fluid motion, and with the help of (16.22), these equations reproduce the quasi-hydrostatic version of (16.19). Conservation of energy follows from the invariance of the action with respect to time shifts. Potential vorticity conservation follows from the invariance of the action with respect to relabelling of Lagrangian particles, cf. Section 3.4.1, and axial angular momentum conservation follows from the invariance of the action with respect to rotations, according to Noether theorem.

16.2.3   Characteristic scales and parameters

Let us consider characteristic scales and define essential dimensionless parameters for the incompressible fluid layer on the rotating sphere. The first parameter is the aspect ratio h_0/L where h_0 is a typical layer thickness and L a typical horizontal scale of the flow. Zero divergence of velocity implies that $W/U = O(h_0/L)$ where U and W are typical horizontal and vertical velocities, respectively, so that neglecting kinetic energy of vertical motion in the Lagrangian (16.24) is allowed if $h_0/L \ll 1$. Once we assume that this hypothesis holds, and use the quasi-hydrostatic Lagrangian (16.24), the aspect ratio disappears from the analysis.

In planar geometry shallowness can be only measured with respect to the proper scale L of the flow, but in spherical geometry shallowness can be also defined with respect to the planet's radius r_0 with the help of parameter ϵ:

$$\epsilon = \frac{h_0}{r_0} \ll 1. \tag{16.29}$$

We use asymptotic expansions in ϵ in what follows. Another global small parameter is the ratio of centrifugal and gravity accelerations on the surface of the sphere rotating with angular velocity Ω:

$$\gamma = \frac{r_0 \Omega^2}{g}. \tag{16.30}$$

Apart from the global parameters ϵ and γ, there are other parameters characterising a given type of the fluid motion. Thus, the planetary Rossby number for the motions with a typical velocity U on the rotating sphere is

$$\mu = \frac{U}{\Omega r_0}. \tag{16.31}$$

If $\mu \ll 1$ the typical fluid velocity is small compared to the velocity due to planetary rotation, which is generally true for large-scale atmospheric and oceanic motions on the Earth. We should stress that this (mild) condition does not necessarily mean that the Rossby number of a given, let us say, synoptic system of typical scale L is small, as generally $L \ll r_0$. Accordingly, the parameter ϵ introduced in (16.29) may be thought of as a product of two parameters: $\epsilon = \frac{h_0}{L} \frac{L}{r_0}$, although we will not introduce these latter. The standard derivation of the shallow-water equations is based on the hypothesis of smallness of $\frac{h_0}{L}$, which is equivalent to the hypothesis of smallness of ϵ for planetary-scale motions. We prefer to use the more general columnar motion hypothesis, which allows us to treat even the situations with $\epsilon = \mathcal{O}(1)$. Although the scale of the thickness of the fluid layer h is given by h_0 it does not define a typical amplitude of the variations $h - h_0$ of h. A nonlinearity parameter, which was repeatedly used in the preceding chapters,

$$\alpha = \frac{||h - h_0||}{h_0} \tag{16.32}$$

is needed for the asymptotic analysis of the shallow-water system (to avoid confusion with longitude, we call it α in this chapter). As usual, the small parameters should be ordered, to render the asymptotic procedure unambiguous. We work under the hypothesis

$$\frac{\epsilon \alpha}{\gamma \mu} = \mathcal{O}(1), \tag{16.33}$$

although we do not introduce α explicitly.

As usual, a bottom topography can be incorporated in shallow-water system. Topography introduces new parameters, such as its typical height. We allow the typical height of the topography to be of the order h_0.

16.2.4   Columnar motion reduction in the variational principle

We start with the variational principle established for primitive equations in the quasi-hydrostatic limit in Section 16.2.2, and systematically apply the columnar motion approximation, which we understand in the spherical sense, that is supposing that the *angular* velocities of the fluid elements in the form of solid angles are independent of the radial position. As applied to the Lagrangian coordinates Λ, Φ, this means that

$$\Lambda = \Lambda(\lambda, \phi, t), \quad \Phi = \Phi(\lambda, \phi, t). \tag{16.34}$$

The incompressibility constraint (16.25) becomes

$$\frac{R^2 \cos \Phi}{r^2 \cos \phi} \frac{\partial (\Lambda, \Phi, R)}{\partial (\lambda, \phi, r)} = \frac{R^2 \partial_r R \cos \Phi}{r^2 \cos \phi} \frac{\partial (\Lambda, \Phi)}{\partial (\lambda, \phi)} = \frac{R^2 \partial_r R}{r^2} \frac{\partial (\Lambda \cos \Phi, \Phi)}{\partial (\lambda \cos \phi, \phi)} = 1. \tag{16.35}$$

This relation can be rewritten as

$$R^2 dR \mathcal{J} = r^2 dr, \tag{16.36}$$

where we introduced the shorthand notation:

$$\mathcal{J} = \mathcal{J}(\Phi, \Lambda; \phi, \lambda) = \frac{\partial (\Lambda \cos \Phi, \Phi)}{\partial (\lambda \cos \phi, \phi)}. \tag{16.37}$$

Due to independence of the angular coordinates on r following from the columnar motion assumption, the relation (16.36) can be immediately integrated:

$$\frac{R^3}{3} = \frac{r^3}{3} \mathcal{J}^{-1} + \mathcal{F}(\lambda, \phi, t), \tag{16.38}$$

where $\mathcal{F}$ is an arbitrary function. From now on we will be considering a fluid layer between two material surfaces: a bottom with an arbitrary topography B, and the top, which is a free surface. In Lagrangian terms,

$$R_b = r_0 + B(\Lambda, \Phi), \quad R_t = r_0 + B(\Lambda, \Phi) + H(\Lambda, \Phi, t). \tag{16.39}$$

Under the Lagrangian mapping they are images of $r_b = r_0 + B(\lambda, \phi)$ and $r_t = r_0 + B + h_I(\lambda, \phi)$, respectively, where $H = h_I$ at $t = 0$. By an additional change of independent variables the initial height of the fluid layer may be rendered uniform, an observation which was already used in Chapters 7 and 8. Hence, without loss of generality we may consider $r_b = r_0$, and $r_t = r_0 + h_0$, with constant h_0 (by doing this we lose the interpretation of λ, ϕ, r as the initial position of a fluid parcel; we trade here transparency for technical convenience). With these assignments $\mathcal{F}$ and $\mathcal{J}$ can be eliminated from (16.38) in favour of $r_0 + B(\Lambda, \Phi)$ and $H(\Lambda, \Phi, t)$:

$$R^3 = \left(r^3 - r_0^3\right) \frac{(r_0 + H + B)^3 - (r_0 + B)^3}{(r_0 + h_0)^3 - r_0^3} + (r_0 + B)^3. \tag{16.40}$$

We thus get, for the Jacobian $\mathcal{J}$,

$$\mathcal{J}^{-1} = \frac{(r_0 + H + B)^3 - (r_0 + B)^3}{(r_0 + h_0)^3 - r_0^3}. \tag{16.41}$$

The relations (16.40) and (16.41) are used to formulate the continuity equation for columnar elements. By rewriting them in the form

$$\frac{(r_0 + H + B)^3 - (r_0 + B)^3}{3} = \frac{(r_0 + h_0)^3 - r_0^3}{3}\mathcal{J}^{-1}, \tag{16.42}$$

differentiating in time, and reusing (16.42) we obtain

$$\frac{d}{dt}\left[\frac{(r_0 + H + B)^3 - (r_0 + B)^3}{3}\right] + \frac{(r_0 + h + B)^3 - (r_0 + B)^3}{3}\frac{d\mathcal{J}/dt}{\mathcal{J}} = 0. \tag{16.43}$$

From the definition of $\mathcal{J}$ (16.37) it follows that

$$\frac{d\mathcal{J}/dt}{\mathcal{J}} = \frac{\partial\left(\dot{\Lambda}\cos\Phi, \Phi\right)}{\partial\left(\Lambda\cos\Phi, \Phi\right)} + \frac{\partial\left(\Lambda\cos\Phi, \dot{\Phi}\right)}{\partial\left(\Lambda\cos\Phi, \Phi\right)} - \tan\Phi\frac{\partial\left(\dot{\Phi}\Lambda\cos\Phi, \Phi\right)}{\partial\left(\Lambda\cos\Phi, \Phi\right)}. \tag{16.44}$$

Using the correspondence between Lagrangian and Eulerian descriptions, we recognise on the right-hand side of this formula the divergence of the (physical) angular velocity vector $\hat{\lambda}\cos\Phi\dot{\Lambda} + \hat{\phi}\dot{\Phi}$ in Lagrangian spherical coordinates. Thus, denoting by $\mathbf{u}$ the physical tangent velocity calculated at some fixed r, cf. Figure 16.3, by $\nabla_{\mathbf{x}}$, the tangent gradient operator along the sphere of the same radius, and $\nabla_{\mathbf{x}} \cdot (...)$ the divergence in the tangent plane, the Eulerian counterpart of (16.43) is

$$(\partial/\partial t + \mathbf{u} \cdot \nabla_{\mathbf{x}}) h + h\nabla_{\mathbf{x}} \cdot \mathbf{u} = 0, \tag{16.45}$$

where

$$h = \frac{(r_0 + H + B)^3 - (r_0 + B)^3}{3r_0^2} \tag{16.46}$$

is the pseudo-thickness of the layer which coincides with the true layer thickness H to the first order in ϵ.

We now consider the variational principle applied to the fluid layer between the bottom and the free surface (top) in the quasi-hydrostatic approximation (16.27). Relation (16.36) allows us to make a change of the radial integration variable $r \to R$. The reduced Lagrangian thus becomes

$$L_r = \int_{-\frac{\pi}{2}}^{\frac{\pi}{2}}\cos\phi\, d\phi \int_0^{2\pi} d\lambda\, \mathcal{J} \int_{R_b}^{R_t} R^2\, dR\left[\frac{1}{2}\left(R^2\cos^2\Phi\dot{\Lambda}^2 + R^2\dot{\Phi}^2\right) + \Omega R^2\cos^2\Phi\dot{\Lambda} - bgR\right], \tag{16.47}$$

where we omitted the already used incompressibility constraint. The integration in R is trivial, and by replacing b by its vertical average $\bar{b}$ we get

$$L_r = \int_{-\frac{\pi}{2}}^{\frac{\pi}{2}} \cos\phi \, d\phi \int_0^{2\pi} d\lambda \, \mathcal{J} \left[\left(\frac{1}{2} \left(\cos^2 \Phi \dot{\Lambda}^2 + \dot{\Phi}^2 \right) + \Omega \cos^2 \Phi \dot{\Lambda} \right) \left(\frac{R_t^5 - R_b^5}{5} \right) \right.$$

$$\left. - \bar{b}g \left(\frac{R_t^4 - R_b^4}{4} \right) \right]. \tag{16.48}$$

Injecting the expressions of R_t, R_b, and $\mathcal{J}$ in terms of h and B obtained from (16.39) to (16.41) in this formula we get a reduced Lagrangian which, after applying the general variational procedure described in Chapter 2, gives the equations of *deep columnar* motion on the sphere, together with (16.45). They may be of interest, yet our goal is to establish *shallow-water* equations. For this, as explained in Section 16.2.3, we introduce the small parameters μ, ϵ, γ, and rescale time and radial coordinate as follows:

$$t \to (\mu\Omega)^{-1} t, \quad R \to r_0(1 + \epsilon z). \tag{16.49}$$

Unless explicitly stated otherwise, we use hereafter dimensionless H, h, and B:

$$H \to \epsilon r_0 H, \quad h \to \epsilon r_0 h, \quad B \to \epsilon r_0 B. \tag{16.50}$$

With this scaling, different terms in the truncated Lagrangian (16.47) scale differently: the kinetic energy terms containing angular velocities squared are proportional to μ^2, the Coriolis term containing Ω is proportional to μ, and the potential energy term containing g is proportional to γ^{-1}. The dimensionless Jacobian (16.41) is expressed as

$$\mathcal{J}^{-1} = \frac{(1 + \epsilon(B + H))^3 - (1 + \epsilon B)^3}{(1 + \epsilon)^3 - 1}, \tag{16.51}$$

and the dimensionless pseudo-thickness (16.46) as

$$h = \frac{(1 + \epsilon)^3 - 1}{3\epsilon} \mathcal{J}^{-1}. \tag{16.52}$$

Up to an inessential constant in front of the whole Lagrangian, the dimensionless kinetic part of the (16.48) is

$$\frac{1}{h} \frac{(1 + \epsilon(H + B))^5 - (1 + \epsilon B)^5}{5\epsilon} \left[\dot{\Lambda} \cos^2 \Phi + \frac{\mu}{2} \left(\dot{\Lambda}^2 \cos^2 \Phi + \dot{\Phi}^2 \right) \right].$$

So far, no approximation has been made. Reintroducing h in the kinetic energy, retaining the terms $\mathcal{O}(1), \mathcal{O}(\mu), \mathcal{O}(\epsilon)$ and omitting higher-order terms, we obtain, for the dimensionless kinetic energy part of (16.48),

$$(1 + \epsilon(h + 2B)) \dot{\Lambda} \cos^2 \Phi + \frac{\mu}{2} \left(\dot{\Lambda}^2 \cos^2 \Phi + \dot{\Phi}^2 \right). \tag{16.53}$$

As to the potential energy part of (16.48), it may or may not be expanded in powers of ϵ, depending on the dynamical regime of interest. Since our focus here is on the effects of Coriolis force, we retain just the dominant term in the potential energy, that is:

$$-\frac{\epsilon}{\gamma\mu}\frac{h+2B}{2}, \tag{16.54}$$

where we supposed, for simplicity, that $\bar{b}$ is constant and equal to unity, without loss of generality. (A generalisation to horizontally varying $\bar{b}(\lambda,\phi)$ is straightforward and leads to non-traditional TRSW equations on the sphere, cf. Chapter 14.)

Retaining the terms (16.53) and (16.54) and omitting higher-order terms, the dimensionless reduced Lagrangian becomes

$$L_{RSW} = \int_{-\frac{\pi}{2}}^{\frac{\pi}{2}} \cos\phi d\phi \int_0^{2\pi} d\lambda \left[\mu \left(\frac{1}{2}\dot{\Phi}^2 + \frac{1}{2}\cos^2\Phi\dot{\Lambda}^2 \right) \right.$$

$$\left. + \left(1 + \underline{\epsilon(h+2B)} \right) \cos^2\Phi\dot{\Lambda} - \frac{\epsilon}{\gamma\mu}\frac{h+2B}{2} \right]. \tag{16.55}$$

The independent dynamical variables in the Lagrangian (16.55) are Λ and Φ. The variable h is not independent, and is expressed in terms of Λ and Φ through (16.37) and (16.52). We underlined in (16.55) the term that engenders non-traditional corrections to the rotating shallow-water equations on the sphere, as will be shown shortly. Under our assumptions, the last, potential energy, term in (16.55) is $\mathcal{O}(1)$. Another $\mathcal{O}(1)$ term is the second one originating from the traditional Coriolis term. The kinetic energy term is $\mathcal{O}(\mu)$, and the correction to the Coriolis term is $\mathcal{O}(\epsilon)$. Therefore, different dynamical regimes can be envisaged depending on the ratio of μ and ϵ. In the essentially shallow situation, all $\mathcal{O}(\epsilon)$ terms are to be omitted, corresponding in spirit to the traditional approximation such as described in Section 16.2.2. This indeed gives the traditional RSW system. If non-shallow corrections are to be taken into account, a relation between μ and ϵ should be prescribed. For $\epsilon \ll \mu \ll 1$, to the leading order only the linear in ϵ correction to the Coriolis term is to be retained, which is the regime we adopt. Although we will not develop it, an essentially deep columnar motion regime is possible, where $\epsilon = \mathcal{O}(1)$ and *all* terms in (16.48)) are to be kept, which results in rather exotic, but still consistent equations of motion. In what follows no further approximation is made and all calculations are exact.

16.2.5   Derivation of the non-traditional rotating shallow-water equations

We require stationarity of the action

$$\delta \int L_{RSW}\left[\Lambda, \dot{\Lambda}, \Phi, \dot{\Phi}\right] dt = 0, \tag{16.56}$$

and, by calculating variations and integrating by parts in time, get

$$\int \cos\phi \, d\phi \, d\lambda \, dt \left[\mu \left[-\left(\ddot{\Phi} + \cos\Phi \sin\Phi \dot{\Lambda}^2 \right) \delta\Phi - \left(\cos^2\Phi \ddot{\Lambda} - 2\cos\Phi \sin\Phi \dot{\Lambda}\dot{\Phi} \right) \delta\Lambda \right] \right.$$

$$+ 2\cos\Phi \sin\Phi \left(\dot{\Phi} \delta\Lambda - \dot{\Lambda} \delta\Phi \right) - \frac{\epsilon}{\gamma\mu} \left(\frac{\delta h}{2} + \delta B \right) - \epsilon\cos^2\Phi \left(\dot{h} + 2\dot{B} \right) \delta\Lambda$$

$$\left. + \epsilon \left[\cos^2\Phi \dot{\Lambda} \left(\delta h + 2\delta B \right) - 2\cos\Phi \sin\Phi \dot{\Lambda} \left(h + 2B \right) \delta\Phi + 2\dot{\Phi}\cos\Phi \sin\Phi \left(h + 2B \right) \delta\Lambda \right] \right] = 0,$$
$$(16.57)$$

where we used the following relations:

$$\delta h = -h \frac{\delta \mathcal{J}}{\mathcal{J}}, \quad \delta B = \delta\Phi \partial_\Phi B + \delta\Lambda \partial_\Lambda B, \qquad (16.58)$$

in order to calculate δh and δB, cf. (16.41) and (16.46). With the help of these formulae the contributions containing δh and δB in (16.57) can be integrated by parts by changing the integration variables from (λ, ϕ) to (Λ, Φ), with the help of the Jacobian $\mathcal{J}$, taken in the leading order in ϵ. Using the fact that the standard integral theorems hold in curvilinear coordinates, if expressed in terms of covariant derivatives, as well as the fact that the covariant derivative of a scalar coincides with the ordinary derivative, we get, for any function F,

$$\int \cos\phi \, d\phi \, d\lambda \, dt \, F \delta h = \int \cos\phi \, d\phi \, d\lambda \, dt \left[\frac{1}{h} \left[\dot{\Lambda} \partial_\Lambda \left(Fh^2 \right) \delta\Lambda + \dot{\Phi} \partial_\Phi \left(Fh^2 \right) \delta\Phi \right] \right]. \quad (16.59)$$

Equations (16.58) written for the time-derivatives

$$\dot{h} = -h \dot{\mathcal{J}} \mathcal{J}^{-1}, \quad \dot{B} = \dot{\Phi} \partial_\Phi B + \dot{\Lambda} \partial_\Lambda B, \qquad (16.60)$$

give for any F:

$$\int \cos\phi \, d\phi \, d\lambda \, dt \, F \left(\dot{h} + 2\dot{B} \right) =$$
$$\int \cos\phi \, d\phi \, d\lambda \, dt \left[-\frac{hF}{\cos\Phi} \left(\partial_\Lambda \left(\cos\Phi \dot{\Lambda} \right) + \partial_\Phi \left(\cos\Phi \dot{\Phi} \right) \right) + 2F \left(\dot{\Lambda} \partial_\Lambda B + \dot{\Phi} \partial_\Phi B \right) \right]. \qquad (16.61)$$

Using these formulae in (16.57) and gathering the terms with Λ and Φ, respectively, we arrive at the equations

$$\mu \left(\ddot{\Lambda} \cos^2\Phi - 2\dot{\Lambda}\dot{\Phi}\cos\Phi \sin\Phi \right) - 2\dot{\Phi} \sin\Phi \cos\Phi = -\frac{\epsilon}{\gamma\mu} \partial_\Lambda \left(h + B \right)$$

$$+ \epsilon \left[2\cos\Phi \sin\Phi (h + 2B)\dot{\Phi} + \frac{1}{h} \partial_\Lambda \left(\cos^2\Phi \dot{\Lambda} h^2 \right) \right. \qquad (16.62)$$

$$\left. + \cos\Phi \left[h\partial_\Lambda \left(\cos\Phi \dot{\Lambda} \right) + h\partial_\Phi \left(\cos\Phi \dot{\Phi} \right) - 2\cos\Phi \dot{\Phi} \partial_\Phi B \right) \right] \right],$$

$$\mu \left(\ddot{\Phi} + \dot{\Lambda}^2 \cos \Phi \sin \Phi \right) + 2 \dot{\Lambda} \sin \Phi \cos \Phi = - \frac{\epsilon}{\gamma \mu} \partial_\Phi (h + B)$$

$$+ \epsilon \left[-2 \cos \Phi \sin \Phi (h + 2B) \dot{\Lambda} + \cos^2 \Phi \dot{\Lambda} \partial_\Phi h + \frac{1}{h} \partial_\Phi (\cos^2 \Phi \dot{\Lambda} h^2) + 2 \cos^2 \Phi \dot{\Lambda} \partial_\Phi B \right].$$

$$(16.63)$$

The derivatives with respect to Lagrangian coordinates in this formulae are understood in the sense $\partial_\Lambda (...) = \dfrac{\partial (..., \Phi)}{\partial (\Lambda, \Phi)}$; $\partial_\Phi (...) = \dfrac{\partial (\Lambda, ...)}{\partial (\Lambda, \Phi)}$, and can be transformed into derivatives with respect to λ, ϕ with the help of the Jacobian $\mathcal{J}$. Together with (16.45) the equations (16.62) and (16.63) constitute the non-traditional RSW equations on the sphere. The underlined terms in (16.62) and (16.63) represent the non-traditional corrections we were seeking. They result from a modification of hydrostatic equilibrium by quasi-hydrostatic terms, non-traditional components of the Coriolis force, and the $O(\epsilon)$ correction to the standard Coriolis force.

It can be immediately shown, either from the invariance of the action with respect to rotations around $\hat{z}$ (shifts in Λ) with the help of the Noether theorem, or by a direct computation, that the axial momentum with the density

$$l_z = r_0^2 \cos^2 \Phi \left(\dot{\Lambda} + \Omega \left(1 + \frac{h + 2B}{r_0} \right) \right),$$

$$(16.64)$$

where we restored dimensions, is conserved. This is the consistent shallow-water reduction of the angular momentum (16.16). Material conservation of potential vorticity can be established from the particle relabelling symmetry of the action, as in Chapter 3.

We can rewrite (16.62) and (16.63) by introducing physical velocity components *at the bottom of the fluid layer*:

$$\mathbf{u} = u \hat{\boldsymbol{\lambda}} + v \hat{\boldsymbol{\phi}} : u = r_0 \cos \phi \dot{\lambda}, \; v = r_0 \dot{\phi}.$$

$$(16.65)$$

It should be stressed that there is an ambiguity in the choice of radial distance (a value between r_0 and $r_0 + B + h$) while transforming angular velocities into physical ones. This is one of the problems with a naive application of the columnar motion hypothesis. Here we make the simplest choice, and restoring dimensions and using dimensional h and B we thus get

$$\partial_t u + \frac{u}{r_0 \cos \phi} \partial_\lambda u + \frac{v}{r_0} \partial_\phi u - \frac{uv \tan \phi}{r_0} - 2\Omega \sin \phi v = - \frac{g}{r_0 \cos \phi} \partial_\lambda (h + B)$$

$$(16.66)$$

$$+ \frac{2\Omega}{r_0} \left(v \sin \phi (h + 2B) + h \partial_\lambda u + u \partial_\lambda h - v \cos \phi \partial_\phi B + \frac{h}{2} \partial_\phi (v \cos \phi) \right),$$

$$\partial_t v + \frac{u}{r_0 \cos\phi} \partial_\lambda v + \frac{v}{r_0} \partial_\phi v + \frac{u^2 \tan\phi}{r_0} + 2\Omega \sin\phi u = -\frac{g}{r_0} \partial_\phi (h + B)$$

$$+ \underline{\frac{2\Omega}{r_0} \left(-u\sin\phi (h + 2B) + u\cos\phi \partial_\phi (h + B) + \frac{h}{2} \partial_\phi (u\cos\phi) \right)}$$

(16.67)

$$\partial_t h + \frac{1}{r_0 \cos\phi} \left(\partial_\lambda (hu) + \partial_\phi (hv \cos\phi) \right) = 0, \tag{16.68}$$

where, again, non-traditional contributions are underlined. If they are omitted, the traditional RSW equations on the sphere are recovered. Traditional equations are usually written in terms of the *position* of the free surface $\tilde{H} = H + B$. Here, non-traditional equations are written in terms of the thickness $h = \tilde{H} - B$. In the f-plane approximation, where the equations (16.66) to (16.68) are considered at fixed reference latitude $\phi = \phi_0$, in the limit $r_0 \to \infty$, the equations of NT RSW on the f-plane (16.12) are recovered from (16.66) and (16.67)after the tangent-plane replacements $\frac{1}{r_0 \cos\phi_0} \partial_\lambda \to \partial_x$, $\frac{1}{r_0} \partial_\phi \to \partial_y$. Retaining the first-order corrections in $\phi - \phi_0$, beta-plane approximation results.

16.3 Example of crucial influence of non-traditional corrections: inertial instability with full Coriolis force

We will now show that corrections due to inclusion of usually neglected components of the Coriolis force lead to considerable changes in growth rates and structure of unstable modes of inertial instability, which was studied in Chapter 10. We start our consideration with theoretical arguments based on the same construction as in Section 10.5 within the rigid-lid two-layer model, and then complement them by a direct linear stability analysis within the framework of the free-surface two-layer model, like in Section 10.4.1.

16.3.1 Inertial instability with full Coriolis force: theoretical considerations

In this section we will be working with the system (16.13), and producing analytic results describing the changes due to NT effects in symmetric inertial instability of barotropic jets.

Background flow and 1.5 dimensional reduction

The first important point is that the difference between the two-layer RSW system in traditional approximation, and with NT terms appears already at the level of stationary jet solutions. Indeed, as follows from (16.13), stationary zonal flow solutions obey the equations which reflect modifications of the geostrophic balance by the NT terms:

$$fu_1 = -\frac{1}{\rho_1}\partial_y\pi + \frac{F}{2}\partial_y u_1(H - h_2),$$

(16.69)

$$fu_2 = -\frac{1}{\rho_2}\partial_y\pi - g'\partial_y h_2 + F\left[\frac{h_2\partial_y u_2}{2} + \partial_y h_2 u_2 + \frac{\rho_1}{\rho_2}\partial_y\left((H - h_2)u_1\right)\right].$$

For the sake of simplification, we will limit ourselves by the approximation, which was already used in Chapter 5, where two densities are close, $\rho_1 \approx \rho_2$, and their difference is significant only in the reduced gravity term $g\frac{\rho_2-\rho_1}{\rho_2} = g'$. Following the lines taken in Section 10.5, we consider configurations which are symmetric with respect to translations in x:

$$\begin{cases} \dfrac{d_1 u_1}{dt} - f v_1 - \dfrac{F}{2}\partial_y v_1 h_1 = 0, \\[2mm] \dfrac{d_1 v_1}{dt} + f u_1 - \dfrac{F}{2}\partial_y u_1 h_1 = -\dfrac{1}{\rho_1}\partial_y\pi, \\[2mm] \dfrac{d_2 u_2}{dt} - f v_2 - \dfrac{F}{2}\partial_y v_2 h_2 = 0, \\[2mm] \dfrac{d_2 v_2}{dt} + f u_2 - \dfrac{F}{2}\left(\partial_y(u_2 h_2) + \partial_y h_2 u_2\right) = -\dfrac{1}{\rho_2}\partial_y\pi - g'\partial_y h_2 + F\dfrac{\rho_1}{\rho_2}\partial_y\left(h_1 u_1\right), \\[2mm] \partial_t h_{1,2} + \partial_y\left(v_{1,2} h_{1,2}\right) = 0, \quad h_1 + h_2 = H_0 = \text{const.} \end{cases}$$

(16.70)

Linearised equations and resulting eigenproblem

We consider barotropic mean flows with $u_{1,2} = U_{1,2}(y)$, entirely defined by distribution of the barotropic pressure π, with no interface displacement $h_1 = H_1 = \text{const}$, $h_2 = H_2 = \text{const}$. Linearisation of the 1.5D equations (16.70) about such solution gives

$$\begin{cases} \partial_t u_1 + (U_1' - f)v_1 - \dfrac{F}{2}\partial_y v_1 H_1 = 0, \\[2mm] \partial_t v_1 + f u_1 - \dfrac{F}{2}\partial_y u_1 H_1 + \dfrac{F}{2}U_1'\eta = -\partial_y\phi, \\[2mm] \partial_t u_2 + (U_2' - f)v_2 - \dfrac{F}{2}\partial_y v_2 H_2 = 0, \\[2mm] \partial_t v_2 + f u_2 - \dfrac{F}{2}\left(\partial_y u_2 H_2 + U_2'\eta + 2\partial_y\eta U_2\right) = -\partial_y\phi - g'\partial_y\eta + F\partial_y\left(-\eta U_1 + H_1\partial_y u_1\right), \\[2mm] \pm\partial_t\eta + \partial_y v_{1,2}H_{1,2} = 0, \end{cases}$$

(16.71)

where we used the weak stratification approximation, and introduced the geopotential $\phi = \pi/\rho$, the perturbation of the interface position η, and the prime notation for ordinary derivatives. The last two equations in (16.71) suggest introduction of a new variable $V = v_1 H_1 = -v_2 H_2$. In terms of this variable we get

$$\begin{cases} \partial_t V + f H_2 u_2 = - H_2 \partial_y \phi - g' H_2 \partial_y \eta \\[2mm] + F H_2 \left[\dfrac{H_2}{2} \partial_y u_2 + \dfrac{U_2'}{2} \eta + U_2 \partial_y \eta + H_1 \partial_y u_1 - \partial_y (U_1 \eta) \right], \\[3mm] - \partial_t V + f H_1 u_1 = - H_1 \partial_y \phi + F H_2 \left(\dfrac{H_1}{2} \partial_y u_1 - \dfrac{U_1'}{2} \eta \right), \\[3mm] H_2 \partial_t u_2 = - \left(U_2' - f \right) V + \dfrac{F}{2} H_2 \partial_y V, \\[3mm] H_1 \partial_t u_1 = + \left(U_1' - f \right) V - \dfrac{F}{2} H_1 \partial_y V. \end{cases} \qquad (16.72)$$

The variable η can be eliminated from the equations (16.72) by time differentiation and the use of the last equation in (16.71). The variables u_1 and u_2 can be then expressed in terms of V using two last equations in (16.72). Finally, geopotential can be eliminated by suitably combining the remaining equations. We arrive in this way at a single equation for V. The whole procedure follows that of Section 10.5 for traditional approximation, but new terms appear in the final equation due to NT contributions. We make a Fourier transformation in time, $V(t, y) = e^{i\omega t} \hat{V}(\omega, y) + \text{c.c.}$, and arrive at a second-order ordinary differential equation for $\hat{V}$, which we write down in the case $H_1 = H_2 = H$, for simplicity:

$$\left[-2\omega^2 + f(f - U_1'(y)) + f(f - U_2'(y)) + \frac{FH}{2} \left(U_1''(y) + U_2''(y) \right) \right] \hat{V}(y) -$$

$$HFU_2'(y)\hat{V}'(y) - H \left[g' + FH(U_1(y) - U_2(y)) + \frac{F^2 H^2}{2} \right] \hat{V}''(y) = 0. \qquad (16.73)$$

Note that at $F \to 0$ this equation coincides, up to the changes $x \leftrightarrow y$, $U \leftrightarrow V$ with the corresponding equation of Section 10.5.

Trapped modes and inertial instability

We introduce the length scale L, the time scale f^{-1}, the velocity scale U and non-dimensional parameters: Rossby number $Ro = U/(fL)$, Burger number $Bu = g'H/(f^2 L^2)$, and $\delta_{NT} = \frac{FH}{fL}$, which measures the strength of NT effects. We then write down (16.73) in non-dimensional form:

$$\left[2(1 - \omega^2) - Ro \left(U_1'(y)) + U_2'(y) \right) + Ro \, \delta_{NT} \frac{(U_1''(y) + U_2''(y))}{2} \right] \hat{V}(y) +$$

$$\delta_{NT} Ro U_2'(y) \hat{V}'(y) - \left[Bu + \delta_{NT} \left((U_1(y) - U_2(y)) \right) + \frac{\delta_{NT}^2}{2} \right] \hat{V}''(y) = 0. \qquad (16.74)$$

The term with the first derivative can be eliminated by the standard transformation of the dependent variable:

$$\hat{V} \to \hat{V} \exp \left(\int dy \delta_{\text{NT}} Ro \frac{U_2'}{2 \left[Bu + \delta_{\text{NT}} (U_1 - U_2) + \frac{\delta_{\text{NT}}^2}{2} \right]} \right). \tag{16.75}$$

The coefficients of thus transformed equation (16.74) are rather cumbersome but we recall that, by definition, in the rotating shallow-water model the vertical scale H is much smaller than the typical horizontal scale L, and hence the parameter δ_{NT} is necessarily small. Thus, we write down the equation resulting from the substitution (16.75) into (16.74) to the leading order in δ_{NT}:

$$[Bu + \delta_{\text{NT}} ((U_1(y) - U_2(y)))] \hat{V}'' +$$

$$\left[2\omega^2 - \left[2 - Ro \left(U_1'(y) \right) + U_2'(y) \right) + Ro \, \delta_{\text{NT}} \frac{(U_1''(y) + U_2''(y))}{2} \right] \right] \hat{V} = 0. \tag{16.76}$$

This equation is a Scrödinger equation with energy $2\omega^2$ and 'potential' $2 - Ro \left(U_1' + U_2' \right) + Ro \, \delta_{\text{NT}} \frac{(U_1'' + U_2'')}{2}$. Under traditional approximation, when $\delta_{\text{NT}} \to 0$, at constant h_2 $U_1 = U_2 = U$, cf. (16.69). As we have seen in Section 10.5, equation (16.76) then has negative eigenvalues ω^2 for strong-enough anticyclonic shears $U'(y) > 0$ of the background flow, and hence provides an instability of standing modes trapped in the minimum of the potential. The NT effects lead to an additional vertical shear of the background flow, compared to traditional approximation. As follows from (16.69) at constant h_2, in non-dimensional terms,

$$U_2 = U + \delta_{\text{NT}} \left(U + U'/2 \right), \quad U_1 = U + \delta_{\text{NT}} U'/2, \tag{16.77}$$

and to the leading order in δ_{NT} the terms with second derivative of U cancel, and (16.76) gives

$$\frac{Bu}{2} \hat{V}''(y) + \left[\omega^2 - \left(1 - Ro(1 + \delta_{\text{NT}}) U'(y) \right) \right] \hat{V}(y) = 0. \tag{16.78}$$

Therefore, at small δ_{NT} the influence of NT effects to the leading order consists in deepening the potential well by rescaling its amplitude, and hence in diminishing eigen-frequencies squared, which leads to an increase in the growth rates of the instability. At the same time, the eigensolutions $\hat{V}(y)$ are distorted with respect to the traditional approximation, according to (16.75).

Thus, we see that the NT corrections tend to enhance the symmetric inertial instability of a barotropic jet, with respect to traditional approximation. Yet, the corrections are small at $\delta_{NT} \to 0$. We will see in Section 16.3.2 that direct linear stability analysis confirms and reinforces this theoretical result.

16.3.2 Inertial instability with full Coriolis force: direct approach to the linear stability analysis

We consider the two-layer rotating shallow-water equations with a free surface (16.14), and, as in Section 16.3.1, study stability of a barotropic, in the sense of absence of vertical shear, Bickley jet evolving in mid-latitudes in the Northern hemisphere, such that the traditional Coriolis parameter is positive $f > 0$ and the non-traditional Coriolis parameter $F = \mathcal{O}(f)$.

Relevant parameters

In the traditional approximation when $F = 0$ a flow with a typical velocity U, horizontal scale L and geopotential gH is characterised by four dimensionless parameters: the Rossby number, the barotropic Burger number, and density and aspect ratios, r and d, respectively. The non-traditional terms in (16.14) have a factor F. After non-dimensionalisation a new characteristic scale FH appears in the equations. It corresponds to the typical scale of vertical variations of the planetary part of absolute angular momentum $\bar{u} - fy + Fh/2$, in a homogeneous layer of height h. The dimensionless parameter associated with this new scale $\delta_{N1} = FH/(fL)$ was already introduced earlier, and is of the order of the aspect ratio of the flow H/L in mid-latitudes. So, in the present context we expect NT effects to be relevant for jets with significant vertical extent.

The system (16.14) has two Lagrangian invariants—potential vorticities of the layers

$$q_1 = \frac{\partial_x v_1 - \partial_y u_1 + f - F\partial_y (h_1/2 + h_2)}{h_1}, \qquad (16.79)$$

$$q_2 = \frac{\partial_x v_2 - \partial_y u_2 + f - F\partial_y h_2/2}{h_2}. \qquad (16.80)$$

Let us recall that inertial instability requires that the product of the (traditional) Coriolis parameter and the absolute vorticity be negative somewhere in the flow. As non-traditional PV (16.79), (16.80) differ from the traditional potential vorticities by terms proportional to F, stability bounds can be altered.

The zonal barotropic Bickley jet has the the velocity

$$\bar{u}(y) = U \cosh^{-2}(y/L), \quad \bar{v}(y) = 0 \qquad (16.81)$$

in both layers. As discussed in Section 16.3.1, the NT terms change the geostrophic balance conditions. So, in order to maintain the zonal velocity (16.81), the thicknesses of the layers have to obey the following modified geostrophic balance equations:

$$\bar{h}_1' = -\frac{F\bar{u}'}{2(g - F\bar{u})(1 - r)} \left(\bar{h}_1 (2r - 1) + \bar{h}_2 \right), \qquad (16.82)$$

$$\bar{h}_2' = \frac{1}{(g - F\bar{u})} \left(-f\bar{u} + \frac{F\bar{u}'}{2(1 - r)} \left(r\bar{h}_1 + \bar{h}_2 \right) \right), \qquad (16.83)$$

where the prime denotes differentiation with respect to y. At given $\bar{u}(y)$, $\bar{h}_{1,2}(y)$ are determined from this system of ordinary differential equations numerically.

In order to perform numerical stability analysis of the jet, we have to fix the values of parameters. We recall that for the Earth $g = 9.81$ m/s^2, $R = 6365km$, and $\Omega - 2\pi$ day^{-1}. We consider a reference configuration with $H \approx 20$ km, $L \approx 450$ km and $U \approx 80$ m/s so that the typical Burger number is not too large, allowing for the symmetric inertial instability. The latitude is $\phi_0 = 30°N$ for which the traditional Coriolis parameter is $f \approx 7 \cdot 10^{-5}$ s^{-1} while its meridional variation across the zonal jet is $2\Omega/r_0 L \approx 9 \cdot 10^{-6}$ s^{-1}. This β term in the traditional Coriolis parameter will be neglected. The NT Coriolis parameter is $F = 2\Omega \cos \phi_0$. Thus, the values of non-dimensional parameters for this reference configuration are

$$Ro = \frac{U}{fL} = 2.7, \quad Bu = \frac{gH}{(fL)^2} = 250, \quad \delta_{NT} = \frac{FH}{fL} = 0.1, \quad r = 0.99, \quad d = 4$$

A jet with these parameters has a region of negative absolute vorticity for $Ro > 3\sqrt{3}/4$, cf. Figure 16.4, and thus is subject to inertial instability .

The numerical solution of (16.82) and (16.83) is plotted in Figure 16.4 for the reference configuration. As follows from the figure, the profiles of thicknesses in the NT background state are slightly modified in comparison with their traditional counterparts, but differences in the potential vorticity profiles are indistinguishable.

Formulation of the linear stability problem

As usual, we add a small perturbation to the background jet: $u_i = \bar{u}(y) + u_i'(x, y, t)$, $v_i = v_i'(x, y, t)$ and $h_i = \bar{h}(y) + h_i'(x, y, t)$, $i = 1, 2$, linearise the system about the jet, and apply

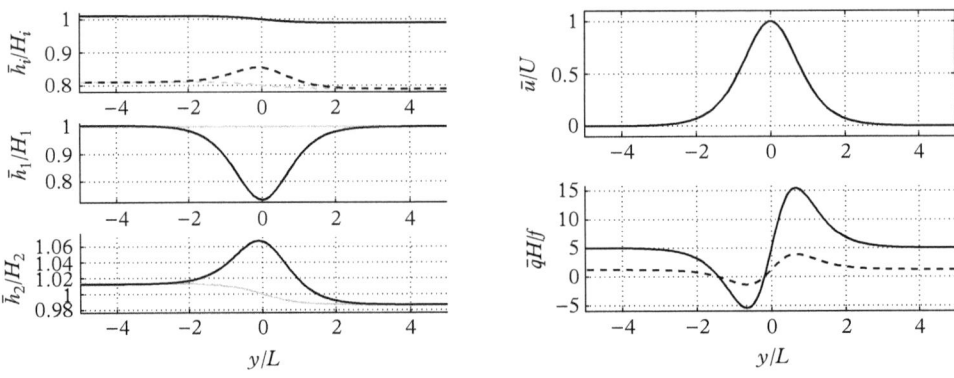

Figure 16.4 *Dimensionless thickness and velocity of the Bickley jet as functions of y/L. Grey: traditional with $\delta_{NT} = 0$, black: NT with $\delta_{NT} = 0.1$ configurations. Top-left panel: normalised thickness of layer 2 (dashed) and the total thickness (solid). Middle-left panel: thickness of the layer 1. Bottom-left panel: thickness of the layer 2, enlarged. Top-right panel: velocity $\bar{u}(y)$ in both layers. Lower-right panel: PV of the layer 2 (dashed) and of the layer 1 (solid).*

Fourier transformations in time and stream-wise direction to the linearised equations of the system. Solutions of the linearised system are sought in the form $(u'_i, v'_i, h'_i) = (\hat{u}_i, \hat{v}_i, \hat{h}_i)e^{i(kx-\omega t)} + \text{c.c.}$. We arrive in this way at the following eigenproblem:

$$\omega \cdot \mathbf{X} = \mathbf{M} \cdot \mathbf{X}, \quad \mathbf{X} = \left(\hat{u}_1(y), \, i\hat{v}_1(y), \, \hat{h}_1(y), \, \hat{u}_2(y), \, i\hat{v}_2(y), \, \hat{h}_2(y), \right)^{\mathrm{T}}, \tag{16.84}$$

where

$$\mathbf{M} = \begin{bmatrix} M_{11} & M_{12} \\ M_{21} & M_{22} \end{bmatrix},$$

with

$$M_{11} = \begin{bmatrix} k\left(\bar{u} - F\bar{h}_1\right) & f - \mathcal{D}\bar{u} - F\left(\mathcal{D}\bar{h}_2 - \dfrac{\bar{h}_1\mathcal{D}}{2}\right) & k\left(g - F\bar{u}\right) \\ f - F\left(\mathcal{D}\left(\bar{h}_1 + \bar{h}_2\right) + \dfrac{\bar{h}_1\mathcal{D}}{2}\right) & k\bar{u} & \left(g - F\bar{u}\right)\mathcal{D} - F\dfrac{\mathcal{D}\bar{u}}{2} \\ k\bar{h}_1 & -\left(\bar{h}_1\mathcal{D} + \mathcal{D}\bar{h}_1\right) & k\bar{u} \end{bmatrix},$$

$$M_{12} = \begin{bmatrix} -Fk\bar{h}_2 & F\left(\mathcal{D}\bar{h}_2 + \bar{h}_2\mathcal{D}\right) & k\left(g - F\bar{u}\right) \\ 0 & 0 & \left(g - F\bar{u}\right)\mathcal{D} \\ 0 & 0 & 0 \end{bmatrix},$$

$$M_{21} = \begin{bmatrix} -Fkr\bar{h}_1 & 0 & kr\left(g - F\bar{u}\right) \\ -rF\left(\mathcal{D}\bar{h}_1 + \bar{h}_1\mathcal{D}\right) & 0 & r\left((g - F\bar{u})\mathcal{D} - F\mathcal{D}\bar{u}\right) \\ 0 & 0 & 0 \end{bmatrix},$$

$$M_{22} = \begin{bmatrix} k\left(\bar{u} - F\bar{h}_2\right) & f - \mathcal{D}\bar{u} + \dfrac{F\bar{h}_2\mathcal{D}}{2} & k\left(g - F\bar{u}\right) \\ f - F\left(\mathcal{D}\bar{h}_2 + \dfrac{\bar{h}_2\mathcal{D}}{2}\right) & k\bar{u} & \left(g - F\bar{u}\right)\mathcal{D} - F\dfrac{\mathcal{D}\bar{u}}{2} \\ k\bar{h}_2 & -\left(\bar{h}_2\mathcal{D} + \mathcal{D}\bar{h}_2\right) & k\bar{u} \end{bmatrix}.$$

This eigenproblem is solved by pseudospectral collocation method with grid stretching. The size of the calculational domain L_y has to be large enough to resolve the eigenmodes trapped in the negative part of PV which has a characteristic length L. Convergence of the results is achieved at $L_y \geq 10L$. Taking $L_y = 10L$, we require the eigenmodes to vanish at $y = \pm 5L$ as we are interested in trapped modes which decay far enough from the jet.

Results of the linear stability analysis of the symmetric problem

We start with symmetric instability, that is $k = 0$. The results for the growth rates $Im(\omega)$ in the reference configuration are plotted as functions of Ro, Bu, r and d, respectively, in Figure 16.5. The meridional structure of the corresponding unstable modes is presented in Figure 16.6. The results can be summarised as follows:

1. *Existence and structure of the unstable modes.* At large enough Ro there is a single unstable mode. It is confined in the region of anticyclonic shear of the jet, and is essentially baroclinic in the sense that perturbations in the two layers have opposite signs. The nature of the instability, and the form of the unstable mode, such as the bell-shaped profile of the meridional velocity, are qualitatively the same as under traditional approximation.

2. *Dependence of the growth rate on parameters Ro, Bu, r, d.* As in Section 10.5, the growth rate $Im(\omega)$ increases with Ro, d, r and decreases with Bu.

3. *Difference between the growth rates with and without NT terms.* The NT growth rate is always higher. At the same Bu, or r, or d the NT growth rates are $\approx 30\%$ higher, which is far from being negligible. The NT terms thus have a destabilising effect upon symmetric inertial instability. There exists a parameter range where the jet is inertially unstable with NT terms, while it is stable in traditional approximation.

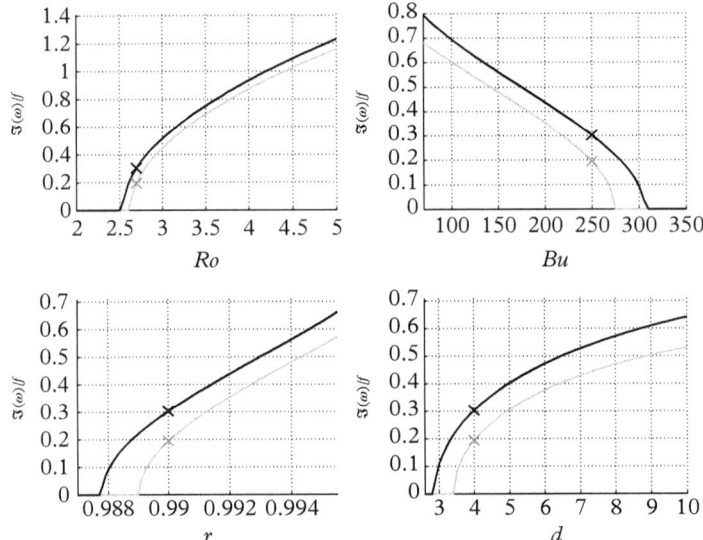

Figure 16.5 *Dependence of the dimensionless growth rate $Im(\omega)/f$ of the symmetric inertial instability on Ro (top-left panel), Bu (top-right panel), r (bottom-left panel), and d (bottom-right panel). Grey: traditional approximation with $\delta_{NT} = 0$. Black: NT with $\delta_{NT} = 0.1$ configuration. Crosses: reference configuration.*

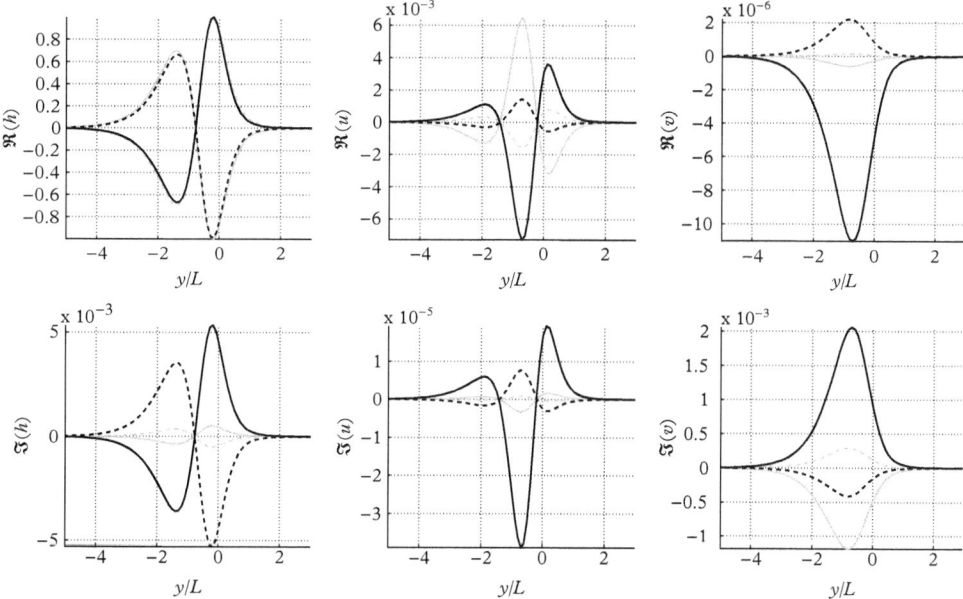

Figure 16.6 *Span-wise structure of the unstable mode of symmetric inertial instability. Dashed: layer 2; Solid: layer 1. Left column: pressure, middle column: zonal velocity, right column: meridional velocity. Upper row: real; bottom row: imaginary part. Grey: traditional approximation with $\delta_{NT} = 0$. Black: NT with $\delta_{NT} = 0.1$.*

Results of the linear stability analysis in the asymmetric configuration

We turn now to the asymmetric problem (16.84) with $k \neq 0$. As in Chapter 10, *asymmetric inertial instability* shows up. In the ageostrophic regime, when $Ro \geq 1$ there are two leading instabilities of the jet: barotropic and baroclinic ones. At sufficiently large Ro the growth rate of the baroclinic instability is dominant in the long-wave sector and has non-zero limit at $k \to 0$. This is, thus, the asymmetric inertial instability, and the situation is qualitatively the same as in traditional approximation. The growth rate of the NT asymmetric inertial instability is larger than in traditional approximation. The difference between the non-traditional and traditional growth rates is largest in the symmetric limit $k = 0$.

Growth rates and phase speeds of the most unstable modes of the asymmetric inertial instability in the reference configuration at different Ro are presented in Figure 16.7. At $Ro = 2.7$ the meridional sections of the three components of the most unstable mode (not shown) are practically indistinguishable from those for the symmetric inertial instability of Figure 16.6, which permits us to conclude that the nature of the instability is the same, in spite of the presence of stream-wise modulations.

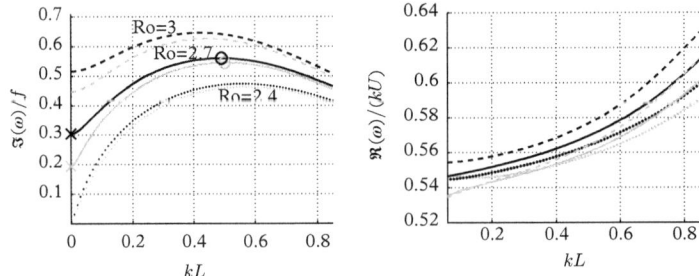

Figure 16.7 *Dimensionless growth rate Im (ω) /f (left panel), and phase velocity Re (ω) /(kU) (right panel) of the most unstable mode of the barotropic Bickley jet in the two-layer model as a function of dimensionless zonal wave number kL. Ro = 2.4 (dashed), Ro = 2.7 (solid, reference configuration) and Ro = 3 (dotted). Grey: traditional approximation with $\delta_{NT} = 0$. Black: NT with $\delta_{NT} = 0.1$. Crosses (circles): symmetric (asymmetric) instability in the reference configuration.*

16.4   Summary, discussion, and bibliographic remarks

The question of 'non-traditional' tangent plane equations was repeatedly discussed in the literature, see the review Gerkema *et al.* (2008) and references therein. The NT effects were shown to have crucial influence on propagation of internal inertia–gravity waves in the ocean in certain configurations (Gerkema and Shrira 2005). It should be stressed that the impact of neglected NT corrections upon long-time simulations of the atmospheric circulation, for example in the context of climate change, was never assessed, although this question was raised and the models incorporating these corrections exist (White *et al.* 2005, Tort and Dubos 2014).

We presented derivations of non-traditional shallow-water equations on the tangent *f* plane and on the whole rotating sphere. While in the first case the straightforward vertical averaging directly leads to the result, the procedure is more subtle in the presence of the effects of curvature, and necessitates the use of variational principle in order to obtain a consistent result. It should be emphasised that the difference between the traditional and the non-traditional systems of rotating shallow-water equations comes from a single new underlined term in the Lagrangian (16.55). The other terms being the standard kinetic and potential energy terms and the Coriolis term in the rotating frame, it seems paradoxical that establishing the form of the non-traditional terms from the equations of motion encounters difficulties. Yet, while the general structure of the first 'non-shallow' correction to the Coriolis term is rather obvious, the problem is to establish the precise values of the coefficients in front of *h* and *B*. As we have seen, they come from the expansion of both the Lagrangian and the Jacobian of the Lagrangian mapping. The coefficients could be determined by imposing the conservation of the axial angular momentum, but the problem is then shifted to a definition of the latter which would be consistent with the approximation.

In order to check the relevance of the approximations used for derivation of the NT RSW equations on the sphere, and also potential importance of the NT corrections, we can make, following Tort *et al.* (2014), some crude estimates. For typical values of parameters of synoptic motions in the Earth's atmosphere we take: $U \sim 10 m/s$, $h_0 \sim 10 km$, and for the meso-scale motions in the ocean: $U \sim 0.1 m/s$, $h_0 \sim 3 km$, together with $\Omega = 7.3 \times 10^{-5} s^{-1}$ and $r_0 \approx 6400 km$ for the Earth. This gives

$$\gamma \approx 3.5 \times 10^{-3}, \tag{16.85}$$

and we get the following estimates: for the atmosphere $\epsilon \approx 1.6 \times 10^{-3}$, $\mu \approx 2.1 \times 10^{-2}$; for the ocean: $\epsilon \approx 5 \times 10^{-4}$, $\mu \approx 2.1 \times 10^{-4}$. This means that the approximations made above are consistent for both Earth's atmosphere and oceans. Unexpectedly, for the ocean the traditional rotating shallow-water approximation, with omission of the $\mathcal{O}(\epsilon)$ terms while retaining $\mathcal{O}(\mu)$ terms, is more problematic than for the atmosphere. In addition, the parameter $\frac{\epsilon}{\mu\gamma}$ appearing in the dimensionless non-traditional RSW equations turns to be large both in the atmosphere and in the ocean. This means that the non-linearity parameter α should be small, for consistency. Then, the non-traditional terms containing derivatives of the position of the free surface $h + B$ can be omitted in this regime in equations (16.66) to (16.68), and $h + B$ can be replaced by its non-perturbed value $h_0 + B$. Further simplifications can be made depending on the smallness of the topography and/or its steepness. We should also mention that higher-order corrections in ϵ are comparable with the corrections due to the deviation of the Earth's form from sphericity and the passage to oblate-spheroidal curvilinear coordinates is then necessary (e.g. White and Wood 2012). Our procedure can be repeated in these coordinates as well.

Concerning other planets, the relevant parameters are available at the NASA website: *http://nssdc.gsfc.nasa.gov/planetary/planetfact.html*. A great uncertainty in the estimates, as with those made earlier, comes from the values of the wind speeds, and of the typical depths of the atmospheres, especially for giant planets. With reasonable guesses for the latter we arrive at values of ϵ that are not much different from the Earth's values. For Jupiter and Saturn we thus get estimates for μ roughly of the same order as for the Earth, while the values of μ are much greater for Venus and Mars due to the slow rotation and small radius, respectively. The close to unity value of μ for Venus means that deep columnar motion corrections to the kinetic energy terms in the reduced Lagrangian could be relevant in this case. Otherwise, the non-traditional approximation as we defined it, is as well justified for the other planets as for the Earth, although the crude estimates above should be refined.

As we have showed, the NT corrections significantly influence the instabilities of jets, changing the structure of the unstable modes, and growth rates, with a difference up to 30%. The inertial instability of the mid-latitude Bickley jet in the two-layer rotating shallow-water model is more vigorous, if considered with full Coriolis force, especially for translationally symmetric perturbations. Nevertheless, its nature remains the same. The overall structure of the unstable modes is similar, reflecting trapping at the anticyclonic side of the jet. The NT corrections are more important at weaker stratifications.

This phenomenon, displayed earlier in the two-layer model, is even more pronounced in the case of continuous stratification, as shown in Tort, Ribstein, and Zeitlin (2016). The corrections to the growth rates, which we considered in the f-plane approximation, are expected to be even more significant at the equator.

The quasi-hydrostatic approximation was introduced in White and Bromley (1995). Derivation of the non-traditional one-layer RSW equations on the f-plane using the variational principle was given in the pioneering paper by Dellar and Salmon (2005). The two-layer free-surface NT RSW equations were derived in Stewart and Dellar (2010). NT RSW model on the sphere was derived in Tort *et al.* (2014), the presentation in Section 16.2 follows this paper. The influence of NT effects upon symmetric inertial instability was studied in Tort, Ribstein, and Zeitlin (2016), and some of the results of this paper are presented in Section 16.3. In this chapter we have derived the NT RSW equations in the f-plane approximation, and on the sphere, and showed how the first follow from the second. The beta-plane approximation with NT effects is not obvious, because of the discussed problem of consistency of the approximation. There exist two variants of non-traditional β-plane equations in the literature. Grimshaw (1975) considered non-traditional equations by retaining the beta effect only in the vertical component Ω_z of the overall rotation. Dellar (2011) extended this result to the case where the horizontal component Ω_y varies with latitude, and the vertical component Ω_z varies with z. The variation of Ω_z with z is captured by the non-traditional extension of the standard Coriolis term in (16.66) and (16.67), while the meridional variation of Ω_y and Ω_z are also recovered from our equations in the tangent plane limit where trigonometric factors are expanded in the first order about the reference latitude.

Figures 16.1, 16.2, and 16.4 to 16.7 are reproduced from Tort, Ribstein, and Zeitlin (2016), and Figure 16.3 from Tort *et al.* (2014), with permissions.

References

Abramowitz, M. and Stegun I.M., 1964, *Handbook of Mathematical Functions*, Dover, NY

Abrashkin, A.A., and Yakubovich, E. I., 2006, *Vortex Dynamics in a Lagrangian Description*, Fizmatlit, Moscow.

Akylas, T.R. Large-scale modulations of edge waves, *J. Fluid Mech.*, **132**, 197–208.

Allen, J.S. and Holm, D.D., Extended quasigeostrophic Hamiltonian models for rotating shallow water motions, *Physica D*, **98**, 229–48.

Annenkov, S.Y., and Shrira, V.I., 2013, Large-time evolution of statistical moments of wind-wave fields, *J. Fluid Mech.*, **726**, 517–46.

Baer, F. and Tribbia, J., 1977, On complete filtering of gravity modes through nonlinear initialization, *Mon. Wea. Rev.*, **105**, 1536–39.

Baey, J.-M. and Carton, X., 2002, Vortex multipoles in two-layer rotating shallow-water flows, *J. Fluid Mech.*, **460**, 151–75.

Balk, A.M. and Nazarenko, S.V., 1990, On the physical realisability of anisotropic Kolmogorov weak-turbulence spectra, *Sov. Phys. JETP*, **97**, 1827–45.

Ball, F.K. 1967, Edge waves in the ocean of finite depth, *Deep-Sea Res.*, **14**, 79–88.

Batchelor, G.K., 1967, *An Introduction to Fluid Dynamics*, CUP, Cambridge.

Batchelor, G.K., 1969, Computation of the energy spectrum in homogeneous two-dimensional turbulence, *Phys. Fluids*, **12**, 233–39.

Bec, J. and Khanin, K., 2007, Burgers turbulence, *Phys. Reports*, **447**, 1–49.

Benilov, E.S., 2004, Stability of vortices in a two-layer ocean with uniform potential vorticity in the lower layer, *J. Fluid Mech.*, **502**, 207–32.

Benilov, E.S. and Reznik, G.M., 1996, The complete classification of large-amplitude geostrophic flows in a two-layer fluid, *Geophys. Astrophys. Fluid Dyn.*, **82**, 1–22.

Benney, D.J. and Newell, A.C., 1969, Random wave closures, *Stud. Appl. Math.*, **48**, 29–35.

Benney, D.J. and Saffman, P.G., 1966, Nonlinear interactions of random waves in a dispersive medium, *Proc. R. Soc. A*, **289**, 301–20.

Betts, A.K. and Miller, M.J., 1986, A new convective adjustement scheme. Part II: Single columns tests using GATE wave, BOMEX, ATEX and arctic air-mass data sets, *Q.J.R. Meteor. Soc.*, **112**, 693–709.

Bouchut, F., 2004, *Nonlinear Stability of Finite-Volume Methods for Hyperbolic Conservation Laws, and Well-Balanced Schemes for Sources*, Birkhauser, Basel.

Bouchut, F., 2007, Efficient numerical finite-volume schemes for shallow water models, 189–256, in *Nonlinear Dynamics of Rotating Shallow Water: Methods and Advances*, V. Zeitlin, ed, Springer, NY.

Bouchut, F., Le Sommer, J., and Zeitlin, V., 2004, Frontal geostrophic adjustment and nonlinear wave phenomena in one-dimensional rotating shallow water. Part 2. High-resolution numerical simulations, *J. Fluid Mech.*, **514**, 35–63.

Bouchut, F., Le Sommer, J., and Zeitlin, V., 2005, Breaking of balanced and unbalanced equatorial waves, *Chaos*, **15**, 013503.

Bouchut, F., Scherer, E., and Zeitlin, V., 2008, Nonlinear adjustment of a front over escarpment, *Phys. Fluids*, **20**, 016602.

Bouchut, F., Lambaerts, J., Lapeyre G., and Zeitlin, V. 2009, Fronts and non-linear waves in a simplified shallow-water model of the atmosphere with moisture and convection, *Phys. Fluids,* **21**, 116–604.

Bouchut, F., Ribstein, B., and Zeitlin, V., 2011, Inertial, barotropic, and baroclinic instabilities of the Bickley jet in two-layer rotating shallow water model, *Phys. Fluids*, **23**, 126601.

Bouchut, F. and Zeitlin, V., 2010, A robust well-balanced scheme for multi-layer shallow water equations, *Discr. Cont. Dyn. Sys. B*, **13**, 739–58.

Boyd, J.P., 1980, Equatorial solitary waves. Part I. Rossby solitons, *J. Phys. Oceanogr.*, **10**, 1–11.

Boyd, J.P., 1987, Orthogonal rational functions on a semi-infinite interval, *J. Comput. Phys.*, **70**, 63–88.

Brannigan, L., 2016, Intense submesoscale upwelling in anticyclonic eddies, *Geoph. Res. Lett.*, **43**, 3360–69.

Caillol, P. and Zeitlin, V., 2000, Kinetic equations and stationary energy spectra of weakly nonlinear internal gravity waves, *Dyn. Atmos. Oceans*, **32**, 81–112; Erratum: **33**, 325–26.

Cairns, R.A., 1979, The role of negative energy waves in some instabilities of parallel flows, *J. Fluid Mech.*, **92**, 1–14.

Carnevale, G.F., Kloosterziel, R.C., and Orlandi, P., 2013, Inertial and barotropic instabilities of a free current in three-dimensional rotating flow, *J. Fluid Mech.*, **725**, 117–51.

Carnevale, G.F., McWilliams, J.C., Pomeau, Y., Weiss, J.B., and Young, W.R., 1991, Evolution of vortex statistics in two-dimensional turbulence, *Phys. Rev. Lett.*, **66**, 2735–37.

Carton, X.J., Flierl, G.R., and Polvani, L.M., 1989, The generation of tripoles from unstable axisymmetric vortex structures, *Europhys. Lett.*, **9**, 339–44.

Cho, J. Y-K., Menou, K., Hansen B.M.S., and Seager, S., 2008, Atmospheric circulations of close-in extrasolar giant planets. I. Global barotropic adiabatic simulations, *Astroph. J.*, **675**, 817–45.

Cullen M.J.P., 2006, *A Mathematical Theory of Large Scale Atmosphere/Ocean Flows*, World Scientific, Singapore.

Cushman-Roisin, B., Sutyrin, G.G., and Tang, B., 1992, Two-layer geostrophic dynamics. Part I: Governing equations, *J. Phys. Oceanogr.*, **22**, 117–27.

Davidson, P.A., 2004, *Turbulence: An Introduction for Scientists and Engineers*, OUP, Oxford.

Dellar, P.J., 2011, Variations on a beta-plane: derivation of non-traditional beta -plane equations from Hamilton's principle on a sphere, *J. Fluid Mech.*, **674**, 174–95.

Dellar, P.J. and Salmon, R., 2005, Shallow water equations with a complete Coriolis force and topography, *Phys. Fluids*, **17**, 106601.

Dempsey, D.P. and Rotunno, R., 1988, Topographic generation of mesoscale vortices in mixed-layer models, *J. Atmos. Sci.*, **45**, 2961–78.

Dritschel, D.G., 1997, Introduction to 'contour dynamics' for the Euler equations in two dimensions, *J. Comp. Phys.*, **135**, 217–19.

Dunkerton, T.J., 1982, A non-symmetric equatorial inertial instability, *J. Atmos. Sci.*, **38**, 807–13.

Eckart, C., 1960, *Hydrodynamics of Oceans and Atmosphere*, Pergamon, NY.

Emanuel, K.A., 1994: *Atmospheric Convection*, OUP.

Esler, J.G., Rump, O.J., and Johnson E.R., 2009, Supercritical rotating flow over topography, *Phys. Fluids*, **21**, 066601.

Falkovich, G.E. and Medvedev, S.B., 1992, Weak turbulence of short inertia-gravity waves, *Europhys. Lett.*, **19**, 279–84.

Farge, M. and Sadourny, R., 1989, Wave-vortex dynamics in rotating shallow water, *J. Fluid Mech.*, **206**, 433–62.

Fedorov, A.V. and Melville, W.K., 1995, On the propagation and breaking of nonlinear Kelvin waves, *J. Phys. Oceanogr.*, **25**, 2519–31.

Flierl, G.R., Larichev, V.D., McWilliams, J.C., and Reznik, G.M., 1980, The dynamics of baroclinic and barotropic solitary eddies, *Dyn. Atmos. Oceans*, **5**, 1–41.

Flor, J.-B., Scolan, H., and Gula, J., 2011, Frontal instabilities and waves in a differentially rotating fluid, *J. Fluid Mech.*, **685**, 532–42.

Ford, R., 1994, The response of a rotating ellipse of uniform potential vorticity to gravity wave radiation, *Phys. Fluids*, **6**, 3694–700.

Frierson, D.M.W., Majda, A.J. and Pauluis, O.M., 2004, Large scale dynamics of precipitation fronts in the tropical atmosphere: a novel relaxation limit, *Commun. Math. Sci.*, **2**, 591–626.

Fukamachi, Y., McCreary, J.P., and Proehl, J.A., 1995, Instability of density fronts in layer and continuously stratified models, *J. Geoph. Res.*, **100**, C2, 2559–77.

Gerkema, T. and Shrira, V.I., 2005, Near-inertial waves in the ocean: beyond the traditional approximation, *J. Fluid Mech.*, **529**, 195–219.

Gerkema, T., Zimmerman, J.T.F., Maas, L.R.M., and van Haren, H., 2008, Geophysical and astrophysical fluid dynamics beyond the traditional approximation, *Rev. Geophys.*, **46**, RG2004

Gill, A.E., 1980, Some simple solutions for heat-induced tropical circulation, *Q.J.R. Met. Soc.*, 106, 447–62.

Gill A.E., 1982, *Atmosphere - Ocean Dynamics*, AP, NY.

Gill, A.E., 1982, Studies of moisture effects in simple atmospheric models: the stable case, *Geophys. Astrophys. Fluid Dyn.*, **19**, 119–52.

Gill, A.E., Davey, M.K., Johnson, E.R., and Linden, P.F., 1986, Rossby adjustment over a step, *J. Marine Res.*, **44**, 713–38.

Goswami, P., and Goswami, B.N., 1991, Modification of n = 0 Equatorial Waves Due to Interaction between Convection and Dynamics, *J. Atmos. Sci.*, **48** 2231–44.

Gouzien, E., Lahaye, N., Zeitlin, V., and Dubos, T., 2017, Instabilities of vortices and jets in thermal rotating shallow water model, 147–52, in *Topical Problems in Fluid Mechanics 2017*, D. Simurda and T. Bodnar, eds., Institute of Thermomechanics, Prague, Prague.

Griffiths, R.W., Killworth, P.D., and Stern, M.E. 1982, Ageostrophic instability of ocean currents, *J. Fluid Mech.*, **117**, 343–77.

Griffiths, R.W. and Linden, P.F., 1982, Laboratory experiments on fronts. Part I: Density-driven boundary currents, *Geophys. Astrophys. Fluid Dyn.*, **19**, 159–87.

Grimshaw, R.H.J., 1975, A note on the beta-plane approximation, *Tellus*, **27**, 351–57.

Gryanik, V.M., 1983, Radiation of sound by linear vortices, *Izv. Atmos. Ocean. Phys.*, **19**, 203–6.

Gula, J., Plougonven, R., and Zeitlin, V., 2009, Ageostrophic instabilities of fronts in a channel in a stratified rotating fluid, *J. Fluid Mech.*, **627**, 485–507.

Gula, J. and Zeitlin, V., 2010, Instabilities of buoyancy-driven coastal currents and their nonlinear evolution in the two-layer rotating shallow-water model. Part 1. Passive lower layer, *J. Fluid Mech.*, **659**, 69–93.

Gula, J. and Zeitlin, V., 2015, Instabilities of shallow-water flows with vertical shear in the rotating annulus, 119–38, in *Modeling Atmospheric and Oceanic Flows: Insights from Laboratory Experiments and Numerical Simulations*, T. von Larcher and P. Williams, eds., AGU-Wiley, Washington.

Gula, J., Zeitlin, V., and Bouchut, F., 2010, Instabilities of buoyancy-driven coastal currents and their nonlinear evolution in the two-layer rotating shallow-water model. Part 2. Active lower layer, *J. Fluid Mech.*, **665**, 209–37.

Gula, J., Zeitlin, V., and Plougonven, R., 2009, Instabilities of two-layer shallow-water flows with vertical shear in the rotating annulus, *J. Fluid Mech.*, **638**, 27–47.

Guza, R. and Bowen, A.J., 1975, The resonant instabilities of long waves obliquely incident on a beach, *J. Geoph. Res.*, **80**, 1529–34.

Hart, J.E., 1972, A laboratory study of baroclinic instability, *Geophys. Astrophys. Fluid Dyn.*, **3**, 181–209.

Hasselmann, K. 1967, Nonlinear interactions treated by the methods of theoretical physics (with application to the generation of waves by wind), *Proc. R. Soc. A*, **299**, 77–100.

Hayashi, Y.Y. and Young, W.R., 1987, Stable and unstable shear modes of rotating parallel flows in shallow water, *J. Fluid Mech.*, **184**, 477–504.

Hide, R., 1958, An experimental study of thermal convection in a rotating liquid, *Phil. Trans. Royal Soc. London*, **A250**, 441–78.

Holm, D.D, Marsden, J.E., and Ratiu, T., 2002, The Euler–Poncaré equations in geophysical fluid dynamics, 251–300, in *Large-Scale Atmosphere-Ocean Dynamics II*, J. Norbury and I. Roulstone, eds., CUP, Cambridge.

Holm, D.D., Marsden, J.E., Ratiu, T., and Weinstein, A., 1985, Nonlinear stability of fluid and plasma equilibria, *Phys. Reports*, **123**, 1–116.

Holton, J.R., 1979, *An Introduction to Dynamic Meteorology*, AP, NY.

Hoskins, B.J. and Bretherton, F.P., 1972, Atmospheric frontogenesis models: mathematical formulation and solution, *J. Atmos. Sci.*, **29**, 11–37.

Huthnance, J.B., 1975, On trapped waves over a continental shelf, *J. Fluid Mech.*, **69**, 689–704.

Iga, K., 2013, Shear instability as a resonance between neutral waves hidden in a shear flow, *J. Fluid Mech.*, **715**, 452–76.

Jeffreys, H. 1926, On the dynamics of geostrophic winds, *Q.J.R. Met.Soc.*, **52**, 85–104.

Kats, A.V. and Kontorivich, V.M., 1973, Symmetry properties of the collision integral and non-isotropic stationary solutions in weak turbulence theory, *Sov. Phys. JETP*, **37**, 80–5; 1974, Anisotropic turbulent distributions for waves with a non-decay dispersion law, *Sov. Phys. JETP*, **38**, 102–7.

Katsman, C.A., van der Vaart, P.C.F., Dijkstra, H.A., and de Ruijet, W.P.M., 2003, Stability of multilayer ocean vortices: a parameter study including realistic Gulf Stream and Agulhas rings, *J. Phys. Oceanogr.*, **33**, 1197–218.

Kenyon, K., 1964, *Notes on the 1964 Summer Study Programme on GFD at WHOI*, **11**, 69

Khouider, B., Majda, A.J., and Stechmann, S.N., 2013, Climate science in the tropics: waves, vortices and PDEs, *Nonlinearity*, **26**, R1.

Killworth, P.D., 1983, On the motion of isolated lenses on a beta-plane, *J. Phys. Oceanogr.*, **13**, 368–76.

Kizner, Z., Reznik, G., Fridman, B., Khvoles, R., and McWilliams, J.C., 2008, Shallow-water modons on the f -plane, *J. Fluid Mech.*, **603**, 305–29.

Klemp, J.B., Rotunno, R., and Skamarock, W.C., 1997, On the propagation of internal bores, *J. Fluid Mech.*, **331**, 81–97.

Kloosterziel, R.C., Carnevale, G.F., and Orlandi, P, 2007, Inertial instability in rotating and stratified fluids: barotropic vortices, *J. Fluid Mech.*, **583**, 379–412.

Kuo, A.C. and Polvani, L.M., 1999, Wave-vortex interactions in rotating shallow water. Part 1. One space dimension, *J. Fluid Mech.*, **394**, 1–27.

Kuo, H.L., 1949, Dynamic instability of two-dimensional non-divergent flows in a barotropic atmosphere, *J. Meteorol.*, **6**, 105–22.

Kuznetsov, E.A., 1972, Turbulene of ion sound in a plasma located in a magnetic field, *Sov. Phys. JETP*, **35**, 310–14.

Kuznetsov, E.A, 2004, Turbulence spectra generated by singularities, *Sov. Phys. JETP Lett.*, 80, 83–9.

Lahaye, N. and Zeitlin, V., 2011, Collisions of ageostrophic modons and formation of new types of coherent structures in rotating shallow water model, *Phys. Fluids*, **23**, 061703.

Lahaye, N. and Zeitlin, V., 2012, Shock modon: a new type of coherent structure in rotating shallow water, *Phys. Rev. Lett.*, **108**, 044502.

Lahaye, N. and Zeitlin, V., 2012, Decaying vortex and wave turbulence in rotating shallow water model, as follows from high-resolution direct numerical simulations, *Phys. Fluids*, **24**, 115106.

Lahaye, N. and Zeitlin, V., 2012, Existence and properties of ageostrophic modons and coherent tripoles in the two-layer rotating shallow water model on the f-plane, *J. Fluid Mech.*, **706**, 71–107.

Lahaye, N. and Zeitlin, V., 2015, Centrifugal, barotropic and baroclinic instabilities of isolated ageostrophic anticyclones in the two-layer rotating shallow water model and their nonlinear saturation, *J. Fluid Mech.*, **762**, 5–34.

Lahaye, N. and Zeitlin, V., 2016, Understanding instabilities of tropical cyclones and their evolution with a moist-convective rotating shallow-water model, *J. Atmos. Sci.*, **73**, 505–23.

Lambaerts, J., Lapeyre, G., and Zeitlin, V., 2011, Moist versus dry barotropic instability in a shallow water model of the atmosphere with moist convection, *J. Atmos. Sci.*, **68**, 1234–52.

Lambaerts, J., Lapeyre, G., and Zeitlin, V., 2012, Moist versus dry baroclinic instability in a simplified two-layer atmospheric model with condensation and latent heat release, *J. Atmos. Sci.*, **69**, 1405–26.

Lambaerts, J., Lapeyre, G., Zeitlin, V., and Bouchut, F., 2011, Simplified two-layer models of precipitating atmosphere and their properties, *Phys. Fluids*, **23**, 046603.

Landau, L.D. and Lifshits, E.M, 1975, *Hydrodynamics*, AP, NY.

Lapeyre, G. and Held, I.M., 2004, The role of moisture in the dynamics and energetics of turbulent baroclinic eddies, *J. Atmos. Sci.*, **61**, 1693–710.

Laplace P.S., 1799, *Traité de la Mécanique céleste*, Ed. Duprat, Paris.

Larichev, V.D. and Reznik, G.M., 1976, Two-dimensional solitary Rossby waves, *Doklady USSR Acad. Sci.*, **231**, 1077–80.

Lavoie, R.L, 1972, A mesoscale numerical model of lake-effect storms, *J. Atmos. Sci.*, **29**, 1025–40

Lax, P.D., 1973, *Systems of conservation laws and the mathematical theory of shock waves*, SIAM, NY.

LeBlond P.H. and Mysak L.A., 1978, *Waves in the Ocean*, Elsevier, Amsterdam.

Le Dizès, S. and Billant, P., 2009, Radiative instability in stratified vortices, *Phys. Fluids*, **21**, 096602.

Legras, B. and Zeitlin, V., 1992, Conformal dynamics for vortex motions, *Phys. Lett. A*, **167**, 265–71.

Le Sommer, J., Medvedev, S.B., Plougonven, R., and Zeitlin, V., 2003, Singularity formation during relaxation of jets and fronts toward the state of geostrophic equilibrium, *Comm. Nonlin. Sci. Num. Sim.*, **8**, 415–42.

Le Sommer, J., Reznik, G.M., and Zeitlin, V., 2004, Nonlinear geostrophic adjustment of long-wave disturbances in the shallow-water model on the equatorial beta-plane, *J. Fluid Mech.*, **515**, 135–170.

Le Sommer, J, Teitelbaum, H., and Zeitlin, V., 2006, Global estimates of equatorial inertia-gravity wave activity in the stratosphere inferred from ERA40 reanalysis, *Geoph. Res. Lett.*, **33**, L07810.

Lighthill, J.M., 1978, *Waves in Fluids*, CUP, NY.

Longuet-Higgins, M. and Gill, A.E., 1967, Resonant interactions of planetary waves, *Proc. R. Soc. A*, **299**, 120–40.

Longuet-Higgins, M.S., 1968, On the trapping of waves along a discontinuity of depth in a rotating occan, *J. Fluid Mech.*, **32**, 417 34.

Longuet-Higgins, M.S., 1968, Double Kelvin waves with continuous depth profiles, *J. Fluid Mech.*, **34**, 49–80.

Lovegrove, A.F., Read, P.L., and Richards, C.J., 2000, Generation of inertia-gravity waves in a baroclinically unstable fluid, *Q.J.R. Met. Soc.*, **126**, 3233–54.

Lynch P., 2006, *The Emergence of Numerical Weather Prediction*, CUP, NY.

Machenhauer, B., 1977, On the dynamics of gravity oscillations in a shallow water model, with application to normal-mode initialization, *Contrib. Atmos. Phys.*, **50**, 253–71.

Malanotte-Rizzoli, P., 1982, Planetary solitary waves in geophysical flows, *Adv. Geophys.*, **24**, 147–224.

Majda, A., 2003, *Introduction to PDEs and Waves for the Atmosphere and Ocean*, AMS, NY.

Marchal, O. and Nycander, J., 2004, Nonuniform upwelling in a shallow-water model of the Antarctic bottom water in the Brazil Basin, *J. Phys. Oceanogr.*, **34**, 2492.

McCartney M.S., 1976, The interaction of zonal currents with topography with applications to the Southern Ocean, *Deep-Sea Res.*, **23**, 413–27.

McCreary, J.P., Kundu, P.K., and Molinari, R.L., 1993, A numerical investigation of dynamics, thermodynamics and mixed-layer processes in the Indian Ocean, *Progr. Oceanogr.*, **31**, 181–244

McWilliams, J.C.,1984, The emergence of isolated coherent vortices in turbulent flow, *J. Fluid Mech.*, **146**, 21–43.

McWilliams, J.C., 1985, Submesoscale, coherent vortices in the ocean, *Rev. Geophys.*, **23**, 165–82.

McWilliams, J.C., 2006, *Fundamentals of geophysical fluid dynamics*, CUP, NY.

Medvedev, S.B., and Zeitlin, V., 2005, Weak turbulence of short equatorial waves, *Phys. Letters A*, **342**, 217–27.

Menelaou, K., Yau, M.K., and Martinez, Y., 2013, On the origin and impact of a polygonal eyewall in the rapid intensification of hurricane Wilma (2005), *J. Atmos. Sci.*, **70**, 3839–58.

Miles, J., 1990, Parametrically excited edge waves, *J. Fluid Mech.*, **214**, 43–57

Minzoni, A.A., and Whitham, G.B., 1977, Nonlinear edge waves and shallow-water theory, *J. Fluid Mech.*, **79**, 273–87.

Monin, A.S. and Obukhov, A.M., 1958, Small oscillations of the atmosphere and adjustment of meteorological fields, *Izvestija—Geophysics*, **11**, 1360–73 (*in Russian*).

Monin, A.S. and Piterbarg, L.I., 1987, A kinetic equation for Rossby waves, *Sov. Phys. Doklady*, **32**, 622–4.

Monin, A.S. and Yaglom, A.M., 2007, *Statistical Fluid Mechanics—Mechanics of Turbulence*, vols. 1 and 2., Dover, NY.

Morse P.M. and Feshbach, H., 1953, *Methods of theoretical physics*, p. I, McGraw-Hill, NY.

Müller, P., Holloway, G., Henyey, F., and Pomphrey, N. Nonlinear interactions among internal gravity waves, 1986, *Rev. Geophys.*, **24**, p. 493–536.

Nazarenko, S., 2011, *Wave Turbulence*, Springer, Berlin.

Neelin, J.D., Held, I.M., and Cook, K.H., 1987, Evaporation-wind feedback and low frequency variability in the tropical atmosphere, *J. Atmos. Sci.*, **44**, 2341–48.

Obukhov, A.M., 1949, On the problem of geostrophic wind, *Izvestija—Geography and Geophysics*, **13**, 281–306 (*in Russian*).

Oliver, M. and Vasylkevich, S., 2016, Generalized large-scale semigeostrophic approximations for the f-plane primitive equations, *J. Phys. A: Math.Theor.*, **49**, 18400.

Ovsyannikov, L.V., 1979, Two-layer shallow water model, *J. Appl. Mech. Tech. Phys.*, **20**, 127 (*in Russian*).

Paldor, N., and Killworth, P.D., 1987, Instabilities of a two-layer coupled front, *Deep-Sea Res.*, **34**, 1525–39.

Pauluis, O., Frierson, D.M.W., and Majda, A.J., 2008, Precipitation fronts and the reflection and transmission of tropical disturbances, *Q.J.R. Meteor. Soc.*, **134**, 913–30.

Pedlosky, J., 1982, *Geophysical Fluid Dynamics*, Springer, NY.

Pennel, R., Stegner, A., and Beranger, K., 2012, Shelf impact on buoyant coastal current instabilities, *J. Phys. Oceanogr.*, **42**, 39–61.

Phillips N.A., 1954, Energy transformations and meridional circulations associated with simple baroclinic wave in a two-level, quasi-geostrophic model, *Tellus*, **6**, 273–86.

Plougonven, R. and Zeitlin, V., 2009, Nonlinear development of inertial instability in a barotropic shear, *Phys. Fluids*, **21**, 106601.

Polvani, L.M., McWilliams, J.C., Spall, M.A., and Ford, R., 1994, The coherent structures of shallow-water turbulence: Deformation-radius effects, cyclone/anticyclone asymmetry and gravity-wave generation, *Chaos*, **4**, 177–86.

Polzin, K. and Lvov, Y.V., 2011, Toward regional characterization of the oceanic internal wavefield, *Rev. Geophys.*, **49**, RG4003.

Poulin, F. and G.R. Flierl, 2003, The nonlinear evolution of barotropic unstable jets, *J. Phys. Oceanogr.*, **33**, 2173–92.

Rayleigh, Lord, 1880, On the stability, or instablity of certain fluid motions, *Proc. Lond. Math. Soc.*, **9**, 57–70.

Rayleigh, Lord, 1917, On the dynamics of revolving fluids, *Proc. R. Soc. Lond. A*, **93**, 148–54.

Reznik, G.M., 1984, On the properties of the equilibrium spectra of weakly nonlinear Rossby waves *Sov. Phys. Oceanology*, **25**, 869–873.

Reznik, G.M., 1992, Dynamics of singular vortices on a beta-plane, *J. Fluid Mech.*, **240**, 405–32.

Reznik, G.M. and Grimshaw R.H.G., 2002, Nonlinear geostrophic adjustment in the presence of a lateral boundary, *J. Fluid Mech.*, **471**, 257–83.

Reznik, G.M. and Soomere, T.E., 1983, On generalized spectra of weakly nonlinear Rossby waves, *Sov. Phys. Oceanology*, **295**, 86–90.

Reznik, G.M. and Zeitlin, V., 2007, Interaction of free Rossby waves with semi-transparent equatorial waveguide. Part 1. Wave triads, *Physica D*, **226**, 55–79.

Reznik, G.M. and Zeitlin, V., 2009, The interaction of free Rossby waves with semi-transparent equatorial waveguide. Wave-mean flow interaction, *Nonlin. Processes Geoph.*, **16**, 381–92.

Reznik, G.M. and Zeitlin, V., 2009, Resonant excitation of coastal Kelvin waves by inertia-gravity waves, *Phys. Lett. A*, **373**, 1019–21.

Reznik, G.M. and Zeitlin, V., 2011, Resonant excitation of trapped waves by Poincaré waves in the coastal waveguides, *J. Fluid Mech.*, **673**, 349–94.

Reznik, G.M., Zeitlin, V., and Ben Jelloul, M., 2001, Nonlinear theory of geostrophic adjustment. Part 1. Rotating shallow water model, *J. Fluid. Mech.*, **445**, 93–120.

Ribstein, B., Gula, J., and Zeitlin, V., 2010, (A)geostrophic adjustment of dipolar perturbations, formation of coherent structures and their properties, as follows from high-resolution numerical simulations with rotating shallow water model, *Phys. Fluids*, **22**, 116603.

Ribstein, B. and Zeitlin, V., 2013, Instabilities of coupled density fronts and their nonlinear evolution in the two-layer rotating shallow-water model: infuence of the lower layer and of the topography, *J. Fluid Mech.*, **716**, 528–65.

Ribstein, B., Zeitlin, V., and Tissier, A.S., 2014, Barotropic, baroclinic, and inertial instabilities of the easterly Gaussian jet on the equatorial -plane in rotating shallow water model, *Phys. Fluids*, **26**, 056605.

Ripa, P., 1991, General stability conditions for a multi-layer model, *J. Fluid Mech.*, **222**, 119–37.

Ripa, P., 1995, On improving a one-layer ocean model with thermodynamics, *J. Fluid Mech.*, **303**, 169–201.

Rossby, C.-G., 1938, On the mutual adjustment of pressure and velocity distributions in certain simple current systems, II, *J. Mar. Res.*, **1**, 239–63.

Rostami, M. and Zeitlin, V., 2017, Influence of condensation and latent heat release upon barotropic and baroclinic instabilities of vortices in a rotating shallow water f-plane model, *Geoph. Astroph. Fluid Dyn.*, **111**, 1–31.

Rozhdestvenskii, B.L. and Janenko, N.N. 1983, *Systems of Quasi-Linear Equations and their Applications to Gas Dynamics*, AMS, RI.

Saffman, P.G. 1992, Vortex Dynamics, CUP, NY

Sagdeev, R.Z. and Galeev, A.A., 1969, *Nonlinear Plasma Theory*, Benjamin, NY.

Sakai, S., 1989, Rossby–Kelvin instability: a new type of ageostrophic instability caused by a resonance between Rossby waves and gravity waves, *J. Fluid Mech.*, **202**, 149–76.

Salby, M.L., 1989, Deep circulations under simple classes of stratification, *Tellus*, **41A**, 48–65.

Salmon, R., 1982, The shape of the main thermocline, *J. Phys. Oceanogr.*, **12**, 1458–1479.

Salmon, R., 1998, *Lectures on Geophysical Fluid Dynamics*, OUP, NY.

Schecter, D.A, and Dunkerton, T. J., 2009, Hurricane formation in diabatic Ekman turbulence, *Q.J.R Meteor. Soc.*, **135**, 823.

Scherer, E. and Zeitlin, V. 2008, Instability of coupled geostrophic density fronts and its nonlinear evolution, *J. Fluid Mech.*, **613**, 309–27.

Stegner, A., 2007, Experimental reality of geostrophic adjustment, 323–79, in *Nonlinear Dynamics of Rotating Shallow Water: Methods and Advances*, V. Zeitlin, ed., Springer, Amsterdam.

Stegner, A. and Zeitlin, V., 1995, What can asymptotic expansions tell us about large-scale quasi-geostrophic anticyclonic vortices?, *Nonlin. Proc. Geophys.*, **2**, 186–93.

Stegner, A. and Zeitlin, V., 1996, Asymptotic expansions and monopolar solitary Rossby vortices in barotropic and two-layer models, *Geophys. Astrophys. Fluid Dyn.*, **83**, 159–94.

Stevens, B. and Bony, S., 2013, What climate models miss?, *Science*, **340**, 1053.

Stewart, A. and Dellar, P.J., 2010, Two-Layer shallow water equations with complete Coriolis force and topography, 1033–38, in *Progress in Industrial Mathematics at ECMI 2008*, A.D. Fitt, J. Norbury, H. Ockendon and E. Wilson, eds., Mathematics in Industry 1, Springer, Berlin.

Stone, P.H., 1966, On non-geostrophic baroclinic instability, *J. Atmos. Sci.*, **23**, 390–400.

Thomas, J., 2016, Resonant fast–slow interactions and breakdown of quasi-geostrophy in rotating shallow water, *J. Fluid Mech.*, **788**, 492–520.

Tort, M. and Dubos, T., 2014, Dynamically consistent shallow-atmosphere equations with a complete Coriolis force, *Q.J.R. Meteorol. Soc.*, **140**, 2388–92.

Tort, M., Dubos, T., Bouchut, F., and Zeitlin, V., 2014, Consistent shallow-water equations on the rotating sphere with complete Coriolis force and topography, *J. Fluid Mech.*, **748**, 789–821.

Tort, M., Ribstein, B., and Zeitlin, V., 2016, Symmetric and asymmetric inertial instability of zonal jets on the f-plane with complete Coriolis force, *J. Fluid Mech.*, **788**, 274–302.

Trefethen, L.N., 2000, *Spectral Methods in MATLAB*, SIAM, NY.

Vallis, G.K, 2006, *Atmospheric and Oceanic Fluid Dynamics: Fundamentals and Large-scale Circulation*, CUP, NY.

Vanneste, J. and Yavneh, I, 2004, Exponentially small inertia - gravity waves and the breakdown of quasigeostrophic balance, *J. Atmos. Sci.*, **61**, 211–23.

Volotsky, S.V., Kats, A.V., and Kontorovich, V.M., 1980, Symmetry transformations of collision integral describing scattering of quasi-particles with a dispersion law close to linear, *Proceedings of the Ukranian Akademy of Science (in Russian)*, **11**, p. 66–9.

Warneford, E.S. and Dellar, P.J., 2013, The quasi-geostrophic theory of the thermal shallow water equations, *J. Fluid Mech.*, **723**, 374–403.

Warneford, E.S. and Dellar P.J., 2014, Thermal shallow water models of geostrophic turbulence in Jovian atmospheres, *Phys. Fluids*, **26**, 016603.

Weiss, J.B., 1991, The dynamics of enstrophy transfer in two-dimensional hydrodynamics, *Physica D*, **48**, 273–94.

White, A. and Bromley, R.A., 1995, Dynamically consistent, quasi-hydrostatic equations for global models with a complete representation of the Coriolis force, *Q.J.R. Met. Soc.*, **121**, 399–418.

White, A.A., Hoskins, B.J., Roulstone, I., and Staniforth, A., 2005, Consistent approximate models of the global atmosphere: shallow, deep, hydrostatic, quasi-hydrostatic and non-hydrostatic, *Q.J.R. Met. Soc.*, **131**, 2081–107.

White, A.A. and Wood, N., 2012, Consistent approximate models of the global atmosphere in non-spherical geopotential coordinates, *Q.J.R. Met. Soc.*, **138**, 980–88.

Whitham, G.B., 1974, *Linear and Nonlinear Waves*, Wiley, NY.

Williams, P.D., Haine, T.W.N., and Read, P.L., 2005, On the generation mechanisms of short-scale unbalanced modes in rotating two-layer flows with vertical shear, *J. Fluid Mech.*, **528**, 1–22.

Williamson, D., Drake, J., Hack, J., Jakob, R., and Swartztrauber, P., 1992, A standard test set for numerical approximations to the shallow water equations in spherical geometry, *J. Comput. Phys.*, **102**, 211–24.

Willoughby, H., Darling, R., and Rahn, M., 2006, Parametric representation of the primary hurricane vortex. Part II: A new family of sectionally continuous profiles, *Mon. Wea. Rev.*, **134**, 1102–20.

Young, W.R., 1994, The subinertial mixed layer approximation, *J. Phys. Oceanogr.*, **24**, 1812–26.

Young, W.R. and Chen, L., 1995, Baroclinic instability and thermohaline gradient alignment in the mixed layer, *J. Phys. Oceanogr.*, **25**, 3172–85.

Yuan, L. and Hamilton, K., 1994, Equilibrium dynamics in a forced-dissipative f-plane shallow-water system, *J. Fluid Mech.*, **280**, 369–94.

Zakharov, V.E., 1984, Kolmogorov spectra in weak turbulence problems, 23–36, in *Handbook of Plasma Physics*, M.N. Rozenbluth and R.Z. Sagdeev, eds, North-Holland, NY.

Zakharov, V.E. and Piterbarg, L.I., 1987, Canonical variables for Rossby waves and drift waves in plasma, *Sov. Phys. Doklady*, **32**, 560.

Zakharov, V.E., L'vov, V.S., and Falkovich, G., 1992, *Kolmogorov Spectra of Turbulence I*, Springer, Berlin.

Zaslavsky G.M. and Sagdeev, R.Z., 1967, Limits of statistical description of a nonlinear wave field, *Sov. Phys. JETP*, **52**, 1081.

Zeitlin, V., 1991, Finite-mode analogs of 2D ideal hydrodynamics: Coadjoint orbits and local canonical structure, *Physica D*, **49**, 353–62.

Zeitlin, V., 1991, On the backreaction of acoustic radiation for distributed two-dimensional vortex structures, *Phys. Fluids A*, **3**, 1677–80.

Zeitlin, V., 1992, Vorticity and waves: geometry of phase-space and the problem of normal variables *Phys. Lett. A*, **164**, 177–83.

Zeitlin, V., 2007, Introduction: fundamentals of rotating shallow water model in the geophysical fluid dynamics perspective, 2–45, in *Nonlinear dynamics of Rotating Shallow Water: Methods and Advances*, V. Zeitlin, ed., Elsevier, Amsterdam.

Zeitlin, V., 2008, Decoupling of balanced and unbalanced motions and inertia-gravity wave emission: small versus large Rossby numbers, *J. Atmos. Sci.*, **65**, 3528–42.

Zeitlin, V., 2013, Resonant excitation of coastal Kelvin waves in the two-layer rotating shallow water model, *Nonlin. Processes Geophys.*, **20**, 993–99.

Zeitlin, V., Reznik, G.M., and Ben Jelloul, M., 2003, Nonlinear theory of geostrophic adjustment. Part 2. Two-layer and continuously stratified primitive equations, *J. Fluid. Mech.*, **491**, 207–28.

Zeitlin, V., Medvedev, S.B., and Plougonven, R., 2003, Frontal geostrophic adjustment, slow manifold and nonlinear wave phenomena in one-dimensional rotating shallow water. Part 1. Theory, *J. Fluid Mech.*, **481**, 269–90.

Index